# Three-Dimensional Imaging, Visualization, and Display

Bahram Javidi · Fumio Okano · Jung-Young Son
Editors

# Three-Dimensional Imaging, Visualization, and Display

*Editors*
Bahram Javidi
Electrical and Computer
Engineering Department
University of Connecticut
USA
bahram@engr.uconn.edu

Fumio Okano
NHK Nippon Hoso Kyokai
Tokyo
Japan
okano.f-la@nhk.or.jp

Jung-Young Son
School of Computer and
Communication Engineering
Daegu University
Korea
sjy@daegu.ac.kr



ISBN: 978-0-387-79334-4 e-ISBN: 978-0-387-79335-1
DOI 10.1007/978-0-387-79335-1

Library of Congress Control Number: 2008939447



Printed on acid-free paper

springer.com

This book is dedicated to the memory of friend
and colleague Dr. Ju-Seog Jang (1961–2004)

# Preface

The history of visual media is characterized by drawing, painting, photo, stereoscope, TV and displays, Holography, and HDTV. Three-dimensional (3-D) imaging, visualization, and display technology is the next phase in this historical development. There is no doubt that 3-D TV and 3-D display are the visual media that will succeed HDTV in near future. The evolutionary merging of communication and broadcast technologies will be enhanced by 3-D visualization, including multimedia communication centers which can respond to all the services and functions provided by the future communication systems.

Comparison of a scene presented by 2-D and 3-D images reveals that the 3-D image provides much more pleasing and realistic information than its 2-D version. The benefits provided by 3-D images result from the depth information they preserve. Depth gives viewers the feeling of being in the place where the 3-D image is captured (i.e., immersive feeling) and the feeling of being present (presence feeling). Depth increases efficiencies in remote site operations, such as remote medical and tele–operations, by making viewers perceive the scene as more real from within their living environments. The accuracy and the presence feeling are the main motives of demands for 3-D images in the areas of communication, broadcasting, entertainment, medical operations, virtual world presentations, advertisement, training, edutainment (education + entertainment), telemarketing, tele-presence, tele-conference, visualization of experimental results, and so on.

The presence feeling will be more enhanced if users can interact with the 3-D images displayed on the screen/display panel. In this case, the images having real object sizes will probably create a better environment for presence feeling, that is, "super presence feeling." The ultimate goal of communication is providing the communication partners with the real feeling of being in the same place and interacting face-to-face with each other. Without 3-D images, it would be difficult to achieve this goal. 3-D images should be user-friendly and provide natural sense of depth. The multi-view 3-D imaging methods are intended to provide a natural sense of depth by equally de-magnifying object images in all three coordinates to fit into the display device, instead of exaggerating the depth sense to maximize visual effect as in the stereoscopic imaging methods.

Our first edition of the book on 3-D TV and Display appeared six years ago by Springer. During the last six years tremendous progress has been made in 3-D imaging technologies. This progress has motivated the major display companies to produce their own 3-D displays. Currently, these displays are in the production stage. The widespread use of mobile phones has created the need to fuse communication and broadcasting technologies. As a consequence, some mobile phones are being equipped with a stereoscopic camera, and 3-D image service has already been provided to the mobile phone company. This is another reason why the major display companies in the world are trying to get into the 3-D visualization market.

The new book provides readers with an introduction as well as an overview of the recent progress made in this field. It consists of 23 chapters describing concepts, applications, and new developments of various areas in the field of 3-D imaging, 3-D visualization, and 3-D display. These include multi-view image acquisition, processing, and display; digital holography; human factors for 3-D display, and recent use of high speed LCDs for 3-D image display. These chapters are written by some of the most active researchers in the field.

We believe that this book provides in-depth coverage of current 3-D imaging technologies and can be a good reference for students, engineers, and researchers who have interests and/or are working on 3-D imaging, 3-D visualization, and 3-D display.

We wish to thank the authors, many of whom are our friends and colleagues, for their outstanding contributions to this book.

This book is dedicated to the memory of our friend and colleague, Dr. Ju Seog Jang (1961–2004).

Storrs, CT — *Bahram Javidi*
Tokyo — *Fumio Okano*
Seoul — *Jung-Young Son*

# Contents

## Part I 3-D Image Display and Generation Techniques Based on I.P

## Part II Multiview Image Acquisition, Processing and Display



## Part III 3-D Image Acquisition, Processing and Display Based on Digital Holography

# Contributors

**Jun Arai** NHK (Japan Broadcasting Corporation) Science & Technical Research Laboratories, 1-10-11, Kinuta, Setagaya-ku, Tokyo 157-8510, Japan, arai.j-gy@nhk.or.jp

**Byung Joon Baek** Division of Mechanical & Aero System Engineering, Chonbuk National University, Jeonju 561-756, Republic of Korea

**Albertina Castro** Instituto Nacional de Astrofísica, Óptica y Electrónica, Apartado Postal 51, Puebla, Pue. 72000, Mexico, betina@inaoep.mx

**Kyung-Hoon Cha** Samsung Electronics Co., Ltd., Suwon, Republic of Korea

**S. Cho** Electronics and Telecommunications Research Institute, 161 Gajeong-dong, Yuseong-gu, Daejeon, Republic of Korea, shee@etri.re.kr

**Seong-Woo Cho** School of Electrical Engineering, Seoul National University, Seoul 151-744, Republic of Korea

**Heejin Choi** School of Electrical Engineering, Seoul National University, Seoul 151-744, Republic of Korea

**Giuseppe Coppola** Istituto di Microelettronica e Microsistemi, Consiglio Nazionale delle Ricerche, Via Campi Flegrei 34, 80078 Pozzuoli (Napoli), Italy

**Mehdi Danesh Panah** Department of Electrical and Computer Engineering, U-2157, University of Connecticut, Storrs, CT 06269-2157, USA

**Masaki Emoto** NHK (Japan Broadcasting Corporation) Science & Technical Research Laboratories, 1-10-11, Kinuta, Setagaya-ku, Tokyo 157-8510, Japan, emoto.m-hy@nhk.or.jp

**C. Fehn** Fraunhofer Institute for Telecommunications, Heinrich-Hertz-Institut, Einsteinufer 37, 10587 Berlin, Germany, Christoph.Fehn@hhi.fhg.de

**Pietro Ferraro** Istituto Nazionale di Ottica Applicata, Consiglio Nazionale delle Ricerche, Via Campi Flegrei 34, 80078 Pozzuoli (Napoli), Italy, pietro.ferraro@inoa.it

**Yann Frauel** Departaments de Ciencias de la Computation, Instituto de investigaciones en Matemáticas Aplicadas y en Sistemas, Universidad Nacional Autónoma de Mexico

**Simonetta Grilli** Istituto Nazionale di Ottica Applicata, Consiglio Nazionale delle Ricerche, Via Campi Flegrei 34, 80078 Pozzuoli (Napoli), Italy

**Yoshio Hayasaki** Center for Optical Research & Education (CORE) Utsunomiya University, Japan

**Bryan M. Hennelly** Department of Computer Science, National University of Ireland, Maynooth, County Kildare, Ireland

**N. Hur** Electronics and Telecommunications Research Institute, 161 Gajeong-dong, Yuseong-gu, Daejeon, Republic of Korea, namho@etri.re.kr

**Seon-Deok Hwang** Samsung Electronics Co., Ltd., Suwon, Republic of Korea

**Y. Iwadate** NHK (Japan Broadcasting Corporation) Science & Technical Research Laboratories, 1-10-11, Kinuta, Setagaya-ku, Tokyo 157-8510, Japan, iwadate.y-ja@nhk.or.jp

**Bahram Javidi** Department of Electrical and Computer Engineering, University of Connecticut, U-2157, Storrs, CT 06269-2157, USA, Bahram@engr.uconn.edu

**P. Kauff** Fraunhofer Institute for Telecommunications, Heinrich-Hertz-Institut, Einsteinufer 37, 10587 Berlin, Germany, Peter.Kauff@hhi.fhg.de

**Dae-Sik Kim** Samsung Electronics Co., Ltd., Suwon, Republic of Korea, daesikkim@samsung.com

**Daesuk Kim** Division of Mechanical & Aero System Engineering, Chonbuk National University, Jeonju 561-756, Republic of Korea, dashi.kim@chonbuk.ac.kr

**Hansung Kim** Knowledge Science Lab, ATR, Kyoto, Japan

**Shin Hwan Kim** School of Computer and Communication Engineering, Daegu University, South Korea, namuri@daegu.ac.kr

**J. Kim** Electronics and Telecommunications Research Institute, 161 Gajeong-dong, Yuseong-gu, Daejeon, Republic of Korea, jwkim@etri.re.kr

**Kyung Tae Kim** Department of Information and Communication Engineering, Hannam University, Daejeon, Korea

**Yongtae Kim** School of Electrical and Electronic Engineering, Yonsei University, Seoul, Republic of Korea

**Yunhee Kim** School of Electrical Engineering, Seoul National University, Seoul, Republic of Korea

**Sung Kyu Kim** Imaging media center, Korea Institute of Science and Technology, Seoul, Korea, kkk@kist.re.kr

**Jae-Phil Koo** Samsung Electronics Co., Ltd., Suwon, Republic of Korea

**B. Lee** Electronics and Telecommunications Research Institute, 161 Gajeong-dong, Yuseong-gu, Daejeon, Republic of Korea, leebh@etri.re.kr

**Byoungho Lee** School of Electrical Engineering, Seoul National University, Seoul 151-744, Republic of Korea, byoungho@snu.ac.kr

**Chaewook Lee** School of Computer and Communication Engineering, Daegu University, Kyungsan, Kyungbuk, Korea

**H. Lee** Electronics and Telecommunications Research Institute, 161 Gajeong-dong, Yuseong-gu, Daejeon, Republic of Korea, hlee2@etri.re.kr

**Osamu Matoba** Department of Computer Science and Systems Engineering, Kobe University, Rokkodai 1-1, Nada, Kobe 657-8501, Japan, matoba@kobe-u.ac.jp

**Conor P. McElhinney** Department of Computer Science, National University of Ireland, Maynooth, County Kildare, Ireland, conormce@cs.nuim.ie

**Sung-Wook Min** Department of Information Display, Kyung Hee University, Seoul, Republic of Korea, mins@khu.ac.kr

**Tomoyuki Mishina** NHK Science & Technical Research Laboratories, 1-10-11, Kinuta, Setagaya-ku, Tokyo 157-8510, Japan, mishina.t-iy@nhk.or.jp; National Institute of Information and Communication Technology, 4-2-1, Nukui-Kitamachi, Koganei, Tokyo 184-8795, Japan, mishina@nict.go.jp

**Kohji Mitani** NHK (Japan Broadcasting Corporation) Science & Technical Research Laboratories, 1-10-11, Kinuta, Setagaya-ku, Tokyo 157-8510, Japan, mitani.k-gw@nhk.or.jp

**Thomas J. Naughton** Department of Computer Science, National University of Ireland, Maynooth, County Kildare, Ireland; University of Oulu, RFMedia Laboratory, Oulu Southern Institute, Vierimaantie 5, 84100 Ylivieska, Finland

**Sergio De Nicola** Istituto di Cibernetica "E. Caianiello", Consiglio Nazionale delle Ricerche, Via Campi Flegrei 34, 80078 Pozzuoli (Napoli), Italy

**Nobuo Nishida** Department of Optical Science and Technology, The University of Tokushima, 2-1 Minamijosanjima, Tokushima 770-8506, Japan

**Takanori Nomura** Department of Opto-Mechatronics, Wakayama University, 930 Sakaedani, Wakayama 640-8510, Japan, nom@sys.wakayama-u.ac.jp

**Fumio Okano** NHK (Japan Broadcasting Corporation) Science & Technical Research Laboratories, 1-10-11, Kinuta, Setagaya-ku, Tokyo 157-8510, Japan, okano.f-la@nhk.or.jp

**Makoto Okui** National Institute of Information and Communication Technology, 4-2-1, Nukui-Kitamachi, Koganei, Tokyo 184-8795, Japan; NHK (Japan Broadcasting Corporation) Science & Technical Research Laboratories, 1-10-11, Kinuta, Setagaya-ku, Tokyo 157-8510, Japan, m-okui@nict.go.jp

**Hiroshi Ono** Centre for Vision Research and Department of Psychology, York University, Toronto, Ontario M3J 1P3, Canada, hono@yorku.ca

**Min-Chul Park** Intelligent System Research Division, Korea Institute of Science and Technology, seoul, South Korea, minchul@kist.re.kr

**Sergey Shestak** Samsung Electronics Co., Ltd., Suwon, Republic of Korea

**Vladmir V. Saveljev** Whole Imaging Laboratory, Hanyang University, Seoul, South Korea, saveljev@hanyang.ac.kr

**Sang Ju Park** School of Electronic and Electrical Engineering, Hongik University, Seoul, South Korea, sjpark@hongik.ac.kr

**Kwanghoon Sohn** School of Electrical and Electronic Engineering, Yonsei University, Seoul, Republic of Korea, khsohn@yonsei.ac.kr

**Jung-Young Son** Center for Advanced Image, School of Information and Communication Engineering, Daegu University, Kyungsan, Kyungbuk, Republic of Korea, sjy@daegu.ac.kr

**Filippo Speranza** Communications Research Centre Canada, 3701 Carling Avenue, Ottawa, Ontario, Canada K2H 8S2, filippo.speranza@crc.ca

**Wa James Tam** Communication Research Centre Canada, 3701 Carling Avenue, Ottawa, Ontario, Canada K2H 8S2, james.tam@crc.ca

**Behnoosh Tavakoli** Department of Electrical and Computer Engineering, U-2157, University of Connecticut, Storrs, CT 06269-2157, USA

**Edward A. Watson** U.S. Airforce Research Laboratories, Wright Patterson Air Force Base, OH 45433, USA, Edward.Watson@wpafb.af.mil

**Hirotsugu Yamamoto** Department of Optical Science and Technology, The University of Tokushima, 2-1 Minamijosanjima, Tokushima, 770-8506, Japan, yamamoto@opt.tokushima-u.ac.jp

**Hirokazu Yamanoue** Science & Technical Research Laboratories, Japan Broadcasting Corporation (NHK), 1-10-11, Kinuta, Setagaya-ku, Tokyo 157-8510, Japan

**Sumio Yano** NHK Science & Technical Research Laboratories, 1-10-11, Kinuta, Setagaya-ku, Tokyo 157-8510, Japan, yano.s-fs@nhk.or.jp; National Institute of Information and Communication Technology, 4-2-1 Nukui-Kitamachi, Koganei, Tokyo 184-8795, Japan, yano.s@nict.go.jp

**K. Yun** Electronics and Telecommunications Research Institute, 161 Gajeong-dong, Yuseong-gu, Daejeon, Republic of Korea, kjyun@etri.re.kr

**Liang Zhang** Communications Research Centre Canada, 3701 Carling Avenue, Ottawa, Ontario, Canada K2H 8S2, liang.zhang@crc.ca

# Part I

## 3-D Image Display and Generation Techniques Based on I.P

# 1

# Three-Dimensional Integral Television Using High-Resolution Video System with 2000 Scanning Lines

Fumio Okano, Jun Arai, Kohji Mitani, and Makoto Okui

## 1.1 Introduction

Use of three-dimensional (3-D) images in broadcasting, communications, and many other areas has been anticipated for some time, but practical applications have shown little progress. One reason is that most 3-D imaging systems fail to simultaneously provide: (1) binocular disparity that can be experienced without special glasses, (2) a convergence point that matches the eye's accommodation point, and (3) motion parallaxes that enable an observer to see different images corresponding to different positions horizontally and vertically (full parallaxes). These capabilities would enable observers to see a 3-D image as though it were a real object. By expanding research on spatial imaging, we aim to develop a 3-D imaging technology that performs all these functions.

Depending on the characteristics of the reconstructed images, 3-D systems can be roughly classified into four types: binocular, multi-view, volumetric imaging, and spatial imaging. The spatial imaging type is able to satisfy the three requirements mentioned above. Holography is a well-known example of this type. The integral method is also a spatial imaging type. Because it can produce 3-D images using natural light (incoherent light), it is considered to be one of the ideal 3-D systems. The integral method was first proposed by

F. Okano
NHK (Japan Broadcasting Corporation) Science & Technical Research Laboratories, 1-10-11, Kinuta, Setagaya-ku, Tokyo 157-8510
e-mail: Japan, okano.f-la@nhk.or.jp

B. Javidi et al. (eds.), *Three-Dimensional Imaging, Visualization, and Display*,
DOI 10.1007/978-0-387-79335-1_1, 

G. Lippmann [1] in 1908 based on a photographic technique. Recently, several attempts have been made to obtain higher quality [2, 3, 4, 5, 6, 7, 8] and moving 3-D images [9, 10].

Figure 1.1 shows the principle of the basic integral method using a single lens array in the capture and display stages. To produce an integral image, a lens array composed of many convex elemental lenses is positioned immediately in front of the capture plate. The integral image is composed of numerous small elemental images that are captured and recorded on the plate. The number of images corresponds to the number of elemental lenses. The integral image is supplied to a transparent display plate. The display plate is placed where the capture plate had been, and is irradiated from behind by an incoherent light source. The light beams passing through the display plate and the

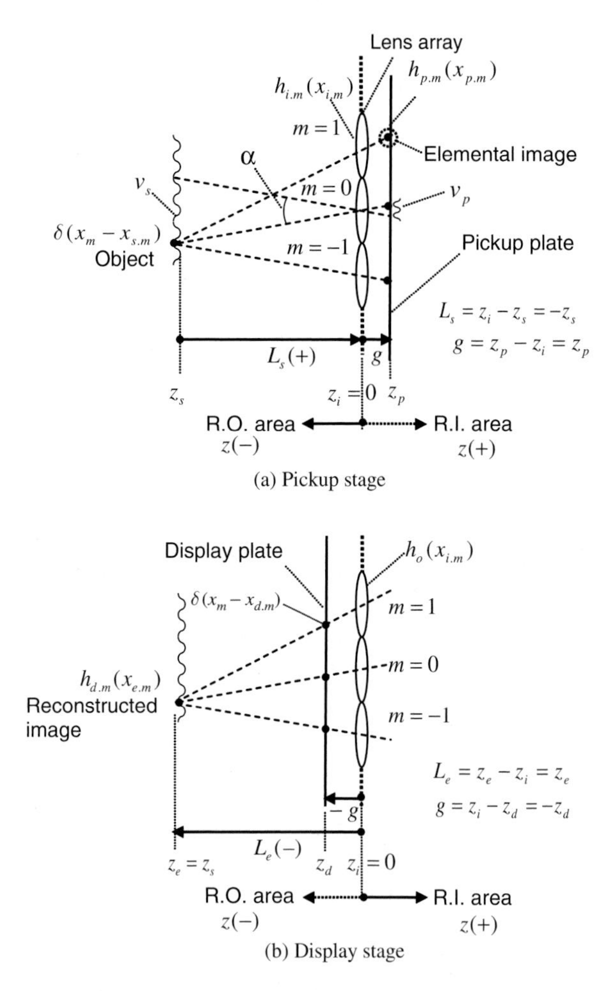


**Fig. 1.1.** Principle of basic integral method

lens array re-trace the original routes and then converge at the point where the object had been, forming an autostereoscopic image.

The total number of pixels, $N_t$, on the capture plate is the product of the number of elemental lenses, $N_m$, and the number of pixels, $N_e$, in each elemental image horizontally and vertically:

$$N_t = N_m N_e. \tag{1.1}$$

The number of elemental lenses determines the upper limit of resolution. The number of pixels in each elemental image affects the resolution of the 3-D image away from the lens array. The capture plate requires $N_e$ times the number of pixels required by conventional television. Thus, extremely high resolution is required for both the capture and display plate. We experimentally developed a 3-D integral television using an extremely high resolution video system with 2,000 scanning lines that provides full color and full parallax 3-D images in real time.

## 1.2 Image Formation

### 1.2.1 Geometric Approach

Though the principle of image formation is described above, a more thorough explanation must include a description of ray tracing by geometrical optics [11, 12]. Figures 1.1(a) and (b) focus on around the $m$-th elemental lenses, many of which constitute an array. A point light source is placed at distance $L_s$ from the lens array and is expressed as a delta function $\delta(x_m - x_{s.m})$, where $x_{s.m}$ represents the object's position. $x_{i.m}$ and $x_{p.m}$ represent the positions in the incident plane of the lens array and capture plate, respectively. Subscript $m$ indicates the position on the coordinates where the intersection of an incident plane and its own optical axis is the origin for each elemental lens. The following $x$ can be obtained by adding $mP_a$ to $x_m$, that is, the distance from the origin of the whole array to the optical axis of the elemental lens:

$$x = x_m + mP_a, \tag{1.2}$$

where $x_{s.m}$, $x_{i.m}$ and $x_{p,m}$ are converted to $x_s$, $x_i$ and $x_p$, respectively, by adding $mP_a$ in the same way. $P_a$ is the pitch between adjacent elemental lenses. Note that $z$ is not assigned a subscript because the coordinates of each elemental lens match those of the whole array. The origin of the $x$ and $z$ coordinates of the whole array is defined as the point where the optical axis crosses the incident plane of the central elemental lens. To simplify calculations, we use the two-dimensional coordinates $(x, z)$, defined by the $x$-axis and optical axis $z$.

Real objects in the capture stage can be located in the space with a negative value of $z$, which is called the real objects area (R.O. area). Real images in the display stage can be located in the space with a positive value of $z$, which is called the real images area (R.I. area). The following calculations can be applied to the three-dimensional coordinates ($x$, $y$, $z$) defined by the optical axis and a plane that crosses it perpendicularly. There is a relationship between $x_{s.m}$ and $x_{p.m}$ in the capture stage shown in Fig. 1.1(a):

$$\frac{x_{s.m}}{z_s} = \frac{x_{p.m}}{g}, \tag{1.3}$$

where $g$ is the gap between the elemental lens and the capture plate. As shown in Fig. 1.1(b), we assume the pitch of the elemental lenses and the gap in the display stage are the same as in the capture stage, respectively. $x_{d,m}$ is the position of the point light source in the display plate and $x_m$ represents the space in which the reconstructed image is formed. There is a similar relationship between $x_{d,m}$ and $x_m$ in the display stage:

$$-\frac{x_{d.m}}{g} = \frac{x_m}{z} \tag{1.4}$$

To avoid the pseudoscopic image with reversed depth [13, 14, 15], each elemental image is converted into a symmetric position for the center of its own, which is obtained by Eq. (1.3):

$$x_{d.m} = -x_{p.m} = -\frac{x_{s.m}}{z_s} g, \tag{1.5}$$

Therefore, we obtain from Eqs. (1.2), (1.4), and (1.5):

$$\begin{aligned} x &= x_m + mP_a = -\tfrac{z}{g} x_{d.m} + mP_a = \tfrac{z x_{s.m}}{z_s} + mP_a = \tfrac{z}{z_s}(x_s - mP) + mP_a \\ &= \tfrac{z}{z_s} x_s + \left(1 - \tfrac{z}{z_s}\right) mP_a \end{aligned} \tag{1.6}$$

Here, we assume $z = z_s$, obtaining the following: $x = x_s$. This $x$ position is independent of elemental lens number $m$ and, therefore, all light rays from the elemental lenses pass the same point $(x_e, z_e)$ where the object is located. Therefore, we obtain:

$$(x_e, z_e) = (x_s, z_s) \tag{1.7}$$

This means the ray from all the elemental lenses converge and form an optical image at the point.

### 1.2.2 Wave Optical Approach

By using wave optics the captured elemental images synthesize an optical image in the display stage [16, 17]. We present the response of the $m$-th elemental lens on the pickup plate shown in Fig. 1.1(a). First, the wave (electric field) entering the elemental lens of the pickup stage is calculated by Fresnel's approximation as

$$\begin{aligned} u_{i.m}(x_{i.m}) &= \frac{1}{j\lambda L_s}\exp\left(-jk\frac{x_{i.m}^2}{2L_s}\right)\int_{object}\delta(x_m - x_{s.m})\exp\left(-jk\frac{x_m^2}{2L_s}\right) \\ &\quad \exp\left(-jk\frac{x_m x_{i.m}}{L_s}\right)dx_m \\ &= \frac{1}{j\lambda L_s}\exp\left(-jk\frac{x_{i.m}^2}{2L_s}\right)\exp\left(-jk\frac{x_{s.m}^2}{2L_s}\right)e\left(-jk\frac{x_{s.m}x_{i.m}}{L_s}\right), \end{aligned} \tag{1.8}$$

where $L_s = Z_i - Z_s$, $k$ is the wave number and equals $2\pi/\lambda$, and $\lambda$ is the wavelength. The output wave from an elemental lens is a product of Eq. (1.8) and the phase shift function of the elemental lens:

$$u_{i.m}(x_{i.m})\exp\left({x_{i.m}^2}/{2f}\right).$$

The wave on the capture plate is obtained by

$$\begin{aligned} h_{p.m}(x_{p.m}) &= \frac{1}{j\lambda g}\exp\left(-jk\frac{x_{p.m}^2}{2g}\right)\int_{-w_a/2}^{w_a/2} u_{i.m}(x_{i.m})\exp\left(\frac{x_{i.m}^2}{2f}\right)\exp\left(-jk\frac{x_{i.m}^2}{2g}\right) \\ &\quad \exp\left(-jk\frac{x_{i.m}x_{p.m}}{g}\right)dx_{i.m}, \end{aligned} \tag{1.9}$$

where $f$ is the focal length of the elemental lens, and $w_a$ is the width of an elemental lens. The amplitude distribution of this equation is obtained by

$$\begin{aligned} \left|h_{p.m}(x_{sp.m})\right| &= \left|\frac{1}{\lambda^2 g L_s}\int_{-w_a/2}^{w_a/2}\exp\left[-jk\left(\tfrac{1}{L_s}+\tfrac{1}{g}-\tfrac{1}{f}\right)\tfrac{x_{i.m}^2}{2}\right]\right. \\ &\quad \left.\exp\left(-jkx_{sp.m}x_{i.m}\right)dx_{i.m}\right| = \left|h_{p.m}(-x_{sp.m})\right|, \end{aligned} \tag{1.10}$$

$$x_{sp.m} \equiv \frac{x_{s.m}}{L_s} + \frac{x_{p.m}}{g}. \tag{1.11}$$

Equation (1.10) is an even function and is symmetric for $x_{sp.m} = 0$. The following equation is obtained from Eq. (1.11):

$$x_{p.m} = -\frac{g}{L_s} x_{s.m}. \tag{1.12}$$

This point is considered the center of the amplitude distribution for each elemental lens. This distribution is the same around each point $x_{p.m}$.

Next, we present the impulse response by the $m$-th elemental lens in the display stage shown in Fig. 1.1(b). To avoid the pseudoscopic image with reversed depth, each elemental image is converted into a symmetric position for the center of its own mentioned above. The symmetric position is obtained by Eq. (1.12):

$$x_{d.m} \equiv -x_{p.m} = \frac{x_{s.m}}{L_s} g = -\frac{x_{s.m}}{z_s} g. \tag{1.13}$$

We assume that the point light source $\delta(x_m - x_{d.m})$ is set in the display plate. The input wave of the elemental lens for the display stage is described by

$$u_{d.m}(x_{i.m}) = \frac{1}{j\lambda g} \exp\left(-jk\frac{x_{i.m}^2}{2g}\right) \int_{elemental\ image} \delta(x_m - x_{d.m}) \exp\left(-jk\frac{x_m^2}{2g}\right) \exp\left(-jk\frac{x_m x_{i.m}}{g}\right) dx_m = \frac{1}{j\lambda g} \exp\left(-jk\frac{x_{i.m}^2}{2g}\right) \exp\left(-jk\frac{x_{d.m}^2}{2g}\right) \exp\left(-jk\frac{x_{d.m} x_{i.m}}{g}\right). \tag{1.14}$$

The output wave is a product of this equation and the phase shift function of an elemental lens:

$$u_{d.m}(x_{i.m}) \exp\left(x_{i.m}^2/2f\right).$$

Therefore, the output wave in the image space is obtained by

$$h_{d.m}(x_m) = \frac{1}{j\lambda|L_e|} \exp\left(-jk\frac{x_m^2}{2L_e}\right) \int_{-w_a/2}^{w_a/2} u_{d.m}(x_{i.m}) \exp\left(\frac{x_{i.m}^2}{2f}\right) \exp\left(-jk\frac{x_{i.m}^2}{2L_e}\right) \exp\left(-jk\frac{x_{i.m} x_m}{L_e}\right) dx_{i.m}, \tag{1.15}$$

where $L_e = z - z_i$ is the distance from the array in the image space. The integral operation is performed in the area of an elemental lens. The amplitude of this equation is obtained by:

$$\begin{aligned} |h_{d.m}(x_m)| &= \left| \frac{1}{\lambda^2 L_e g} \int_{-w_a/2}^{w_a/2} \exp\left(-jk\frac{1}{2L_f}x_{i.m}^2\right) \exp\left[-jk\left(\frac{x_{s.m}}{L_s}+\frac{x_m}{L_e}\right)x_{i.m}\right] dx_{i.m} \right| \\ &= \left| \frac{1}{\lambda^2 L_e g} \int_{-w_a/2}^{w_a/2} \exp\left\{-jk\frac{1}{2L_f}\left[x_{i.m}+L_f\left(\frac{x_{s.m}}{L_s}+\frac{x_m}{L_e}\right)\right]^2\right\} dx_{i.m} \right|, \end{aligned} \tag{1.16}$$

$$\frac{1}{L_f} \equiv \frac{1}{g} + \frac{1}{L_e} - \frac{1}{f}. \tag{1.17}$$

Equation (1.16) is the absolute value of the definite integral of a complex Gaussian function and can be modified as

$$\begin{aligned} |h_{d.m}(x_{se})| &= \left| \frac{1}{\lambda^2 L_e g} \int_{-w_a/2}^{w_a/2} \exp\left\{-jk\frac{1}{2L_f}\left[x_{i.m}+L_f\left(\frac{x_s-mP_a}{L_s}+\frac{x-mP_a}{L_e}\right)\right]^2\right\} dx_{i.m} \right| \\ &= |h_{e.m}(-x_{se})|, \end{aligned} \tag{1.18}$$

$$x_{se} \equiv \frac{x_s - mP_a}{L_s} + \frac{x - mP_a}{L_e}, \tag{1.19}$$

where $x_{s.m}$ and $x$ are rewritten by the coordinates of the whole array given in the same way as in Eq. (1.1). The amplitude of the light wave exiting any elemental lens is spread according to Eq. (1.18). The spread is symmetric for the point $x_{se} = 0$, and we obtain the $x$ coordinate from Eq. (1.19):

$$x = mP_a\left(1 + \frac{L_e}{L_s}\right) - \frac{L_e}{L_s}x_s. \tag{1.20}$$

We can say that this equation shows the light ray from the elemental image.

Here, we assume $L_e = -L_s$, and the following is obtained:$x = x_s$ This $x$ position is independent of the elemental lens number $m$, so all waves from the elemental lenses are symmetric for point $(x_e, z_e)$, where the waves from all the elemental lenses converge and form an optical image:

$$(x_e, z_e) = (x_s, -L_s) = (x_s, z_s). \tag{1.21}$$

The image is located at point $(x_s, z_s)$ where the object is located. This result corroborates previous findings on geometrical analysis described above.

The obtained wave at the image point is given by the sum of the squared amplitudes of the waves from each elemental lens, since the light waves emitted from all points $x_{d.m}$ (defined by Eq. (1.13)) in the display plate are incoherent. Equation (1.18) is modified at the image plane,

$$\left|h_{d.m}(x_e)\right| = \left| \frac{1}{\lambda^2 L_e g} \int_{-w_a/2}^{w_a/2} \exp\left\{ -jk\frac{1}{2L_f}\left[x_{i.m} + \frac{L_f}{L_s}(x_s - x_e)\right]^2 \right\} dx_{i.m} \right|. \tag{1.22}$$

This equation shows that all amplitudes of the waves from each elemental lens are the same, since they are independent of elemental lens number $m$. The synthesized wave is obtained as

$$\sum \left|h_{d.m}(x_e)\right|^2 = M \left|h_{d.m}(x_e)\right|^2, \tag{1.23}$$

where the summation is performed for all elemental lenses and $M$ is the number of elemental lenses. Therefore, the MTF of the synthesized wave is equal to that of an elemental lens and can be calculated as the Fourier transform of the squared amplitude of the point spread function for an elemental lens.

$$MTF_d(v_s) \equiv F\left(\left|h_{d.m}(x_{e.m})\right|^2\right), \tag{1.24}$$

where $F$ shows the Fourier transform operation. $v_s$(cycles/mm) is the spatial frequency in the image and the object planes.

The MTF of the pickup stage is also given by

$$MTF_p(v_p) \equiv F\left(\left|h_{p.m}(x_{p.m})\right|^2\right). \tag{1.25}$$

$v_p$(cycles/mm) is the spatial frequency of the elemental image in the capture and display plates. The overall MTF, including the pickup and display plate, is a product of the MTFs of both stages:

$$MTF_t = MTF_p \bullet MTF_d, \tag{1.26}$$

Pixel structure in the capture and display devices affects the MTFs [18]. In this discussion, we assume that the pixel pitches are sufficiently fine or that the devices do not have a pixel structure. Therefore, we do not consider the sampling effects of the pixel structure.

The description given here shows that a 3-D optical image generated by the integral method is positioned at the same point as the object point. The MTF of an optical image formed by the whole array is the same as that of an element lens. These results, produced by the Fresnel regime, corroborate previous findings related to the resolution characteristics of the integral method [19, 20].

## 1.3 Resolution

### 1.3.1 Derivation of MTF in Capture and Display Stages

As shown in Fig. 1.1, spatial frequency $\alpha$ is normalized by the distance [19]. These have the following relationship between $v_s$, $v_p$, and $g$:

$$\alpha = \frac{1}{\tan^{-1}\left[1/(v_s\,|z_s|)\right]} \cong v_s\,|z_s| = v_p g. \tag{1.27}$$

The MTF for the capture stage can be expressed as the product of the elemental lens's MTF and the capture device's MTF. Let $MTF_p(\alpha)$ represent the MTF for the capture stage. The MTF of the display stage can also be expressed as the product of the elemental lens's and display device's MTFs. Let $MTF_d(\alpha)$ represent the MTF for the display stage. In this section, we assume that the numbers of pixels of these capture and display devices are infinite, meaning the MTF of the elemental lens is the sole factor affecting the resolution. These MTFs are obtained by Eqs. (1.24) and (1.25) and are rewritten by $\alpha$.

MTF can be calculated as the Fourier transform of the squared amplitude of the point spread function. It is equal to calculating the autocorrelation function of the pupil function [20, 21], as is well-known. It is assumed that the pupil of each elemental lens is a two-dimensional circle. The pupil function $P_{fp}$ of the elemental lens for the capture stage, which includes the effect of the defocusing, is expressed as follows:

$$P_{fp} = \exp\left[i\pi(x_{i.m}^2 + y_{i.m}^2)E_p(z_s)/\lambda\right], \tag{1.28}$$

where

$$E_p(z_s) = \left|{}^1/_g - {}^1/_{z_s} - {}^1/_f\right|, \tag{1.29}$$

$\lambda$ is the wavelength, $f$ is the focal length of the elemental lens for capture and display, and $z_s$is object distance or image distance, mentioned above. Coordinate $(x_{i.m}, y_{i.m})$ is applied to the plane of the pupil. The pupil function $P_{fd}$ of the elemental lens for display is expressed as follows:

$$P_{fd} = \exp\left[i\boldsymbol{\pi}(x_{i.m}^2 + y_{i.m}^2)E_d(z_s)/\boldsymbol{\lambda}\right], \tag{1.30}$$

where

$$E_d(z_s) = \left|{}^1/_{z_s} + {}^1/_g - {}^1/_f\right|, \tag{1.31}$$

$MTF_p$ in the capture stage and $MTF_d$ in the display stage are given by

$$
\begin{aligned}
MTF_p(\alpha) &= \frac{1}{S}\iint\limits_{pupil} P_{fd}(x_{i.m}+\frac{\alpha\lambda}{2})P^*_{fd}(x_{i.m}-\frac{\alpha\lambda}{2})dx_{i.m}dy_{i.m} \\
&= \frac{1}{S}\iint\limits_{pupil} \exp\left[i2\pi\alpha x_{i.m}E_p(z_s)\right]dx_{i.m}dy_{i.m},
\end{aligned}
\tag{1.32}
$$

$$
\begin{aligned}
MTF_d(\boldsymbol{\alpha}) &= \frac{1}{S}\iint\limits_{pupil} P_{fd}(x_{i.m}+\frac{\boldsymbol{\alpha\lambda}}{2})P^*_{fd}(x_{i.m}-\frac{\boldsymbol{\alpha\lambda}}{2})dx_{i.m}dy_{i.m} \\
&= \frac{1}{S}\iint\limits_{pupil} \exp\left[i2\boldsymbol{\pi\alpha} x_{i.m}E_d(z_s)\right]dx_{i.m}dy_{i.m},
\end{aligned}
\tag{1.33}
$$

where the asterisk denotes a complex conjugate, and $S$ is the areas of the pupil of the elemental lens for the capture and display stages.

Here, we set $g = f$ and obtain:

$$E_p(z_s) = E_d(z_s) = \left|\frac{1}{z_s}\right| \equiv E(z_s). \tag{1.34}$$

Therefore,

$$MTF_p(\boldsymbol{\alpha}) = MTF_d(\boldsymbol{\alpha}) \equiv MTF(\boldsymbol{\alpha}) \tag{1.35}$$

where

$$MTF(\alpha) = {}^{4R}/_{SB} \cdot (D_1\cos\varphi - D_2\sin\varphi), \tag{1.36}$$

$$
\begin{aligned}
D_1 = {} & 2[J_1(RB)(\frac{\theta}{2}+\frac{\sin 2\theta}{4}) - J_3(RB)(\frac{\sin 2\theta}{4}+\frac{\sin 4\theta}{8}) \\
& + J_5(RB)(\frac{\sin 4\theta}{8}+\frac{\sin 6\theta}{12}) - \cdots],
\end{aligned}
\tag{1.37}
$$

$$
\begin{aligned}
D_2 = {} & J_0(RB) - 2[J_2(RB)(\frac{\sin\theta}{2}+\frac{\sin 3\theta}{6}) - J_4(RB)(\frac{\sin 3\theta}{6}+\frac{\sin 5\theta}{10}) \\
& + J_6(RB)(\frac{\sin 5\theta}{10}+\frac{\sin 7\theta}{14}) - \cdots],
\end{aligned}
\tag{1.38}
$$

$B = 2\boldsymbol{\pi\alpha}E(z_s)$, $\varphi = B\alpha\lambda/2, J_n(\cdot)$ is a Bessel function of the $n$-th order, $R$ is the radius of the pupil of the elemental lens, and $\theta = \cos^{-1}(\alpha\lambda/2R)$. Here, we assume that the aberration of each elemental lens is negligible.

### 1.3.2 Examples of MTF

The spatial frequency measured from the observer's position, i.e., visual spatial frequency $\beta$ (cpr), is defined here to clarify the argument. Spatial frequencies $\alpha$ and $\beta$ have the following relationship [19]:

$$\beta = \alpha \left( L_{OB} - z_s \right) / \left| z_s \right| , \tag{1.39}$$

where $L_{OB}$ is the viewing distance between the lens array and the observer. This $\beta$ is originally defined in the display stage. It can be expanded in the capture stage and considered as a spatial frequency when an object is viewed from the observer's position.

When the observer views the reconstructed image, it is being sampled at the elemental lens, as shown in Fig. 1.2. The maximum spatial frequency of reconstructed images is limited to the Nyquist frequency. With $P_a$ representing the pitch between elemental lenses, the Nyquist frequency can be expressed as follows based on the visual spatial frequency:

$$\beta_{NL} = L_{OB} / 2P_a . \tag{1.40}$$

The sampling effect is conspicuous if the elemental lenses and observer's pupil are pinholes. It is also clear when the image is located on the lens array. We assume the Nyquist frequency limitation is expanded when the elemental lenses and the pupil are lenses, not pinholes, or when the image is not located on the lens array.

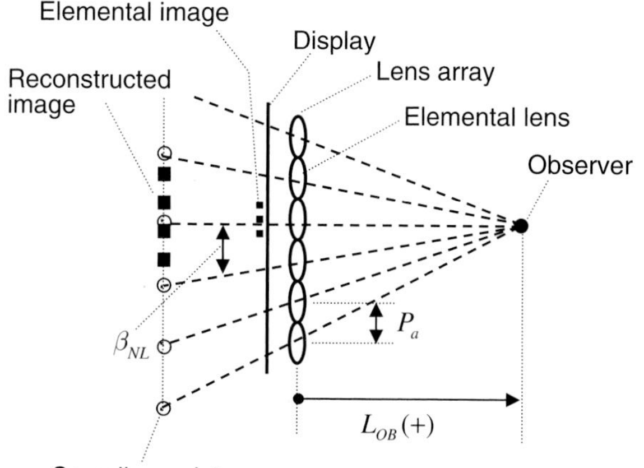


**Fig. 1.2.** Sampling by lens array

Here, we assume a standard observer with 20/20 eyesight [22]. The observer is usually introduced to evaluate television systems. The eyesight corresponds to a resolution of 1.0 min in a visual angle. The spatial frequency corresponding to the observer's eyesight, called the visual-limit spatial frequency (VLSF), is given by

$$\beta_{VL} = \frac{360 \times 60/2}{2\pi} = 1720 \quad \text{(cycles/rad.: cpr)}. \tag{1.41}$$

The observer cannot perceive an elemental lens arrangement with a visual spatial frequency greater than the VLSF. We can set the Nyquist frequency to be greater or less than the VLSF, depending on viewing distance. Here, we set the distance to obtain:

$$\beta_{NL} = \beta_{VL} \tag{1.42}$$

In addition, actual resolution deteriorates due to an aberration in the elemental lens. However, MTF was calculated using only the influence of diffraction with focus defects because it was assumed that ideal lenses were used as elemental lenses.

Figure 1.3 plots the MTFs in relation to the visual spatial frequencies. The diameters of the lenses are 0.25, 0.5, and 1.0 mm. Viewing distances that give the VLSF are 0.86, 1.72, and 3.44 m, respectively. The normalized object distance is one-quarter of the viewing distance. As the graph shows, the smaller the diameter of the elemental lens, the smaller the MTF is. Figure 1.4 shows the MTFs and the normalized object distance is –1/4 of the viewing distance. Other parameters are the same as those in Fig. 1.3. The graph also shows that the smaller the diameter of the elemental lens,

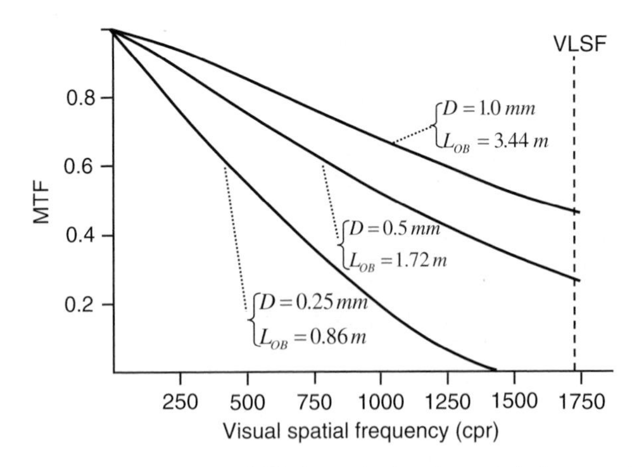


**Fig. 1.3.** Relationship between MTFs and visual spatial frequencies. Object is located at a distance of $0.25L_{OB}$ (R.I. area)

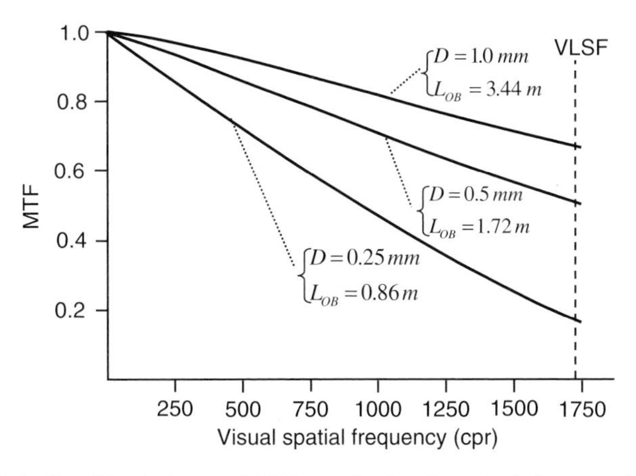


**Fig. 1.4.** Relationship between MTFs and visual spatial frequencies. Object is located at a distance of $-0.25L_{OB}$ (R.O. area)

the smaller the MTF is. The MTFs are improved compared with Fig. 1.3, because the effect of defocusing is less than the MTFs in Fig. 1.3. Figure 1.5 shows the MTF responses in the VLSF. The responses decrease as the object is located at a distantance from the lens array. Because of diffraction, this tendency is greater when the radius of the elemental lens is smaller. When the lens is 1.0 mm in diameter, the MTF is not zero even if the object is placed sufficiently far away, at a distance of 100 $L_{OB}$, enabling high quality images to be observed.

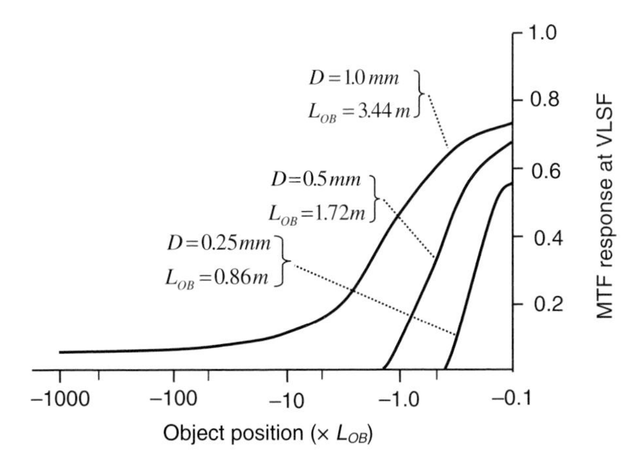


**Fig. 1.5.** Relationship between MTF responses at VLSF and object positions

### 1.3.3 Effect of Pixel Pitch

Here, it is assumed the elemental image has a pixel structure, and the pitch is given by $P_{NP}$. The elemental image is sampled by the pixels on the elemental images. The maximum projected spatial frequency of the object is limited by the Nyquist frequency:

$$\alpha_{NP} \cong \frac{g}{2P_{NP}}. \tag{1.43}$$

Visual spatial frequency $\beta_{NP}$ according to $\alpha_{\mathrm{NP}}$, is given by Eq. (1.39):

$$\beta_{NP} = \alpha_{NP} \frac{L_{OB} - Z_s}{|Z_s|} = \frac{g}{2P_{NP}} \frac{L_{OB} - Z_s}{|Z_s|} \tag{1.44}$$

Here, assuming $-z_s$ is infinite, we obtain:

$$\beta_{NP} = \frac{g}{2P_{NP}}. \tag{1.45}$$

Maximum visual spatial frequency $\beta_{MX}$ is given by the following equation:

$$\boldsymbol{\beta}_{MX} = \min(\boldsymbol{\beta}_{NP}, \boldsymbol{\beta}_{DF}, \boldsymbol{\beta}_{VL}, \boldsymbol{\beta}_{NL}) \tag{1.46}$$

where $\beta_{DF}$ is a spatial frequency that gives the null response by diffraction or the defocus described in the former section. When the diameter of the elemental lens is more than about 1.0 mm, $\beta_{DF}$ is more than $\beta_{VL}$ under the condition in Eq. (1.42). If the pixel pitch of each elemental image $P_{NP}$ is too large, the MTF response depends mainly on the Nyquist frequency $\beta_{NP}$ by the pixel pitch.

The viewing zone is given by [19]

$$\boldsymbol{\phi}_{VZ} \cong \frac{P_a}{g} \tag{1.47}$$

A wide viewing zone requires small g, but small $g$ degrades $\beta_{NP}$. To compensate for the degradation, $P_{NP}$ needs to be smaller.

## 1.4 Experimental System

To obtain moving pictures, an electronic capture device, such as a CCD or CMOS image sensor, is set on the capture plate and takes elemental images. For reconstruction, a display device such as an LCD panel or an EL panel is placed behind the lens array. Video signals are transmitted from the capture device to the display device.

The size of the lens array for capturing needs to be large enough to obtain a large parallax; however, it is difficult to develop a large capture device for moving pictures. In the actual system, a television camera using a pickup lens is set to capture all elemental images formed by the elemental lenses. In the future, a capture device of the same size as the lens array will need to be developed and set immediately behind the lens array.

Figure 1.6 shows an experimental real-time integral imaging system. A depth control lens, a GRIN lens array [14, 15], a converging lens, and an EHR (extremely high resolution) camera with 2,000 scanning lines [23] are introduced for capturing. The depth control lens [10] forms the optical image of the object around the lens array, and the GRIN lens array captures the optical image. Many elemental lenses (GRIN lenses) form elemental images near the output plane of the array. An elemental GRIN lens acts as a specific lens forming an erect image for the object in the distant R.O. area to avoid pseudoscopic 3-D images. The converging lens [10], which is set close to the GRIN lens array, leads the light rays from elemental GRIN lenses to the EHR camera. The converging lens uses light rays efficiently, but is not an essential part of the system. In principle, the camera is focused on the elemental images, which are formed around the output plane of the GRIN lens.

Table 1.1 shows the experimental specifications of the capture setup. Figure 1.7 shows the two-dimensional arrangement of the GRIN lens array used in the experiment. The pitch between the adjacent elemental lenses corresponds to 21.3 pixels of the EHR camera. The arrangement has a delta structure, which is more efficient than a grid structure. The horizontal pitch is considered 21.3/2 and the vertical one is considered $21.3 \times \sqrt{3}/2$, equivalently.

The video signal of the camera is led directly to a display device. Table 1.2 shows the experimental specifications of the display setup. The electronic display device combines a color LCD panel and an array of convex lenses. The lens array has the same delta structure as the GRIN lens array. The pitch between the adjacent elemental lenses corresponds to 21.3 pixels of the

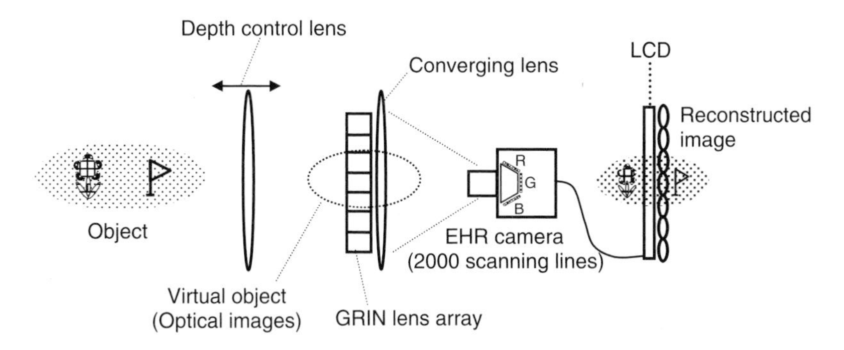


**Fig. 1.6.** An experimental 3-D integral television system

**Table 1.1.** Specifications of capture setup

| | | |
|---|---|---|
| Camera | Sensor | 3 CMOS s |
| | Number of pixels | 3,840(H)×2,160(V) |
| | Size of pickup device | 16.1(H)×9.1(V) mm$^2$ |
| | Frame frequency | 60 Hz |
| | Focal length of lens | 90 mm |
| | Color system | Color separation prism (R, G, B) |
| GRIN lens array | Number of GRIN lenses | 160 (H) × 125 (V) |
| | Diameter | 1.085 mm (20.5 pixels) |
| | Pitch between lenses | 1.14 mm (21.3 pixels) |
| | Lens length | 20.25 mm |
| | Focal length | –2.65 mm |
| | Size | 182.4 (H) × 123.5 (V) mm$^2$ |
| | Structure | Delta (Line offset) |
| Depth control lens | Focal length | 225 mm |
| Converging lens | Focal length | 650 mm |

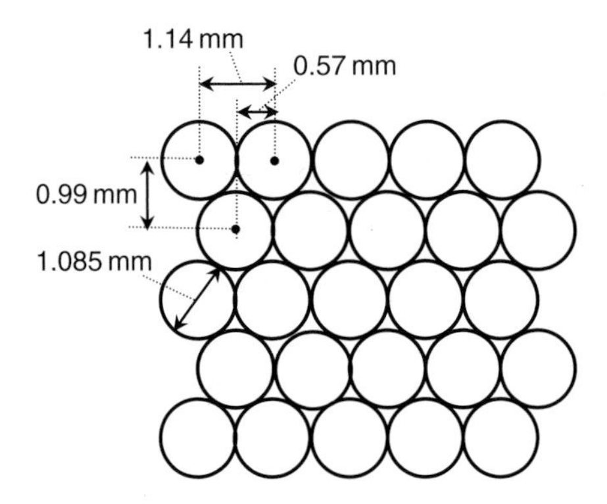


**Fig. 1.7.** Elemental lenses arrangement of capture stage

LCD panel. The gap between the lens array and display device is about the same as the focal length of the elemental convex lens.

Figure 1.8 shows calculated maximum visual spatial frequency$\boldsymbol{\beta}_{MX}$ in the horizontal and vertical directions, which degrades for the objects that are distant from the lens array in the R.O. and R.I. areas. There is difference of $\boldsymbol{\beta}_{NL}$ between horizontal and vertical directions because the lens array has a delta structure. In this experimental system, the limitation of the pixel pitch on the elemental image is dominant compared to the diffraction and defocus by the elemental lens mentioned in the previous section because the pixel pitch in the experimental system is too large. The depth control lens shifts the object around the lens array as an optical image. This prevents distant objects from being degraded. Figure 1.9(a)–(e) shows a reconstructed 3-D image, "balloon

**Table 1.2.** Specifications of display setup

| | | |
|---|---|---|
| Display panel | Device | LCD |
| | Number of active pixels | 3,840 (H) X 2,160 (V) |
| | Frame frequency | 60 Hz |
| | Color system | R, G, B Stripe |
| Lens array | Number of lenses | 160 (H) × 125 (V) |
| | Diameter | 2.64 mm (21.3 pixels) |
| | Pitch between lenses | 2.64 mm (21.3 pixels) |
| | Focal length | 8.58 mm |
| | Size | 422.4 (H) × 286.1 (V) mm$^2$ |
| | Structure | Delta (Line offset) |
| Viewing area | Angle | 12 degree (measured value) |
| | Width (at 2 m) | 594 mm |

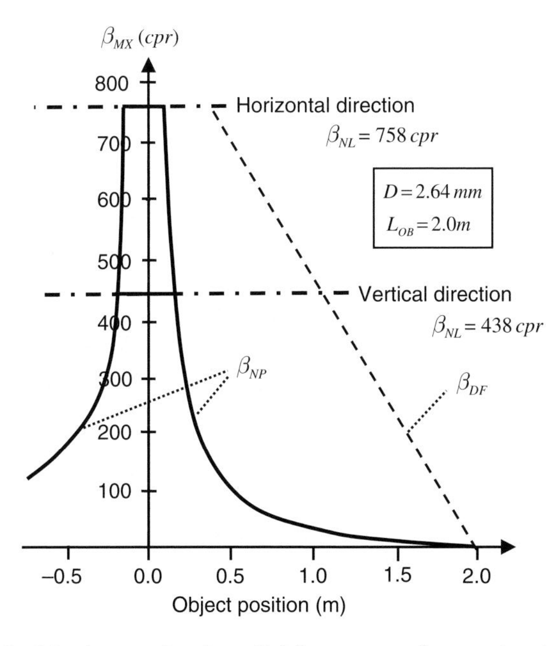


**Fig. 1.8.** Maximum visual spatial frequency of reconstructed images

and lady," taken from upper, center, lower, left, and right viewpoints. The obtained figures differ according to the viewpoints. Differences between the figures can clearly be seen around the lady's neck, as shown by the white circles in the figures. The reconstructed 3-D image is formed with full parallax in real-time.

(a) Upper viewpoint

(d) Left viewpoint

(b) Center viewpoint

(e) Right viewpoint

(c) Lower viewpoint

**Fig. 1.9.** Reconstructed image viewed from different viewpoints

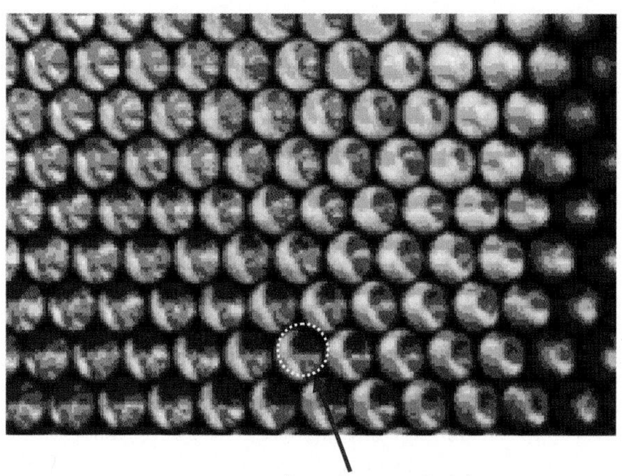


**Fig. 1.10.** Enlarged portion of elemental images

Figure 1.10 shows an enlarged portion of the elemental images that are slightly different from the neighboring elemental images. These elemental images are isolated from each other. The number of active elemental images is 160(H)×125(V), which is insufficient for television. The viewing zone angle is about 12 degrees as a value measured horizontally and vertically in an actual

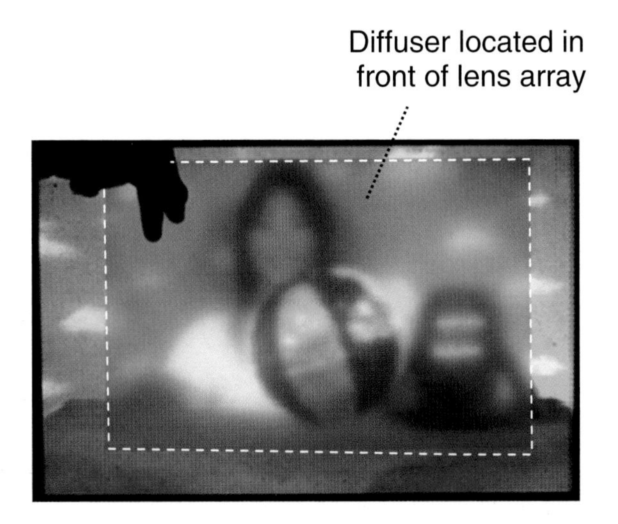


**Fig. 1.11.** An example of reconstructed image projected on diffuser

system, which is not sufficient for a television system. Figure 1.11 shows the images projected on the diffuser set in front of the array. The diffuser corresponds to the position of the balloon, so the balloon is imaged and the lady is less clear than in Fig. 1.9. It is evident that the integral imaging reconstructs optical 3-D images as holography does.

## 1.5 Conclusion

An experimental system using an EHR video system with 2,000 scanning lines was developed and shown. The system produces full-color and full-parallax 3-D images in real-time. The number of elemental lenses rules the resolution of the images located around the lens array. This corresponds to the upper limit of resolution of reconstructed images. Its number should be increased to match the number of pixels of conventional television (i.e., 480(V)×640(H) pixels) in the future. The pitch of the elemental lens or its diameter should be small enough to obtain the desired resolution.

However, because of diffraction, too small a diameter of the elemental lens degrades the resolution of images distant from the lens array. The diameter should be around 1.0 mm. The pixel pitch in an elemental image also affects the resolution of images distant from the lens array. A fine pixel pitch is also effective in obtaining a wide viewing zone and is desired to match the limiting resolution by diffraction of the elemental lens. As a result, there are more than 100 vertical and horizontal pixels of an elemental image. The total number of pixels of an elemental image exceeds 10,000. If the system is limited to capture an object near the lens array, the diameter of each elemental lens can be smaller than about 1.0 mm. The diffraction limits on resolution are relieved under this condition. The number of the pixels in an elemental image can be reduced.

To produce high quality 3-D images for television, this system requires an extremely large number of pixels for the capture and display stages. Although this problem must be overcome, we confirmed that our real-time system can be applied to 3-D television. Furthermore, because elemental images contain 3-D information, this approach can also be applied to many other applications.

## References

[1] G. Lippmann, Comptes-Rendus, 146, pp. 446–451 (1908).
[2] A. P. Sokolov, "Autostereoscopy and integral photography by Professor Lippmann's method," Izd. MGU, Moscow State University Press (1911).
[3] H. E. Ives, "Optical properties of a Lippmann lenticulated sheet," J. Opt. Soc. Am., 21, pp. 171–176 (1931).
[4] R. L. DeMontebello, "Wide-angle integral photography – The integral system," in Proc. of SPIE Annu. Tech. Conf., San Diego, Seminar 10, No. 120–08, Tech. Digest, pp. 73–91 (1977).
[5] H. E. Ives, "Optical device," U.S. Patent 2 174 003, Sept. 26 (1939).
[6] N. Davies, M. McCormick, and M. Brewin, "Design and analysis of an image transfer system using microlens arrays", Opt. Eng., 33(11), pp. 3624–3633 (1994).
[7] J. Y. Son, V. Saveljev, B. Javidi, and K. Kwack, "A method of building pixel cells with an arbitrary vertex angle," Opt. Eng., 44(2), pp. 0240031–0240036 (2005).
[8] J. S. Jang and B. Javidi, "Depth and size control of three-dimensional images in projection integral imaging," Opt. Exp., 12(6), pp. 3778–3790 (2004).
[9] F. Okano, H. Hoshino, J. Arai, and I. Yuyama, "Real-time pickup method for a three-dimensional image based on integral photography," Appl. Opt., 36(7), pp. 1598–1603 (1997).
[10] F. Okano, J. Arai, H. Hoshino, and I. Yuyama, "Three-dimensional video system based on integral photography," Opt. Eng., 38(6), pp. 1072–1077 (1999).
[11] F. Okano and J. Arai, "Optical shifter for a three-dimensional image by use of a gradient-index lens array," Appl. Opt., 40(20), pp. 4140–4147 (2002).
[12] F. Okano, J. Arai, and M. Okui, -Amplified optical window for three-dimensional images,, Opt Lett., 31(12), pp. 1842–1844 (2006).
[13] J. Hamasaki, M. Okada, and S. Utsunomiya, "Lens-plate 3D camera using orthoscopic-pseudoscopic-image-conversion optic," Monthly Journal of Institute of Industrial Science, University of Tokyo, 40(3), pp. 127–136 (1988).
[14] F. Okano, J. Arai, and H. Hoshino, "Stereoscopic image pickup device and stereoscopic display device," Japan Patent Appl. No. 08-307763 (1996).
[15] J. Arai, F. Okano, H. Hoshino, and I. Yuyama, "Gradient-index lens array method based on real-time integral photography for three-dimensional images," Appl. Opt., 37(11), pp. 2034–2045, (1998).
[16] F. Okano, J. Arai, and M. Kawakita, -Wave optical analysis of integral method for three-dimensional images,- Opt Lett., 2007, 32 (4), pp. 364–366 (2007).

[17] F. Okano; J. Arai, and M. Okui, "Visual resolution characteristics of an afocal array optical system for three-dimensional images, 46(02), p. 023201 (2007).
[18] A. Stern and B. Javidi, "Shannon number and information capacity of three-dimensional integral imaging," J. Opt. Soc. Am. A, 21, pp. 1602–1612 (2004).
[19] H. Hoshino, F. Okano, H. Isono, and I. Yuyama, "An analysis of resolution limitation of integral photography," J. Opt. Soc. Am. A, 15(8), pp. 2059–2065 (1998).
[20] J. Arai, H. Hoshino, M. Okui, and F. Okano, "Effects of focusing on the resolution characteristics of integral photography," J. Opt. Soc. Am. A, 20 (6), pp. 996–1004 (2003).
[21] H. H. Hopkins, "The frequency response of a defocused optical system," Proc. Roy. Soc., A231, pp. 91–103 (1955).
[22] "Relative quality requirements of television broadcast systems," Rec. ITU-R BT.1127.
[23] I. Takayanagi, M. Shirakawa, K. Mitani, M. Sugawara, S. Iversen, J. Moholt, J. Nakamura, and E. Fossum, "A 1 $^1/_4$ inch 8.3 M Pixel Digital Output CMOS APS for UDTV Application," in Proc. of ISSCC 2003, vol. 1, pp. 216–217 (2003).

# 2

# High Depth-of-Focus Integral Imaging with Asymmetric Phase Masks

Albertina Castro, Yann Frauel, and Bahram Javidi

**Abstract** In this chapter, we address the problem of the limited depth-of-field and depth-of-focus of integral imaging systems. We first describe the origin of the problem in both the pickup and the reconstruction stages. Then we show how the depth-of-field/depth-of-focus can be significantly improved by placing an asymmetric phase mask in front of each lenslet. We apply this technique in the pickup as well as in the reconstruction stages, and we demonstrate that very out-of-focus objects can be resolved, at the price of a slight diffusion effect. Moreover, since the use of a phase mask preserves all the spectrum information within its passband, it is possible to apply a digital restoration step in order to eliminate the diffusion effect and retrieve clean images over a large range of distances.

## 2.1 Introduction

Recently, three-dimensional (3-D) imaging has attracted a lot of interest. It has many potential applications in medicine, in security and defense, in manufacturing and entertainment industries among others. Integral imaging has shown to be one of the most promising techniques for 3-D displays [1, 2, 3, 4, 5, 6, 7, 8, 9, 10, 11, 12, 13, 14, 15, 16, 17, 18, 19, 20]. Contrary to holography, it uses regular—polychromatic and incoherent—light [11], and does not require the use of special glasses. Another advantage is that it allows the viewer to

A. Castro
Instituto Nacional de Astrofísica, Óptica y Electrónica, Apartado Postal 51, Puebla, Pue. 72000, Mexico
e-mail: betina@inaoep.mx

B. Javidi et al. (eds.), *Three-Dimensional Imaging, Visualization, and Display*,
DOI 10.1007/978-0-387-79335-1_2, 

freely move to see different perspectives of the scene and that several people can simultaneously see the display. However, as traditional optical systems, integral imaging suffers from a limited depth-of-focus and a limited depth-of-field that cause in-depth scenes to partly appear out-of-focus. There have been several approaches to increasing the depth-of-field of integral imaging systems. For instance Jang, et al. introduced the concept of real and virtual image fields [12]. Another technique uses a curved device for capturing and displaying the elemental images in the pickup and the reconstruction stages, respectively [13]. A different approach utilizes an array of lenslets with different focal lengths [14], and its diffractive version uses an array of binary zone plates with different focal lengths [15]. Annular binary amplitude filters have also been proposed to extend the depth-of-field [16, 17, 18]. In this chapter we present the use of odd symmetry phase masks [21, 22, 23, 24, 25, 26, 27]. Unlike amplitude filters these phase-only masks extend the depth-of-field/depth-of-focus of the integral imaging system without reducing the amount of collected light [3,6]. The extension of the depth-of-field/depth-of-focus is obtained at the price of a slight diffusion effect. However, this degrading effect can be reversed through a proper digital restoration process.

In Section 2.2 we present a brief review of the principles of an integral imaging system. We explain both the limited depth-of-field obtained with a regular lenslet array during pickup and the limited depth-of-focus during reconstruction. Section 2.3 presents the phase-only masks used to increase the depth-of-field/depth-of-focus. Sections 2.4 and 2.5 present the results of the modified integral imaging system that incorporates an asymmetric phase mask in the pickup and in the reconstruction stage, respectively. Lastly, the final remarks are given in Section 2.6.

## 2.2 Principles of Integral Imaging

The idea of integral imaging [19] is to capture 3-D information of a scene through the use of a two-dimensional array of imaging elements such as pinholes or lenslets. During the *pickup stage*, each element forms an *elemental image* obtained from a particular point of view. The set of elemental images is recorded by either a photographic medium or a digital camera [20]. During the *reconstruction stage*, all the previously recorded images are displayed in their original positions. When these images are viewed through the same imaging array that was used during pickup, a perception of 3-D is obtained. In the following subsections we separately detail the pickup stage and the reconstruction stage.

### 2.2.1 Pickup Stage

Let us first consider the simpler case of integral imaging with a pinhole array. During the pickup stage, each pinhole projects a particular object point onto

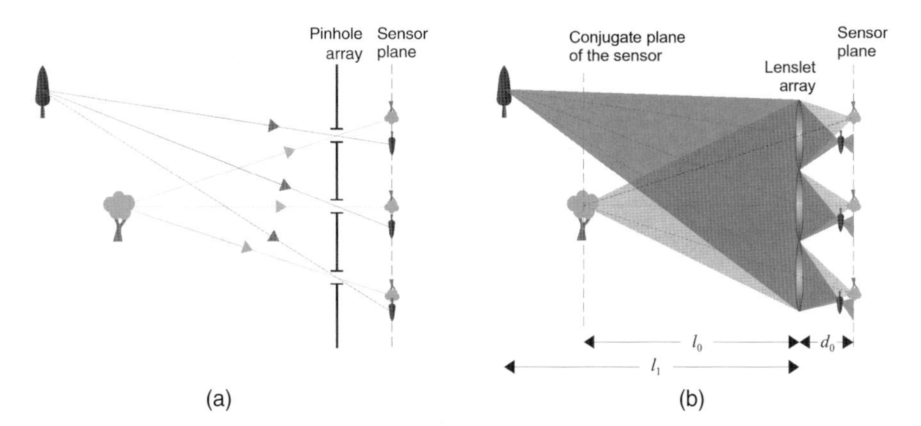


**Fig. 2.1.** Pickup stage (**a**) with a pinhole array and (**b**) with a lenslet array

the elemental image plane. The locations of the projections depend on both the lateral and the longitudinal position of the original object (Fig. 2.1(a)). In particular, the distance between the different projections of the same point is related to the distance of this point. The depth information is thus captured through the relationships between elemental images. Because a single ray is selected by each pinhole, the projected points are equally obtained for any distance of the object.

Now pinholes are not actually usable since they strongly reduce the amount of collected light. In practice, the elemental images are obtained with an array of lenslets. We consider here the case of paraxial optics without aberrations other than defocus. Each lenslet collects a whole beam of rays emitted by an object point and concentrates it to form an image point (Fig. 2.1(b)). The relation between the object distance $l$ and the conjugate image distance $d$ is given by the lens law

$$\frac{1}{l} + \frac{1}{d} = \frac{1}{f}, \tag{2.1}$$

where $f$ is the focal length of the lenslet. Unfortunately, image distance $d$ does not necessarily correspond to the distance of the sensor $d_0$. In that case, the projection of the object point onto the sensor plane is spatially expanded, meaning that the obtained elemental image is blurred. Only objects at distance $l_0$, conjugate of $d_0$, will exactly appear in focus. Figure 2.2 presents the intensity image of a point object—or *Intensity Point Spread Function* IPSF—for various object distances corresponding to defocuses of $W_{20} = 0$, $W_{20} = 3\lambda$, $W_{20} = 5\lambda$ and $W_{20} = 7\lambda$, respectively (see Section 2.3). It can be seen that the blurring effect is all the stronger as the object is farther away from the in-focus plane. The depth-of-field can be defined as the range of distances $l$ around $l_0$ such that the defocus effect is barely perceivable, which can be expressed as.

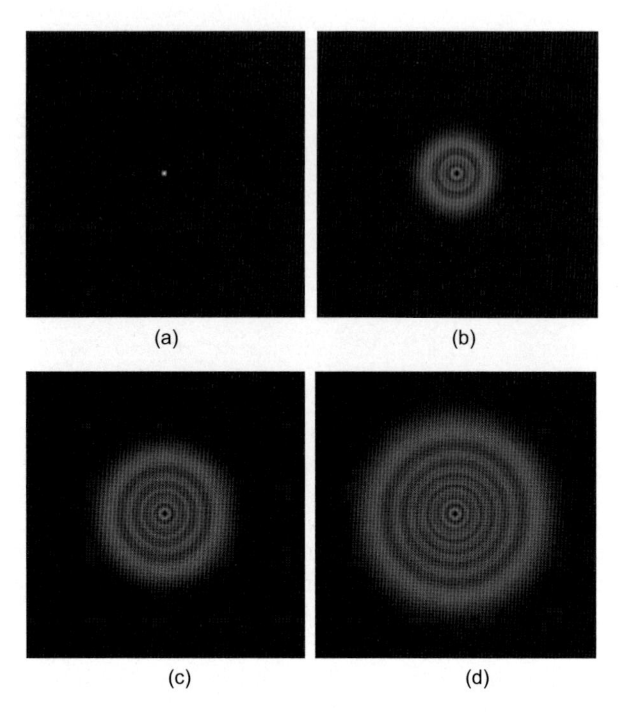


**Fig. 2.2.** Image of a point object located at various distances from the array of lenslets. (**a**) $l = 100$ mm (**b**) $l = 143$ mm (**c**) $l = 200$ mm (**d**) $l = 333$ mm

$$H_l(0,0) \geq \frac{1}{\sqrt{2}} H_{l_0}(0,0), \qquad (2.2)$$

where $H_l(x, y)$ and $H_{l_0}(x, y)$ are the IPSFs for an object at distance $l$ and $l_0$, respectively [32]. Figure 2.3 gives the evolution of the peak value $H_l(0, 0)$ of the normalized IPSF versus object distance $l$ for a regular lenslet of diameter 2 mm and a focal length of 5 mm. The depth-of-field in this case is about 6 mm.

In order to see the effect of defocus, it is possible to digitally simulate an integral imaging setup as shown in Fig. 2.4. The imaging array contains $7 \times 7$ lenslets having a diameter of 2 mm and a focal length of 5 mm. The sensor plane—where the elemental images are captured – is located at distance $d_0 = 100/19$ mm, in a way that the plane located at $l_0 = 100$ mm is in focus. The object scene contains four E-chart objects at distances $l_0 = 100$ mm, $l_1 = 143$ mm, $l_2 = 200$ mm and $l_3 = 333$ mm from the lenslet array, respectively. Figure 2.5 shows a sample of three elemental images provided by different lenslets of the array. It is clear from this figure that not all objects are sharply imaged. Since the purpose of integral imaging is precisely to capture in-depth objects or scenes, this impossibility to obtain sharp images for all distances at

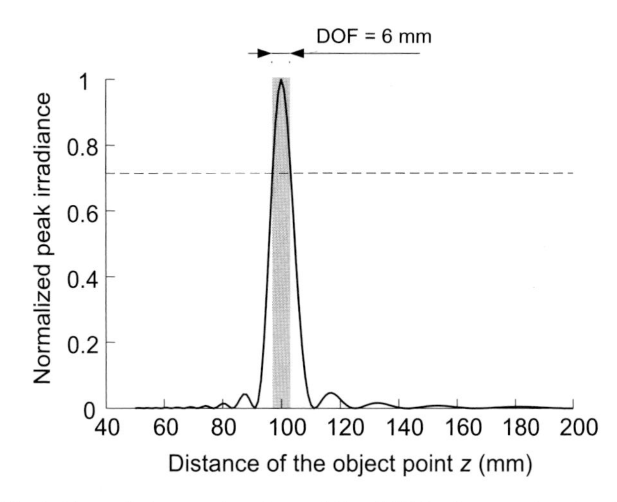


**Fig. 2.3.** Evolution of the peak value of the IPSF of a regular lenslet. The gray area represents the depth-of-field

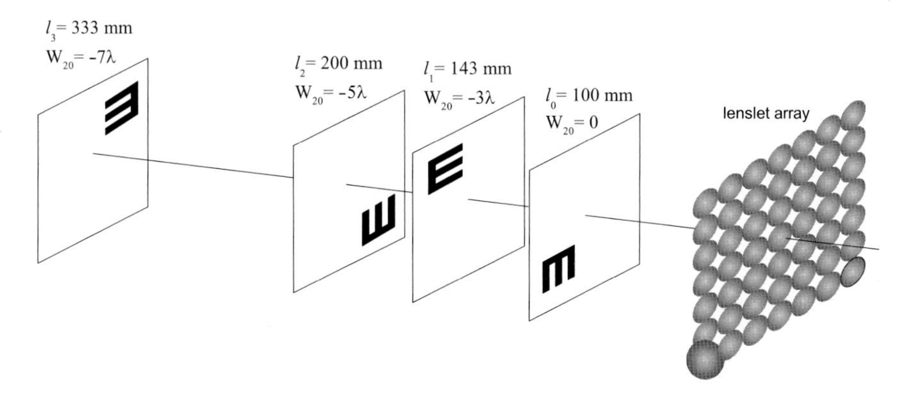


**Fig. 2.4.** Simulated integral imaging setup

the same time is a critical issue. Obviously, if the captured elemental images are blurred the 3-D object reconstructed at a later stage will also be blurred.

### 2.2.2 Reconstruction Stage

For the reconstruction stage, we set aside the limitation of the depth-of-field explained in the previous section. We thus assume that the elemental images are all sharp. For instance, they may have been obtained through a pinhole array or computationally generated. These elemental images are displayed at a distance $d_0$ from the imaging array. When the reconstruction is done using a pinhole array, the rays coming from different elemental images recombine at the 3-D location of

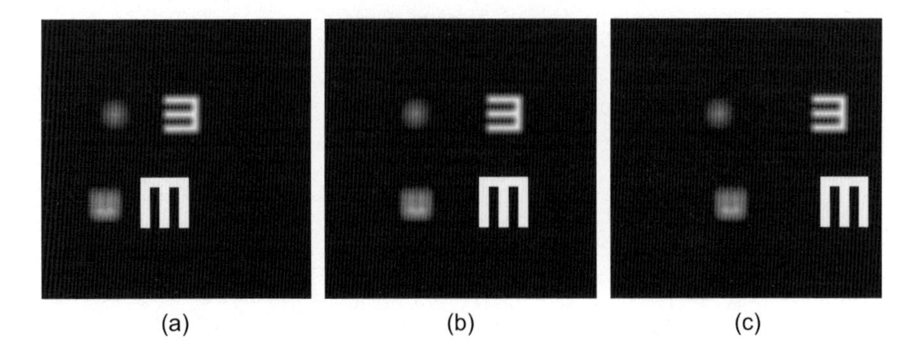


**Fig. 2.5.** Elemental images obtained through various lenslets of the array. The objects are located at 100 mm, 143 mm, 200 mm and 333 mm, respectively. (**a**) Lenslet (0,-2) (**b**) Central lenslet (0,0) (**c**) Lenslet (0,+2)

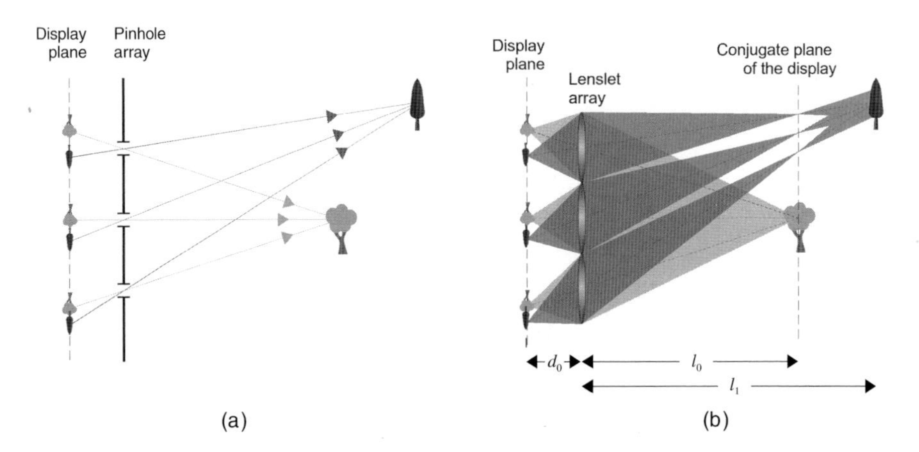


**Fig. 2.6.** Reconstruction stage (**a**) with a pinhole array and (**b**) with a lenslet array

the original object point, at a distance $l$ from the array (Fig. 2.6(a)). The viewer, therefore, perceives that the light is emitted from this location.

When the reconstruction is performed with a lenslet array, again each lenslet conveys a whole beam of light (rather than a single ray) for each elemental image point. Here two competing processes take place. On the one hand, the beams transmitted by different lenslets still recombine at the location of the original object, at distance $l$. On the other hand, the beam conveyed by a particular lenslet converges in the conjugate plane of the elemental image display plane, at distance $l_0$. This latter distance only depends on the position of the display plane (see Eq. 2.1) and is fixed, while distance $l$ only depends on the position of the original objects. Both distances are, in general, different. The result of this discrepancy is that the object is still reconstructed in its correct location, but appears blurred because the individual beams are not

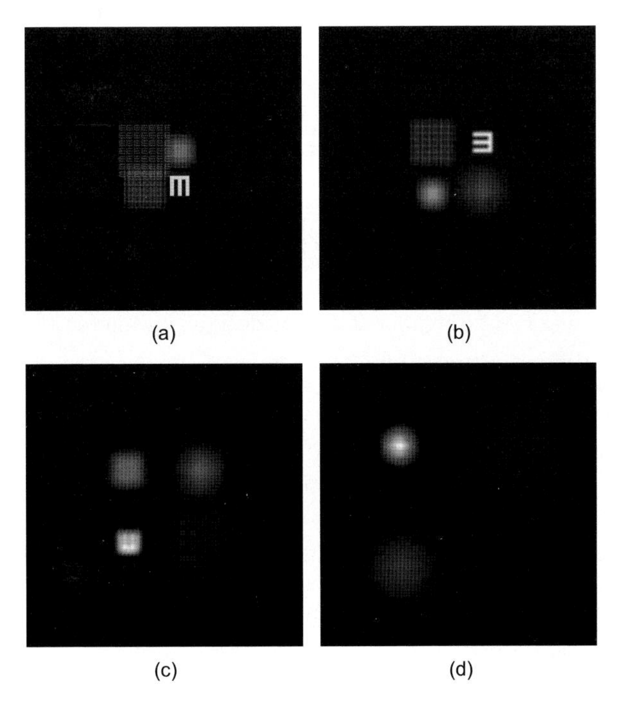

**Fig. 2.7.** Reconstructed scene obtained with an array of lenslets for various reconstruction distances (**a**) $l = 100$ mm (**b**) $l = 143$ mm (**c**) $l = 200$ mm (**d**) $l = 333$ mm

focused to a point at this location (see Fig. 2.6(b)). Only objects that were originally located at distance $l = l_0$ are sharply reconstructed. Figure 2.7 provides images of the reconstruction of the scene in Fig. 2.4 in various planes. As mentioned above, the objects are reconstructed in their correct locations, but are all the more blurred as this location is far away from the conjugate plane of the elemental image display plane $l_0 = 100$ mm.

There exists an interval distance around the in-focus plane $l_0$ where the defocus is barely noticeable. This range is called *depth-of-focus* and is defined similarly to the depth-of-field described in Section 2.2.1.

## 2.3 Asymmetric Phase Masks

It has been shown that the depth-of-field and depth-of-focus can be extended by introducing an amplitude or phase modulation in the pupil of the system [21, 22, 23, 24, 25, 26, 27, 28, 29, 30, 31]. The advantage of phase modulation over amplitude modulation is that the light gathering power of the system is preserved. In particular, a family of odd symmetric phase masks—or asymmetric phase masks—has been demonstrated to be useful in extending the depth-of-field [21–27]. These masks have a rectangularly separable profile given by

$$\varphi(x) = \exp\left[i2\pi\alpha \text{sgn}(x)\left|\frac{x}{w}\right|^k\right], \tag{2.3}$$

where the spatial coordinate in the pupil plane $x$ is normalized by the semi-width of the physical extent of the pupil $w$, sgn represents the signum function and $\alpha$ is an adjustable factor that represents the maximum phase delay introduced by the mask. The exponent $k$ is a design parameter that defines the phase mask order.

In order to easily analyze the effect of pupil modulation on the depth-of-field, it is possible to define a *generalized pupil function* [33] as follows

$$P(x, y) = Q(x, y)M(x, y)D(x, y), \tag{2.4}$$

where $x$ and $y$ are the spatial coordinates, $Q(x, y)$ stands for the pupil aperture (circular in our case), $M(x, y) = \varphi(x)\varphi(y)$ is the phase modulation introduced by the phase mask and $D(x, y)$ is a quadratic phase factor that accounts for the defocus:

$$D(x, y) = \exp\left[i2\pi\frac{W_{20}}{\lambda}\left(\frac{x^2 + y^2}{w^2}\right)\right], \tag{2.5}$$

with $\lambda$ the wavelength, $w$ the radius of the lenslet and

$$W_{20} = \frac{1}{2}\left(\frac{1}{d_0} - \frac{1}{l} - \frac{1}{f}\right)w^2 = \frac{1}{2}\left(\frac{1}{l_0} - \frac{1}{l}\right)w^2, \tag{2.6}$$

as Hopkins' defocus coefficient [34].

In the pickup stage (see Fig. 2.1), $l$ represents the distance of the object and $l_0$ corresponds to the location of the conjugate plane of the sensor. In that case, both distances refer to the object space. In the reconstruction stage (see Fig. 2.6), $l$ represents the distance of the reconstructed object and $l_0$ corresponds to the location of the conjugate plane of the display device. Both distances then refer to the image space.

While the IPSF reveals the behavior of the optical system in the spatial domain, its *Optical Transfer Function (OTF)* characterizes its behavior in the frequency domain. Both functions can be computed for a particular value of defocus using the generalized pupil function of Eq. (2.4). The IPSF is the square modulus of the Fourier transform of $P(x, y)$, and the OTF is obtained as the autocorrelation of the properly scaled $P(x, y)$. The modulus of the OTF is also known as *Modulation Transfer Function* or MTF.

For the asymmetric phase masks of Eq. (2.3), it was found that the OTF has very interesting properties: (i) it is mostly invariant for a large range

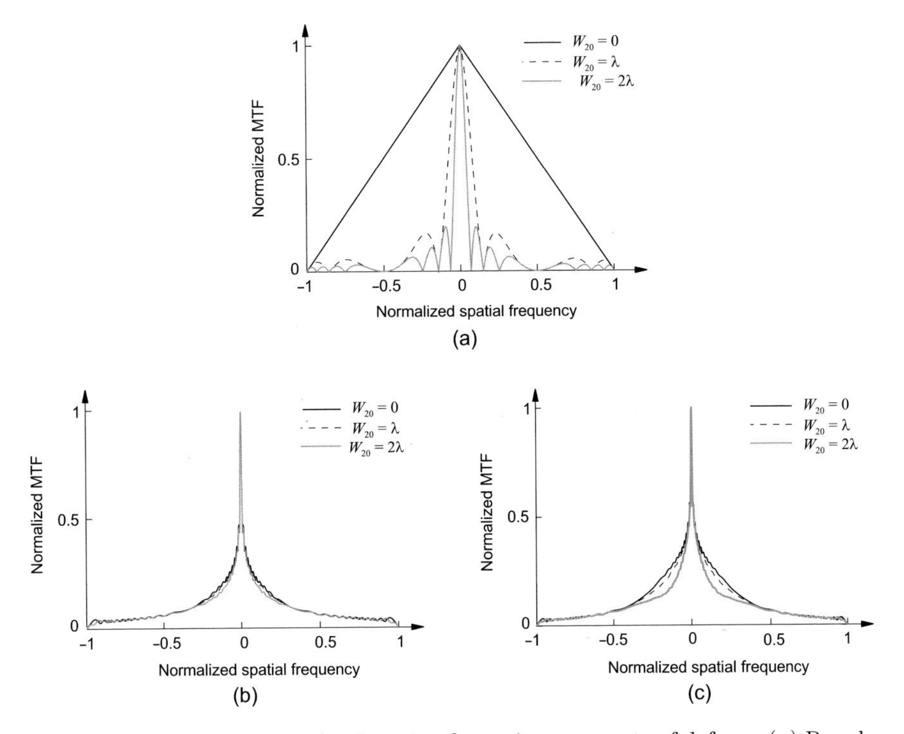


**Fig. 2.8.** Modulation Transfer Function for various amounts of defocus (**a**) Regular lenslet (**b**) Lenslet with a phase mask of order $k = 4$. (**c**) Lenslet with a phase mask of order $k = 5$

of defocus and (ii) it does not lose any information (it is always different from zero). Figure 2.8 presents the MTF of a lenslet for various amounts of defocus given in terms of the defocus coefficient $W_{20}$. Figure 2.8(a) corresponds to a regular lenslet, while Fig. 2.8(b)–(c) corresponds to a lenslet with an asymmetric phase mask of order $k = 4$ and $k = 5$, respectively. Let us recall that a defocus of $\lambda$ is considered severe. For the case of the quartic phase mask, the low values in the MTF cause a degradation of image quality. Nonetheless, since there is no zero value in the MTF it is possible to perform a digital restoration by applying a deconvolution process. In the absence of noise, this restoration consists of an inverse filtering and is done by dividing the Fourier spectrum of the image by the OTF of the phase mask. When noise is present a more cautious process, such as Wiener filtering, must be used [35]. Thanks to the invariance of the OTF with respect to defocus, the same restoration filter simultaneously recovers objects at several distances.

In some cases—for instance when an optical reconstruction is needed — the digital restoration step might not be available. For such eventualities, it was found that an asymmetric phase mask of order $k = 5$ produces images with acceptable contrast, even without any digital enhancement. In this case, the invariability of the OTF is not needed anymore. Figure 2.8(c) presents MTF curves for various amounts of defocus for a lenslet with an asymmetric phase mask of order $k = 5$. From this figure, it can be seen that, again, there is no loss of information.

## 2.4 Pickup with an Asymmetric Phase Mask

We computationally simulate an imaging array where a quartic asymmetric phase mask is placed in front of each lenslet. Parameter $\alpha$ is empirically set to $35/\pi$ (see Eq. 2.3). Figures 2.9(a)–(d) show the images given by the central lenslet for a point object located at various distances from the imaging elements. Comparing this figure with Fig. 2.2, one can see that the phase mask provides invariance to the distance of the object. However the IPSF is not as sharp as with a regular lenslet for an in-focus object. The evolution of the IPSF peak versus object distance is given in Fig. 2.10. The depth-of-field is now about 54 mm, nine times greater than for a regular lenslet.

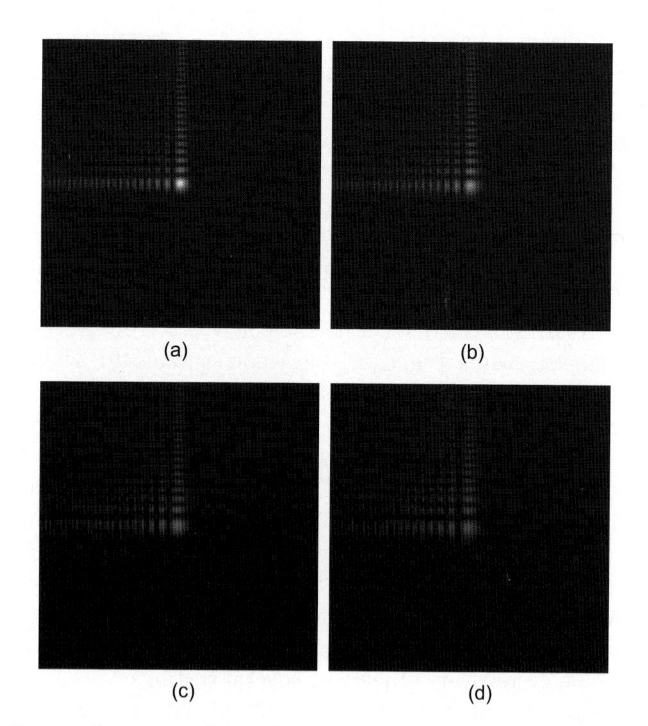


**Fig. 2.9.** Image of a point object located at various distances from the array of lenslets with quartic phase masks (**a**) $l = 100$ mm (**b**) $l = 143$ mm (**c**) $l = 200$ mm (**d**) $l = 333$ mm

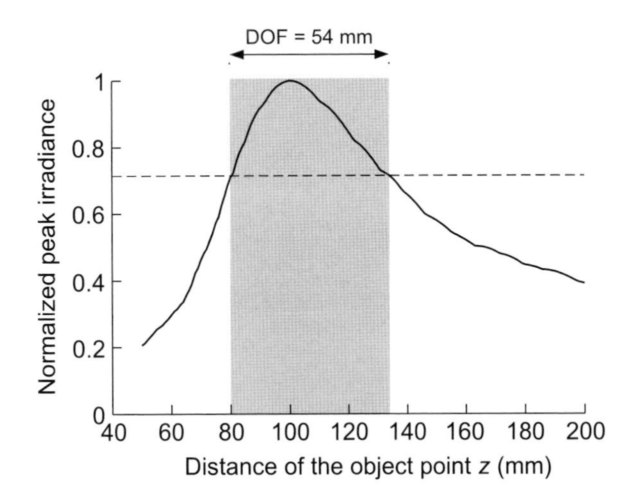


**Fig. 2.10.** Evolution of the peak value of the PSF of a lenslet with a quartic phase mask. The gray area represents the depth-of-field

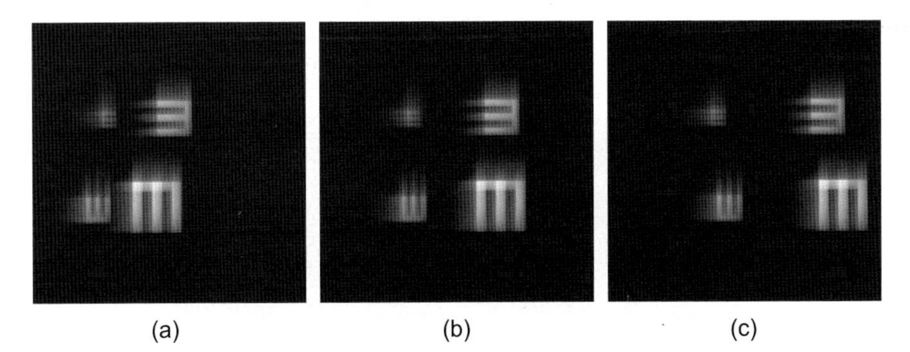


**Fig. 2.11.** Elemental images obtained through various lenslets with quartic phase masks. The objects are located at $l_0 = 100$ mm, $l_1 = 143$ mm, $l_2 = 200$ mm and $l_3 = 333$ mm, respectively. (**a**) Lenslet (0,-2) (**b**) Central lenslet (0,0) (**c**) Lenslet (0,+2)

Now, as in Section 2.2.1, we consider the integral imaging setup described in Fig. 2.4. Figure 2.11 shows a few elemental images obtained without digital restoration, while Fig. 2.12 shows the same images after restoration. The inverse filter used for the restoration is obtained by averaging five OTFs corresponding to defocuses of $W_{20} = -2\lambda$, $W_{20} = -\lambda$, $W_{20} = 0$, $W_{20} = +\lambda$ and $W_{20} = +2\lambda$, respectively . It can be noted that the objects are resolved in all cases, even for distance well outside the depth-of-field obtained in Fig. 2.10. Using digital restoration, the visual quality is equivalent to an in-focus regular lenslet.

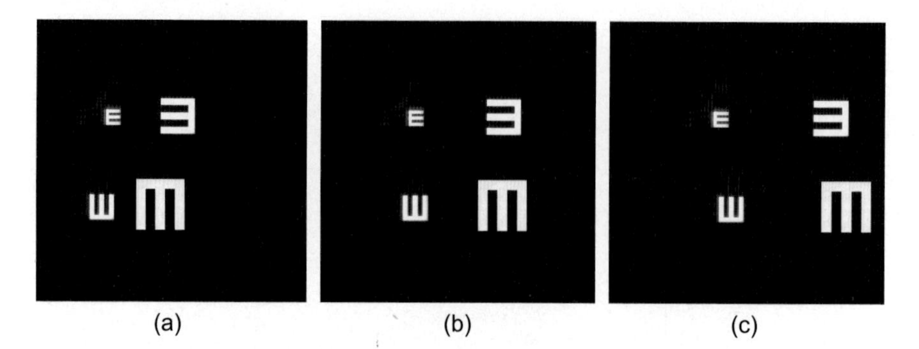

Fig. 2.12. Elemental images obtained through various lenslets with quartic phase masks and digital restoration. The objects are located at $l_0 = 100$ mm, $l_1 = 143$ mm, $l_2 = 200$ mm and $l_3 = 333$ mm, respectively. (**a**) Lenslet (0,-2) (**b**) Central lenslet (0,0) (**c**) Lenslet (0,+2)

## 2.5 Reconstruction with an Asymmetric Phase Mask

In this section, we assume that we already have a set of elemental images of the scene of Fig. 2.4, and that these elemental images do not present any defocus effect. Such images can be obtained computationally, by capture through a

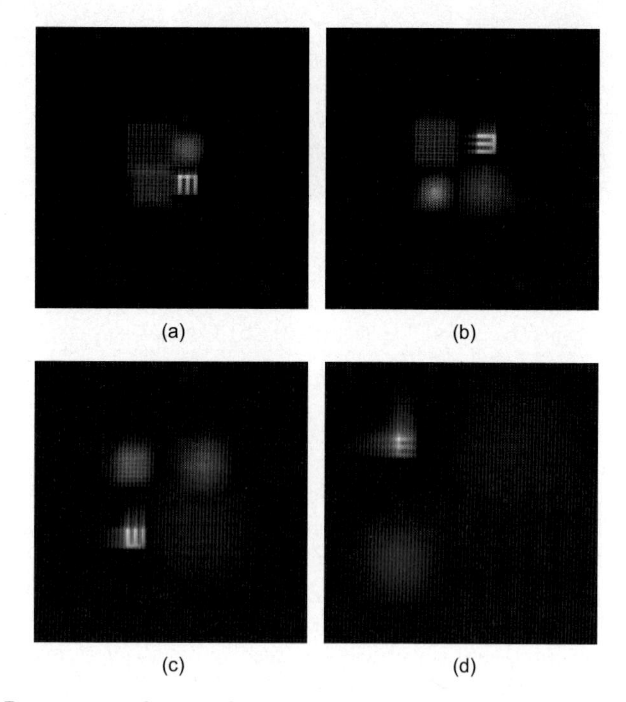

Fig. 2.13. Reconstructed scene obtained with an array of lenslets with quintic phase masks. (**a**) $l = 100$ mm (**b**) $l = 143$ mm (**c**) $l = 200$ mm (**d**) $l = 333$ mm

pinhole array, or by digital restoration of elemental images captured with an array of lenslets with phase masks (see Section 2.4). Now the reconstruction is performed using an array of lenslets with phase masks of order $k = 5$. In this case, we prefer the quintic mask over the quartic one because the most common use of the system would be to perform an optical reconstruction in order for an observer to directly see the 3-D reconstruction. A digital restoration step is thus not possible. Figure 2.13 presents the reconstructed images at various distances. From this figure it is evident that none of the object is perfectly sharp, but all of them have a similar and reasonably in-focus aspect. The farthest object is well resolved, contrary to what was obtained with a regular lenslet (see Fig. 2.7).

## 2.6 Conclusion

In this chapter, we showed that integral imaging systems based on a lenslet array are subject to limited depth-of-field and depth-of-focus. To alleviate this problem we suggest placing an asymmetric phase mask in front of each lenslet. We considered the effect of a phase mask in both the pickup and the reconstruction stages. For pickup, we proposed to use asymmetric phase masks of order four and we showed that the point spread function of each lenslet becomes mostly invariant over a large range of object distances. We estimated that the depth-of-field increases about nine times compared to a regular integral imaging system. In addition, the introduction of a phase mask results in avoiding the presence of zeros in the optical transfer function of the lenslets, which allows us to digitally restore the elemental images. The obtained images are sharp over a large range of distances. For the reconstruction stage, we suggested the use of phase masks of order five that provide better viewing qualities than quartic masks in the absence of any digital restoration. This property is desirable in the case of an all-optical reconstruction.

**Acknowledgment** The authors acknowledge financial support from Consejo Nacional de Ciencia y Tecnología of Mexico (CONACyT) under grant CB05-1-J49232-F.

## References

[1] Javidi B, Okano F, eds (2002) Three Dimensional Television, Video, and Display Technologies, Springer Berlin
[2] Arai J, Kawai H, Okano F (2007) Microlens arrays for integral imaging system. Appl Opt 45:9066–9078
[3] Castro A, Frauel Y, Javidi B (2007) Integral imaging with large depth-of-field using an asymmetric phase mask. Opt Express 15:10266–10273

[4] Yoo H, Shin D-H (2007) Improved analysis on the signal property of computational integral imaging system. Opt Express 15:14107–14114
[5] Tavakoli B, Panah MD, Javidi B, et al. (2007) Performance of 3D integral imaging with position uncertainty. Opt Express 15:11889–11902
[6] Castro A, Frauel Y, Javidi B (2007) Improving the depth-of-focus of integral imaging systems using asymmetric phase masks. AIP Conf Proc 949:53–58
[7] Hwang D-C, Shin D-H, Kim ES (2007) A novel three-dimensional digital watermarking scheme basing on integral imaging. Opt Commun 277:40–49
[8] Yeom S, Javidi B, Watson E (2007) Three-dimensional distortion-tolerant object recognition using photon-counting integral imaging. Opt Express 15:1533–1533
[9] Hong S-H, Javidi B (2005) Three-dimensional visualization of partially occluded objects using integral imaging. J Disp Technol 1:354–359
[10] Park J-H, Kim H-R, Kim Y (2005) Depth-enhanced three-dimensional-two-dimensional convertible display based on modified integral imaging. Opt Lett 29:2734–2736
[11] Frauel Y, Tajahuerce E, Matoba O, et al. (2004) Comparison of passive ranging integral imaging and active imaging digital holography for three-dimensional object recognition. Appl Opt 43:452–462
[12] Jang J-S, Jin F, Javidi B (2003) Three-dimensional integral imaging with large depth-of-focus by use of real and virtual fields. Opt Lett 28:1421–1423
[13] Jang J-S, Javidi B (2004) Depth and lateral size control of three-dimensional images in projection integral imaging. Opt Express 12:3778–3790
[14] Jang J-S, Javidi B (2003) Large depth-of-focus time-multiplexed three-dimensional integral imaging by use of lenslets with nonuniform focal lens and aperture sizes. Opt Lett 28:1925–1926
[15] Hain M, von Spiegel W, Schmiedchen M, et al. (2005) 3D integral imaging using diffractive Fresnel lens array. Opt Express 13:315–326
[16] Martínez-Corral M, Javidi B, Martínez-Cuenca R, et al. (2004) Integral imaging with improved depth-of-field by use of amplitude-modulated microlens arrays. Appl Opt 43:5806–5813
[17] Martínez-Cuenca R, Saavedra G, Martínez-Corral M, et al. (2004) Enhanced depth-of-field integral imaging with sensor resolution constraints. Opt Express 12:5237–5242
[18] Martínez-Cuenca R, Saavedra G, Martínez-Corral M, et al. (2005) Extended depth-of-field 3-D display and visualization by combination of amplitude-modulated microlenses and deconvolution tools. J Disp Technol 1:321–327
[19] Lippmann G (1908) Épreuves réversibles. Photographies intégrales. C R Acad Sci 146:446–451
[20] Okano F, Hoshino H, Arai J, et al. (1997) Real-time pickup method for a three-dimensional image based on integral photography.Appl Opt 36:1958–1603
[21] Dowski ER, Cathey WT (1995) Extended depth-of-field through wave-front coding. Appl Opt 34:1859–1865
[22] Ojeda-Castañeda J, Castro A, Santamaría J (1999) Phase mask for high focal depth. Proc SPIE 3749:14
[23] Castro A, Ojeda-Castañeda J (2004) Asymmetric phase mask for extended depth-of-field. Appl Opt 43:3474–3479
[24] Sauceda A, Ojeda-Castañeda J (2004) High focal depth with fractional-power wave fronts. Opt Lett 29:560–562

[25] Mezouari S, Harvey AR (2003) Phase pupil functions reduction of defocus and spherical aberrations. Opt Lett 28:771–773
[26] Castro A, Ojeda-Castañeda J (2005) Increased depth-of-field with phase-only filters: ambiguity function. Proc SPIE 5827:1–11
[27] Castro A, Ojeda-Castañeda J, Lohmann AW (2006) Bow-tie effect: differential operator. Appl Opt 45:7878–7884
[28] Mino M, Okano Y (1971) Improvement in the OTF of a defocused optical system through the use of shade apertures. Appl Opt 10:2219–2224
[29] Hausler G (1972) A method to increase the depth-of-focus by two step image processing. Opt Commun 6:38–42
[30] Ojeda-Castañeda J, Andrés P, Díaz A (1986) Annular apodizers for low sensitivity to defocus and to spherical aberration. Opt Lett 5:1233–1236
[31] Ojeda-Castañeda J, Berriel-Valdos LR, Montes E (1987) Bessel annular apodizers: imaging characteristics. Appl Opt 26:2770-2772
[32] Siegman AE (1986) Lasers. University Science, Sausalito, CA
[33] Goodman JW (1996) Introduction to Fourier optics, 2nd Ed. Chap. 6 McGraw-Hill, New York, NY
[34] Hopkins HH (1951) The frequency response of a defocused optical system. Proc Roy Soc Lond Series A 231:91–103
[35] Castleman KR (1996) Digital image processing. Prentice-Hall, Upper Saddle River, NJ

# 3

# Integral Imaging Using Multiple Display Devices

Sung-Wook Min, Yunhee Kim, Byoungho Lee, and Bahram Javidi

**Abstract** We introduce an integral imaging system using multiple display devices in this chapter. This system improves viewing characteristics such as image depth, viewing angle and image resolution. We implemented and tested this system which combines two integral imaging devices and uses a beam splitter to show these enhancements.

## 3.1 Introduction

Autostereoscopy is an interesting three-dimensional (3-D) display method that does not require the observer to use any special aids, such as polarizing glasses [1, 2, 3]. Instead, special optical components are placed in front of the display device to make the binocular disparity, which is one of the important cues of 3-D perception, visible to the observer. The autostereoscopic 3-D display system can be classified by the optical components it uses. These components include a lenticular lens, a parallax barrier, holographic optical elements and

S.-W. Min
Department of Information Display, Kyung Hee University, Seoul, Republic of Korea
e-mail: mins@khu.ac.kr

Y. Kim and B. Lee
School of Electrical Enginnering, Seoul National University, Seoul, Rebublic of Korea

B. Javidi
Electrical and Computer Enginnering Department, University of Connecticut, Storrs, CT, USA

B. Javidi et al. (eds.), *Three-Dimensional Imaging, Visualization, and Display*, DOI 10.1007/978-0-387-79335-1_3, 

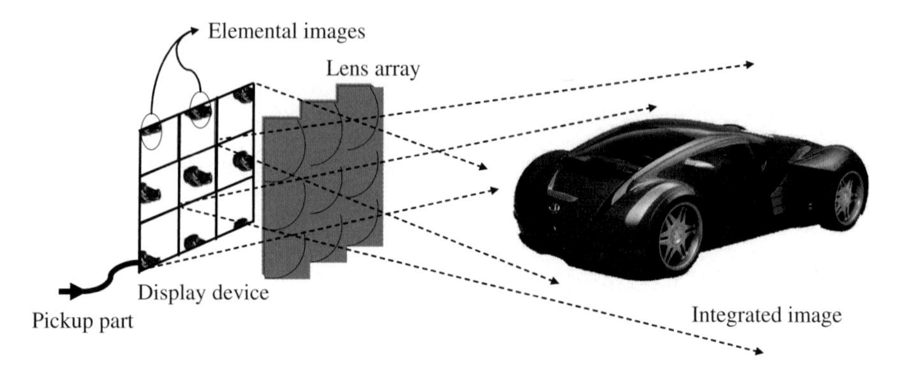


**Fig. 3.1.** The integral imaging display device

a lens array. The observer using the autostereoscopic system does not need to wear bothersome glasses, but has to stay at a restricted position where the image can represent a 3-D effect. In other words, the autostereoscopic display system has a limited viewing region for its 3-D image display.

Integral imaging is one of the promising autostereoscopic 3-D display techniques because it can display autostereoscopic 3-D images which have full color and both horizontal and vertical parallaxes [4, 5, 6, 7]. The integral imaging display system consists of a display device and a lens array which is an optical component to make full parallaxes. Figure 3.1 shows the basic concept of the integral imaging display system. In the system, the display device presents the elemental images which are generated from the pickup part. The elemental images are the snap images of the object which are captured from slightly different viewpoints at the same time and have different perspectives of the object. The elemental images are reconstructed through the lens array into a 3-D integrated image.

The integral imaging display system has special viewing characteristics which can be analyzed using viewing parameters such as image resolution, viewing angle and image depth [8]. These viewing parameters are related to each other. There is a trade-off among the viewing parameters. Therefore, to improve one viewing parameter without the degradation of the others, we must use a display device which has very high resolution or special methods to improve the viewing characteristics, such as the time-variant, the spatial-variant, or the polarization techniques [9, 10, 11, 12, 13, 14, 15]. In this chapter, we introduce a method using multiple display devices to improve the viewing characteristics of the integral imaging system [16, 17]. We explain the method to increase the image depth that uses two integral imaging devices combined by a beam splitter. We also explain a method to widen the viewing angle. In addition, we describe a method to enhance the viewing resolution. We prove the feasibilities of the systems with some basic experiments.

## 3.2 A Method to Improve Image Depth

### 3.2.1 Image Depth

As mentioned above, the integral imaging system has viewing parameters which represent the quality of a 3-D image. The image depth is defined as the thickness of the integrated image. Figure 3.2 explains the image depth in the integral imaging system. The integrated image is thrown by each elemental lens; so, the reconstructed images of the integral imaging system are located around the central depth plane which is determined by the lens law. The position of the central depth plane can be obtained by [18]

$$l = \frac{fg}{g - f}, \tag{3.1}$$

where $l$ is the distance between the central depth plane and the lens array, $f$ is the focal length of lens array, and $g$ is the gap between the lens array and the display panel as shown in Fig. 3.2. When $g$ is bigger than $f$, $l$ has a positive value. This positive value means the central depth plane is located in front of the lens array and the integrated image is displayed in the real mode. On the other hand, when $g$ is smaller than $f$, $l$ has a negative value and the integrated image and the central depth plane are located behind the lens array, which is the virtual mode.

Figure 3.3 shows an example of the integrated images and their elemental images which are in the real mode and the virtual mode, respectively. The character "A" shown in Fig. 3.3(a) is located at 80 mm in front of the lens array while that in Fig. 3.3(c) is at 80 mm behind the lens array. The integral

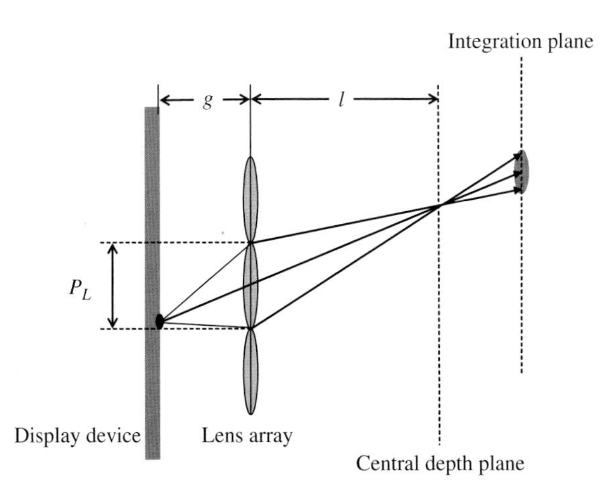


**Fig. 3.2.** The image depth of the integral imaging system

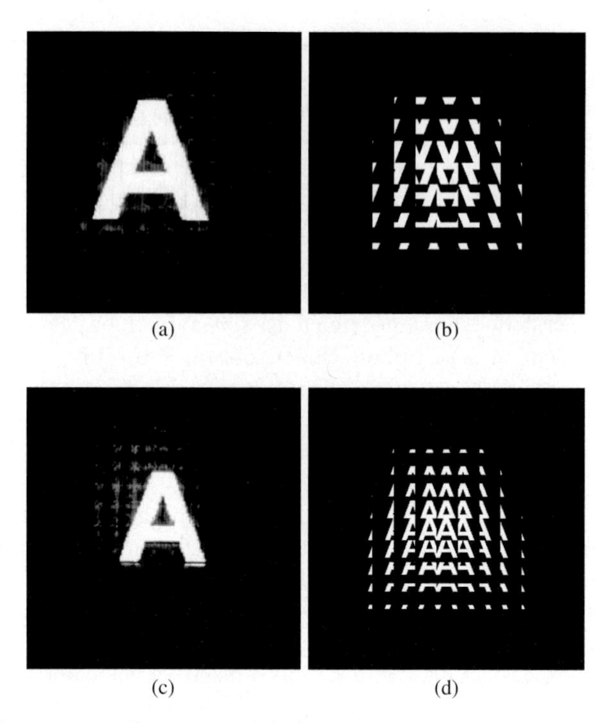

**Fig. 3.3.** The integrated images and their elemental images for the real and virtual modes: (**a**) the integrated image in the real mode, (**b**) elemental images for the real mode image, (**c**) the integrated image in the virtual mode, and (**d**) elemental images for the virtual mode image

imaging system of Fig. 3.3 is assembled with a lens array where the elemental lens size is 10 mm, the focal length is 22 mm, and a display panel where the pixel size is 0.264 mm.

The elemental image presented on the display device is focused on the central depth plane by each elemental lens. Because the pupil of the elemental lens is limited to being small enough, the whole image cannot be observed on the central depth plane and the part of the image makes a cross section on the integration plane. Therefore, the observer feels as if the integrated image, the sum of these cross sections, is located on the integration plane. The integrated image is increasingly broken and distorted if it is located farther from the central depth plane. The distance between the central depth plane and the integration plane, where the integrated image can be observed without severe brokenness and distortion, is named the marginal image depth of the integral imaging system. The marginal image depth is related to the size of the elemental lens, the position of the central depth plane, and the pixel size of the integrated image. For a longer marginal image depth, it is necessary to increase the pixel size of the integrated image, to push the central depth plane farther from the lens array, and to reduce the size of the elemental lens. However, these conditions incur a lower image resolution and a narrower viewing angle which make the viewing condition badly deteriorated. For this reason, we need an enhanced system using a new and unique method to increase the marginal image depth without deterioration of the viewing condition.

### 3.2.2 A System to Extend Image Depth

There are several methods [9, 11, 13, 17] proposed to improve the image depth without the degradation of the other viewing parameters. Such methods are the multiple integral imaging devices method [17], the moving lens array method [9], the optical path controller method [11], and so on. In this section, we introduce a method using two integral imaging devices to improve the image depth.

It is easy to assemble a system using two integral imaging devices by using a beam splitter. Figure 3.4 shows the basic concept of the system to extend the image depth. As shown in Fig. 3.4, the central depth plane of device 2 is located in front of that of device 1 at a certain distance; then, the marginal image depth of the system is extended about twice without any degradation.

Figure 3.5 shows the experimental results of the system using two integral imaging devices. The specification of the integral imaging system for the experiments is the same as that of the system used in Fig. 3.3. Figure 3.5(a) shows images of a conventional integral imaging system where the gap is adjusted to make the central depth plane located at 80 mm. The integration plane of the character "A" is located at 80 mm, while that of the character "B" is at 160 mm. Since the longitudinal position of the character "B" is too far from the central depth plane, the image is severely broken. Figure 3.5(b) shows images of the system using two integral imaging devices. The character "A" is displayed on integral imaging device 1 where the central depth plane is set to be located at 80 mm; while the character "B" is reconstructed by

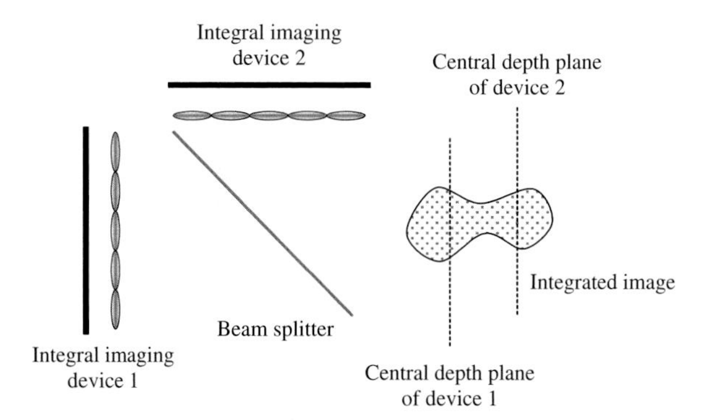


**Fig. 3.4.** A system using two integral imaging devices to improve image depth

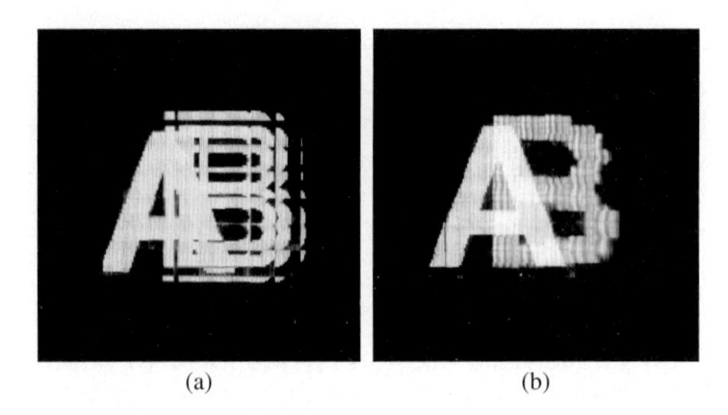

(a) (b)

**Fig. 3.5.** Integrated images from (**a**) a conventional system and (**b**) a system using two integral imaging devices

integral imaging device 2 where the central depth plane is located at 160 mm. As shown in Fig. 3.5(b) both of the images can be observed clearly, so that improved image depth is achieved by the system using two integral imaging devices.

The system, using two integral imaging devices, can be assembled easily without any mechanical moving parts and any complicated optical instruments. Ease of assembly is the main merit of this method. The system, however, is bulky and this burdensomeness worsens when the system uses more than three integral imaging devices because the system needs a complex optical beam combiner.

## 3.3 A Method to Improve the Viewing Angle

### 3.3.1 Viewing Angle

The 3-D image using the autostereoscopic system can be observed within a limited viewing region. The integral imaging system has a viewing angle, which means the angular region where the integrated image can be observed, without cracking and flipping.

Figure 3.6 shows the scheme for the viewing angle of the integral imaging system. The viewing angle is induced by the elemental lens and the elemental images. Each elemental image has a corresponding elemental lens through which the image can be observed correctly. When the elemental image is observed through the neighboring elemental lens, the flipping image occurs which has the wrong perspective. As shown in Fig. 3.6, each elemental lens has a viewing angle which is calculated by [19]:

$$\Omega = 2\arctan\left(\frac{P_L}{2g}\right), \tag{3.2}$$

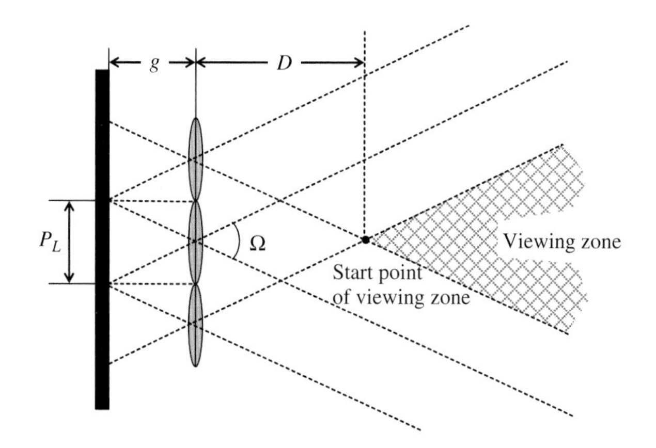


**Fig. 3.6.** Viewing angle of an integral imaging system

where $\Omega$ is the viewing angle, $P_L$ is the size of the elemental lens, and $g$ is the gap between the lens array and the display panel. The viewing zone of the integral imaging system is the area where all the viewing angles of the elemental lenses are overlapped as shown in Fig. 3.6. The start point of the viewing zone is obtained by [19]

$$D = \frac{P_L}{2\tan\left(\frac{\Omega}{2}\right)}(N-1) = g(N-1), \tag{3.3}$$

where $D$ is the position of the start point of the viewing zone and $N$ is the number of the elemental lenses. Accordingly, the viewing zone seems narrower for an observer at a certain position as the integral imaging system is expanded, which confines the size of the lens array and the whole system.

The viewing angle is very important because the integrated image can be observed with full parallaxes only within the viewing zone which is made by the viewing angle. To improve the viewing angle, the *f*-number of the elemental lens should be small. But, an elemental lens with a small *f*-number degrades the image resolution and the image depth. Therefore, a special method is necessary to improve the viewing angle without the degradation of the other viewing parameters.

### 3.3.2 A System to Widen the Viewing Angle

Many methods have been reported [14, 17, 20, 21] to widen the viewing angle. In this section, we explain a method using two integral imaging devices [17]. Figure 3.7 shows a scheme of a system using two integral imaging devices to widen the viewing angle. The system to widen the viewing angle is very similar to the system that extends image depth. Two integral imaging systems are

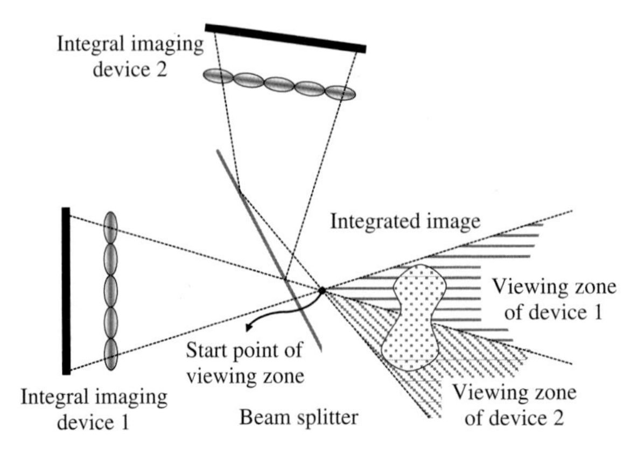


**Fig. 3.7.** A system using two integral imaging devices to widen the viewing angle

combined by a beam splitter with tilting. Tilting is the difference compared with the extended system.

As shown in Fig. 3.7, the reference point of the combination is the start point of the viewing zone. The tilting angle is about the viewing angle. In this way, the viewing angle of the system can be doubled without the degradation of the image resolution and the image depth.

Figure 3.8 shows experimental results from the system that widens the viewing angle by using two integral imaging devices. The specification of the experimental setup is the same as that of the system used for the experiment of Fig. 3.5. The viewing angle of the system is about 18° which can be calculated using Eq. (3.2). As shown in Fig. 3.8(b), the perspective of the integrated image is increased and the viewing angle of the system is doubled.

The flipping image is still shown in Fig. 3.8(b). In the system depicted in Fig. 3.7, the flipping image appears in the entire viewing region because the viewing region of device 1 is located out of the viewing region of device 2. Therefore, the flipping image must be eliminated using another method. One

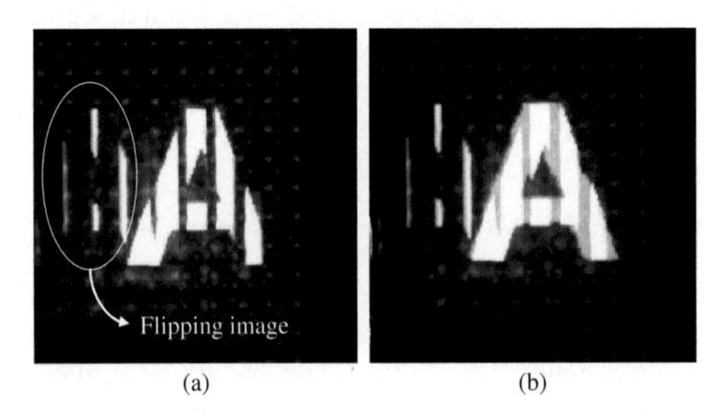


**Fig. 3.8.** (**a**) Integrated and flipping images of the conventional integral imaging system (**b**) An integrated image of the system using two integral imaging devices

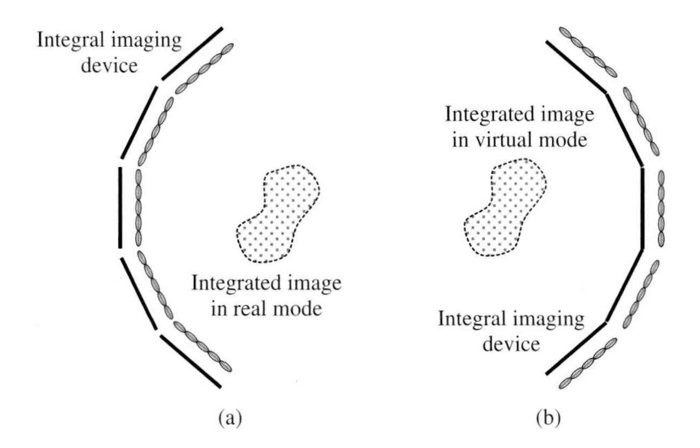


**Fig. 3.9.** A system using multiple integral imaging devices: (**a**) for a real mode, and (**b**) for a virtual mode

of the methods to eliminate the flipping image is to set the vertical barriers between each display panel and lens array along the edges of the elemental lenses [20].

We can implement the system using multiple devices by advancing the system using two integral imaging devices. Figure 3.9 shows the system using multiple integral imaging devices to improve the viewing angle [17]. The system shown in Fig. 3.9(a) is designed using the integral imaging devices in the real mode; while the system shown in Fig. 3.9(b) uses virtual mode. Using the curved structure, the viewing angle can be widened [21]. For the continuous viewing region, the curved structure for the real mode is set to be concave, while that for the virtual mode is set to be convex as shown in Fig. 3.9. The integral imaging devices in Fig. 3.9 also need to remove their flipping image like the system using two integral imaging devices.

## 3.4 A Method to Enhance Viewing Resolution

### 3.4.1 Viewing Resolution

In an integral imaging system, a 3-D image is reconstructed from the sampled elemental images through the lens array (or a pinhole array) as mentioned above. The viewing resolution of the reconstructed 3-D image depends on many system parameters [18, 22, 23] – the size and the pitch of the lens array, the resolution of the display device, etc. The resolution of the display device is important because the elemental image should be displayed with sufficient resolution. However, the primary factor that limits the viewing resolution of the 3-D image is the pitch of the lens array. Here, the pitch of the lens arrays (or the distance between the pinholes) determines the sampling rate of the elemental images in the spatial dimension. A pixel of a 3-D image is determined by the pitch of the lens array.

The viewing resolution of a 3-D image is limited fundamentally because of the sampling rate. The image quality is degraded in a 3-D display. In this regard, to enhance the viewing resolution, the pitch of the lens arrays should be small. But, the small lens pitch for the viewing resolution makes the viewing angle narrow. It is necessary to improve the viewing resolution without the degradation of the other viewing parameters.

### 3.4.2 A System to Enhance Viewing Resolution

Some methods have been proposed to overcome the low resolution problem such as using non-stationary micro-optics [24] and rotating a pair of prism sheets [25]. However, these methods involve rapid mechanical movement that causes some problems such as air resistance and noise. In this section, we

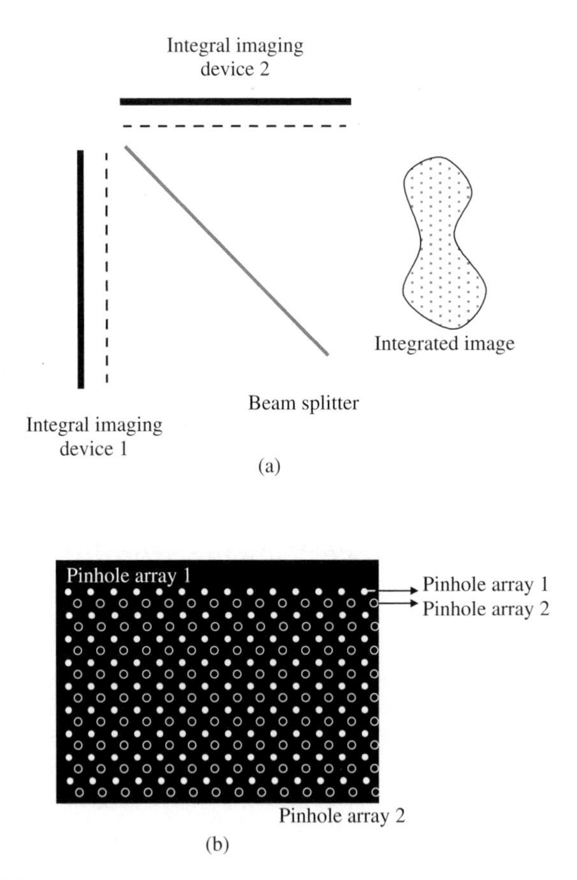


**Fig. 3.10.** (**a**) A system using double integral imaging devices to enhance the viewing resolution (**b**) Pinhole array combined by a beam splitter

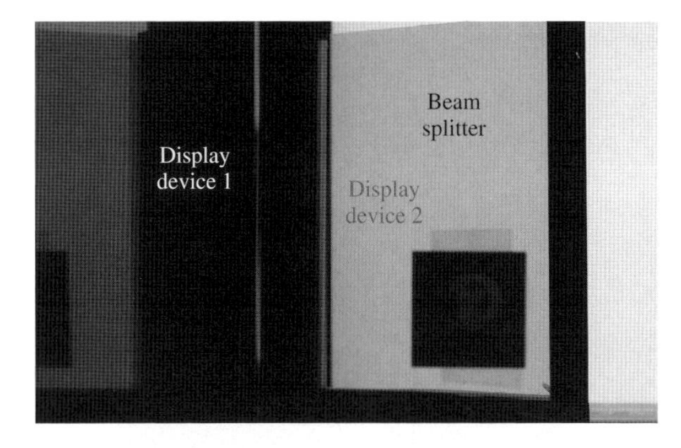


**Fig. 3.11.** The experimental setup

describe a possible resolution improvement using double display devices without any mechanically moving device.

Figure 3.10 (a) shows a configuration using double devices to enhance the viewing resolution [26]. The images of display device 2 and pinhole array 2 are reflected by the beam splitter. Here, we shift pinhole array 2 by half of the pixel period in the diagonal direction with respect to the transmitted image of pinhole array 1. This is the main characteristic of the proposed resolution enhanced system. Figure 3.10(b) shows this situation in detail.

For displaying 3-D images, elemental images according to pinhole array 1 are displayed in display device 1. The 3-D image is generated through pinhole array 1. Similarly, elemental images according to the shifted pinhole array 2

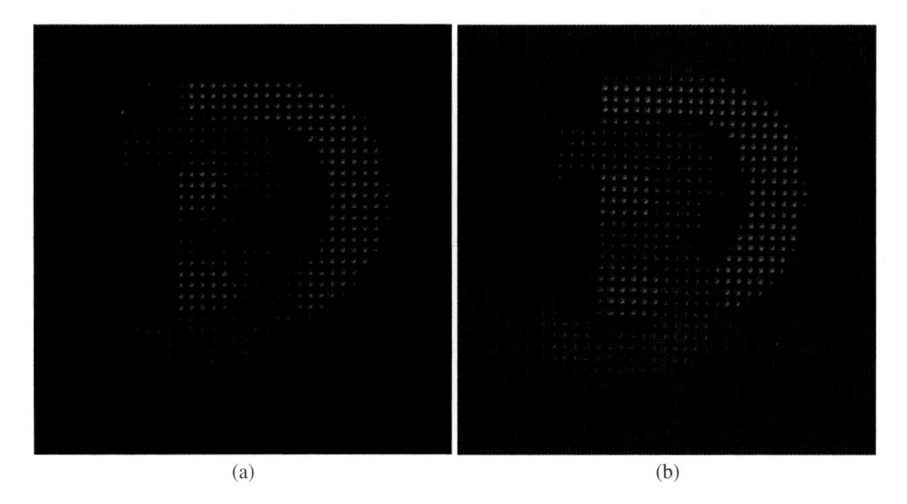


**Fig. 3.12.** Integrated images (**a**) by integral imaging device 1 and (**b**) by integral imaging device 2

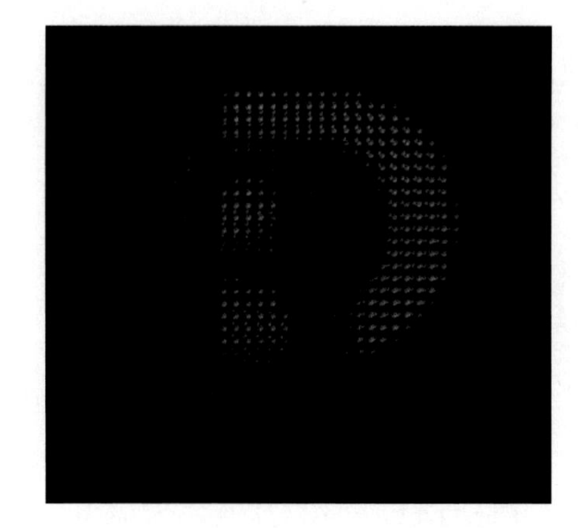

**Fig. 3.13.** Integrated images using integral imaging devices

are displayed and the corresponding 3-D image is generated by pinhole array 2. These two 3-D images, that express different pixels from the same image, are reconstructed separately by the two integral imaging systems and combined with a beam splitter to reconstruct the whole 3-D image with enhanced resolution. The resolution of the combined 3-D image is twice that of the resolution of the conventional integral imagine system. Figure 3.11 shows the experimental setup.

Figure 3.12 shows experimental results illustrating images generated by each integral imaging system. We combined the images in Figs. 3.12(a) and (b) by a beam splitter. Figure 3.13 shows integrated images using the two integral imaging systems. The pixels are not uniformly distributed because of the difficulty in aligning the beam splitter, the two pinhole planes, and the elemental images. However, as the results indicate, resolution is enhanced using the two integral imaging devices.

## 3.5 Summary

In this chapter, we introduced a system using multiple integral imaging devices to improve viewing characteristics. To extend the image depth, we combined two integral imaging devices using a beam splitter. The marginal image depth of the system was doubled. To widen the viewing angle, the integral imaging devices were combined with a tilted angle. We found that the viewing region of the system could be enlarged if the flipping image of each device was effectively blocked. To enhance the viewing resolution, images that express different pixels in the same image were reconstructed separately by the two integral imaging devices and combined to reconstruct entire images with enhanced resolution. The feasibility for a system using multiple integral imaging devices was also investigated. The methods and the systems explained in this chapter can be applied to advance research of 3-D display systems.

## References


[1] T. Okoshi, *Three-Dimensional Imaging Techniques*, Academic Press, New York, 1976.

[2] B. Javidi and F. Okano (editors), *Three Dimensional Television, Video, and Display Technologies*, Springer Verlag, Berlin, 2002.

[3] J. Y. Son, B. Javidi, and K. Kwack, "Methods for displaying three-dimensional Images," *Proc. IEEE*, vol. 94, no. 3, pp. 502–523, 2006.

[4] G. Lippmann, "La photographie integrale," *Comptes-Rendus*, vol. 146, pp. 446–451, 1908.

[5] H. E. Ives, "Optical properties of a Lippmann lenticulated sheet," *J. Opt. Soc. Am.*, vol. 21, pp. 171–176, 1931.

[6] F. Okano, H. Hoshino, J. Arai, and I. Yuyama, "Real-time pickup method for a three-dimensional image based on integral photography," *Appl. Opt.*, vol. 36, pp. 1598–1603, 1997.

[7] A. Stern and B. Javidi, "Three-dimensional image sensing, visualization and processing using Integral Imaging," *Proc. IEEE,* vol. 94, no. 3, pp. 591–607, 2006.

[8] S.-W. Min, J. Kim, and B. Lee, "New characteristic equation of three-dimensional integral imaging system and its applications," *Japan. J. Appl. Phys.*, vol. 44, pp. L71–L74, 2005.

[9] B. Lee, S. Jung, S.-W. Min, and J.-H. Park, "Three-dimensional display by use of integral photography with dynamically variable image planes," *Opt. Lett.*, vol. 26, no. 19, pp. 1481–1482, 2001.

[10] B. Lee, S. Jung, and J.-H. Park, "Viewing-angle-enhanced integral imaging by lens switching," *Opt. Lett.*, vol. 27, no. 10, pp. 818–820, 2002.

[11] J. Hong, J.-H. Park, S. Jung, and B. Lee, "Depth-enhanced integral imaging by use of optical path control," *Opt. Lett.*, vol. 29, no. 15, pp. 1790–1792, 2004.

[12] R. Martinez, A. Pons, G. Saavedra, M. Martinez-Corral, and B. Javidi, "Optically-corrected elemental images for undistorted integral image display," *Opt. Express*, vol. 14, no. 22, pp. 9657–9663, 2006.

[13] R. Martnez-Cuenca and G. Saavedra, M. Martnez-Corral, and B. Javidi, "Extended depth-of-field 3-D display and visualization by combination of amplitude-modulated microlenses and deconvolution tools," *IEEE J. Display Technol.*, vol. 1, no. 2, pp. 321–327, 2005.

[14] J. S. Jang and B. Javidi, "Very-large scale integral imaging (VLSII) for 3-D display," *Opt. Eng.*, vol. 44, no. 1, article 01400, 2005.

[15] J. S. Jang and B. Javidi, "Depth and size control of three-dimensional images in projection integral imaging," *Opt. Express*, vol. 12, no. 16, pp. 3778–3790, 2004.

[16] B. Lee, S.-W. Min, and B. Javidi, "Theoretical analysis for three-dimensional integral imaging systems with double devices," *Appl. Opt.*, vol. 41, no. 23, pp. 4856–4865, 2002.

[17] S.-W. Min, B. Javidi, and B. Lee, "Enhanced three-dimensional integral imaging system by use of double display devices," *Appl. Opt.*, vol. 42, pp. 4186–4195, 2003.

[18] J.-H. Park, S.-W. Min, S. Jung, and B. Lee, "Analysis of viewing parameters for two display methods based on integral photography," *Appl. Opt.*, vol. 40, pp. 5217–5232, 2001.

[19] H. Choi, Y. Kim, J.-H. Park, S. Jung, and B. Lee, "Improved analysis on the viewing angle of integral imaging," *Appl. Opt.*, vol. 44, no. 12, pp. 2311–2317, 2005.

[20] H. Choi, S.-W. Min, S. Jung, J.-H. Park, and B. Lee, "Multiple-viewing-zone integral imaging using a dynamic barrier array for three-dimensional displays," *Opt. Express*, vol. 11, no. 8, pp. 927–932, 2003.

[21] Y. Kim, J.-H. Park, H. Choi, S. Jung, S.-W, Min, and B. Lee, "Viewing-angle-enhanced integral imaging system using a curved lens array," *Opt. Express*, vol. 12, no. 3, pp. 421–429, 2004.

[22] C. B. Burckhardt, "Optimum Parameters and Resolution Limitation of Integral Photography," *J. Opt. Soc. Am.* vol. 58, no. 1, pp. 71–76, 1968.

[23] T. Okoshi, "Optimum design and depth resolution of lens-sheet and projection-type three-dimensional displays," *Appl. Opt.* vol. 10, no. 10, pp. 2284–2291, 1971.

[24] J.-S. Jang and B. Javidi, "Improved viewing resolution of three-dimensional integral imaging by use of nonstationary micro-optics," *Opt. Lett.* vol. 27, no. 5, pp. 324–326, 2002.

[25] H. Liao, T. Dohi, and M. Iwahara, "Improved viewing resolution of integral videography by use of rotated prism sheets," *Opt. Express,* vol. 15, no. 8, pp. 4814–4822, 2007.

[26] Y. Kim, J.-H. Jung, J.-M. Kang, Y. Kim, B. Lee, and B. Javidi, "Resolution-enhanced three-dimensional integral imaging using double display devices," *2007 IEEE Lasers and Electro-Optics Society Annual Meeting*, Orlando, Florida, USA, Oct. 2007, paper TuW3, pp. 356–357.

4

# 3-D to 2-D Convertible Displays Using Liquid Crystal Devices

Byoungho Lee, Heejin Choi, and Seong-Woo Cho

**Abstract** Three-dimensional (3-D)/two-dimensional (2-D) convertibility is an important factor in developing 3-D displays for wide commercialization. In this chapter we review current 3-D/2-D convertible display technologies that use liquid crystal devices. These technologies include the parallax barrier method, the lenticular lens method and integral imaging.

## 4.1 Background

The three-dimensional (3-D)/two-dimensional (2-D) convertible display is a new technique, in the autostreoscopic 3-D display field, now attracting significant attention. Ideally, a 3-D display is a method showing all aspects of a virtual object as if it really existed. With the tremendous improvement of electronic devices and display technologies, the 3-D display techniques are coming to the real world and various 3-D products have been developed and even commercialized. However, in spite of this progress, full 3-D display systems are barely in the early pioneer stage of adoption.

As a result, the 3-D/2-D convertible display is proposed as a new 3-D display solution that could penetrate even into the 2-D flat panel display (FPD) market. Currently 3-D content is limited, compared to the abundant

B. Lee, H. Choi, and S.-W. Cho
School of Electrical Engineering, Seoul National University, Seoul 151-744, Republic of Korea
e-mail: byoungho@snu.ac.kr

B. Javidi et al. (eds.), *Three-Dimensional Imaging, Visualization, and Display*,
DOI 10.1007/978-0-387-79335-1_4, 

amount of 2-D content. Most 3-D dynamic image display systems have low resolution. Hence, the notion of a 3-D/2-D convertible system is that of adding a 3-D display function to a normal 2-D display system. In other words, from the viewpoints of 2-D systems users, the 3-D/2-D convertible display has additional value, i.e., 3-D display function, that users can easily select for specific 3-D applications. For this purpose, the 3-D/2-D convertible system is required to have the following properties.

(1) Minimized 2-D display quality degradation in the normal 2-D mode
(2) Electrical changeability between the 3-D mode and 2-D mode for the convenience of the user
(3) Comparability to conventional 2-D FPDs with respect to the thickness of the systems

Property (1) concerns the general use of a 3-D/2-D convertible system. For most of the operating time, because of lack of 3-D content and low 3-D resolution, the 3-D/2-D convertible system is expected to be used in the 2-D mode and, therefore, the quality of the 2-D mode is an essential issue. It is impossible, of course, to eliminate degradation (in most cases) in the quality of the 2-D mode by the additional 3-D functional device. As a consequence, achieving property (1) is at the top of the most important issues in 3-D/2-D convertible displays. Property (2) is about the reliability and durability of the system. In most cases, a system without mechanical movement can have a longer warranty because of its simpler structure. Property (3) concerns the marketability of a 3-D/2-D convertible system. The system should not be much bulkier than 2-D FPD systems.

We recognize that the above requirements are not for the 3-D performance of the system, but for the marketability or commercial success of such a system. From this viewpoint, the 3-D/2-D convertible display system has more restrictions than the normal 2-D and even 3-D display systems. However, an optimized 3-D/2-D convertible technique can result in a high value-added product and thus claim a new market.

From the technical viewpoint of realizing a 3-D/2-D convertible display, the material for implementing 3-D/2-D convertible display systems is expected to have the following properties.

(1) The material should have some optical anisotropy for 3-D/2-D conversion.
(2) The material of the devices stacked or added for 3-D/2-D conversion functionality should be transparent (low light loss) for visible light.
(3) The material should be controlled electrically.

For the above requirements, liquid crystal (LC) is a useful material which can satisfy all of these properties. Regarding properties (1) and (3), the LC is an appropriate birefringence material that can be controlled with bias voltage. As a result, most of the 3-D/2-D convertible techniques adopt an LC display (LCD) technology for their realization. In this chapter, three kinds of 3-D/2-D convertible techniques using the LCD technique will be explained and summarized. Some of them are already commercialized and the others are in development. However, each technique has its own advantages and bottlenecks so we need to review them all.

## 4.2 LC Parallax Barrier Techniques

### 4.2.1 The LC Parallax Barrier Method

The LC parallax barrier uses the basic principle of the parallax barrier method. The parallax barrier is one of the simplest autostreoscopic methods. It is widely used for many applications. The parallax barrier is composed of black and white patterns. The LC parallax barrier is realized by using a simple LCD panel which can display only white and black stripe patterns. Since the white area is transparent in the LCD, the light can be transmitted through the white stripe. On the other hand, the black stripe blocks the light and acts like an optical barrier which can be generated or eliminated by the applied voltage. With this principle, 3-D/2-D convertible display systems have been developed that consist of LCD panels and parallax LC barriers [1, 2, 3]. The location of the LC barrier can be in front of or behind the LCD panel. The

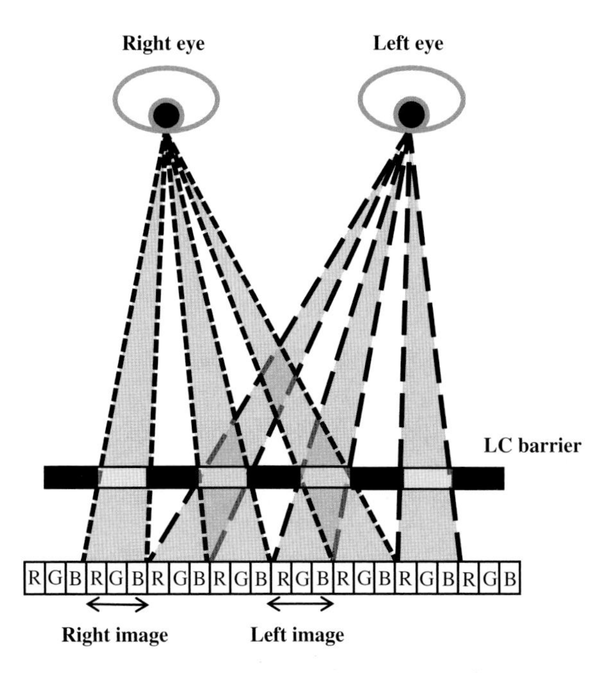


**Fig. 4.1.** The structure of a LC parallax barrier system (R: red, G: green, B: blue sub-pixels)

structure of the LC parallax barrier system is shown in Fig. 4.1. Through the open windows two eyes at specific positions observe different images displayed on interleaved different areas of pixels of the display panel. Hence, the stereo images for these two eyes are displayed in the interleaved format.

With the above principle and structure, the LC parallax barrier system has the following advantages and disadvantages.

### *Advantages*

(1) Simple Structure: The LC parallax barrier has the simplest structure when compared with other methods. In fact, the LCD system is not a highly efficient system as far as transmittance is concerned. The transmittance of an LCD system is mainly decided by factors such as the aperture ratio, color gamut (thickness of the color filter) and resolution. The pattern of electrodes in the LC barrier is a simple stripe pattern and the driving has only two statuses – black and white. These simple structures may provide a high aperture ratio and can increase the transmittance. For the color gamut, the LC parallax barrier does not need to display color images and, therefore, has no color filter on it. Moreover, the parallax barrier may have a relatively lower resolution than most LCD panels.

(2) Low Cost: The LC parallax barrier is basically a kind of LC panel. With the improved technologies for designing and manufacturing LC panels, the LC panel has become one of the cheapest electronic display devices. As a result, the LC parallax barrier is an economical device that can be fabricated easily and cheaply.

(3) Thickness: The thickness of the 3-D/2-D converting device is also an important issue, especially for mobile applications such as cellular phones and notebook computers. Compared with other 3-D/2-D converting techniques, the LC parallax barrier 3-D/2-D convertible system can have a thin structure composed of only two pieces of glass with LC.

### *Disadvantages*

(1) Low Brightness in the 3-D Mode: Basically the LC parallax barrier is a kind of blocking device composed of optical barriers. Although these optical barriers play the most important role in the 3-D mode, they can also block some portion of light rays which are emitted from the display panel and decrease the brightness of the 3-D display system. This problem becomes even worse if the system supports multiple 3-D views. With the increase of the 3-D views, we need to block more areas of display panel (i.e., the slits should be narrower to spatially multiplex the display panel at more viewing locations) because of the principle of the parallax barrier. Also, the brightness of the 3-D mode decreases. Ideally, if the 3-D/2-D convertible system has $N$ views in 3-D mode, the brightness of the system is decreased by a factor of $1/N$. Therefore, it is

essential to enhance the brightness of the 3-D mode in LC parallax barrier systems. To mitigate the reduction in brightness, some parallax barrier 3-D/2-D systems increase the backlight intensity at the 3-D mode.

(2) Color Separation in 3-D Mode: In common display devices, each pixel is composed of R(red)/G(green)/B(blue) dots for the color display. These dots are arranged in a stripe pattern. In the 2-D mode, these stripe-patterned R/G/B dots are combined into one pixel and observed at once. However, in the 3-D mode with a parallax barrier, the observer can only see a designated part of the display panel and the R/G/B dots may not appear to merge to a single pixel to an observer. As a result, the observed color information may be distorted with the location of the observer. This phenomenon is called color separation (or color dispersion) and is shown in Fig. 4.2.

(3) Reduced Resolution and Black Stripe Patterns in the 3-D Mode: In 3-D mode, the LC barrier plays an important role in inducing the binocular disparity, but it also blocks more than half of the displayed image. As a result, there is a black stripe pattern in the 3-D image that severely degrades the 3-D image resolution and quality.

Summarizing from the above analysis, the LC parallax barrier method has advantages in cost and productivity, but it also has disadvantages in display quality when in 3-D mode. As a result, this method is commonly used in mobile applications which require low cost and medium 3-D image qualities. For example, Sharp and Samsung SDI have introduced 3-D/2-D convertible cellular phones based on the parallax barrier technique [1, 3]. Some developers of large sized displays are also adopting this technique. The most advanced example is that of the LG Electronics/LG. Philips LCD system, 42-inches in size with a 1,920×1,080 pixel panel [2]. The luminance for the 3-D display is

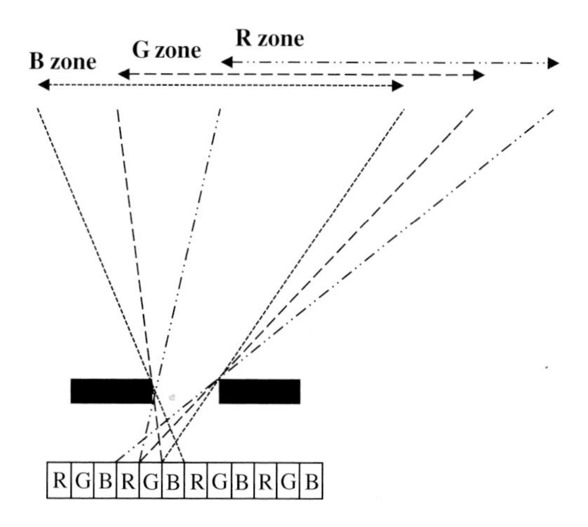


**Fig. 4.2.** The color separation phenomenon in the LC parallax barrier system

450 cd/m$^2$ and for 2-D is 200 cd/m$^2$. The optimum observation position for the 3-D image is 4 m from the panel.

### 4.2.2 The Time Multiplexing Method

The time multiplexing method applies the latest and most improved version of the LC parallax barrier. It is composed of two phases as shown in Fig. 4.3.

Figure 4.3(a) shows the same principle as the conventional LC parallax barrier. The left/right eye observes the odd/even set of R/G/B pixels, respectively. Since only half of the total pixels are shown to each eye, the resolution of the observed 3-D image is decreased by a factor of 1/2. In Fig. 4.3(b), we show a configuration similar to Fig. 4.3(a) except the black/white patterns have been exchanged. The left/right eye observes the opposite (even/odd) set of R/G/B pixels, respectively. As a result, by switching the system with twice the driving speed of the conventional driving method (mostly 60 Hz) between the two modes of Figs. 4.3(a) and (b) to induce the afterimage effect, each eye can observe all of the pixels of the display panel. The field sequential method is similar to this technique except that at each phase all light rays are directed to a single (left or right) eye only.

With the above principle, this time multiplexing method has the following advantages/disadvantages.

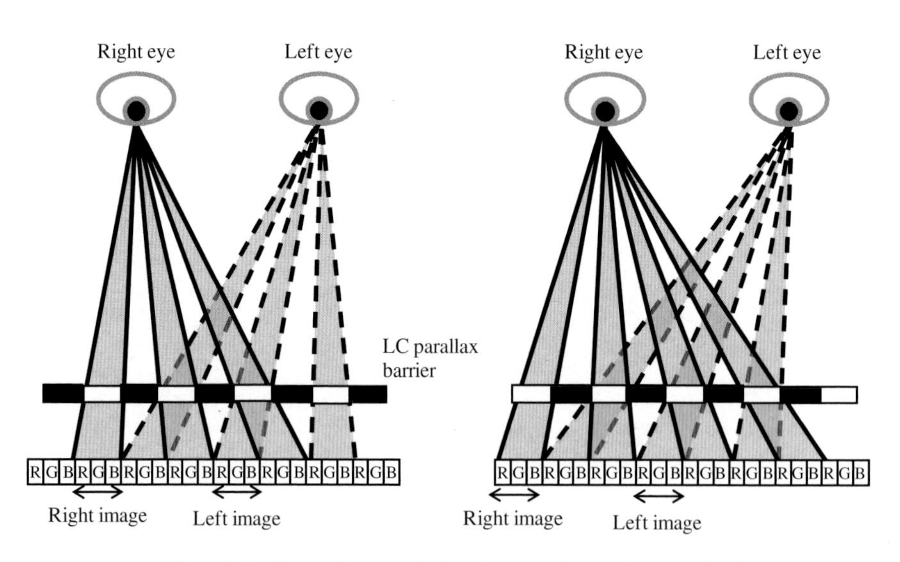


**Fig. 4.3.** Two phases of the time multiplexing method

***Advantages***

(1) High Quality 3-D Mode: In the time multiplexing method, the number of pixels per eye is increased two-fold, compared with the conventional LC parallax barrier method. As a result, there is no decrease of resolution in the 3-D mode. Moreover, for the black stripe pattern of the parallax barrier, the time-sequential switching averages the luminance differences between the white/black stripes and prevents the stripes from being observed. This results in improved 3-D image quality.

(2) Simultaneous Display of 3-D and 2-D Images (partial 3-D technique): It is also possible to display 3-D and 2-D images simultaneously (partial 3-D). The principle of this partial 3-D can also be applied to the conventional parallax barrier method. Only some parts of the whole images are provided in 3-D by the use of binocular disparity, while the 2-D image portion is generated by providing the same images to both eyes. This partial 3-D technique can be used for emphasizing some special and important parts of images with the 3-D mode. It is expected to be an effective 3-D display method for commercial applications such as 3-D advertisement or 3-D shopping malls.

***Disadvantages***

(1) High Driving Speed in the 3-D Mode: To induce the afterimage effect between the two phases, the time multiplexing parallax barrier system requires a high driving speed (mostly 120 Hz). Since the LCD is not a suitable device for high speed driving, the high driving speed is a major bottleneck of this method. Although it was recently reported that some commercial LCD display systems also supported a 120 Hz operation, the high speed driving still requires improved driving techniques and can cause an increase in cost.

(2) Low Brightness in the 3-D Mode: Even with a time multiplexing technique, half of the display panel is always blocked by the black stripes, which results in a decrease in the brightness.

Samsung SDI and the Samsung Advanced Institute of Technology are pursuing this technique [4]. Samsung SDI is using an optically compensated birefringence mode of LC barriers [3].

## 4.3 The LC Lenticular Lens Technique

LC lenticular lenses use the optical birefringence of the LC. There are two kinds of commercialized LC lenticular lens methods. One uses fixed-state LC lenses and the other changes the state of the LC.

### 4.3.1 The Solid Phase LC Lenticular Lens Method

In the solid phase method developed by Ocuity Ltd., the state of the LC lenticular lens does not change and the different polarization of light is chosen so that the light feels or does not feel the existence of the lens. The structure of the solid phase method and its principle in 2-D and 3-D modes are shown in Fig. 4.4 [5, 6].

The solid phase structure is composed of a display panel, LC lenses, a refractive medium, an electrical polarization switch, and a polarizer. In the 2-D mode, the vertical polarization component of the backlight is selected. This makes the light through the lens feel the same refractive index as that of the outside medium. Hence, no refraction occurs. In this situation, the polarization of the output light is not matched with the axis of the polarizer. The polarization switching is needed to change the polarization. On the contrary, in the 3-D mode, the horizontal polarized light component is used. It feels the higher refractive index of the LC lens than that of the outside refractive medium. In this situation, there is refraction at the boundary of the LC lens. The images are displayed by the principle of a lenticular 3-D display.

With the above principles, the solid phase LC lenticular method has the following advantages/disadvantages.

#### *Advantage*

(1) High Optical Efficiency: Although light efficiency depends on system structure, the solid phase system realizes a high optical efficiency with the high transmittance of the LC lens.

#### *Disadvantages*

(1) High Cost: High cost is the common disadvantage of lenticular methods and can be considered as a trade-off with its high performance.

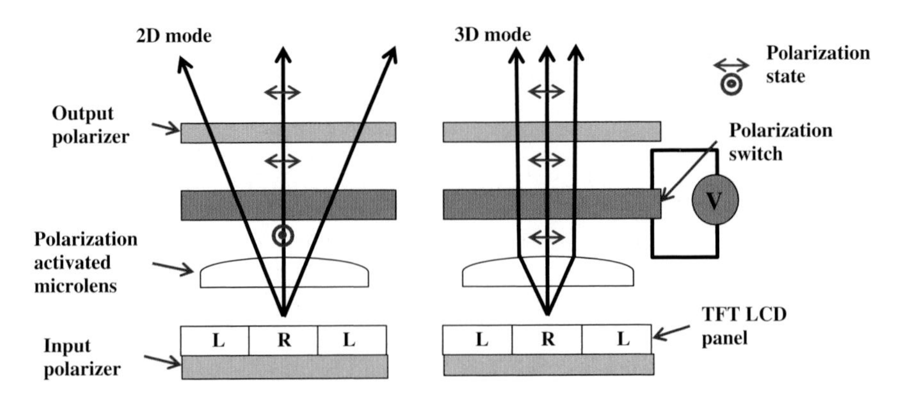


**Fig. 4.4** The principle and structure of the solid phase LC lenticular lens

(2) Thickness: Although the LC lens system can be fabricated within a thin system size, it still requires a thicker space than the parallax barrier system because of the lens structure.

### 4.3.2 The LC Active Lenticular Lens Method

The LC active lenticular lens, developed by Philips, is an active device which can function as a lens or not [7]. The principle uses the optical property of the LC of having different refractive indexes depending on its state. The structure and principle of the LC active lenticular lens method is shown in Fig. 4.5. In Fig. 4.5, each lenticular lens is filled with LC. The indium tin oxide (ITO) electrodes are positioned around it to provide the control voltage. Outside of the lens, there is a replica which has the same refractive index as that of the LC when the bias voltage is applied. In the 3-D mode, no voltage is applied to the electrode and there is a difference between the refractive indexes of the LC and the replica. In this case, polarization of the light is used that feels the extraordinary refractive index of the LC in the lens. As a result, the 3-D image can be formed through the LC active lenticular lens as shown in Fig. 4.5(a). In the 2-D mode, with appropriate voltage, the difference of refractive indexes

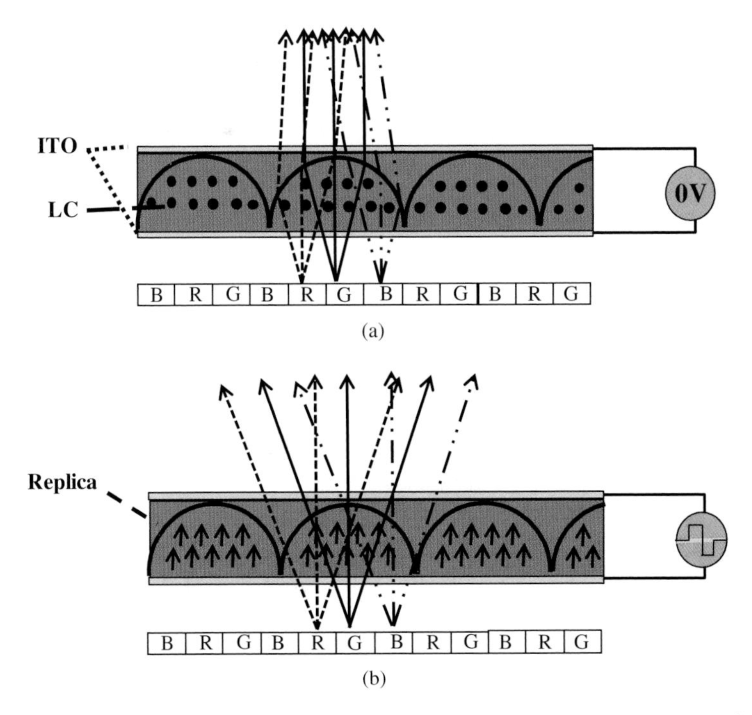


**Fig. 4.5.** The principle of the structure of the LC active lenticular lens: (**a**) 3-D mode and (**b**) 2-D mode

between the LC and replica vanishes and the LC active lenticular lens cannot perform as a lens. Therefore, the observer can see the images on the display panel without any distortion. The LC active lenticular lens array method has a similar principle as the solid phase LC lenticular lens, and also has the same advantages and disadvantages.

### 4.3.3 The Slanted Lenticular Lens Method

Although the slanted lenticular lens technique is not an issue directly related to the 3-D/2-D convertible display, it is an important method that can be used to display 3-D images which are more natural and have better quality. Therefore, we need to review the slanted lenticular lens method. In conventional 3-D systems with parallax barriers or normal lenticular lenses, only the horizontal parallax is provided to satisfy the binocular disparity of 3-D images. As a result, only the horizontal resolution of the 3-D display system is decreased, whereas the vertical resolution remains unchanged. Moreover, the periodical structure of the parallax barrier and normal lenticular lens is only in the

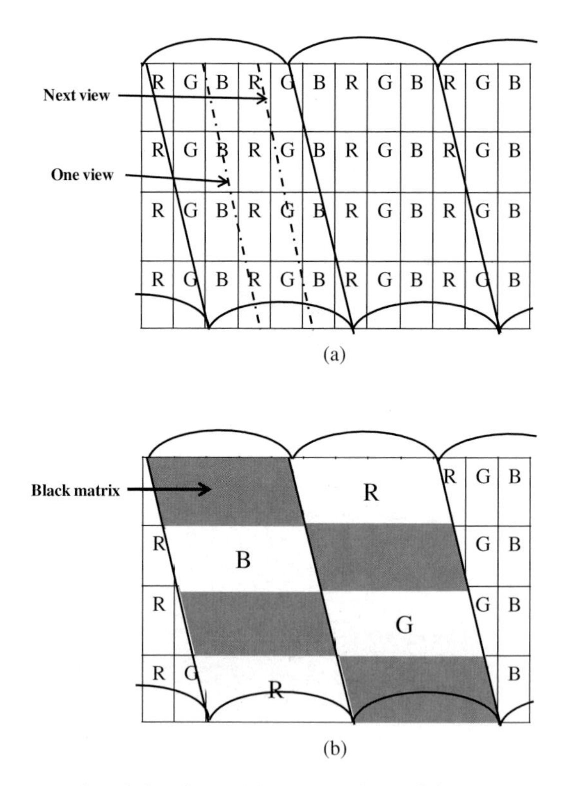


**Fig. 4.6.** The principle of the slanted lenticular lens: (**a**) the structure and (**b**) the observed color sub-pixels

horizontal direction. This structure results in a black stripe pattern and color separation which comes from the sub-pixel structure of the display panel. The pattern and color separation degrade the quality of the 3-D image.

The slanted lenticular lens technique is the most advanced method which can mitigate some of the disadvantages of the current parallax barrier/normal lenticular lens method by tilting the lenticular lens [7, 8, 9]. The structure and principle of the slanted lenticular lens are shown in Fig. 4.6. With the slanted structure of the lenticular lens, the 3-D pixel is composed of R/G/B sub-pixels which are located in a different column and row as shown in Fig. 4.6(a). As a result, the resolution degradation in the 3-D mode is divided into both horizontal and vertical directions. Moreover, with the slanted structure, there is no black stripe, as shown in Fig. 4.6(b), since the R/G/B sub-pixels are observed between the black matrices. Therefore, the quality of the 3-D image can be enhanced.

Philips has adopted this technique and demonstrated a 42-inch 3-D display system. It recently extended this technique and demonstrated a 3-D/2-D convertible 42-inch display system [9].

## 4.4 3-D/2-D Convertible Integral Imaging

The integral imaging [10, 11, 12] is a technique which is now working its way through necessary research. It has not yet been developed as part of any commercial products although some demonstration systems have been developed by a few companies like Hitachi (in cooperation with the University of Tokyo), Toshiba and NHK [12, 13, 14]. In this section, we categorize and review various 3-D/2-D convertible integral imaging systems. In integral imaging, there are mainly two kinds of 3-D/2-D convertible methods. One method adopts a point light source array, while the other uses a lens array. These two methods have different properties and, therefore, are suitable for different applications.

### 4.4.1 3-D/2-D Convertible Integral Imaging with Point Light Sources

In a common integral imaging system, the 3-D image is formed from elemental images through a lens array. However, the lens array is a fixed device and is an obstacle for 3-D/2-D conversion. Therefore, various 3-D/2-D convertible methods which do not use a lens array have been developed. In an integral imaging system, a lens array can be replaced with a point light source array or pinhole array [15, 16]. The principle to display a 3-D image, using a point light source array and elemental images, is illustrated in Fig. 4.7 [15]. The lens array has better optical properties than the point light source array or pinhole array. However, the point light source array is easier to generate/eliminate and is suitable for 3-D/2-D convertible integral imaging.

### Integral Imaging with Point Light Source Array and the Polymer-Dispersed Liquid Crystal

The earliest 3-D/2-D convertible method uses a point light source array and a polymer-dispersed liquid crystal (PDLC) [15, 17]. The PDLC is a kind of active diffuser that can be both diffusive and transparent according to the voltage application. In the PDLC, the liquid crystal droplets are randomly arranged in the polymer. With no voltage applied, the directions of LCs in LC droplets do not have uniformity. The light through the PDLC is scattered by those droplets. If voltage is applied, the LCs in LC droplets are arranged into a uniform direction and the PDLC becomes transparent. The earliest 3-D/2-D convertible methods used a collimated light [15]. A 2-D lens array is used to form a point light source array in Fig. 4.7. If the PDLC plane, that is attached to the lens array, scatters light, then it destroys the collimated light and no point light source array is formed. This case can be used for the 2-D display mode. Therefore, electrical 3-D/2-D conversion is possible by controlling the PDLC.

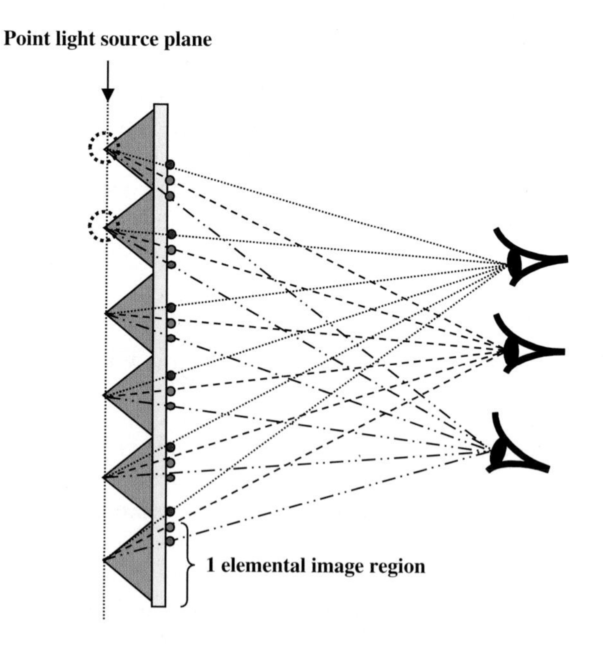


**Fig. 4.7.** The principle of the 3-D display with a point light source array

With the above principle, 3-D/2-D convertible integral imaging with the PDLC method has the following advantages/disadvantages.

### *Advantage*

(1) High Optical Efficiency: The point light source array is formed by the lens array which has a transparent structure and can provide high optical efficiency.

### *Disadvantage*

(1) Thickness: To collimate the light rays, the thickness of the proposed system should be more than 10 cm for a 1.8-inch screen size. Therefore, the system has use in only limited applications and the system is not suitable for mobile devices.

### Integral Imaging Using a LED Array

In this section, we introduce a 2-D/3-D convertible system using a light-emitting diode (LED) array. LEDs are currently being used for backlight units of 2-D LCD systems. LEDs can also be used for 2-D/3-D convertible integral imaging systems with reduced thickness as compared with the previous method. The basic concept of the system is that each mode of either 2-D or 3-D uses different combinations of LED arrays [18]. Figure 4.8 shows the

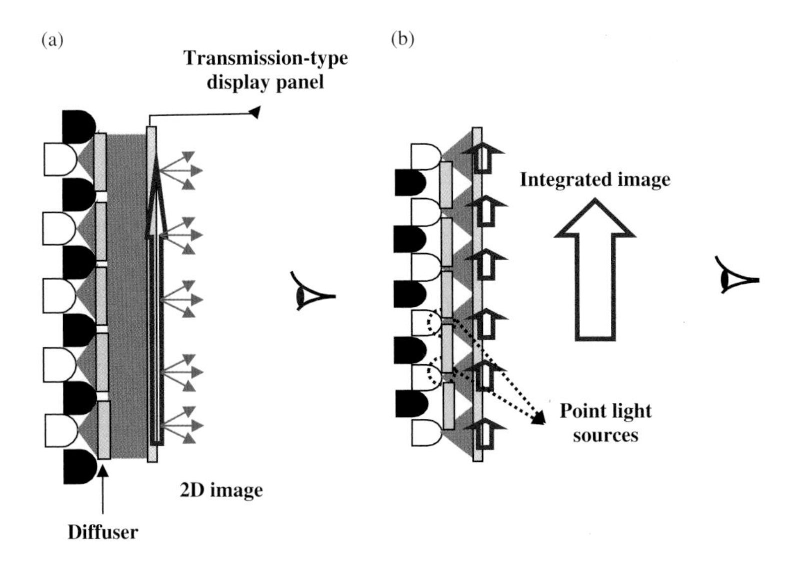


**Fig. 4.8.** The structure of the LED array method: (**a**) the 2-D mode and (**b**) 3-D mode

concept of the proposed system. The backlight is constructed by two kinds of LEDs – one kind for the 3-D mode and the other for the 2-D mode. The diffuser is placed in front of each LED used for the 2-D mode. The backlight for the 3-D mode comes out from the apertures (not passing through the diffuser) in front of the LEDs used in the 3-D mode. An LCD is used to modulate the backlight in the form of a 2-D image or elemental images for a 3-D display. In the 3-D display mode, the LEDs for the 3-D mode are emitting lights. The lights diverge through the apertures and go through the display panel. A lens array can also be inserted between the LEDs and LCD so that the number of point light sources becomes larger than the number of 3-D mode LEDs. For example, $2N \times 2N$ uniformly distributed point light sources can be generated using a lens array consisting of $N \times N$ elemental lenses. In the 2-D display mode, the LEDs for the 2-D mode are illuminating and the light goes through the diffusers which are placed in front of the LEDs. Then, the light is diffused and used as a backlight for the LCD that displays 2-D images.

***Advantage***

(1) High Optical Efficiency: Because there is no light blocking structure, the optical efficiency is high.

***Disadvantage***

(1) Thickness: The number of point light sources determines the 3-D image resolution. To increase the number of point light sources, a lens array should also be inserted, which increases the thickness of the system. The thickness should be 7 cm for a 1.8-inch screen size. Therefore, the system has only limited applications and the system is not suitable for mobile devices.

## Integral Imaging Using a Pinhole Array on a Polarizer

A proposed improvement over the above methods, especially for the thickness of a system, is a method using a pinhole array fabricated on a polarizer (PAP) [19]. In the PDLC method, the point light sources are generated through the lens array and high optical efficiency can be acquired. However, a gap is needed to form the point light source. This gap increases the thickness. The method explained here reverses the advantages and disadvantages of the PDLC method – it has a low optical efficiency, but thin system thickness. The key device of the method is the PAP. According to the polarization of the light, the PAP can become either a pinhole array for the 3-D mode or a transparent sheet for the 2-D mode. Using this property, a point light source array is generated from the pinholes in the 3-D mode; whereas, it vanishes in the 2-D mode. 3-D/2-D conversion is possible by changing the polarization state of the backlight unit. The structure of the PAP system is shown in

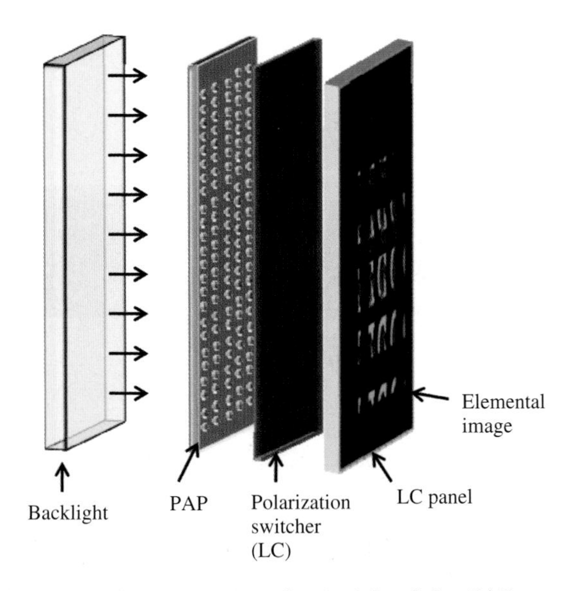


**Fig. 4.9.** The structure and principle of the PAP system

Fig. 4.9 [19]. As shown in the figure, the system is composed of an LCD panel, the PAP, a polarization switcher which is a single LCD cell, and a backlight unit. If the polarization switcher is adjusted so that the polarization of light incident on the PAP is orthogonal to the polarizer direction of the PAP panel, only the light that comes through the pinholes survives and forms a point light source array. On the other hand, if the polarization of light is set to the polarizer direction of the PAP panel, then the light transmits through the whole PAP plane and forms a planar light source for the 2-D display. Figure 4.10 shows experimental results.

With the above principle, the PAP method has the following advantages and a disadvantage.

### *Advantages*

(1) Thin System Size: In the PAP system, only a small gap is needed between the LCD panel and the PAP and, therefore, it is possible to design a thin system which is suitable for various applications.

(2) Low Cost: The PAP can be fabricated in the rear polarizer of the display LCD panel and does not require an additional structure. In addition, the polarization switcher is a single LC cell and has a simpler structure than the LC parallax barrier. Therefore, the PAP system can be realized with low cost which can compete with the LC parallax barrier system and even conventional 2-D display systems.

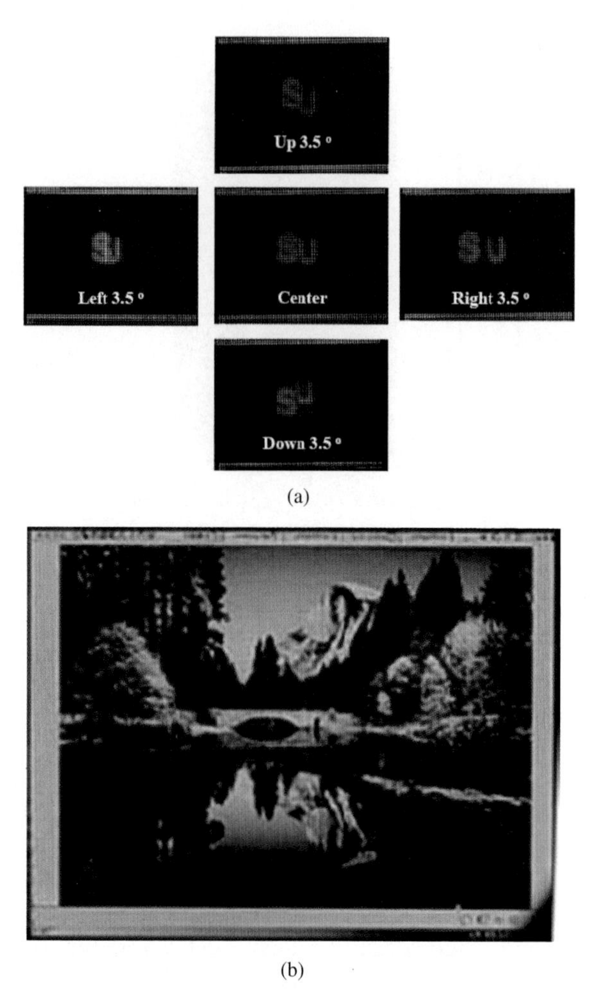


(a)

(b)

**Fig. 4.10.** Experimental results: (**a**) 3-D mode and (**b**) 2-D mode

(3) High Optical Efficiency in the 2-D Mode: In the 2-D mode of the PAP system, the PAP becomes a transparent sheet and ideally induces no optical loss. Our experiments show that 15~20 percent loss occurs due to the reflection and absorption of the polarizer. In the 2-D mode, it is possible to provide high luminance.

***Disadvantage***

(1) Low Optical Efficiency in the 3-D Mode: In the 3-D mode of the PAP system, the PAP becomes a pinhole array and can only have a very low transmittance which is normally below 10 percent. As a result, the optical efficiency of

the 3-D mode becomes very low and a special technique for compensating the luminance decrease is needed. However, since the 3-D mode is an additional function, the PAP method is useful for simple 3-D/2-D convertible systems.

### Integral Imaging Using a Pinhole Array on a Liquid Crystal Panel

The latest version of the PAP method uses a pinhole array on an LC panel (PALC) for a partial 3-D display [20]. In the PALC method, all principles are the same as those of the PAP method, except the PAP and the polarization switcher are combined and together become the PALC. Since the PALC is an active device, it is possible to form pinholes only in designated regions and to realize a partial 3-D display.

### Integral Imaging Using a Fiber Array

In 3-D/2-D convertible integral imaging with a point light source array, the above methods both have big advantages and disadvantages. The method using a fiber array to generate point light sources is in a medium position between the above extremes – it has a high optical efficiency and medium system size. In the fiber array method, the point light source is generated from a stack of fibers which is arranged in order as shown in Fig. 4.11 [21]. Since the fiber is a good waveguide, it can deliver the light from the source to the designated position with accuracy. The other advantage of the fiber array is that it is bendable, which can reduce the system size by not locating the light source for 3-D behind the LCD panel. As shown in Fig. 4.11, the system consists of a display LCD panel, a fiber array fixed on a transparent plate, light source 1 for the 2-D mode, and light source 2 for the 3-D mode. In 2-D mode, light source 1 is used to illuminate the LCD panel and the images on the LCD panel are shown. In the 3-D mode, light source 2 is turned on. This source generates point light sources at the location of each fiber and 3-D images can be formed. Since the fiber array has a higher optical efficiency than the pinhole array, the luminance in both the 2-D and the 3-D modes can be sufficiently high.

With the above principle, the fiber array method has the following advantages and a disadvantage.

#### *Advantages*

(1) High Optical Efficiency: This system can deliver high luminance for both 2-D and 3-D modes.

(2) Medium System Size: Since the fiber array can be bent, there is some freedom in the location of light source 2. Although the system thickness is increased when compared with the PAP and PALC methods, it can still be thinner than the PDLC method. Therefore, the fiber array method can be

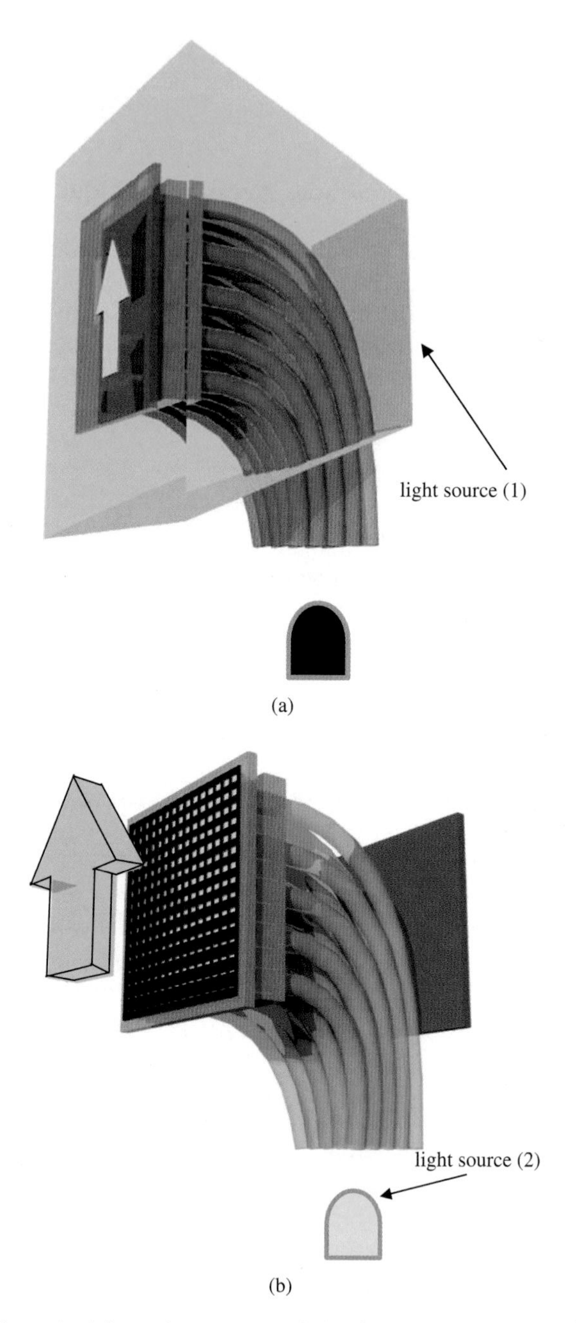


**Fig. 4.11.** The principle and structure of the fiber array method: (**a**) 2-D mode and (**b**) 3-D mode

used for the devices such as monitors and TVs which require high luminance and a medium system size.

#### *Disadvantage*

(1) The fibers deteriorate the uniformity of the light from light source 1 for a 2-D display.

### 4.4.2 3-D/2-D Convertible Integral Imaging with a Lens Array – A Multilayer Display System

The lens array is a key device in integral imaging which provides both 3-D views and high optical efficiency, but decreases the resolution of the image and can be an obstacle for a 2-D display. There is a method which avoids this disadvantage by using the optical property of the LCD [22].

The basic principle of the multilayer display system is preparing independent devices for the 2-D and the 3-D modes. Since the LCD panel is a transmission type display which displays the image by controlling the transmittance of the light, it is possible to stack multiple LCD panels and drive them independently. In the proposed multilayer structure, an additional LCD panel (panel 2) is located in front of the lens array of the conventional integral imaging system as shown in Fig. 4.12. In the 3-D mode, LCD panel 2 is set to be transparent by displaying a white screen; whereas, the elemental images are displayed on LCD panel 1 and 3-D images are formed through the lens array. In the 2-D mode, on the other hand, LCD panel 1 displays a white screen to become a light source for LCD panel 2 and the 2-D image on the LCD panel 2 is shown to the observer. Because the panel 2 can be attached on the lens array, the total thickness of the system can be thin as shown in Fig. 4.13.

Using the above principle, this multilayer method has the following advantages and disadvantages.

#### *Advantages*

(1) High Resolution in the 3-D Mode: The system can have a higher resolution than the methods which use a point light source array.

(2) High Contrast Ratio in the 2-D Mode: Since LCD panel 1 is used as a backlight source in the 2-D mode, an advanced technique to control the luminance of the backlight can be used and the contrast ratio of the 2-D image can be increased by decreasing the black luminance.

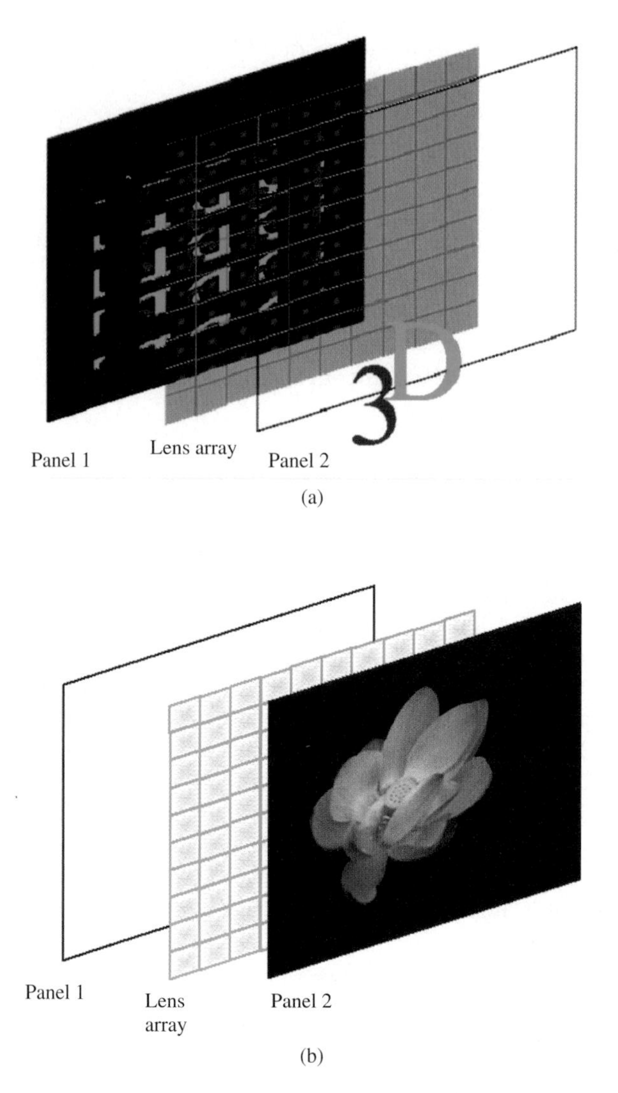


**Fig. 4.12.** The principles of the multilayer method: (**a**) 3-D mode and (**b**) 2-D mode

### *Disadvantages*

(1) Low Optical Efficiency: Although the LCD panel is a transmission type display, the optical efficiency of the LCD device is not so high and stacking of the multiple LCD devices would result in a low optical efficiency in both the 2-D and 3-D modes.

(2) Lens Array Recognition: Undesirable recognition of the lens array is also an important problem in integral imaging with a lens array. In the

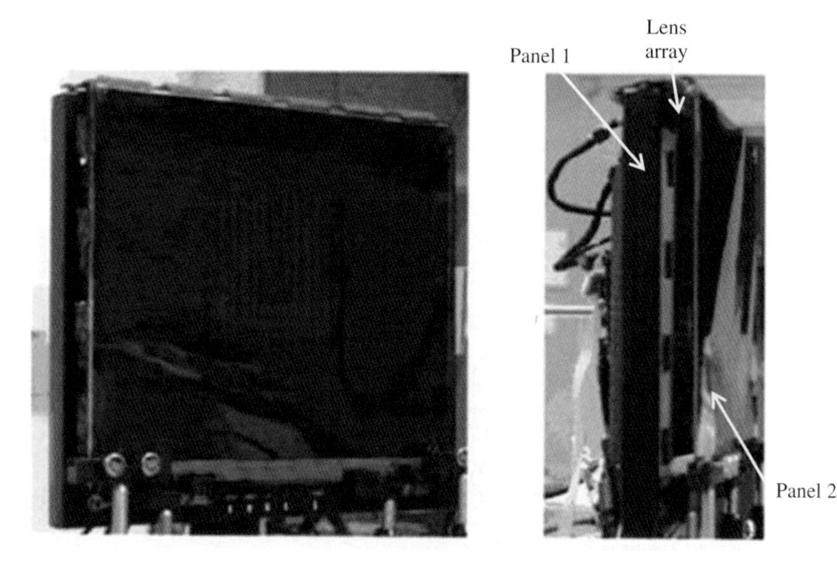


**Fig. 4.13.** Picture of the multilayer 3-D/2-D convertible integral imaging system

multilayer method, the seam lines of the lens array can be observed in the 3-D mode and even in the 2-D mode through LCD panel 2.

(3) High Cost: The multilayer method adopts two LCD panels and, thus, the cost of the system would be increased.

## 4.5 Conclusion

The 3-D/2-D convertibility described in this chapter is an essential factor in the commercialization of a 3-D display. The LC is a suitable material with its optical anisotropy and electric properties. It is widely used for various 3-D/2-D convertible techniques. There are three kinds of autostereoscopic 3-D/2-D convertible techniques – the LC parallax barrier, the LC lenticular lens, and integral imaging. The first two have been commercialized. The third technique has been developed to a certain level of technology and is close to commercialization. Each of these techniques has its own advantages and disadvantages

**Table 4.1.** A comparison of 3-D/2-D convertible techniques with LC

| | Brightness | Multi-view 3-D | 3-D Quality | Cost |
|---|---|---|---|---|
| LC parallax barrier | ▲ | ▲ | ▲ | ○ |
| LC lenticular lens | ○ | ○ | ○ | ▲ |
| Integral imaging | ○ | ○ | ▲ | ▲ |

as shown in Table 4.1. It is important to adopt an appropriate technique for a specific application with consideration of the techniques' characteristics.

**Acknowledgements** This work was supported by the Korea Science and Engineering Foundation and the Ministry of Education, Science and Technology of Korea through the National Creative Research Initiative Program (#R16-2007-030-01001-0).

## References

[1] A. Jacobs, J. Mather, R. Winlow, D. Montgomery, G. Jones, M. Willis, M. Tillin, L. Hill, M. Khazova, H. Stevenson, and G. Bourhill, "2-D/3-D switchable displays," Sharp Technical Journal, no. 4, 2003. http://www.sharp-world.com/corporate/info/rd/tj4/4-2-3.html

[2] H. Kang, M. K. Jang, K.J. Kim, B.C. Ahn, S. D. Yeo, T. S. Park, J. W. Jang, K. I. Lee, and S. T. Kim, "The development of 42" 2-D/3-D switchable display," Proc. of The 6th International Meeting on Information Display and The 5th International Display Manufacturing Conference (IMID/IDMC 2006), Daegu, Korea, Aug. 2006, pp. 1311–1313.

[3] H. J. Lee, H. Nam, J. D. Lee, H. W. Jang, M. S. Song, B. S. Kim, J.S. Gu, C. Y. Park, and K. H. Choi, "A high resolution autostereoscopic display employing a time division parallax barrier," Society for Information Display 2006 International Symposium, San Francisco, CA, USA, June 2006, vol. 37, book 1, pp. 81–84.

[4] D.-S. Kim, S. D. Se, K. H. Cha, and J. P. Ku, "2-D/3-D compatible display by autostereoscopy," Proc. of the K-IDS Three-Dimensional Display Workshop, Seoul, Korea, Aug. 2006, pp. 17–22.

[5] G. J. Woodgate and J. Harrold, "A new architecture for high resolution autostereoscopic 2-D/3-D displays using free-standing liquid crystal microlenses," Society for Information Display 2005 International Symposium, vol. 36, 2005, pp. 378–381.

[6] J. Harrold, D. J. Wilkes, and G. J. Woodgate, "Switchable 2-D/3-D display – solid phase liquid crystal microlens array," Proc. of International Display Workshops, Niigata, Japan, Dec. 2004, pp. 1495–1496.

[7] S. T. de Zwart, W. L. IJzerman, T. Dekker, and W. A. M. Wolter, "A 20-in. switchable auto-stereoscopic 2-D/3-D display," Proc. of International Display Workshops, Niigata, Japan, Dec. 2004, pp. 1459–1460.

[8] O. H. Willemsen, S. T. de Zwart, M. G. H. Hiddink, and O. Willemsen, "2-D/3-D switchable displays," Journal of the Society for Information Display, vol. 14, no. 8, pp. 715–722, 2006.

[9] M. G. H. Hiddink, S. T. de Zwart, and O. H. Willemsen, "Locally switchable 3-D displays," Society for Information Display 2006 International Symposium, San Francisco, CA, USA, June 2006, vol. 37, book 2, pp. 1142–1145.

[10] B. Lee, J.-H. Park, and S.-W. Min, "Three-dimensional display and information processing based on integral imaging," in Digital Holography and Three-Dimensional Display (edited by T.-C. Poon), Springer, New York, USA, 2006, Chapter 12, pp. 333–378.

[11] A. Stern and B. Javidi, "Three-dimensional image sensing, visualization, and processing using integral imaging," Proc. of the IEEE, vol. 94, no. 3, pp. 591–607, 2006.
[12] J. Arai, M. Okui, T. Yamashita, and F. Okano, "Integral three-dimensional television using a 2000-scanning-line video system," Applied Optics, vol. 45, no. 8, pp. 1704–1712, 2006.
[13] T. Koike, M. Oikawa, K. Utsugi, M. Kobayashi, and M. Yamasaki, "Autostereoscopic display with 60 ray directions using LCD with optimized color filter layout," Stereoscopic Displays and Applications XVIII, Electronic Imaging, Proc. SPIE, vol. 6490, Paper 64900T, pp. 64900T-1–64900T-9, San Jose, CA, USA, Jan. 2007.
[14] http://techon.nikkeibp.co.jp/english/NEWS_EN/20050418/103839/
[15] J.-H. Park, H.-R. Kim, Y. Kim, J. Kim, J. Hong, S.-D. Lee, and B. Lee, "Depth-enhanced three-dimensional-two-dimensional convertible display based on modified integral imaging," Optics Letters, vol. 29, no. 23, pp. 2734–2736, 2004.
[16] J.-Y. Son, Y.-J. Choi, J.-E. Ban, V. Savelief, and E. F. Pen, "Multi-view image display system," U. S. Patent No. 6,606,078, Aug. 2003.
[17] J.-H. Park, J. Kim, Y. Kim, and B. Lee, "Resolution-enhanced three-dimension/two-dimension convertible display based on integral imaging," Optics Express, vol. 13, no. 6, pp. 1875–1884, 2005.
[18] S.-W. Cho, J.-H. Park, Y. Kim, H. Choi, J. Kim, and B. Lee, "Convertible two-dimensional-three-dimensional display using an LED array based on modified integral imaging," Optics Letters, vol. 31, no. 19, pp. 2852–2854, 2006.
[19] H. Choi, S.-W. Cho, J. Kim, and B. Lee, "A thin 3-D-2-D convertible integral imaging system using a pinhole array on a polarizer," Optics Express, vol. 14, no. 12, pp. 5183–5190, 2006.
[20] H. Choi, Y. Kim, S.-W. Cho, and B. Lee, "A 3-D/2-D convertible display with pinhole array on a LC panel," Proc. of the 13th International Display Workshops, Otsu, Japan, vol. 2, Dec. 2006, pp. 1361–1364.
[21] Y. Kim, H. Choi, S.-W. Cho, Y. Kim, J. Kim, G. Park, and B. Lee, "Three-dimensional integral display using plastic optical fibers," Applied Optics, vol. 46, no. 29, pp. 7149–7154, 2007.
[22] H. Choi, J.-H. Park, J. Kim, S.-W. Cho, and B. Lee, "Wide-viewing-angle 3-D/2-D convertible display system using two display devices and a lens array," Optics Express, vol. 13, no. 21, pp. 8424–8432, 2005.

# 5

# Effect of Pickup Position Uncertainty in Three-Dimensional Computational Integral Imaging

Mehdi DaneshPanah, Behnoosh Tavakoli, Bahram Javidi, and Edward A. Watson

**Abstract** This chapter will review the relevant investigations analyzing the performance of Integral Imaging (II) technique under pickup position uncertainty. Theoretical and simulation results for the sensitivity of Synthetic Aperture Integral Imaging (SAII) to the accuracy of pickup position measurements are provided. SAII is a passive three-dimensional, multi-view imaging technique that, unlike digital holography, operates under incoherent or natural illumination. In practical SAII applications, there is always an uncertainty associated with the position at which each sensor captures the elemental image. In this chapter, we theoretically analyze and quantify image degradation due to measurements' uncertainty in terms of Mean Square Error (MSE) metric. Experimental results are also presented that support the theory. We show that in SAII, with a given uncertainty in the sensor locations, the high spatial frequency content of the 3-D reconstructed images are most degraded. We also show an inverse relationship between the reconstruction distance and degradation metric.

## 5.1 Introduction

There has been an increasing interest in three-dimensional (3-D) image sensing and processing as well as 3-D visualization and display systems recently [1, 2, 3, 4, 5]. Among different techniques of 3-D imaging and display, Integral

---

B. Javidi
Department of Electrical and Computer Engineering, U-2157 University of Connecticut, Storrs, CT 06269-2157, USA
e-mail: Bahram@engr.uconn.edu

B. Javidi et al. (eds.), *Three-Dimensional Imaging, Visualization, and Display*,
DOI 10.1007/978-0-387-79335-1_5, 

Imaging (II) is based on the idea of integral photography [6, 7, 8, 9, 10] in which the scene is conventionally imaged with a lens from multiple perspectives. Each 2-D image, known as elemental image, is a 2-D intensity image that is acquired from different locations with respect to a 3-D scene. This is in contrast with the holographic techniques of wavefront recording and reconstruction which are based on interferometric measurement of magnitude and phase [2, 3]. In II, perspective views can be generated by microlens array or by translating a conventional camera over a synthetic aperture. The latter is known as SAII and performs better for applications requiring longer imaging range or better 3-D reconstruction resolution; we refer the reader to refs [11, 12, 13, 14, 15, 16] for excellent work on this technique. Elemental images convey the directional and intensity information and can be used for optical or computational visualization of the scene visual information [17, 18, 19, 20]

There have been efforts to improve the resolution, viewing angle, depth of focus and eye accommodation for the II technique in both aspects of pickup and display [21, 22, 23]. On the pickup side, methods such as time multiplexing computational integral imaging [21] and Synthetic Aperture Integral Imaging (SAII) [12,13] are proposed to increase the resolution of optical or computational reconstructions. In addition, techniques such as non-stationary micro-optics [23], optical path control [24], use of embossed screen [25] or amplitude modulation of individual lenslets [26] are investigated to improve resolution or depth-of-focus. Computational reconstruction of elemental images has proven to be promising in a variety of applications including 3-D object recognition, occlusion removal and multiple viewing point generation [27, 28, 29].

Synthetic aperture imaging includes collecting information from a moving platform. Therefore, practical issues related to systems involving translation, including the inaccuracy in position measurement, need to be studied for this imaging system. This raises the question of the sensitivity of the 3-D image reconstruction results to the amount of position measurement inaccuracy at the pickup stage. In this chapter, we provide an overview of our investigations using statistical analysis on the performance of 3-D computational reconstruction of SAII technique to the random pickup position errors [30]. Section 5.2 briefly overviews the concepts of integral imaging technique and computational reconstruction. Section 5.3 models the position measurement errors mathematically and inspects the vulnerability of the system to such errors, while in Section 5.4 a Monte Carlo simulation for a point source imaged with SAII is studied. Experimental results with real 3-D objects are presented in Section 5.5 and we conclude in Section 5.6.

## 5.2 Integral Imaging and Computational Reconstruction

Conventional Integral Imaging (II) systems use an array of small lenses mounted on a planar surface called lenslet array to capture elemental images on a single optoelectronic sensor. Each lens creates a unique perspective view of the scene at its image plane. As long as elemental images do not overlap in the image plane of the lenslet, one can capture all elemental images on a an opto-electronic sensor array such as Charge-Coupled Device (CCD) at once. This technique has the merits of simplicity and speed. However, for objects close to the lenslet array, the elemental images may overlap in the image plane which requires one to use additional optics to project the separated elemental images on a sensor. Also, the typically small aperture of lenslets creates low resolution or abberated elemental images. In addition, the pixels of the imaging device must be divided between all elemental images, which leads to the low number of pixels per elemental image.

Synthetic Aperture Integral Imaging (SAII) is a method which can potentially resolve some of the problems associated with conventional II. In this technique, an imaging device such as a digital camera scans a planner aperture in order to capture two dimensional images from different perspectives. These elemental images can be captured with well corrected optics on a large optoelectronic sensor; therefore the resolution and aberration of each elemental image can be dramatically enhanced, compared to lenslet based II.

The pickup process of Synthetic Aperture Integral Imaging is illustrated in Fig. 5.1. The point of view of each camera depends on its location with respect to the 3-D scene. Therefore, directional information is also recorded in an integral image as well as the 2-D intensity information. Specifically, depth information is indicated in the change of the parallax of various views.

Since capturing the perspective views requires multiple acquisitions, SAII in this form is suitable for imaging the relatively immobile targets. However, [12] proposes a solution to this problem by introducing an array of imaging devices on a grid which can be used for imaging dynamic objects.

One of the possible approaches in the 3-D II reconstruction computationally is to simulate reversal of the pickup process using geometrical optics [17]. In this method, a 2-D plane of the 3-D scene located at a particular distance from the pickup plane is reconstructed by back propagating the elemental images through the simulated pinhole array. The back projection process consists of shifting and magnifying each elemental image with respect to the

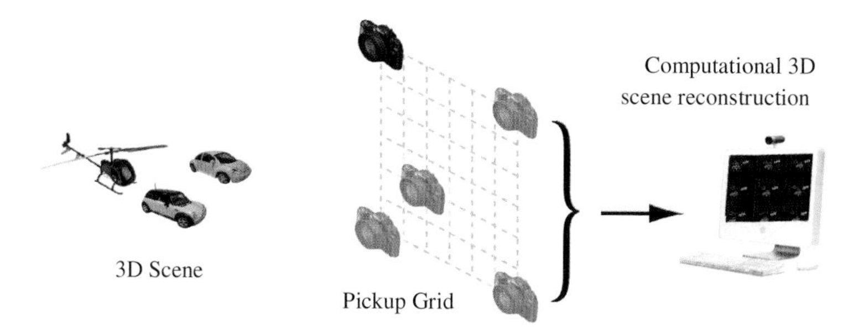


**Fig. 5.1.** Pickup process of three-dimensional synthetic aperture integral imaging

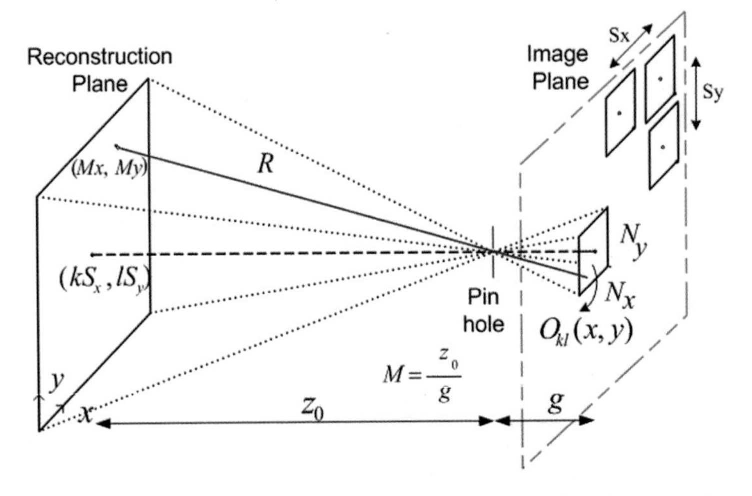


**Fig. 5.2.** Schematic of II reconstruction process consists of back projection of each elemental image to the reconstruction plane

location of its associated imaging device on the pickup plane and the desired distance of reconstruction. The rays coming from different elemental images intersect to create an image of the objects located originally at the distance of reconstruction (See Fig. 5.2).

Alternatively, one can shrink the spatial grid, on which elemental images are taken, proportional to the desired reconstruction distance while keeping the size of elemental images constant. In effect, the pickup grid separation is divided by the magnification factor ($z_0/g$). This approach is more appropriate for SAII in which each elemental image is captured on a full sensor and has a better resolution, compared to lenslet based elemental images. An additional merit of this approach on the computational side is that there would be no need to handle magnified images (very large matrices) for the reconstruction process. This greatly reduces the reconstruction time, required resources and computational burden.

According to the latter method, original elemental image, $O_{kl}(x,y)$, is shifted according to the following expression:

$$I_{kl}(x, y, z_0) = O_{kl}\left(x + \left(\frac{1}{M}\right) S_x k,\, y + \left(\frac{1}{M}\right) S_y l\right), \tag{5.1}$$

in which $S_x$ and $S_y$ denote the separation of sensors in $x$ and $y$ directions at the pickup plane, respectively, whereas subscripts $k$ and $l$ signify the location of elemental image $O_{kl}$ in the pickup grid. The magnification factor $M$, is given by $z_0/g$ where $z_0$ is the distance between the desired plane of reconstruction and the sensor along the optical axis, and $g$ denotes the distance of the image plane from each lens. The final reconstructed plane consists of partial overlap of the shifted elemental images as:

$$I(x, y, z_0) = \sum_{k=1}^{K} \sum_{l=1}^{L} I_{kl}(x, y, z_0)/R^2(x, y), \tag{5.2}$$

where $K$ and $L$ denote the number of elemental images acquired in the $x$ and $y$ directions; also $R$ compensates for intensity variation due to different distances from the object plane to elemental image $O_{\mathrm{kl}}$ on the sensor and is given by:

$$R^2(x, y) = (z_0 + g)^2 + \left[(Mx - S_x k)^2 + (My - S_y l)^2\right] \times \left(\frac{1}{M} + 1\right)^2 \tag{5.3}$$

In essence, Eq. (5.3) is the square of the distance of each pixel in the elemental image from the corresponding pixel in the reconstruction plane.

However, for most cases of interest such as long range 3-D imaging, Eq. (5.3) is dominated by the term $(z_0+g)^2$ because, for reconstruction, we always have $|x| \leq \frac{S_x k}{M} + \frac{N_x}{2}$ or equivalently $|Mx - S_x k| \leq \frac{MN_x}{2}$. So, if the viewing field is smaller than the reconstruction distance, $MN_x < z_0$, then the term $(z_0+g)^2$ would dominate in Eq. (5.3). This condition is equivalent to having an imaging sensor (CCD) which is small in dimensions compared to the effective focal length of the imaging optics.

In the computational reconstruction the adjacent shifted elemental images overlap such that for objects close to the reconstruction plane, the overlap of all elemental images would be aligned, whereas for objects located away from the reconstruction plane the overlap would be out of alignment resulting in a blurred reconstruction. Thus, with computational reconstruction one is able to get an in-focus image of an object at the correct reconstruction distance, while the rest of the scene appears out of focus.

The collection of 2-D scenes reconstructed at all distances can then provide the 3-D visual clues of the scene [17].

## 5.3 Sensitivity Analysis of Synthetic Aperture Integral Imaging (SAII)

A practical concern in the most synthetic aperture methods is the questionable accuracy of spatial localization of the information collecting apparatus and its (usually) degrading effect on the final processed data. However, the sensitivity of different synthetic aperture methods to such errors depends on the parameters of the system and thus is highly application specific. Similarly, the sensitivity of SAII to uncertainty in pickup position measurements is a natural question that is of prominent practical importance in system design.

Note that in the case of the SAII system, the lens and the optoelectronic sensor are aligned while the position of the whole image acquisition device (camera) is uncertain. In order to analyze the effect of positioning uncertainty at the pickup stage on the quality of the 3-D reconstructed images, a random position error, $\Delta p$, is introduced in the position from which elemental images of an accurate SAII pickup are acquired. Let the accurate position of the camera at each node in a planar pickup grid with $K$ nodes in $x$ and $y$ directions be $\{S_x\, k,\, S_y\, l\}$ where $k$, $l$=1,2,...,$K$.. Thus the distorted position of the sensor for capturing the elemental image $O_{kl}$ is represented by $\left\{S_x k + \Delta p^x, S_y l + \Delta p^y\right)\}$ where ($\Delta p^x$, $\Delta p^y$) are modeled as two independent random variables.

According to geometrical optics, applying distorted position in the computational reconstruction is equivalent to shifting the $i$th elemental image, $O_i(\mathbf{p})$, from its original location by $\Delta p_i/M$ where $\mathbf{p}$=($x$,$y$) denotes the position variable in the sensor plane and $M$ is the magnification associated with the plane of reconstruction. To avoid unnecessary complications in the equations, a single index is assigned to each elemental image following a raster pattern (see Fig. 5.2). Note that the position of each elemental image in the 2-D grid is still used for reconstruction. According to Eq. (5.2), the reconstruction with dislocated elemental images can be written as:

$$I_e(x, y, z) = \sum_{i=1}^{K \times K} I_i(x + \frac{\Delta p_i^x}{M}, y + \frac{\Delta p_i^y}{M})/R^2, \tag{5.4}$$

where K denotes the number of elemental images taken in each direction. We define the difference between Eqs. (5.2) and (5.4) as the error metric:

$$err(\mathbf{p}, z) = I(\mathbf{p}, z) - I_e(\mathbf{p}, z). \tag{5.5}$$

To quantify the image degradation, we use the well-known Mean Square Error (MSE) metric. This metric provides an estimate of average error per pixel at the reconstruction plane that should be expected for a particular measurement uncertainty.

For clarity, we initially assume that all positioning errors for all elemental images captured in a rectangular grid of $K \times K$ images lie along the same direction, for instance $x$. Extension of the results to the more realistic case of two-dimensional random position errors with independent components in $x$ and $y$ directions is also provided. Since we assume a stochastic process is governing the position measurement process, the expected value of the error associated with a specific location error probability distribution needs to be studied. The average error resulting from different realizations of the location error distributions is given by:

$$E\left|err(x, y, z)\right|^2 = \frac{1}{R^4} E\left\{\left|\sum_{i}^{K \times K} I_i(x, y) - I_i(x + \frac{\Delta p_i}{M}, y)\right|^2\right\}, \tag{5.6}$$

in which $E(.)$ is the expectation operator; $R$ is a function of system parameters and is dominated by $(z+g)$ in Eq. (5.3). Also, $\Delta p_i$ is the error associated with $i$th sensor. We proceed with the analysis in the Fourier domain where the stochastic spatial shifts of elemental images appear as phase terms. Thus $err(\mathbf{p},z)$ can be written as:

$$\begin{aligned} Err(\mathbf{f},z) = F.T\left\{err(\mathbf{p},z)\right\} &= \frac{1}{R^2}\sum_{i=1}^{K\times K}\left[\tilde{I}_i(f_x,f_y) - \tilde{I}_i(f_x,f_y)e^{-jf_x\frac{\Delta p_i}{M}}\right] \\ &= \frac{1}{R^2}\sum_{i=1}^{K\times K}\left[\tilde{\mathbf{I}}_i - \tilde{\mathbf{I}}_i e^{-jf_x\frac{\Delta p_i}{M}}\right] \end{aligned} \quad (5.7)$$

The symbol $\sim$ denotes the Fourier transformation; also $\mathbf{f}=(f_x, f_y)$ stands for the couplet of spatial frequencies in the $x$ and $y$ directions, i.e.; $\tilde{\mathbf{I}} = I(f_x, f_y) = F.T\left\{I(\mathbf{p})\right\}$.

As we assume that the random displacements are all in the $x$ direction, the shift appears as a phase term $\exp(-jf_x\Delta p_i/M)$ in Eq. (5.7) only incorporating $f_x$. Henceforth, we use the term error spectrum for $|Err\,(\mathbf{f},\, z)|^2$.

We show in the Appendix how one gets the following expression for the expected value of the error spectrum:

$$E\left|Err(\mathbf{f},z)\right|^2 = \frac{1}{R^4}\left[(2-\gamma-\gamma^*)\sum_{i}^{K\times K}\left|\tilde{\mathbf{I}}_i\right|^2 + |1-\gamma|^2\sum_{i\neq j}^{K\times K}\sum_{j\neq i}^{K\times K}\tilde{\mathbf{I}}_i\tilde{\mathbf{I}}_j^*\right], \quad (5.8)$$

where $\gamma = E\left\{\exp(-jf_x\Delta p/M)\right\}$ is related to the moment generating function of random variable $\Delta p$ and can be derived for all basic probability distributions, i.e., Gaussian, uniform, Laplacian and, etc [31]. Hereafter, we assume that the camera location error follows a zero mean Gaussian distribution with variance $\sigma^2$, i.e., $N(0,\, \sigma^2)$. However, it is straightforward to assume other distributions.

For a Gaussian random variable $X\sim N(\mu,\sigma^2)$ the moment generating function is $M_X(t) = E\left\{e^{tX}\right\} = \exp\left(\mu t + \sigma^2 t^2/2\right)$ [31], thus for $\Delta p\sim N(0,\sigma^2)$, $\gamma$ in Eq. (5.8) becomes a real number, $\gamma = E\{\exp(-jf_x\Delta p/M)\} = \exp(-f_x^2\sigma^2/2M^2)$. Essentially this parameter depends on the reconstruction distance and the characteristics of the random measurement errors and is a function of frequency. Equation (5.8) can be further simplified to:

$$E\left|Err(\mathbf{f},z)\right|^2 = \frac{2(1-\gamma)}{R^4}\times\sum_{i}^{K\times K}\left|\tilde{\mathbf{I}}_i\right|^2 + \frac{(1-\gamma)^2}{R^4}\times\sum_{i\neq j}^{K\times K}\sum_{j\neq i}^{K\times K}\tilde{\mathbf{I}}_i\tilde{\mathbf{I}}_j^*. \quad (5.9)$$

For the general case when the dislocation has both $x$ and $y$ components, Eq. (5.9) can be adjusted to the following form:

$$E\left|Err(\mathbf{f},z)\right|^2 = \frac{2(1-\gamma^2)}{R^4} \times \sum_{i}^{K\times K}\left|\tilde{\mathbf{I}}_i\right|^2 + \frac{(1-\gamma^2)^2}{R^4} \times \sum_{i\neq j}^{K\times K}\sum_{j\neq i}^{K\times K}\tilde{\mathbf{I}}_i\tilde{\mathbf{I}}_j^*. \quad (5.10)$$

The total expected error is the area under the error spectrum for all spatial frequencies. According to Parseval's theorem we have:

$$\int \left|err(\mathbf{p},z)\right|^2 d\mathbf{p} = \int \left|Err(\mathbf{f},z)\right|^2 d\mathbf{f}, \quad (5.11)$$

and since the expectation and integration operations can be exchanged, one has the following relationship for the MSE at distance $z$ on the reconstruction plane:

$$MSE(z) = \int E\left|err(\mathbf{p},z)\right|^2 d\mathbf{p} = \int E\left|Err(\mathbf{f},z)\right|^2 d\mathbf{f}$$
$$= \frac{1}{(z+g)^4}\int\left[2(1-\gamma^2)\sum_{i}^{K\times K}\left|\tilde{\mathbf{I}}_i\right|^2\right]d\mathbf{f} + \int\left[(1-\gamma^2)^2\sum_{i\neq j}^{K\times K}\sum_{j\neq i}^{K\times K}\tilde{\mathbf{I}}_i\tilde{\mathbf{I}}_j^*\right]d\mathbf{f} \quad (5.12)$$

For a given position measurement error distribution and at a certain distance, $\gamma$ is a function of frequency. Thus, the term $(1-\gamma^2)$ in the first integral in the right hand side of Eq. (5.12) acts as a weighting coefficient for the energy spectrum of the shifted elemental images. As discussed earlier, in the case of a Gaussian positioning error, $(1-\gamma^2) = [1-\exp(-f_x^2\sigma^2/M^2)]$, is a high pass filter function with respect to $f_x$. This means that the higher spectral components of elemental images contribute more significantly to the MSE of each plane compared to the energy contained in the low spatial frequencies.

In addition, at larger reconstruction distances, i.e., larger $M$, the stop band of this filter becomes wider and thus diminishes more of the spectral components of the elemental images, and consequently reduces the resulting MSE. The bandwidth of this high pass filter depends solely on the ratio of variance of the positioning error probability distribution function, $\sigma^2$, and the magnification factor $M$. In particular, for a Gaussian sensor positioning error, i.e., $\Delta p \sim N(0, \sigma^2)$, the Full Width Half Maximum (FWHM) is given by:

$$\mathrm{FWHM} = 2M\sqrt{\ln(2)}/\sigma = 1.66z/g\sigma. \quad (5.13)$$

The inverse relationship of FWHM with $\sigma$ is confirmed by the fact that with better positioning accuracy (small $\sigma$), more spatial frequencies are going to be suppressed in Eq. (5.12) and, thus, the MSE would be reduced. This is the primary reason that we can expect to see more degradation near the edges of the objects in focus at the reconstruction plane and less degrading effect in areas containing low spatial frequencies or objects that are at a distance away from the reconstruction plane. Also, FWHM increases with reconstruction distance $z$, which means one should expect less degradation when reconstructing far objects. This important outcome has been verified and demonstrated in the experimental results. Likewise, the second integral in the right hand side of Eq. (5.12) also filters out the low spatial frequencies from the cross correlation of two distinct shifted elemental images. This has the same effect as described for the first term with the only difference that $(1-\gamma^2)^2$ has a larger FWHM compared to $(1-\gamma^2)$, which acts as the filter in the first term, and thus the elemental image cross correlation contents contribute less in the total MSE.

## 5.4 Degradation Analysis for a Point Source

Monte Carlo simulation for the case of imaging a point source is used in order to estimate the degradation of reconstructed images due to sensor position measurement uncertainty. We simulate integral imaging of a point source from a rectangular pickup grid using geometrical optics; however, for clarity, we first carry out the analysis for a one-dimensional linear array with $K$ elemental images. Extension to 2-D pickup arrays is straightforward. Without loss of generality, we assume the point source to be located at distance $z$ from the pickup baseline on the optical axis of the first elemental image, i.e., $\delta(p)$; the distance between image plane and the lens is $g$ (see Fig. 5.2); also the camera is linearly translated with pitch $S_{\mathrm{p}}$. Thus, according to Eq. (5.1), the $k$th shifted elemental image is:

$$I_k(p) = \delta(p + kS_p(g/z)), \tag{5.14}$$

where $p$ is the position variable.

According to Eq. (5.5) the following expression for the error spectrum is:

$$E|Err(f_x, z)|^2 = \frac{2(1-\gamma)}{(z+g)^4}K + \frac{2(1-\gamma)^2}{(z+g)^4}\sum_{k\neq l}^{K}\sum_{l\neq k}^{K}\exp[-jf_xS_p(g/z)(k-l)] \tag{5.15}$$

Equation (5.15) can be rewritten as:

$$E|Err(f_x, z)|^2 = \frac{2\gamma(1-\gamma)}{(z+g)^4}K + \frac{2(1-\gamma)^2}{(z+g)^4}\frac{\sin^2(Kf_xS_p(g/z))}{\sin^2(f_xS_p(g/z))}. \tag{5.16}$$

In particular, when the distribution governing the displacement of sensors is $\Delta p \sim N(0, \sigma^2)$, then $\gamma = \exp(-f_x^2\sigma^2 g^2/2z^2)$ and Eq. (5.16) can be rearranged to:

$$E|Err(f_x, z)|^2 = \frac{2(1 - e^{-f_x^2\sigma^2 g^2/2z^2})}{(z+g)^4}\Bigg[Ke^{-f_x^2\sigma^2 g^2/2z^2} + \ldots$$
$$(1 - e^{-f_x^2\sigma^2 g^2/2z^2})\frac{\sin^2(Kf_x S_p(g/z))}{\sin^2(f_x S_p(g/z))}\Bigg] \quad (5.17)$$

Note that the total MSE expected from such a random displacement of sensors is the total energy of error spectrum as in Eq. (5.12).

We perform Monte Carlo simulation to statistically compute the MSE in the case of imaging a point source with a one-dimensional linear array in which the position of the sensor is altered about a perfect linear grid of pinholes according to the random variable $\Delta p{\sim}N(0,\ 1)$. The simulation is repeated for 1,000 trials with different samples of random position errors. At each trial, the set of elemental images are used to reconstruct the point source at its respective distance and the results are compared with the result of reconstruction using correctly positioned sensors. In each trial, MSE is computed according to Eqs. (5.6) and (5.11) and all the 1,000 MSEs for each reconstruction distance are averaged. The total MSE computed from Monte Carlo simulation is compared to the MSE calculated using Eq. (5.17) with similar parameters as $K = 16$, $S_p = 2.5$ mm and $\Delta p{\sim}N(0,1)$ for the point source located at different distances from $z = 24$ to 40 cm. Figure 5.3 shows the agreement of simulation results and that of the mathematical analysis.

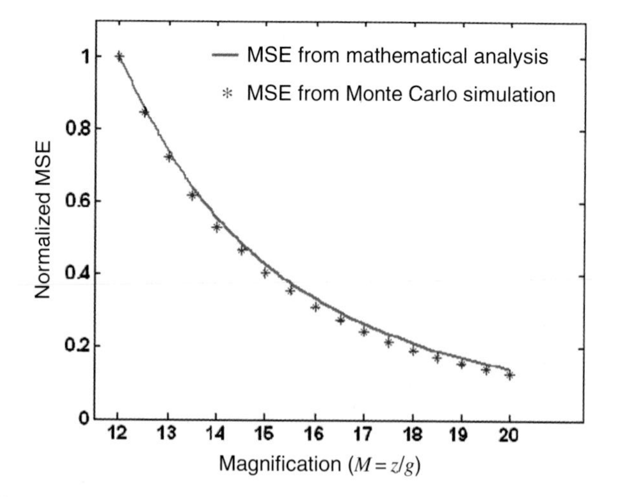


**Fig. 5.3.** Comparison of the MSE results of the Monte Carlo simulation and Eq. (5.17) for a point source located at z = 24–40 cm (M = 12–20) at g = 2 cm

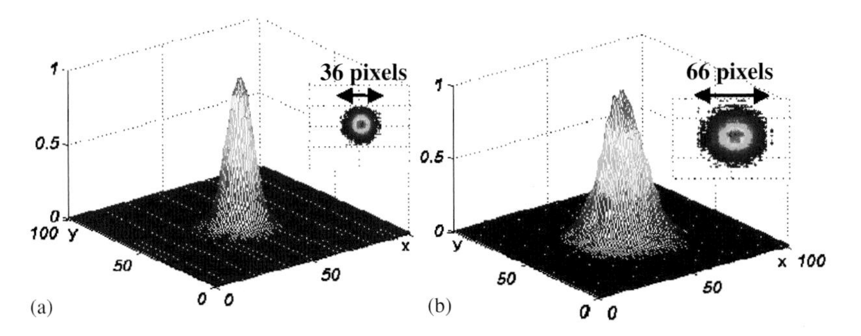


**Fig. 5.4.** Result of the Monte Carlo simulation for the point source reconstruction

For the two-dimensional Monte Carlo simulation, the same process is repeated using a 16×16 rectangular grid, pitch $S_x = S_y = 2.5$ mm, the position displacements follow $\Delta p \sim N(0, 1)$ and the point source is located at (a) $z =$ 40 cm and (b) 24 cm. Figure 5.4 shows the combination of the reconstruction results from all 1,000 trials at 24 and 40 cm. Notice that, for zero position errors, the plots would have been a single point at the origin.

The simulation results in Fig. 5.4 approximate the average impulse response at the corresponding reconstruction distance. The bell shape impulse response in Fig. 5.4 resembles the standard blurring by optics and detectors' finite dimensions and has the similar effect on the reconstructed images. The blurring associated with the Airy disk of imaging optics is given in radians by $\alpha_{diff} = 1.21\lambda/d$ in which $d$ is the radius of the input aperture and $\lambda$ is the wavelength of the illumination. On the other hand, the blurring associated with the pixilation of the sensor in radians is $\alpha_{pxl} = c/f$, where $c$ and $f$ are the sensor pixel size and focal length of imaging optics, respectively. The expression for blurring due to diffraction yields 17 and 29 $\mu$m for object distance of 24 and 40 cm, respectively, at $\lambda = 600$ nm and $d = 10$ mm. For pixel size of $c = 10$ $\mu$m and lens focal length of $f = 2$ cm, the blurring due to pixilation is 120 and 200 $\mu$m at the same object distances.

However, when positioning error is introduced, the resulting blurring radii can be measured as 2.5 and 1.4 mm by projecting the half maximum radii of the bell shapes in Fig. 5.4 to their respective reconstruction distances at 24 and 40 cm.

This comparison suggests that improved positioning systems are more effective to reduce the total MSE of the reconstructed images, compared to improvements related to the MTF of imaging optics or the sensor pixel pitch. As mentioned earlier, the exact impulse response depends on the plane in which the elemental images are reconstructed. Figure 5.4 shows that for a point source located 24 cm from the sensor, the impulse response is wider than that of the case of a point source at 40 cm, i.e., the blurring effect of sensor position uncertainty is more degrading for objects closer to the pickup plane.

## 5.5 Experimental Results

The results of SAII experiments are presented to demonstrate the quantitative and qualitative degradation of computational 3-D reconstructions by introducing sensor position uncertainty in the pickup process and also to verify the mathematical analysis performed earlier. Figure 5.5(a) shows a 2-D image of the 3-D scene used in the experiments. The scene is composed of two toy cars and a model helicopter located at different distances from the sensor. The depth of the scene is 16 cm for which the closest and farthest objects are located 24 and 40 cm away from the center of the pickup grid. The dimension of the cars is about 5×2.5×2 cm, whereas the helicopter is 9×2.5×2 cm. The scene is illuminated with diffused incoherent light.

The experiment is performed by moving a digital camera transversely in an $x$–$y$ grid with the pitch of 5 mm in both $x$ and $y$ directions. At each node, an elemental image is captured from the scene. The active CCD area of the imaging sensor is 22.7×15.6 mm with 10 $\mu$m pixel pitch and the effective focal length of the camera lens is about 20 mm. In Fig. 5.5(b)–(d) we show the 3-D reconstruction of the scene in three different distances of the objects according to Eq. (5.2). At each distance one of the objects is in focus while the others appear washed out.

The 16×16 set of elemental images are captured in a planar grid pattern (see Fig. 5.1). A subset of the captured elemental images can be seen in Fig. 5.6, each convening different perspective information from the 3-D scene.

**Fig. 5.5.** (**a**) A 2-D image of the 3-D scene, (**b**) reconstruction at the distance $z =$ 24 cm, (**c**) $z = 30$ cm and (**d**) $z = 36$ cm

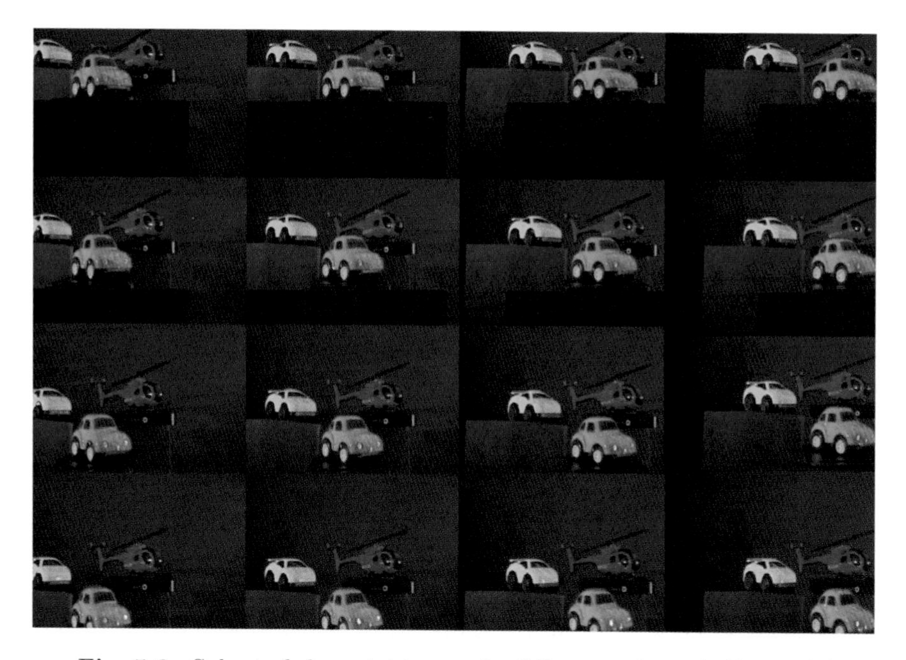

**Fig. 5.6.** Subset of elemental images for 3-D scene shown in Fig. 5.5(**a**)

Monte Carlo simulation is used to study the degradation effect of sensor position uncertainty during the pickup process on the reconstructed images. Position errors in $x$ and $y$ directions, $(\Delta p^{\mathrm{x}},\Delta p^{\mathrm{y}})$, are modeled as two independent zero mean Gaussian random variables. In the experiments, a rectangular grid with equal spacing in $x$ and $y$ directions is used, i.e., the pitch of $S_{x,y} = S_p$ $= 5$ mm. Since $\Delta p^{\mathrm{x,y}}$ follows a $N(0,\sigma^2)$, we define the fraction $100\sigma^2/S_{\mathrm{p}}$ to be the pitch error percentage for degradation evaluation. Note that $\Delta p^{\mathrm{x,y}}/S_{\mathrm{p}}$ represents a normalized positioning error metric with respect to the gap between elemental image acquisition.

To perform a Monte Carlo simulation, the random position error for each camera is randomly chosen 500 times in the form of a couplet $(\Delta p^{\mathrm{x}},\Delta p^{\mathrm{y}})$, and utilized to computationally reconstruct one plane of the 3-D scene located at distance $z$ using Eq. (5.4). As a result, we have 500 reconstructed images of the same plane which is compared with the reconstruction using correct positions obtained from Eq. (5.2). Mean Square Error metric is computed using Eq. (5.6) to measure the difference of these images quantitatively. This simulation is done for distances from 24 to 40 cm which corresponds to magnification, $M$, from 12 to 20. The pitch error is chosen to be 30 percent in both directions, i.e., $\sigma^2/S_p = 0.3$. Since $\Delta p^{\mathrm{x,y}}$ is a Gaussian random variable, such an error means that 70 percent of the time $\Delta p^{\mathrm{x,y}}$ remains less than 0.3 of the pitch $(0.3S_p)$.

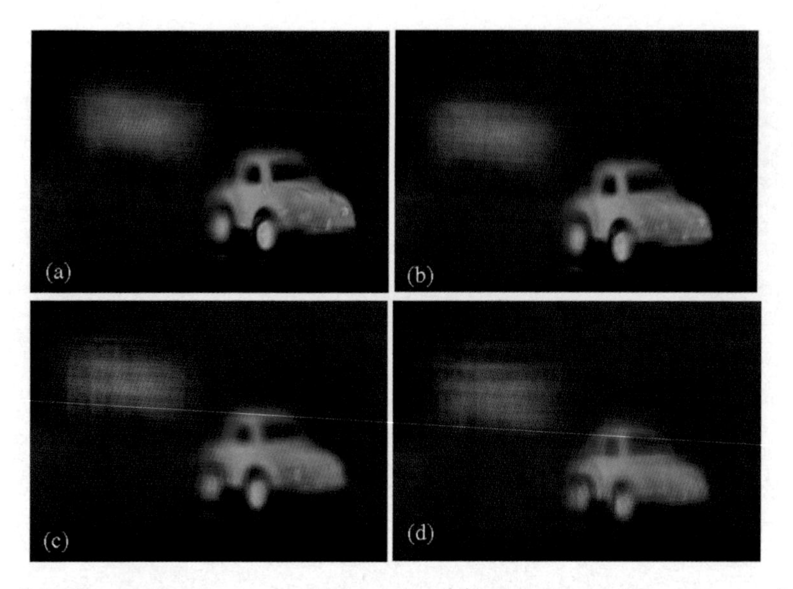


**Fig. 5.7.** Reconstruction at $z = 24$ cm using (**a**) original camera position, and (**b**) using distorted camera position with 20 percent pitch error, (**c**) 30 percent pitch error, (**d**) 50 percent pitch error

Figure 5.7(a) shows the results of reconstruction at $z = 24$ cm without introducing the position errors, while in Fig. 5.7(b)–(d) we illustrate the reconstruction results for three different error percentages, i.e., 20, 30, 50 percent position error, respectively. This figure compares the degradation effect of using distorted camera positions on the particular plane of reconstruction qualitatively. One can clearly observe that the degradation is more severe for larger pitch errors.

Figure 5.8, shows the box-and-whisker diagram of the MSEs to demonstrate their statistical properties at different reconstruction distances. The blue box shows the upper and lower quartiles of MSEs which is related to their variance, and the dotted blue line shows the smallest and largest computed MSEs. According to Monte Carlo simulation the average of these 500 MSEs is computed at each plane, illustrated by the solid red line in Fig. 5.8.

This average for each particular plane of the scene is a reasonable estimation of the error one can expect due to a 30 percent camera positioning error. Figure 5.8 also illustrates the decreasing trend of the average MSE with increasing magnification. With the constant $g$, magnification increase is identical to the reconstruction distance increase. Note that the variance of the error at each plane decreases when reconstruction distance increases, while its rate of decrease is greater than the rate of decrease of the average MSE. This fact can be explained using Eq. (5.12) which shows MSE is inversely proportional to $z$.

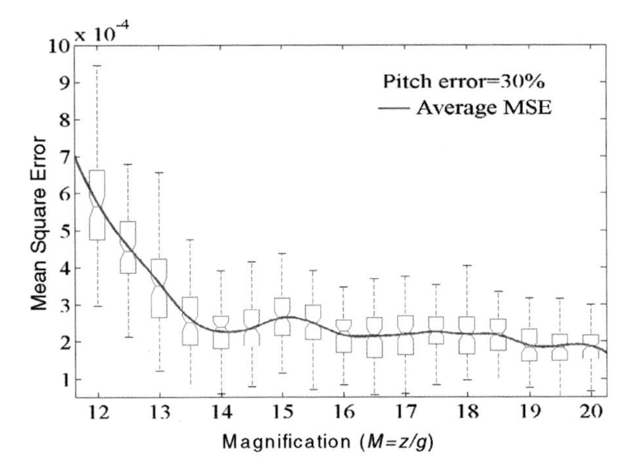


**Fig. 5.8.** Box-and-whisker diagram of the MSEs for z = 24 cm (M = 12) to z = 40 cm (M = 20) when the pitch error is 30 percent

The simulation done for 30 percent pitch error is repeated for 10, 20 and 50 percent pitch errors. According to each of these pitch errors a random position is generated by adding a random Gaussian variable with appropriate variance to the original position. We repeat the process 500 times and each time the MSE of the degraded and original images is computed. As a result, for each pitch error, we have 500 MSEs at various planes of the scene. The average of

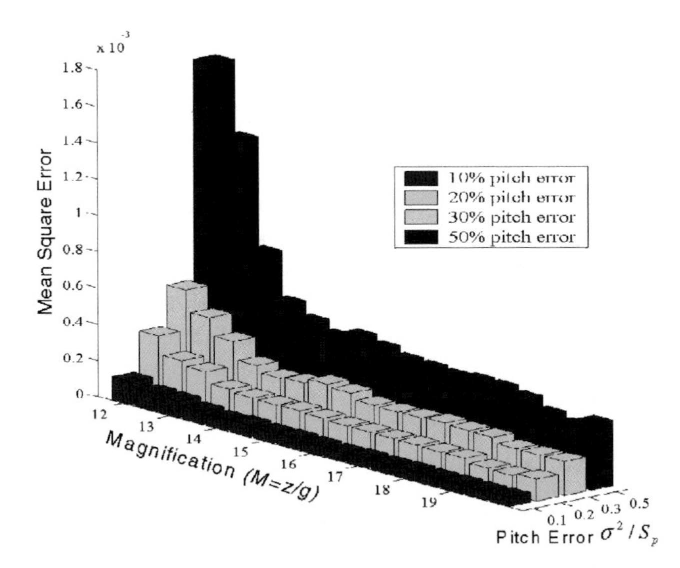


**Fig. 5.9.** Mean of MSE corresponding to 10, 20, 30 and 50 percent pitch errors for the distances from $z = 24$ to 40 cm, i.e., $M = 12$–20

**Fig. 5.10.** Total error for the range of distances from (**a**) z = 35–38 cm, (**b**) z = 29–32 cm and (**c**) z = 24–27 cm

MSE of 500 trials is computed for each plane at different pitch errors and the combined result is shown in Fig. 5.9.

As discussed in Section 5.4, the contribution of all spatial frequencies is not the same in the total MSE. To illustrate this fact, the total MSE in three different ranges, i.e., 35–38 cm, 29–32 cm and 24–27 cm for the helicopter, yellow car and green car, is computed respectively while the pitch error is 50 percent. The results in the Fig. 5.10 show that the object in focus which exhibits more detail and higher spatial frequencies is distorted more significantly by the introduced position error for the cameras. Clearly, the maximum error occurs around edges. This observation confirms the mathematical analysis result in Eq. (5.12) which suggests that the term $(1-\gamma^2)$ emphasizes contributions of higher spatial frequencies of the elemental images in MSE, while suppressing the low spatial frequencies' contributions. Figure 5.10 also verifies the fact that error is larger for objects closer to the pickup plane (sensor).

## 5.6 Conclusion

An introduction to the Synthetic Aperture Integral Imaging (SAII) technique for 3-D imaging and computational reconstruction is explained in this chapter. In addition, an overview on the effect of position measurement accuracy at the pickup stage of SAII on the quality of 3-D computational reconstructions is presented [30]. Both theoretical analysis and experimental results are provided. Mean Square Error (MSE), calculated through a Monte Carlo simulation, is chosen as the statistical degradation metric, and Fourier transform has been used in theoretical analysis. The results suggest that, for a given position uncertainty, the degradation in terms of MSE decreases with increasing reconstruction distance. In addition, theoretical analysis suggests that low spatial frequency components in the energy spectrum of the elemental images contribute less in the resultant MSE, whereas high spectral components are the more significant cause of degradation for a fixed positioning error distribution. We provide experimental results that support the theoretical predictions for image degradation. From this study one can optimize the parameters of a SAII system's hardware in order to achieve a tolerable degradation in the presence of sensor positioning error.

## Appendix

The error spectrum follows the expression (5.18). Note that in the Eq. (5.18) only Δp has a random nature. Thus, the expected value for the error spectrum depends on the behavior of this variable, i.e., the distribution governing the spatial dislocation of the sensors during pickup.

$$
\begin{aligned}
\left|Err(\mathbf{f},z)\right|^2 &= \frac{1}{R^4}\left|\sum_{i}^{K\times K}\tilde{\mathbf{I}}_i(1-e^{-jf_x\frac{\Delta p_i}{M}})\right|^2 \\
&= \frac{1}{R^4}\left[\sum_{i}^{K\times K}\sum_{j}^{K\times K}\tilde{\mathbf{I}}_i\tilde{\mathbf{I}}_j^*(1-e^{-jf_x\frac{\Delta p_i}{M}})(1-e^{+jf_x\frac{\Delta p_j}{M}})\right].
\end{aligned}
\tag{5.18}
$$

Since expectation is a linear operator, one can break down the error expectation as follows:

$$
\begin{aligned}
E\left\{\left|Err(\mathbf{f},z)\right|^2\right\} &= \frac{1}{R^4}\left[\sum_{i}^{K\times K}\sum_{j}^{K\times K}\tilde{\mathbf{I}}_i\tilde{\mathbf{I}}_j^*E\left\{(1-e^{-jf_x\frac{\Delta p_i}{M}})(1-e^{+jf_x\frac{\Delta p_j}{M}}\right\}\right] \\
&= \frac{1}{R^4}\left[\sum_{i}^{K\times K}\sum_{i}^{K\times K}\tilde{\mathbf{I}}_i\tilde{\mathbf{I}}_i^*E\left\{2-e^{-jf_x\frac{\Delta p_i}{M}}-e^{+jf_x\frac{\Delta p_i}{M}}\right\}\right]+\ldots \\
&\quad \frac{1}{R^4}\left[\sum_{i\neq j}^{K\times K}\sum_{j\neq i}^{K\times K}\tilde{\mathbf{I}}_i\tilde{\mathbf{I}}_j^*E\left\{(1-e^{-jf_x\frac{\Delta p_i}{M}})(1-e^{+jf_x\frac{\Delta p_j}{M}})\right\}\right]
\end{aligned}
\tag{5.19}
$$

Now, let $\gamma = E\left\{\exp(-jf_x\Delta p/M)\right\}$ which is the moment generating function of random variable $\Delta p$ [31]. Equation (5.19) reduces to:

$$
\begin{aligned}
E\left\{\left|Err(\mathbf{f},z)\right|^2\right\} &= \frac{1}{R^4}\left[\sum_{i}^{K\times K}\sum_{i}^{K\times K}\tilde{\mathbf{I}}_i\tilde{\mathbf{I}}_i^*(2-\gamma-\gamma^*)\right]+ \\
&\quad \frac{1}{R^4}\left[\sum_{i\neq j}^{K\times K}\sum_{j\neq i}^{K\times K}\tilde{\mathbf{I}}_i\tilde{\mathbf{I}}_j^*E\left\{(1-e^{-jf_x\frac{\Delta p_i}{M}})(1-e^{+jf_x\frac{\Delta p_j}{M}})\right\}\right].
\end{aligned}
\tag{5.20}
$$

Note that $\Delta p_i$ denotes the sensor location error associated with $i$th sensor which is a random variable and the expected value of the random variable or a function of the random variable is constant which can come outside of the summation. In addition the sensor location errors are assumed to be independent, thus we have:

$$
\begin{aligned}
E\left\{(1-e^{-jf_x\frac{\Delta p_i}{M}})(1-e^{+jf_x\frac{\Delta p_j}{M}})\right\} &= E\left\{1-e^{-jf_x\frac{\Delta p_i}{M}}\right\} E\left\{1-e^{+jf_x\frac{\Delta p_j}{M}}\right\} \\
&= (1-\gamma)(1-\gamma^*) = |1-\gamma|^2 .
\end{aligned}
\tag{5.21}
$$

Consequently, the error spectrum expectation can be written as in Eq. (5.9).

## References


[1] Benton S A (2001) Selected Papers on Three-Dimensional Displays. SPIE Optical Engineering Press, Bellingham, WA
[2] Javidi B, Okano F (2002) Three Dimensional Television, Video, and Display Technologies. Springer, Berlin
[3] Okoshi T (1976) Three-Dimensional Imaging Technique. Academic Press, New York
[4] Javidi B, Hong S H, Matoba O (2006) Multi dimensional optical sensors and imaging systems. Appl. Opt. 45:2986–2994
[5] Levoy M (2006) Light fields and computational imaging. IEEE Comput. 39(8):46–55
[6] Lippmann M G (1908) Epreuves reversibles donnant la sensation durelief. J. Phys. 7:821–825
[7] Dudnikov Y A (1974) On the design of a scheme for producing integral photographs by a combination method. Sov. J. Opt. Technol. 41:426–429
[8] Ives H E (1931) Optical properties of a Lippmann lenticuled sheet. J. Opt. Soc. Am. 21:171–176
[9] Sokolov P (1911) Autostereoscpy and Integral Photography by Professor Lippmann's Method. Moscow State Univ. Press, Moscow, Russia
[10] Okano F, Hoshino H, Arai J et al (1997) Real time pickup method for a three dimensional image based on integral photography. Appl. Opt. 36:1598–1603
[11] Stern A, Javidi B (2006) Three-dimensional image sensing, visualization, and processing using integral imaging. Proc. IEEE 94:591–607
[12] Wilburn B, Joshi N, Vaish V, Barth A et al (2005) High performance imaging using large camera arrays. Proc. ACM 24:765–776
[13] Jang J S, Javidi B (2002) Three-dimensional synthetic aperture integral imaging. Opt. Lett. 27:1144–1146
[14] Burckhardt B (1968) Optimum parameters and resolution limitation of integral photography. J. Opt. Soc. Am. 58:71–76
[15] Hoshino H, Okano F, Isono H et al (1998) Analysis of resolution limitation of integral photography. J. Opt. Soc. Am. A 15:2059–2065
[16] Wilburn B, Joshi N, Vaish V et al (2005) High performance imaging using large camera arrays. ACM Trans. Graph. 24(3): 765–776
[17] Hong S H, Jang J S, Javidi B (2004) Three-dimensional volumetric object reconstruction using computational integral imaging. Opt. Express 12:483–491

[18] Stern A, Javidi B (2003) 3-D computational synthetic aperture integral imaging (COMPSAII). Opt. Express 11:2446–2451

[19] Igarishi Y, Murata H, Ueda M (1978) 3-D display system using a computer-generated integral photograph,. Jpn. J. Appl. Phys. 17:1683–1684

[20] Erdmann L, Gabriel K J (2001) High resolution digital photography by use of a scanning microlens array. Appl. Opt. 40:5592–5599

[21] Kishk S, Javidi B (2003) Improved resolution 3-D object sensing and recognition using time multiplexed computational integral imaging. Opt. Express 11:3528–3541

[22] Martínez-Cuenca R, Saavedra G, Martinez-Corral M et al (2004) Enhanced depth of field integral imaging with sensor resolution constraints. Opt. Express 12:5237–5242

[23] Jang J S, Javidi B (2002) Improved viewing resolution of three-dimensional integral imaging by use of nonstationary micro-optics. Opt. Lett. 27:324–326

[24] Hong J, Park J H, Jung S et al (2004) Depth-enhanced integral imaging by use of optical path control. Opt. Lett. 29:1790–1792.

[25] Min S W, Kim J, Lee B (2004) Wide-viewing projection-type integral imaging system with an embossed screen. Opt. Lett. 29:2420–2422

[26] Martínez-Corral M, Javidi B, Martínez-Cuenca R et al (2004) Integral imaging with improved depth of field by use of amplitude modulated microlens array. Appl. Opt. 43:5806–5813

[27] Hwang Y S, Hong S H, Javidi B (2007) Free view 3-D visualization of occluded objects by using computational synthetic aperture integral imaging. J. Display Technol 3:64–70

[28] Yeom S, Javidi B, Watson E (2005) Photon counting passive 3-D image sensing for automatic target recognition. Opt. Express 13:9310–9330

[29] Frauel Y, Javidi B (2002) Digital three-dimensional image correlation by use of computer-reconstructed integral imaging. Appl. Opt. 41:5488–5496

[30] Tavakoli B, Danesh Panah M, Javidi B et al (2007) Performance of 3-D integral imaging with position uncertainty. Opt. Express 15:11889–11902

[31] Mukhopadhyay N (2000) Probability and Statistical Inference. Marcel Dekker, Inc. New York

# 6

# 3-D Image Reconstruction with Elemental Images Printed on Paper

Daesuk Kim and Bahram Javidi

## 6.1 Introduction

There has been much interest in 3-D visualization [1, 2, 3, 4, 5, 6, 7, 8, 9, 10, 11, 12, 13, 14, 15, 16, 17, 18, 19, 20, 21, 22, 23, 24, 25, 26, 27, 28]. Among various approaches, lenticular technique has been a promising candidate in most 3-D display industries since it can provide high quality 3-D effect with easy setup. However, the lenticular approach has an inherent drawback because it cannot provide autostereoscopic images without using a supplementary tool such as polarizing glasses. Likewise, despite being able to provide an autostereoscopic 3-D image, holography also has many constraints which make it impractical when used in commercial 3-D display systems. The rainbow hologram, invented by S. Benton [1], has been positioning its practicability in the 3-D display industry due to the benefit of 3-D image reconstruction capability based on white light illumination. However, the rainbow hologram also sacrifices one-dimensional parallax information for practicability. The ability to view holographic images in white light was a vital step toward making holography suitable for display applications.

Recently, progress in spatial light modulator (SLM) and dramatically enhancing computer technology has been motivating the use of Integral Imaging (II) technique for full parallax 3-D TV and visualization. The basic idea

D.Kim
Division of Mechanical & Aero System Engineering, Chonbuk National University, Jeonju 561-756, Republic of Korea
e-mail: dashi.kim@chonbuk.ac.kr

B. Javidi et al. (eds.), *Three-Dimensional Imaging, Visualization, and Display*, DOI 10.1007/978-0-387-79335-1_6, © Springer Science+Business Media, LLC 2009

of II was originated from Lippmann in 1908 [2]. However, II was not one of the major approaches in 3-D visualization until recent outstanding advanced studies examined the enormously enhanced practicability of the II based 3-D display [4, 5, 6, 7, 8, 9, 10, 11, 12, 13, 14, 15, 16, 17, 18, 19, 20, 21, 22, 23, 24, 25, 26, 27, 28]. II has inherent benefits and distinct characteristics compared with the lenticular technique in terms that it captures full parallax of a 3-D object at the imaging step by use of a lenslet array and reconstructs 3-D images with full parallax capability simply by employing another lenslet array. II can provide a real volumetric display technique that forms 3-D images in free space. And the scope of II has been extended to the object recognition and the depth estimation [29]. One advantage of II for 3-D image recognition lies in its compactness of multiple perspective imaging. Multi-view scenes of 3-D objects are recorded by a single shot without using multiple sensors or changing the position of the sensor. For dynamic 3-D full parallax 3-D visualization, we need the whole set of imaging and display tools which include CCD and SLM. However, in some specific display applications that require static 3-D picture based advertisement, entertainment and information capability, we need a simple and compact approach. In this chapter, we propose a novel full parallax static 3-D display technique based on elemental images printed on paper. We experimentally show that the proposed printed II based 3-D display can provide the capability of full parallax with moderate resolution. The proposed approach could potentially be applied in practical use within various 3-D visualization industries.

## 6.2 Dynamic Integral Imaging

For dynamic 3-D display applications, a series of two-dimensional elemental images that have different perspectives for a given 3-D object are generated using a lenslet array and then recorded in a two-dimensional CCD camera as depicted in Fig. 6.1(a). To reconstruct the dynamic 3-D object

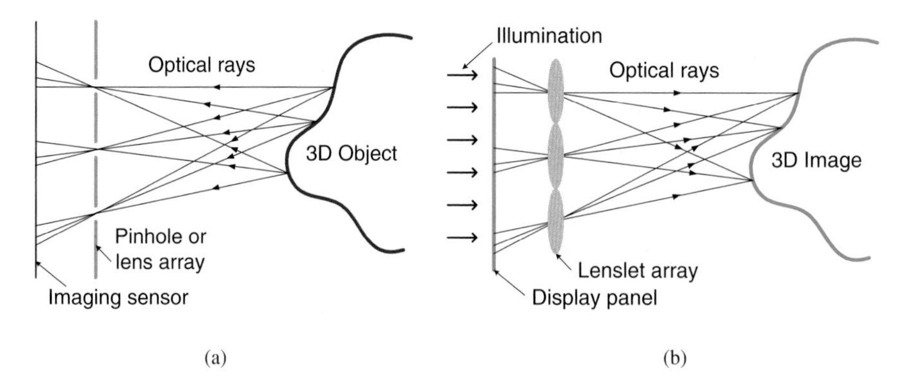


**Fig. 6.1.** Schematics of dynamic integral imaging scheme: (**a**) II imaging step and (**b**) II display step

movement in free space from the consecutively captured elemental images, a SLM such as an LCD is used to create continuously varying elemental images. Finally, by employing another lenslet array in combination with the SLM, the real dynamic 3-D object with full parallax information can be visualized as described in Fig. 6.1(b). For a given observation point, one can see a group of sub-images sampled from their own elemental images with different perspectives through the display lenslet array. For this kind of dynamic 3-D full parallax 3-D visualization, we need the whole set of imaging and display tools which include CCD and SLM.

## 6.3 Static Integral Imaging

Figure 6.2(a) shows the schematic of the proposed static II based 3-D display method implemented with only a lenslet array and a set of elemental images printed on paper. This method is as simple as conventional lenticular technique in implementation. The only thing that needs to be taken into account is to align a lenslet array at a correct position as designed. First, elemental images with full parallax information are obtained by commonly used integral imaging setup. We need only one frame integral image for 3-D reconstruction. Then, for the static 3-D visualization, we print out the elemental images. Next, we reconstruct the 3-D image by directly attaching a lenslet array on the printed elemental images. No special glasses are required and white light is used for visualizing the 3-D object. Figure 6.2(b) shows the printed elemental images obtained by using the integral imaging system experimentally. We used 30 $\times$ 30 elemental images printed with a state-of-the-art color laser printer to maintain high quality elemental images. The resolution of the printed elemental images is 400 dpi. It is estimated that the printing resolution of the elemental images used is sufficient for high quality 3-D visualization. However, due to the small number of elemental images we can get from the integral image data and some dimensional mismatches between the elemental images and the lenslet array, the results we obtained were at a slightly lower resolution.

## 6.4 Experimental Results

Figure 6.3 illustrates the 3-D objects used in our experiments. They consist of two dice of different size and a round button. We can visualize a 3-D object in free space just by combining a lenslet with the printed thin paper in white light environment as in Fig. 6.2(a). Figure 6.2(b) shows the elemental images printed on high quality laser paper. The elemental images are captured in the following experimental conditions: A die with a linear dimension of 15 mm is used as a main 3-D object. Also, a smaller die with a linear dimension of 4 mm is located on the big die and a round button shaped object with a diameter of 8 mm is put around 5 mm ahead of the big die. Those 3-D objects are

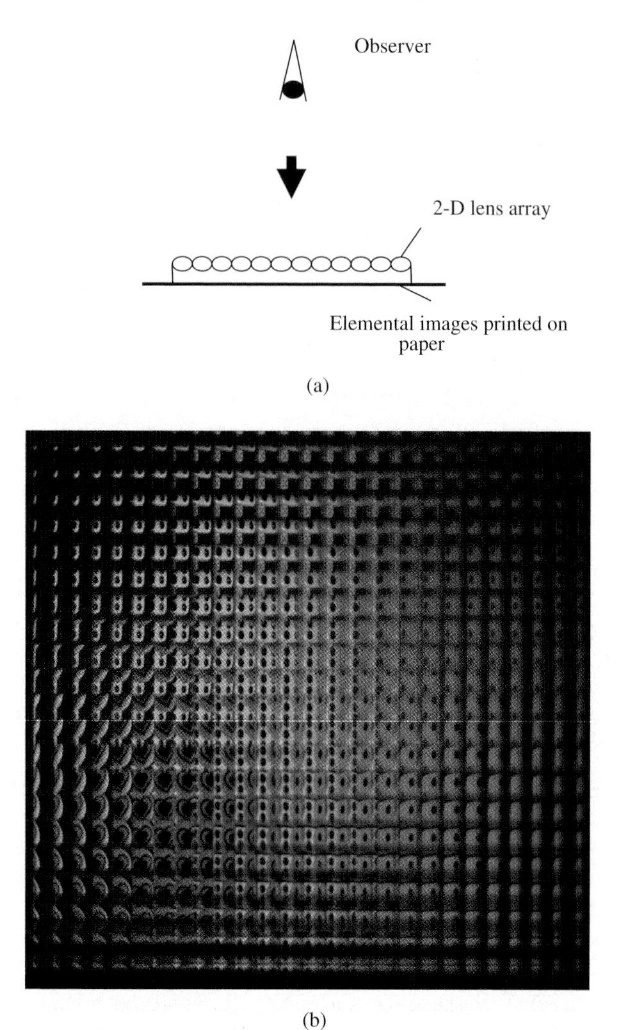


**Fig. 6.2.** A schematic on static 3-D display setup that consists of elemental images printed on paper and 2-D lens array: (**a**) schematic of the full parallax 3-D display and (**b**) elemental images printed on paper

illuminated by spatially incoherent white light. The lenslet array is placed in front of them to form the elemental images. The lenslet array has 30 × 30 circular refractive lenses in a 50 mm square area. The distance between the objects and the lenslet array is around 50 mm. The image array of the objects is formed on CCD camera by inserting a camera lens with a focal length of 50 mm between the CCD camera and the lenslet array.

**Fig. 6.3.** Photographs of the 3-D objects used in the experiment

For the static 3-D visualization, we put the lenslet array, which is made through the pre-described integral imaging process. In order to maintain the distance between the paper and the lenslet back surface at 3 mm, several gap spacers are positioned between them. With this simple setup, an observer can see 3-D multiple perspectives without glasses. Figure 6.4 shows the reconstructed full parallax 3-D images captured by a CCD camera from the observer

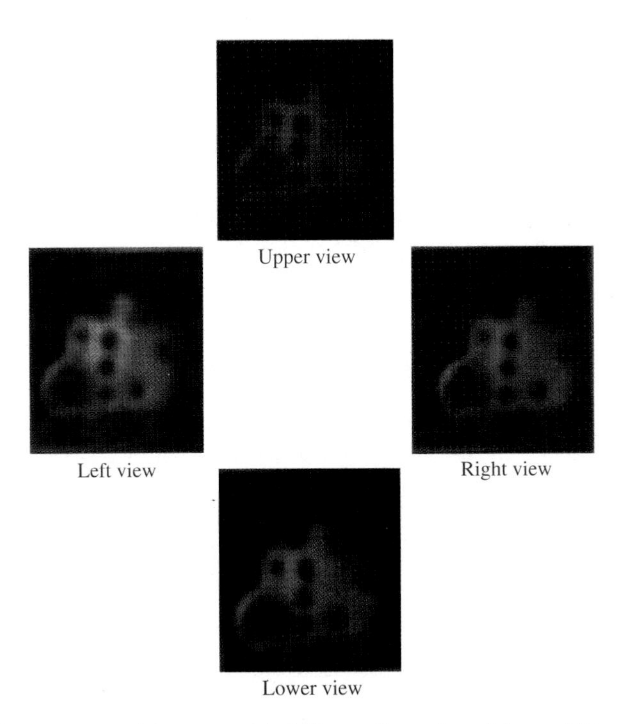


**Fig. 6.4.** Reconstructed images with full parallax viewed from different angles in horizontal and vertical axes

position. It represents four different perspectives in both horizontal and vertical axes. The view angle difference between the upper view and the lower one is around 5~6 degrees. Although the multi-perspective angle difference is not big enough to distinguish it with ease, we can clearly see that full parallax multiple perspective capability can be performed. Some white spots are caused by the alignment error between the elemental images printed on paper and the lenslet array placed upon the elemental images. However, we expect that the resolution of the reconstructed 3-D object may be potentially improved as the higher number of elemental images is employed for this static 3-D display technique, and such white spot noise would be eliminated perfectly through a more accurate approach by employing precision mechanisms.

## 6.5 Conclusion

We have described a novel static 3-D display method based on elemental images printed on paper. It has a significant advantage over conventional commercial static 3-D display techniques in that it can provide a real full parallax autostereoscopic capability while maintaining high practicability without need of additional apparatus like glasses. We expect that the proposed simple and compact scheme may have a high potential for use in various industrial applications that require static 3-D display, such as entertainment and advertisement fields, in the near future.

## References

[1] S. A. Benton, "On a method for reducing the information content of holograms," Journal of the Optical Society of America, 59: 1545, 1969.
[2] G. Lippmann, "La photographie integrale," Comptes-Rendus Academie des Sciences, 146: 446–451, 1908.
[3] F. Okano, H. Hoshino, J. Arai, and I. Yuyama, "Real-time pickup method for a three-dimensional image based on integral photography," Applied Optics, 36: 1598–1603, 1997.
[4] J.-S. Jang and B. Javidi, "Improved viewing resolution of three-dimensional integral imaging by use of nonstationary micro-optics," Optics Letters, 27: 324–326, 2002.
[5] J.-S. Jang and B. Javidi, "Three-dimensional projection integral imaging using micro-convex-mirror arrays," Optics Express, 12: 1077–1083, 2004.
[6] B. Javidi and F. Okano eds, *Three Dimensional Television, Video, and Display Technologies*, Springer Verlag, Berlin, 2002.
[7] A. Stern and B. Javidi, "3-D image sensing, visualization, and processing using integral imaging," Proceedings of the IEEE Journal, 94: 591–608, March 2006. (Invited Paper).
[8] J. S. Jang and B. Javidi, "Three-dimensional TV and display with large depth of focus and improved resolution," Optics and Photonics News Magazine, 36–43, April 2004.

[9] A. Castro, Y. Frauel, and B. Javidi, "Integral imaging with large depth of field using an asymmetric phase mask," Journal of Optics Express, 15(16): 10266–10273, 2007.
[10] R. Martinez-Cuenca, G. Saavedra, A. Pons, B. Javidi, and M. Martinez-Corral, "Facet braiding: a fundamental problem in integral imaging," Optics Letters, 32(9): 1078–1080, 2007.
[11] R. Martinez, A. Pons, G. Saavedra, M. Martinez-Corral, and B. Javidi, "Optically-corrected elemental images for undistorted integral image display," Journal of. Opt. Express, 14(22): 9657–9663, 2006.
[12] B. Javidi, S. H. Hong, and O. Matoba, "Multidimensional optical sensor and imaging system," Journal of Applied Optics, 45: 2986–2994, 2006.
[13] B. Javidi and S. Hong, "Three-dimensional visualization of partially occluded objects using integral imaging," IEEE Journal of Display Technology, 1(2): 354–359, 2005.
[14] M. Hain, W. von Spiegel, M. Schmiedchen, T. Tschudi, and B. Javidi, "3D integral imaging using diffractive Fresnel lens arrays," Journal of Optics Express, 13(1): 315–326, 2005.
[15] R. Martinez-Cuenca, G. Saavedra, M. Martinez-Corral, and B. Javidi, "Extended depth-of-field 3-D display and visualization by combination of amplitude-modulated microlenses and deconvolution tools," IEEE Journal of Display Technology, 1(2): 321–327, 2005.
[16] S. Hong and B. Javidi, "Improved resolution 3D object reconstruction using computational integral imaging with time multiplexing," Journal of Optics Express, 12(19): 4579–4588, 2004.
[17] O. Matoba and B. Javidi, "Three-dimensional polarimetric integral imaging," Journal of Optics-Letters, 29(20): 2375–2377, 2004.
[18] J. S. Jang and B. Javidi, "Very-large scale integral imaging (VLSII) for 3D display," Journal of Optical Engineering, 44(1): 01400 to 01400–6, 2005.
[19] J. S. Jang and B. Javidi, "Depth and size control of three-dimensional images in projection integral imaging," Journal of Optics Express, 12(16): 3778–3790, 2004.
[20] J. S. Jang and B. Javidi, "Three-dimensional projection integral imaging using micro-convex-mirror arrays," Optics Express, on-line Journal of the Optical Society of America, 12(6): 1077–1083, 2004.
[21] J. S. Jang and B. Javidi, "Three-dimensional Integral Imaging of Micro-objects," Journal of Optics Letters, 29(11): 1230–1232, 2004.
[22] F. Jin, J. Jang, and B. Javidi, "Effects of device resolution on three-dimensional integral imaging," Journal of Optics Letters, 29(12): 1345–1347, 2004.
[23] A. Stern and B. Javidi, "3D Image sensing and reconstruction with time-division multiplexed computational integral imaging (CII)," Journal of Applied Optics, 42(35): 7036–7042, 2003.
[24] J. S. Jang and B. Javidi, "Large depth-of-focus time-multiplexed three-dimensional integral imaging by use of lenslets with nonuniform focal lengths and aperture sizes," Journal of Optics Letters, 28(20): 1924–1926, 2003.
[25] A. Stern and B. Javidi, "3-D computational synthetic aperture integral imaging (COMPSAII)," Optics Express, on-line Journal of the Optical Society of America, 2003.
[26] S. W. Min, B. Javidi, and B. Lee, "Enhanced 3D integral imaging system by use of double display devices," Journal of Applied Optics-Information Processing, 42(20): 4186–4195, 2003.

[27] J. S. Jang and B. Javidi, "Three-dimensional integral imaging with electronically synthesized lenslet arrays," Journal of Optics Letters, 27(20): 1767–1769, 2002.

[28] S. Jung, J. H. Park, B. Lee, and B. Javidi, "Viewing-angle-enhanced integral imaging using double display devices with masks," Journal of Optical Engineering, 41(10): 2389–2391, 2002.

[29] S. Yeom, B. Javidi, and E. Watson, "Photon counting passive 3D image sensing for automatic target recognition," Optics Express, 13: 9310–9330, 2005.

# Part II

# Multiview Image Acquisition, Processing and Display

# 7

# Viewing Zones of IP and Other Multi-view Image Methods

Jung-Young Son

**Abstract** Multi-view, three-dimensional imaging methods use a different set of view images (multi-view) to create three-dimensional images. In these methods, the images can be projected with an array of projectors (projection type) or displayed on a display panel (contact type). However, the methods are basically based on both binocular and motion parallaxes as their depth cue. For the parallaxes, the viewing zone should be divided into many viewing regions and each of these regions allows viewers to perceive an individual view image or a mixed image composed of parts from more than two different view images in a multi-view image set. The number of viewing regions and the composition of the image at each of the regions can be predicted by the number of different view images in the multi-view image and of pixels in a pixel cell. When the pixel cell is composed of non-integer number pixels, more regions are created than an integer number and the compositions become more complicated. This is because a number of pixel cells are involved in defining the viewing regions.

In this chapter, the viewing zones for the multi-view 3-D imaging systems are analyzed and the image's composition at each viewing region of the zone is defined. For the contact type, the analysis is extended for both integer and non-integer number pixels in the pixel cell, and it is shown that the depth cues in IP are parallaxes.

J.-Y. Son
Center for Advanced Image, School of Information and Communication Engineering, Daegu University, Kyungsan, Kyungbuk, Republic of Korea
e-mail: sjy@daegu.ac.kr

B. Javidi et al. (eds.), *Three-Dimensional Imaging, Visualization, and Display*,
DOI 10.1007/978-0-387-79335-1_7, 

## 7.1 Introduction

To generate a three-dimensional (3-D) image from a set of different view images, i.e., a multi-view image for providing (a) depth cue(s), a display panel for displaying the image and optics to form a viewing zone (VZ), i.e., viewing zone forming optics (VZFO), are required. VZ is defined as a spatial region where viewers can perceive images through the entire surface of a display panel. For the plane images, the VZ covers the front space of the display panel in most cases, but for the multi-view 3-D images a confined space in front of the panel, defined by VZFO, is based on geometrical optics principle. The VZFO makes the different view images in a multi-view image set be seen separately, or different images can be synthesized and then be seen separately at different regions of the VZ. The size and shape of the VZ are defined mostly by the mechanisms of displaying and loading the multi-view image on the display panel or screen. There are currently three known mechanisms: contact, projection and scanning type [1]. The contact type comprises the 3-D imaging methods that have the structure of superposing the VZFO under or above a display panel. This type mechanism is typical for the methods using a flat panel display as the panel. The methods include the multi-view, IP and some eye tracking based methods [2]. The projection type comprises the methods which use the VZFO as the image projection screen. This mechanism type is typical for methods using a high-speed projector or many spatially aligned projectors. The methods are based on a projector array or a high-speed shutter array. The scanning type is typical for methods employing a moving element such as a scanner to construct a frame of image time sequentially. Focused light array (FLA) [3] is currently the only known method for the scanning type. Hence, the VZs for the contact and the projection are formed by the imaging action of each elemental optic consisting of the VZFO and the imaging action of the optics itself, respectively. Due to the imaging action, the VZs for the contact and the projection are determined by the sizes of imaging objects of the VZFO. For the scanning, the VZ is defined by the multiplexing angle of the multi-view image and the scanning range of the scanner. To include many separated viewing regions for different view images, the VZs for the contact and the projection are divided into many segments. Viewers are getting depth sense by locating their eyes in two different segments. Hence, the main depth cue for these types are both binocular and motion parallaxes. To eliminate the chance of directing each viewer's eyes to the same region, the horizontal widths of these regions should not be more than viewers' interocular distance. Since the imaging objects in the contact and projection type are the image cell under each elemental optic, which consists of a number of pixels and exit pupil of each projector, respectively, VZs are inherently divided. Hence, there is almost no gap or overlapping between regions in the contact, but there can be a large gap or overlapping between regions in the projection. For the scanning type, the depth cue is no different from the contact, but there is no division in the VZ. However, viewers will perceive different images at different locations in the VZ because the image perceived by viewers at a specific location of the viewing zone is formed by specific pixels from many different view images. The size of VZs in the contact and projection types are proportional to that of the image cell and the number of projectors, respectively. But, for the contact type, each image cell can be imaged by many elemental optics, hence many VZs can be created. Furthermore, the number of viewing regions is more than that of different view images employed and the images perceived at each of the regions are different from each other. This is a major difference between the contact and the projection. The viewing zone can also be shifted in the eye tracking based contact type.

## 7.2 Basic Viewing Zone Forming Principle of Multi-view Imaging Methods

The VZs in the projection type are defined by the exit pupil(s) of the projector objective(s). However, the geometries of viewing zone forming are slightly different for the multiplexing method employed by the type. There are basically two multiplexing methods such as spatial and time multiplexing in projection type [1]. The spatial multiplexing is employed when the projectors have a frame speed not exceeding 60 Hz, but for the time multiplexing, the frame speed exceeds more than 60 Hz. The geometry for the spatial multiplexing method is depicted in Fig. 7.1. The optical axis of each projector's objective is aligned in a horizontal plane in parallel or in radial with those of other projectors as in Fig. 7.1(a). A viewing region for each image is defined as the exit pupil image of each projector's objective, formed by the projection screen. The distance between two neighboring viewing regions can be adjusted to be optimum by changing that between their corresponding exit pupils. Depending on the distance between exit pupils of two adjacent projector's objectives, the viewing regions can be (1) separated, (2) joined and (3) overlapped as shown in Fig. 7.1(b). For the separated case, the viewing region for each view image is separated from its neighbors with a certain gap. At this gap area, no image will be seen, i.e., blackened stripes appear as the viewer moves his/her eyes in a horizontal direction. For the joined case, the edge of each image viewing region is joined with that of its neighbor's viewing regions together. This is the optimum case. Viewing region is continuous and clear image is seen at any position in VZ. For the third, a part of each viewing region is overlapped to its neighbors. Hence, two adjacent view images will be seen simultaneously at the superposed part, i.e., the images will be blurred. The maximum size of the superposed part should not exceed the half distance between the two adjacent viewing regions. For all three cases, the distance between two adjacent viewing regions should not be greater than the viewers' interocular distance to insure that a viewer's two eyes will not be in the same viewing region. For the case of time multiplexing, a high-speed projector with an objective with a large

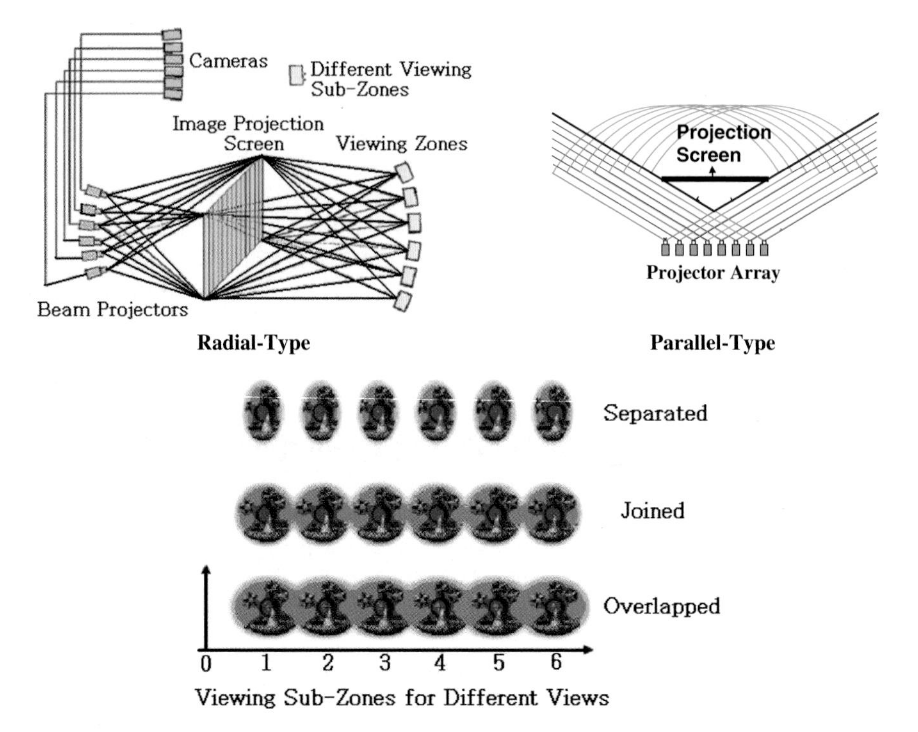


**Fig. 7.1.** Projector and viewing zone configurations in projection type multi-view 3-D imaging systems; (**a**) projector arrays; (**b**) viewing zone configurations

exit pupil is used [4]. In this case, the projector should have the frame speed exceeding more than $60N$, where $N$ is a number of different view images to be displayed and $N$ strip shutters with On/Off speed of $1/60N$ second should be placed on the exit pupil of the projector's objective. Hence, the viewing zone is determined by the exit pupil size and each viewing region by the width of each strip in the shutter. In this case, the viewing regions are all joined with their neighbors. The strip shutter's On/Off sequence is synchronized with the projection sequence of different view images and the shutter covers the entire pupil area to eliminate mixed viewing regions. Otherwise, all different view images will be seen through the uncovered pupil region.

For the contact type methods, the VZ forming geometry is different from the projection type due to many elemental optics composing the VZFO. In these methods, VZ should be a common volume shared by each image volume formed by each elemental optic. Figure 7.2 shows the basic optical geometry of viewing zone forming in multi-view 3-D imaging methods. The geometry is formed based on a point light source (PLS) array. However, this geometry will be equally applied for the methods based on lenticular, parallax barrier

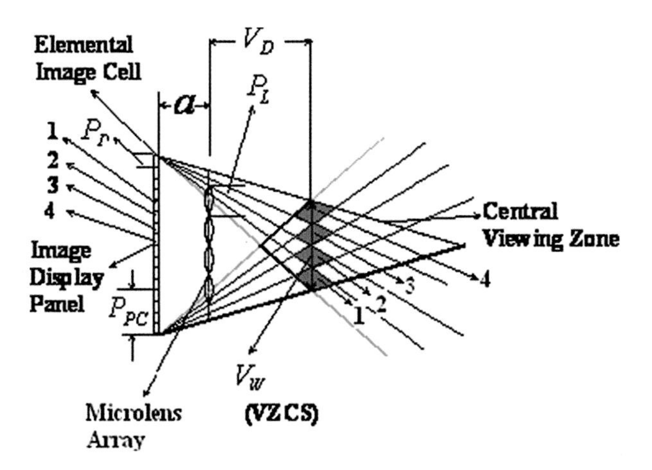


**Fig. 7.2.** Viewing zone forming geometry of contact type multi-view 3-D imaging system

and microlens array plates as their VZFO [5]. Two basic components forming the geometry are the image display panel and the PLS array. The panel is composed of image cells of the same size and shape, called pixel cells. The pixel cell works like a pixel in the plane image, and is a basic unit of displaying the multi-view image in the type methods. The number and array size of the cell are the same as those of PLSs in the array. The geometry is configured such that a ray from each PLS in the array, which passes through the upper right corner of its corresponding pixel cell, meets together with others at the upper right corner of VZCS. Rays passing other corners of their corresponding pixel cells are also meeting at their corresponding corners of the VZCS. The extensions of these rays create an empty space common to all pixel cells, in front of the display panel, and determine (1) achievable depth range, (2) relative position of display panel and various VZFO, and (3) relative sizes of pixel cell and the pitch of various VZFO [6]. From the geometry, it can be seen that the VZCS is the place where all the magnified pixel cells by their corresponding PLSs are completely matched. Hence, if all pixel cells have the same pixel pattern, the VZCS will also have the same pattern, i.e., be divided into segments corresponding to the pixel number in the pixel cell, and the viewing zone will be divided into smaller regions according to the pattern. From the geometry, the pixel cell width $P_C$ and the PLS array pitch $P_L$ satisfies the relationship $P_L/P_C = (a + d_V)/d_V$ where $a$ and $d_V$ are the distances between the PLS array and the display panel, and between the panel and the VZCS, respectively. $P_L - P_C = \Delta P$ makes the magnified images of all pixel cells in the panel superpose completely with others at the VZCS. Hence, $\Delta P$ cannot be neglected though it is very small compared with $P_L$ and $P_C$.

## 7.3 Image Composition in the Multi-view

There are two different setup conditions to align different view images in the multi-view methods: (1) the multi-view image and each pixel cell are the same dimension, i.e., the multi-view image is consisted of $I \times J$ different view images and the pixel cell $I \times J$ pixels, and the resolution of each view image $K \times L$ is the same as the number of the pixel cells in the panel, and (2) the multi-view image and the pixel cells in the display panel have the same dimension of $I \times J$, and the resolution of each view image and the number of pixels in the pixel cell are the same dimension of $K \times L$. The first is typical of the contact type multi-view 3-D imaging methods (henceforth designated by the Multi-view), and the second for IP, when the relationship above is equally applied to both cases. For the first case, the multi-view image is aligned such that $k \times l$th, where $k = 1, 2, 3, \ldots, K$ and $l = 1, 2, 3, \ldots, L$, pixel from each view image is aligned at $k \times l$th pixel cell in the order of cameras for different view images, in the multi-view camera array. For the second case, $k \times l$th image is aligned to $k \times l$th pixel cell. Hence, in $i \times j$th segment in the VZCS, $i \times j$th (where $i = 1, 2, 3, \cdots\cdots, I$ and $j = 1, 2, 3, \cdots\cdots, J$) pixel from all pixel cells will be seen. Since the $i \times j$th pixel represents $i \times j$th view image for the Multi-view, $i \times j$th view image is seen at $i \times j$th segment. But, in the IP, $i \times j$th pixel in each view image, seen at the $i \times j$th segment will synthesize a new image. This indicates that the image arrangements and the image compositions in IP and the Multi-view have a conjugate relationship. To increase the resolution of images seen at different viewing regions in VZCS, the conditions $K > I$ and $L > J$ for the Multi-view, and $K \leq I$ and $L \leq J$ for IP should be met. Due to the segmentation of VZCS, the viewing zone is also divided into different regions. This is shown in Fig. 7.3 for the case when four different view images have a resolution of $5 \times 1$. Figure 7.3 is just a plane view of the viewing zone forming geometry. The actual viewing zone should have a diamond shape formed by joining two pyramids of different heights, but of the same base shape and size through their bases when each side of VZCS is shorter than its corresponding side of the display panel. When one side of VZCS is not shorter than its corresponding side of the display panel, the distant sides of the viewing zone cannot be defined, i.e., the shape cannot be formed [7]. In this case, the width of each viewing region can work as a limiting factor of the viewing zone because the width is larger than that for the shorter VZCS case. In Fig. 7.3, the zone is divided into 16 different viewing regions. The same number of viewing regions is formed in a triangle corresponding to one of the pyramids. Each viewing region is also divided into four subregions corresponding to the number of point light sources minus 1. These regions are specified by a combination of five numbers. These numbers are specifying the pixel order in each pixel cell from the left side. For example, the region specified with number 11,111 indicates that the first pixel in each pixel cell is seen at this region, and the region specified with number 22,334 indicates that the second pixel from the first two pixel

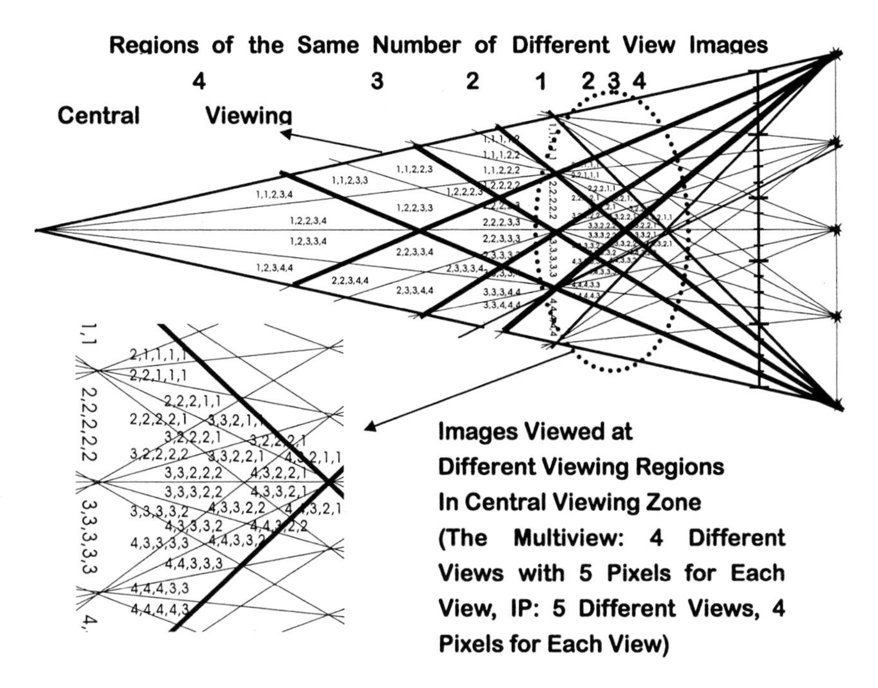


**Fig. 7.3.** Viewing regions and image compositions

cells, third pixel from the next two pixel cells and fourth pixel from the fifth pixel cell are seen at the region. This means that the image seen at the region specified by 22,334 is synthesized by patching the first two one-fifth portions of view image number 2, 3 and 4 one-fifth portions of view image number 3 and 5 one-fifth portion of view image number 4, together. So these numbers represent the image compositions seen at the regions. Hence, in the regions specified by the same order number combinations, a complete view image specified by the number will be seen, and in the regions specified by different number combinations, synthesized images from a part of each of the different view images specified by the different numbers are seen. The four subregions in each region indicate that the compositions of synthesized images become different for different subregions. However, the difference between the images is the portion size of each view image composing them. It is doubtful that this creates a distinctive difference between two adjacent subregion images. In Fig. 7.3, the image compositions in different viewing regions have distinctive differences: (1) four regions along the VZCS specified as region group 1 are for each view image; (2) three regions in front and behind the VZCS specified as region group 2 for synthesized images from two different view images; (3) two regions away from the three regions specified as region group 3 for synthesized images from three different view images, and (4) the region at the top of each triangle specified as region group 4 for synthesized images from four different view images. The compositions are specified in Fig. 7.3. The images seen at region group 1 consisted of the same numbers. For region group 2, the mixed images are 1 and 2, 2 and 3, and 3 and 4 from the top region. The image compositions are 11,112, 11,122, 11,222 and 12,222 for the subregions from the top in the top region. If there is $K$ PLSs, $K-1$ subregions will be formed in each region belonging to region groups 2 to 4. The image compositions in these subregions are $K-1$ 1s and 2, $K-2$ 1s and 22, ⋯, 11 and $K-2$ 2s and 1 and $K-1$ 2s, from the top. As $K$ becomes bigger, the subregion size becomes smaller. Hence, there will be no distinction between subregions if the sum of subregion widths is less than a viewer's pupil size. In this case, many subregion images will be projected simultaneously to each of the viewer's eyes. Hence, whether these images are viewed separately or mixed together as an image is not known.

The same analysis can also be applied to the vertical direction. Hence, it is not difficult to estimate the total number of different viewing regions to be formed for $I \times J$ different view images. The number will be $(I \times J)^2$. However, this number is not all the viewing regions formed in the contact type multi-view imaging methods due to an existing side viewing zone, the zones between viewing zones. Since each elemental optic is not only imaging the pixel cell at its front, but also the neighboring cells to the cell in front, and each viewing zone has a diamond shape, there can be many side viewing zones and zones formed between adjacent viewing zones. The viewing zone formed by a front pixel cell of each elemental optic is designated as the central viewing zone. The existence of the side viewing zones is introducing pseudoscopic viewing areas because the image projected to viewers' left and right eyes are reversed when viewers locate their two eyes at different viewing zones. In this case, viewers' left eyes get a view image with a higher number than the image for their right eyes. The viewing areas appear along the boundaries of viewing zones [7]. At the boundary regions located at the zones between viewing zones, the synthesized images can also have the reversed number order, i.e., a part of higher order view image is patched to the left side of the lower order. This means that the synthesized image itself can have a pseudoscopic nature. Hence, the visual quality of images perceived at these regions will be worse than that of images at the regions in viewing zones. Despite the presence of pseudoscopic viewing regions, the combined size of the viewing zones is big enough to include many simultaneous viewers. Since there are more than two different viewing regions, the viewers can also shift their eyes to different viewing regions because this shifting allows them to see other view images; hence, motion parallax is provided. In the head or eye tracking based multi-view imaging methods, either the viewing zone [8] is moved or the viewing region in the viewing zone is shifted [9] in accordance with the viewer head/eye movement, and different view images seen at the viewer's new eye positions are displayed. The viewing zone is moved by shifting the viewing zone forming optics relative to the multi-view images displayed under the optics and, for the opposite case, the viewing region is shifted. The advantages of the head or eye tracking methods are (1) the resolution of each view image can be reduced to one half of its original resolution instead of $1/I$ because only two different view images are displayed at every moment, and (2) no pseudoscopic viewing region can appear. These advantages are obtained with the cost of tracking devices and algorithms for real time-detection of viewer's eye or head movement, and a mechanism of shifting the viewing zone forming optics. However, the methods are, in general, for one viewer and the dynamic range of tracking is limited to the central viewing zone. Hence, the effected size of the obtainable viewing zone will be smaller than that of the non-tracking multi-view 3-D image methods.

Figure 7.3 also shows that the boundaries of the VZ are defined by the two outside pixel cells, and viewing regions by the number of different view images. Since the size of VZCS can be at most $(I \times J)^2$ multiplied by the square of the smallest interocular distance of viewers, as the number of pixel cells increases, the VZ become smaller and, consequently, the viewing regions also become smaller by the geometrical reasons. If the number of different view images is increased $v$ times without changing the VZ size, the number of viewing regions will be increased in proportion to the $v^2$ and the number of viewing subregions to $1/v$ because, as the number of different view images increases, the resolution of each view image is reduced to the same ratio. As a consequence, the sizes of the viewing region and subregions are reduced in proportion to $v^2$ and $v$. Increasing the number of different view images can reduce the VZ size, due to the decrease in the visual quality of the synthesized images. As the number increases, there are an increased number of different view images that can be involved in the images perceived at the viewing regions located further away from VZCS. The visual qualities of the images should be reduced as the involved number of different view images increases. Hence the quality of perceived images will deteriorate as the viewers move away from the VZCS. If the quality is properly defined, it can be used to define the VZ size, other than the geometrical optics. The visual quality can be quantified to help define the VZ more accurately, but its validity should be determined by considering the human factor [7]. Since the disparity difference between first and last view images in the multi-view image will be more as the number of different view images increases, the relative size of the viewing zone can be reduced with the increase.

According to Fig. 7.2, the far end vertex of each viewing zone, i.e., the tail of each viewing zone, can diverge to infinity. This means that the diverged space can also create many different viewing regions. Though these viewing regions are excluded from the VZ definition, they can still give disparity between images in different viewing regions. The images in these regions are synthesized by repeated pixel order intermixing with different view images in the multi-view image. The difference between the images of these regions is the starting pixel numbers. The visual quality of images in these regions will not be good, but there still can be disparity between images of different viewing regions. As the number of different view images increases, the disparity can no longer be perceived because the starting pixel number difference may not be recognizable between the images in the different viewing regions. This is the probable reason of perceiving no depth sense at the far distance of the display panel in the multi-view 3-D imaging methods.

## 7.4 Viewing Zones for Non-integer Number of Pixels in a Pixel Cell

The previous section discussed the viewing regions and images seen when the number of pixels in each pixel cell has an integer value. However, in practice, it is hard to make the pixel cell pitch correspond to an integer number of pixels due to the geometrical relationship between pitches of the pixel cell and the PLS array as depicted in Fig. 7.2 and the tolerance in pitch value of the VZFO. Hence, it is typical for a pixel cell to comprise non-integer number pixels in both horizontal and vertical directions. The impact of the non-integer pixel cell is that (1) the viewing zone is divided into more regions than the integer case in Fig. 7.3, (2) assigning a specific view image to the shared pixel can change image compositions in parts of the viewing regions, and (3) the mixed image regions also appear along the VZCS. In these mixed image regions, the perceived images are composed of two adjacent view images intermixed with a pixel by pixel order. The images are just like those displayed on the display panel for a 2×2 multi-view 3-D image system for the case when $q = 0.5$. In general, the width of a pixel cell $P_C$ can be represented as $P_C = (K + q)P_P$, where $K$ is the number of integer pixels, $P_P$ pitch of a pixel and $q$ fraction of a pixel. For the pixel cell with integer number pixels in each direction, $q$ will be zero. However, for the non-integer case, $q$ can be represented as $q = n/m$, where $n$ and $m$ are integer numbers satisfying the relationship $m > n$. $q$ can be an irrational number, but in this case the closest rational number to $q$ can be assigned by approximation. Since a pixel is a basic unit in the display panel, it cannot be divided. A pixel can only be shared by two adjacent pixel cells for the case of a non-integer pixel cell. The pixel can be shared by four adjacent pixel cells in vertical directions. The ratio of sharing is different for different adjacent pixel cells within the $m$ pixel cells. Hence, the first and last pixels are the fractional pixels for the second to $(m - 1)$th cells in the $m$ pixel cells, if the first pixel in the first pixel cell is a complete pixel. By changing the sharing ratio, the pixel arrangements for $m$ consecutive pixel cells become different to each other. After $m$ pixel cells, the same pixel arrangement is repeated for the next $m$ pixel cells. Hence, the viewing zone is defined by the $m$ pixel cells and, consequently, the width of each viewing region along the VZCS becomes $1/m$ of that for the integer case. If the display panel has a very high resolution, the non-integer pixel cell case can be eliminated because it is possible to assign the number of pixels in $m$ pixel cells to a pixel cell. Within $m$ pixel cells, there are $m - 1$ shared pixels on average. The shared pixels can be left empty or assigned a specific view image. Leaving them empty will be uneconomical in the point of utilizing limited number of pixels in a display panel. However, assigning a specific view image is also not desirable because it can increase the pseudoscopic viewing areas. Figure 7.4 shows the viewing regions formed when the pixel cell comprises 4.5 pixels, and each view image is composed of 6×1 pixels in a horizontal direction. Since $q = 0.5$, $m = 2$ and $n = 1$. The viewing zone is defined by two pixel cells and the width of each viewing region along the VZCS becomes one-half of that for the integer case. The pixel cells can be arranged two different ways: (1) the fractional

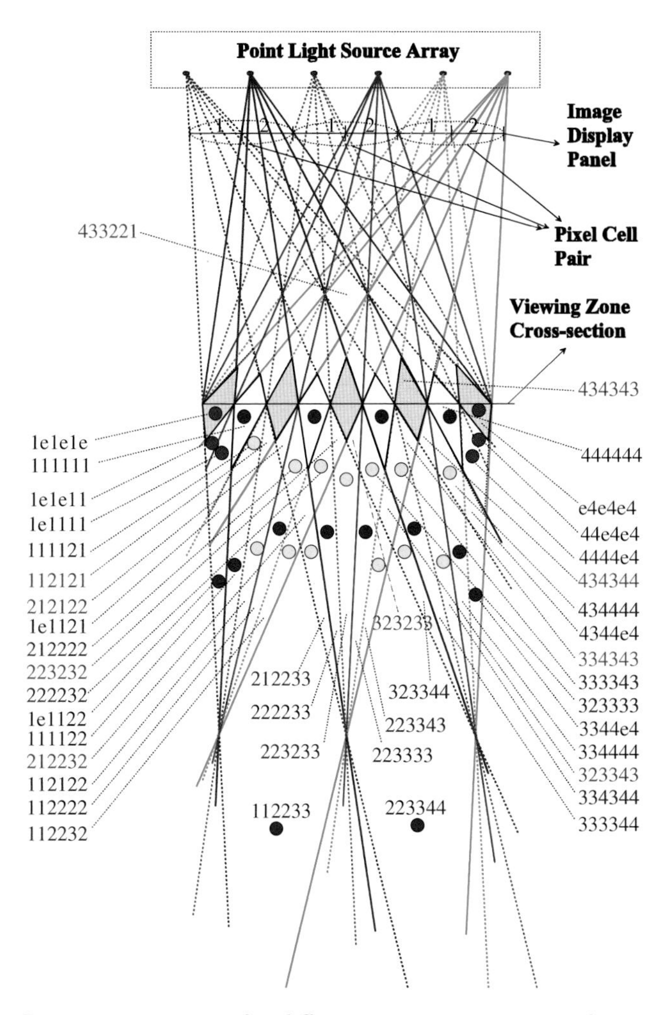


**Fig. 7.4.** Image compositions for different viewing regions in the viewing zones formed when a pixel cell is composed of 4.5 pixels; (**a**) when the fractional pixel is between two pixel cells; (**b**) when the fractional pixels are in the outside

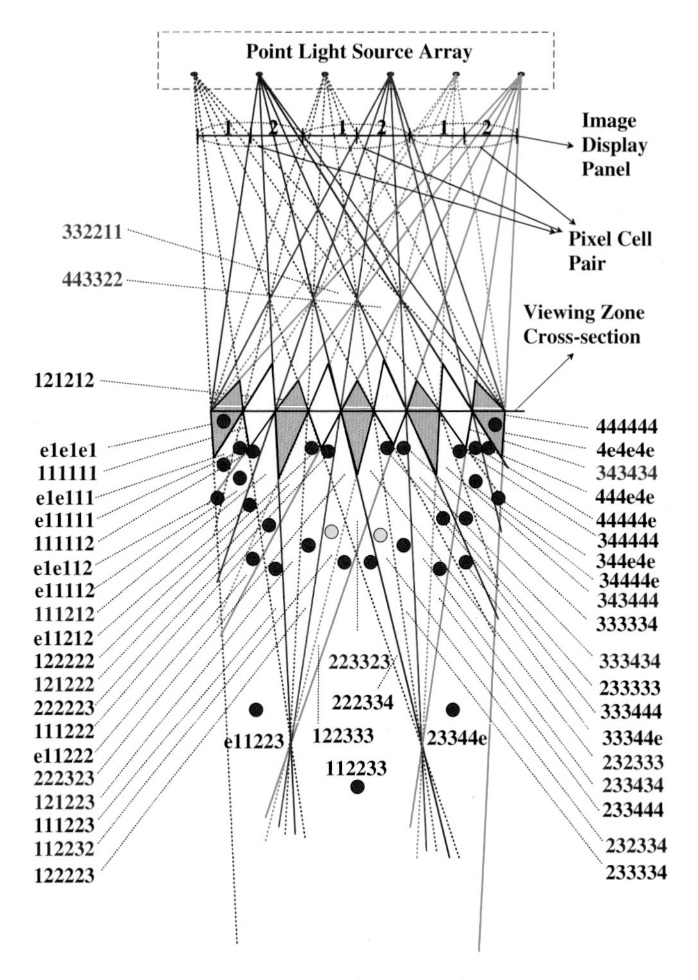


**Fig. 7.4.** (continued)

pixels specified by e are either in between two cells so that they are joined to be a complete pixel as in Fig. 7.4(a), or outside of two cells so that they are joined with other pixel cells as in Fig. 7.4(b). In Fig. 7.4(a), the viewing regions along the VZCS and images seen at these regions are depicted in Fig. 7.5(a). The mixed image viewing regions are appearing in-between each view image. The width ratio of each view image to mixed image viewing region is equal to $q/(1-q)$. When $q = 0.5$, i.e., $m = 2$, the ratio is 1. The image compositions in the inner three regions in Fig. 7.4(a) – 212,121, 323,232 and 434,343 –are exactly the same as the image arrangements in the display panel of a 2×2 multi-view 3-D imaging system, and the compositions in the two outmost regions – 1e1e1e and e4e4e4 – are comprising the fractional pixel. The images in the three inner regions look reversed because the higher numbered

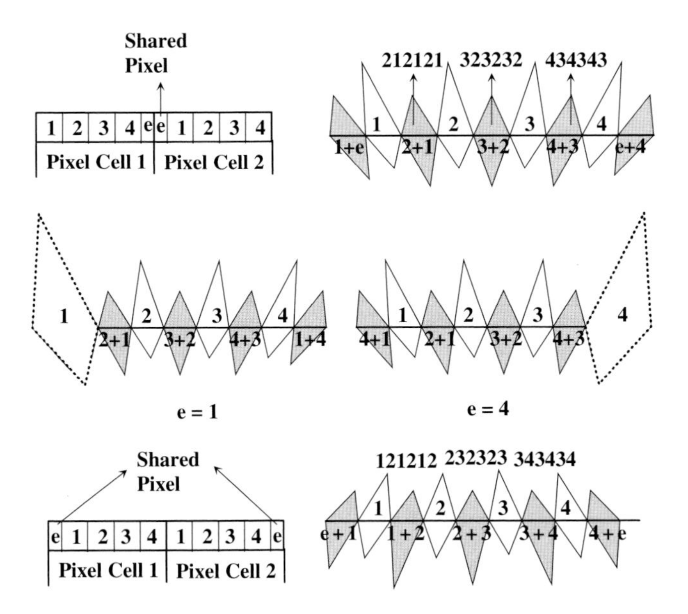


**Fig. 7.5** Image compositions in viewing regions along viewing zone cross section; (**a**) when the fractional pixel is in between two pixel cell; (**b**) two fractional pixels are outside of the pixel cells

view images are in front of the lower numbered ones. This indicates that the images themselves are not balanced and will probably show a pseudoscopic feature. When $m > 2$, the compositions are no longer showing the feature of multi-view image arrangement in a $2{\times}2$ multi-view 3-D imaging system. When $m = 3$, three pixel cell are involved in defining the viewing regions, Hence the view number $i = 2$ region is divided into three different regions for viewing images of 211,211···, 212,212···, and 222,222, and four subregions of having compositions 211,121(11)···, 211,221(12)···, 212,221(22)··· and 222,222(22) for $m = 4$. In general for $m = m$, the view number $i$ region is divided into $i\ m - 1$ each $(i - 1)$, $i\ m - 2$ each $(i - 1)\ i - 1$, ······, $i\ i - 1\ m - 2$ each $i$ and $m$ each $i$.

There are four possible ways of loading images in the fractional pixel, i.e., no image, image 1 or 2, or image 5. When no image is loaded, the fractional pixel works as a dummy pixel. Hence, each view 1 and 4 images will be interlaced with the dummy pixel. The interlaced images will be individual view images, but the qualities are degraded. When image 1 or 4 is loaded, the viewing regions for image 1 or 4 are extended as shown in Fig. 7.5(a). When image 5 is loaded, the image composition in the left most viewing region is 151,515 and in the right most 545,454. Since the 151,515 viewing region will create a pseudoscopic viewing region with 111,111 viewing region, the pseudoscopic viewing region is effectively increased by the view number 5. So the new image 545,454 viewing region is created by sacrificing the size of orthoscopic viewing region. Figure 7.4(a) also shows three tails in the viewing zone, two large viewing regions between them and no region for the mixed image of four different view images. These are major differences between Figs. 7.3 and 7.4(a). The two large viewing regions cannot work as real viewing regions because their widths are mostly greater than the interocular distance. For the Fig. 7.4(b), the image compositions are mostly in forward direction, and the viewing zone has only two tails. Hence, it shows a large viewing region and two viewing regions that are about half the size of the large one. However, these three regions will also be too big to work as real viewing regions. The image compositions in viewing regions along VZCS are specified in Fig. 7.5(b).

## 7.5 Viewing Zones in IP

Theoretically, the viewing zone forming geometry of IP is not different from that of contact type multi-view 3-D imaging methods as shown in Fig. 7.2. However, in practice, there are several differences between IP and the Multi-view: (1) the image arrangement and (2) the dimension of elemental optics array in VZFO and the number of pixels in each pixel cell as explained in Section 7.2, and (3) the pixel cell size. In IP, the pixel cell size is approximated as the pitch size of the elemental optics. This doesn't change the geometry, but the viewing regions are defined as in the non-integer pixel cell in the Multi-view and image compositions in the regions become unique for the IP.

The typical setup conditions of IP are such that (1) the display panel where the pixel cell is placed is parallel to VZFO which is composed by an array of microlenses, (2) all microlenses in the VZFO have the same optical characteristics, (3) the distance between the panel and VZFO is variable, (4) the elemental image cell has a square shape and its side length is the same as the microlens diameter, i.e., $\Delta P = 0$, and (5) the parallax differences between adjacent images are determined by the pitch of each microlens. However, none of these conditions supports that the 3-D image displayed in IP reveals real depth as in a hologram. The reasons for this are as follows: Conditions (1) and (2) inform that the magnified image of each elemental image by its corresponding microlens is focused at a plane parallel to the array. The distance between the focused plane and the microlens array is determined by the lens law. When considering the depth of focus of each microlens, the magnified image has a certain volume because the magnified image can also be formed in space surrounding the plane. This image volume can have be shaped either as a pyramid with a flat top or a circular cone with a flat top, depending on the microlens shape because the image size in the volume increases continuously as the distance from the microlens array increases. The images from other microlenses will also have the same shape as above. Since each volume is formed by a plane image of increasing size, it is difficult to say that the volume forms a volumetric image. When these volume images are overlapped, only a common dissecting plane of all the volumes, where the focused image of each elemental image by its corresponding microlens is overlapped orderly with others to completely reconstruct the original object image. The images appearing behind and before the plane will be blurred because each elemental image is not focused and it is not overlapped properly with others. This is shown in Fig. 7.6 for six horizontally arranged microlenses. Figure 7.6 also depicts possible FOVs defined optically and electronically. Note that the electronically defined one has no optical ground. It is virtual and described here for comparison purposes only. Since the FOV defines the viewable angle range of a scene in front of a lens, it is determined by the size of the image sensor under each microlens and its focal length. The optically defined FOV is much wider than the electronically defined one because the sensor size is much bigger than that of the elemental image cell, satisfying condition 4. As a result, the optically defined FOV of each microlens overlaps with the others and the electronic one is within the range of the optically defined one. This explains why

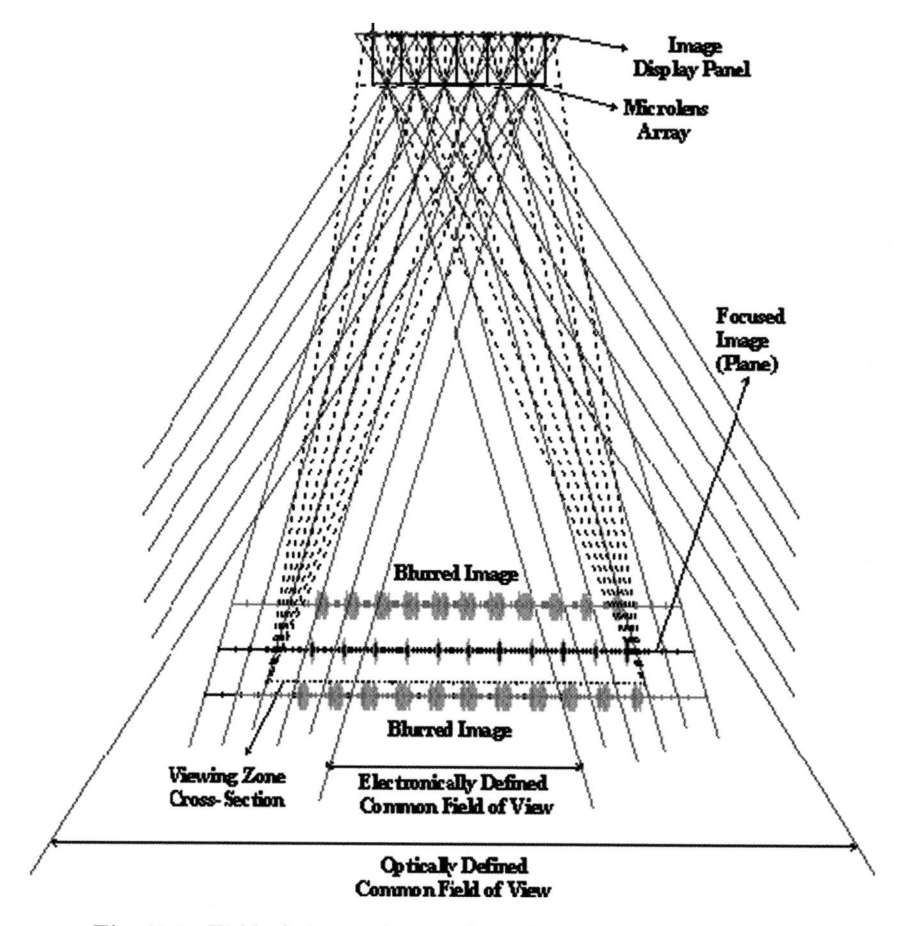


**Fig. 7.6.** Field of view and image formed in IP optical configuration

side viewing zones are formed, the electronically defined FOV is within the range of the optically defined one and the images seen at the central space are within the electronically defined FOV. Since the pitches of the pixel cell and microlens are the same and the optics axis of the microlens is passing the center of the cell, the electronically defined FOV of each microlens is parallel to each other and two adjacent FOVs are aligned with a distance equal to a pitch of the microlens array. The FOVs cannot completely overlap to each other and only a part of each FOV can overlap each other and form a common FOV. Since this common FOV is representing only a part of each elemental image cell, the focused image formed from a part of each elemental image can be seen in it as shown in Fig. 7.6. Figure 7.6 also shows the images of pixel grids in each volume image. They should appear if a flat panel display is used to display images, though they may not be visible due to its pixel grids which are smaller than other elemental images. In any case, the pixel boundaries are not seen in IP; additional evidence . that the image lacks a real depth and is visually unstable.

When the display panel is divided into segments with an equal size and shape without a gap between them to make the images of segments match completely to each other at a plane located at a certain distance from the lens array, the plane is always within the common FOV of the optically defined FOVs because it is wide enough. This defines the viewing zone forming geometry in the Multi-view. The plane is optically defined without regard to the distance between the microlens array plate and the display panel in the optical geometry as shown in Fig. 7.6 [10]. The plane is denoted as a viewing zone cross section (VZCS) and the segments represent the pixel cells in the Multi-view.

The second condition raises the question of what is the optimum array dimension of the microlens array? If each elemental image contributes piece by piece to complete the depth compressed object image, the dimension cannot be too big to avoid the redundant overlapping of elemental images in the common FOV region. The higher dimension and resolution of each elemental image required by IP probably intends to increase the number and resolution of images projected to viewers' eyes to provide highly resolved 3-D images with smooth parallax [11] as in the multi-view.

The third condition raises other puzzling questions like (1) where does the focused image plane appear, if the elemental image plane is at the focal plane of the microlens array, and (2) how are both virtual and real images simultaneously viewed [12, 13] when the distance between the microlens plate and the elemental image plane is fixed? For a fixed distance, the image formed by each microlens should be either virtual or real, but not both. This means that the 3-D image projected to viewers' eyes in IP is not a lens image, but a parallax based image.

In IP, the recorded image under each microlens is reversed for orthoscopic image display. In this process, the recorded image is redefined to have the same size as the microlens for simplicity. Hence, the size of the elemental image has the dimension of a microlens [14]. But this is a nominal value because there is no way to limit the FOV of each microlens, except the image sensor size, and the thickness and focal length of the microlens. Since the image sensor covers an array of the microlenses and the height is not high enough to limit the normal ray of the microlens to the size of the microlens, the actual FOV of each microlens should be much wider than that determined by the elemental image size. Hence, each elemental image is most likely projected simultaneously by its corresponding microlens and its adjacent microlenses. This implicates that the image projected to the image plane will be much wider than the elemental image size used in practice. This is supported by the nominal specs provided by flat panel TV manufacturers. For example, a typical LCD TV provides the viewing angle of more than 170° [15]. Hence, each pixel has a beam divergence angle of more than 170°. This angle covers the width $\approx 22.8a$ at the surface of the microlens array. This width will correspond to several 10 microlens pitches.

Since the FOVs of all microlenses in the array are parallel to each other, the focused image can always be seen through entire elemental image cells in the display panel only in the common FOV of all microlenses in the array. The common FOV of all microlenses is starting from the cross point of two boundary lines defining the FOVs of microlenses in both edges of the array. The focused image plane should be in the common FOV, but its position should be as far as possible from the cross point so that the perceived image is composed of the most pixels from each elemental image. From the viewer point of view, the image formed in the image plane may not be seen in full because each point of the image can be seen within the angle defined approximately by $(\tan^{-1} P_L/2a)/\xi$, where $\xi$ is the image magnification factor, due to the imaging forming principle of a lens. The beam originated from an emitting point under a lens can have a maximum beam angle of $(\tan^{-1} P_L/2a)$. When the microlens f-number and $\xi$ are assumed as 2 and 50, respectively, the angle becomes 0.56°. This angle can be too small to view the entire image at a short distance from the image, especially when the image size is big. This explains why the image focused by a lens is rapidly shifting as a viewer changes his view positions. Hence, it is hard to see the entirely focused image in a highly magnified imaging system. However, rapid image shifting is not observed in IP even though it is a highly magnified imaging system. This implies that the image projected to viewers' eyes is not the focused image.

The viewer's position and the width of viewing zone are defined as the focused image plane [14] and the distance between two points which are created by crossing the extended lines of connecting two edges of each pixel cell to the center of its corresponding microlens as shown in reference [16], respectively. Furthermore it indicates the existence of side viewing zones. When viewers are positioned at the focused image plane, they cannot see any image; only highly blurred light points originated from a portion of the focused image determined by pupil sizes of the viewers. This means that the focused image is not the image projected to each of the viewer's eyes, but viewers still perceive depth sense by plane images displayed on the full surface of the display panel. This implies that the depth sense is obtained by a parallax difference between images projected to each viewer's two eyes. This is a binocular parallax. The existence of the side viewing zone manifests that each elemental image is diverging with the angle greater than $(\tan^{-1} P_L/2a)$, i.e., the FOV of each microlens is greater than the one defined by condition 4. Furthermore, when the width of the viewing zone is defined as the reference [14], the optical geometry of IP becomes exactly the same as that of the Multi-view as shown in Fig. 7.2.

This evidence confirms that the projected image in IP is not a hologram-like 3-D image, but like the images in the Multi-view. Hence, the depth sensing mechanism of the IP is both binocular and motion parallaxes. The further proof on the parallaxes based depth sensing mechanism of IP and the images projected to viewers' eyes in the VZCS can be found by applying the viewing zone forming principle of the Multi-view [6]. From the analysis above, it is concluded that IP and the Multi-view are sharing the same viewing zone forming geometry and the depth cues of IP are both binocular and moving parallaxes.

## 7.6 Image Composition in IP

The conclusion drawn in Section 7.5 enables us to estimate the compositions of images seen at different viewing regions of IP. The image composition at different viewing regions when $\Delta P \neq 0$, i.e., when the difference between IP and the Multi-view is specified by the image arrangements in the pixel cell, is derived in Section 7.3. When $\Delta P$ is assumed as 0, $\Delta P$ portion from the second pixel cell is shifted to the first pixel cell. As a consequence, the second pixel cell is starting with a fraction pixel with a width $P_P - \Delta P$. Hence, for the $k$th pixel cell, $(k - 1)\Delta P$ portion of the pixel cell becomes shifted to $(k - 1)$th pixel cell. In Fig. 7.7 , the VZCS formed by pixel cells for $P_C = nP_P$ and $\Delta P = P_P/l$ are depicted. Since the VZCS is formed by the images of pixel cells, the VZCS is divided into $n$ regions for n pixels and a small region corresponding to $\Delta P$. However the second pixel cell divides again the VZCS such that a region corresponding to $P_P - \Delta P$, $n-1$ regions corresponding to $n - 1$ pixels and a region corresponding to $2\Delta P$. The VZCS is divided into regions corresponding to $P_P - (l - 1)\Delta P$ for the first region, $P_P$ for next $n$ regions. Hence, every $l$ pixel cells, the same pattern repeats. However, the starting region (from left) is for view number 1 image for the first $l$ pixel cells, but for the following $l$ multiple pixel cells, the starting region changes for the view number 2, 3, 4, ····, images. Since the width of starting regions in the $l$ multiple pixel cells, the width is reduced by the amount corresponding to $\Delta P$, $M$th and $(M + 1)$th view images are simultaneously viewed in the region corresponding to the first pixel of the $Ml$th pixel cell within the $Ml$th to $\{(M + 1)l - 1\}$th pixel cells. This is shown in Fig. 7.7. In Fig. 7.7, the compositions of images perceived at $i$th viewing regions can be predicted as $i$th pixel from the first pixel cell, $l$ − 1ea of

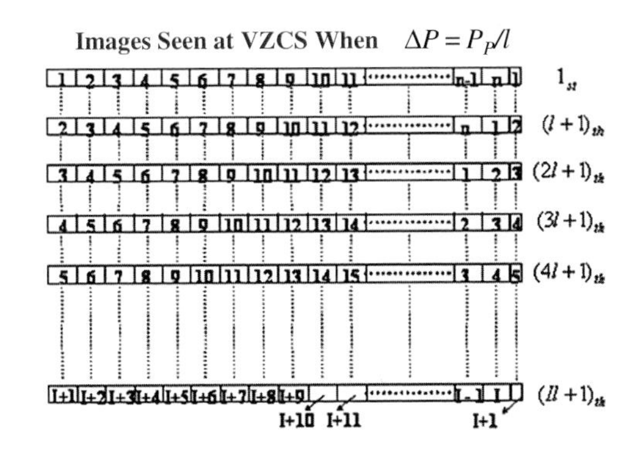


**Fig. 7.7.** Image compositions in VZCS when $\Delta P = P_P/l$

$i$th or $(i + 1)$th pixel from the next $l - 1$ pixel cells, $(i + 1)$th pixel from $(l + 1)$th pixel cell, $(i + 1)$th or $(i + 2)$th pixels from $(l + 2)$th to $2l$th pixel cells, $(i + 2)$th pixel from $(2l + 1)$th pixel cells, $(i + 2)$th or $(i + 3)$th pixels from $(2l + 2)$th to $3l$th pixel cells, • • • • • • • •, $(I + i - 1)$th from $\{(I - 1)l + 1\}$th pixel cell, $(I + i - 1)$th or $(I + i)$th pixels from $\{(I - 1)l + 2\}$th to $Il$th pixel cells when there are $Il$ pixel cells. By the pixel shifting for every $l$ pixel cells, there appears $n - l + 1$(1 is attributed by fractional pixel in the first pixel cell) different images composed of forward pixel orders from $Il$ pixel cells and $l$ images of mixed order. There should be a pseudoscopic viewing region between the $n - l + 1$ and $l$ images. The visual quality of the images synthesized from different number pixels from different pixel cells will be worse than that from the images in the Multi-view, due to the disparity jump for every $l$ pixels in the images.

## 7.7 Image Compositions in Other Multi-view Image Methods

The multi-view image methods described so far are based on time and spatial multiplexing methods in presenting the multi-view image. The multi-view images can also be presented by angular multiplexing. In this multiplexing method, the multi-view images are arranged the same as the Multi-view and IP. However, each pixel cell image is either combined as a point and each view pixels are directed to a specified angle direction as in the focused light array (FLA) method, or directed as a collimated image to a common image plane with a specified angle as in the multiple image [17]. The complete multi-view image is displayed by time-wise scanning of all pixel cell images in the pixel order in the FLA, and all collimated pixel cell images are simultaneously displayed with angles corresponding to their image order for the multiple image. Since the same number pixel in each pixel cell is directing the same direction with those in other pixel cells, these pixels are creating a collimated image.

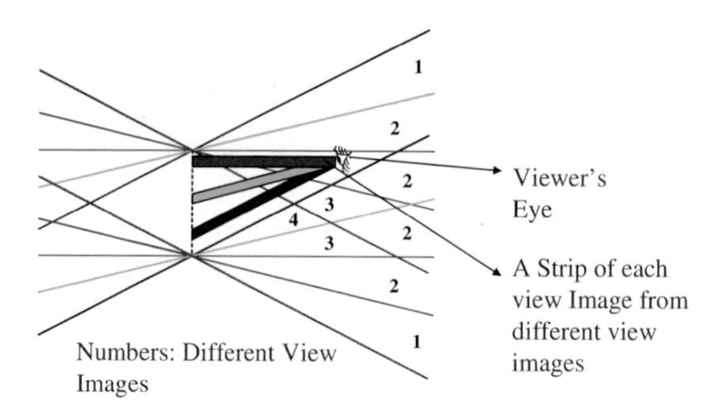


**Fig. 7.8.** Image forming in the angular multiplexed imaging methods

Hence, these two methods are based on the same image forming principle. In these methods, the image projected to each of the viewer's eyes is defined by the pupil size and position of the eye. As shown in Fig. 7.8, the image is composed of a strip from each collimated image which passes through the pupil. The width of the strip is determined by the pupil size. To make a full size image with the strips from images which are passing the pupil direction, the angle distance between images and the number of different view images in the multi-view image should be chosen carefully. Otherwise, there will be a gap between the strips and a full size image cannot be generated.

## 7.8 Conclusions

The multi-view 3-D dimensional imaging methods provide both binocular and motion parallaxes as their depth cue, but their mechanisms of forming viewing zones are different for projection, contact and scanning types. For the contact type, IP and the Multi-view have the same viewing zone forming geometry and their depth cues are the same; both binocular and motion parallaxes. The image compositions of the Multi-view and IP are basically holding a conjugate relationship, but neglecting the pitch difference between the pixel cell and the elemental optics in IP makes the quality of images perceived in IP worse than that in the Multi-view.

## References

[1] J.-Y. Son, B. Javidi, and K.-D. Kwack, "Methods for Displaying 3 Dimensional Images" (Invited Paper), Proceedings of the IEEE, Special Issue on: 3-D Technologies for Imaging & Display, Vol. 94, No. 3, pp 502–523, 2006

[2] J.-Y. Son and B. Javidi, "3-Dimensional Imaging Systems Based on Multi-view Images," (Invited Paper), IEEE/OSA. J. Display Technol., Vol. 1, No. 1, pp 125–140, 2005

[3] Y. Kajiki, H. Yoshikawa, and T. Honda, "Ocular Accommodation by Super Multi-View Stereogram and 45-View Stereoscopic Display," IDW'96, Proceedings of the 11th International Display Workshops, pp 489–492, 1996
[4] J.-Y. Son, V. G. Komar, Y.-S. Chun, S. Sabo, V. Mayorov, L. Balasny, S. Belyaev, M. Semin, M. Krutik, and H.-W. Jeon, "A Multiview 3 Dimensional Imaging System With Full Color Capabilities," SPIE Proc., Vol. 3295A, pp. 218–225, 1998.
[5] J.-Y. Son, V. V. Saveljev, Y.-J. Choi, J.-E. Bahn, and H.-H. Choi, "Parameters for Designing Autostereoscopic Imaging Systems Based on Lenticular, Parallax Barrier and IP Plates," Opt. Eng., V 42, No. 11, pp 3326–3333, 2003
[6] J.-Y. Son, V. V. Saveljev, M.-C. Park, and S.-W. Kim, "Viewing Zones in PLS Based Multiview 3 Dimensional Imaging Systems," IEEE/OSA J. Display Technol., Vol. 4, No. 1, pp 109–114, 2008.
[7] J.-Y. Son, V. V. Saveljev, J.-S. Kim, S.-S. Kim, and B. Javidi, "Viewing Zones in 3-D Imaging Systems Based on Lenticular, Parallax Barrier and Microlens Array Plates", Appl. Opt., Vol. 43, No. 26, 4985–4992, 2004 (Appl. Opt. Cover)
[8] R. Boerner, "Three Autostereoscopic 1.25m Diagonal Rear Projection Systems with Tracking Feature," IDW'97 Proc. pp. 835–838, Japan, 1997
[9] G. J. Woodgate, D. Ezra, J. Harrold, N. S. Holliman, G. R. Jones, and R. R. Moseley, "Observer Tracking Autostereoscopic 3D Display Systems," Proc. of SPIE, Vol. 3012, pp 187–198, 1997
[10] J.-Y. Son, "Autostereoscopic Imaging System Based on Special Optical Plates" in Three-Dimensional Television, Video, and Display Technology, B. Javidi and F. Okano (eds.), Springer-Verlag, New-York, 2002
[11] M. C. Forman, N. Davies, and M. McCormick, "Continuous Parallax in Discrete Pixelated Integral Three-Dimensional Displays," J. Opt. Soc. Am. A, Vol. 20, No. 3, pp 411–420, 2003
[12] M. Martinez-Corral, B. Javidi, R. Martinez-Cuenca, and G. Saavedra, "Multifacet Structure of Observed Reconstructed Integral Imaging," J. Opt. Soc. Am. A, Vol. 22, No. 4, pp 597–603, 2005
[13] J. S. Jang and B. Javidi, "Large Depth-of-Focus Time-Multiplexed Three-Dimensional Integral Imaging by Use of Lenslets with Nonuniform Focal Lengths and Aperture Sizes," Opt. Lett., Vol. 28, No. 20, pp 1924–1926, 2003.
[14] F. Okano, H. Hosino, J. Arai, M. Yamada, and I. Yuyama, "Three Dimensional Television System Based on Integral Photography," in Three-Dimensional Television, Video, and Display Technique, B. Javidi and F. Okano (eds.), Springer, Berlin, Germany, 2002
[15] http://www.Projectorpeople.com
[16] S. S. Kim, V. Saveljev, E. F. Pen, and J. Y. Son, "Optical Design and Analysis for Super-Multiview Three-Dimensional Imaging Systems," SPIE Proc., Vol. 4297, pp 222–226, January 20–26, San Jose, 2001
[17] Y. Takaki and H. Nakanuma, "Improvement of Multiple Imaging System Used for Natural 3D Display Which Generates High-Density Directional Images," Proc. SPIE, Vol. 5243, 2003, pp 43–49

# 8

# Rich Media Services for T-DMB: 3-D Video and 3-D Data Applications

BongHo Lee, Kugjin Yun, Hyun Lee, Sukhee Cho, Namho Hur, Jinwoong Kim, Christoph Fehn, and Peter Kauff

**Abstract** Mobile TV applications have gained a lot of interest in recent years. The ability to provide the user with a personal viewing experience as well as the flexibility to add new services have made this technology attractive for media companies and broadcasters around the world. In this chapter, we present a novel approach for mobile 3-D video and data services that extend the regular 2-D video service of the Terrestrial Digital Multimedia Broadcasting (T-DMB) system. We demonstrate the main concepts and features of our *3-D DMB* approach and discuss our latest research on the efficient compression of stereoscopic 3-D imagery.

## 8.1 Introduction

Many view three-dimensional television (3-D TV) as one of the most promising applications for the next generation of home video entertainment. Although the basic principles of stereoscopic acquisition and reproduction were demonstrated in the 1920s by John Logie Baird [2], the introduction of 3-D TV, so far, has been hampered by the lack of high quality autostereoscopic 3-D display devices, i.e., devices that don't require the viewer to wear special glasses in order to perceive the 3-D effect. Fortunately, this situation

B. Lee
Electronics and Telecommunications Research Institute, 161 Gajeong-dong, Yuseong-gu, Daejeon, Republic of Korea
e-mail: leebh@etri.re.kr

B. Javidi et al. (eds.), *Three-Dimensional Imaging, Visualization, and Display*,
DOI 10.1007/978-0-387-79335-1_8, 

**Fig. 8.1.** Mobile devices equipped with autostereoscopic 3-D displays. A wide range of different consumer products are already available on the market today

has changed beneficially in the recent past and especially in the commercially highly attractive area of small mobile devices, where usually only a single user is watching and interacting with the content, a number of promising autostereoscopic 3-D display solutions (see for example [14, 23] have been developed by major electronics companies such as Samsung and Sharp (see Fig. 8.1). These 2-D/3-D switchable displays could provide the basis introducing 3-D into the mobile TV market, and especially in the Republic of Korea, where the Terrestrial Digital Multimedia Broadcasting (T-DMB) system [25] has already been established for 2-D mobile television, 3-D video seems to be a very promising future service enhancement [3, 12].

This chapter focuses on the development of a stereoscopic 3-D extension to the basic 2-D video service provided by T-DMB. First, we present a general overview about the T-DMB system including its service framework, its protocol stack, and the applications it is able to support. Thereafter, we explain in detail our concept for a future *3-D DMB* system as well as the two different types of 3-D services that it shall provide: (a) a 3-D video service, and (b) a 3-D data service. This is followed by a description of our latest research on the efficient compression of stereoscopic 3-D imagery. Finally, we conclude the chapter with a brief summary and an outlook on future work.

## 8.2 The T-DMB System

Terrestrial Digital Multimedia Broadcasting (T-DMB) is a method for sending multimedia content (audio, video, and data) to mobile devices such as mobile phones. It has been developed as a compatible extension to the Eureka-147 Digital Audio Broadcasting (DAB) standard, which had originally been designed for providing digital radio services [10]. Eureka-147 DAB is very well suited for mobile reception and provides very high robustness against multipath propagation. It enables the use of single frequency networks (SFNs) for high frequency efficiency. Besides high quality digital audio services (monophonic, two-channel, or multichannel stereophonic), DAB allows the transmission of program associated data as well as a multiplex of other data services (i.e., travel and traffic information, still and moving pictures, etc.) [15]. In order to cope with the specific demands of mobile reception, DAB utilizes the orthogonal frequency division multiplex (OFDM) technique, a well-proven technology that is already widely used in wireless applications.

Technically, T-DMB extends DAB by adding multimedia protocol stacks that enable the provision of mobile TV services. It incorporates the latest high performance media coding technologies such as MPEG-4 Part 3 BSAC (Bit Sliced Arithmetic Coding) and HEAAC v2 (High Efficiency Advanced Audio Coding) [20] for the compression of audio, and defines MPEG-4 Part 10 H.264/AVC (Advanced Video Coding) [19] for the compression of video. In order to guarantee tight synchronization as well as efficient multiplexing of the encoded audio and video streams both the MPEG-2 [16] and MPEG-4 [18] system tools are utilized. Furthermore, T-DMB supports interactive data services by utilizing MPEG-4 BIFS (BInary Format for Scenes) as well as a powerful, Java-based middleware engine that exhibits an execution environment for native and downloadable applications.

In terms of distributing the data, T-DMB has the key advantage over mobile telecommunication networks, which are primarily being conducted by point-to-point interaction, in that it is able to deliver massive multimedia content using the more efficient concept of point-to-multipoint transmission. In the anticipated convergence scenario, where the telecommunication and broadcasting networks are weaved together, T-DMB could thus play a key role in disseminating advanced audio/visual and data services that could not be conveyed economically via the telecommunication network.

### 8.2.1 Service Framework of T-DMB

The service framework of T-DMB is based on the Eureka-147 DAB system. As illustrated in Fig 8.2, the video data of the DMB service is conveyed in the so-called *stream mode* of DAB, which is specifically designed for the delivery of streaming contents. The other paths of the conceptual transmission architecture are fully compatible with DAB. The stream mode allows a service application to accept and deliver data transparently from source to destination at a data rate that must be a multiple of 8 kbit/s [10]. An important difference between the stream mode and the so-called *packet mode* is that for the stream mode only one service component can be mapped into a single DAB sub-channel, while the packet mode enables several data service components within the same sub-channel.

Taking a look at the complete video delivery chain (see again Fig 8.2), the encoded video stream is first fed into the video multiplexer. This module outputs Transport Stream (TS) packets, which are then protected using both equal and unequal error protection (EEP/UEP) featuring convolutional encoding and time interleaving. The so-protected packets are finally inserted in the main service (MSC) multiplexer, where they are combined with data from other services. The necessary side information from the DMB video

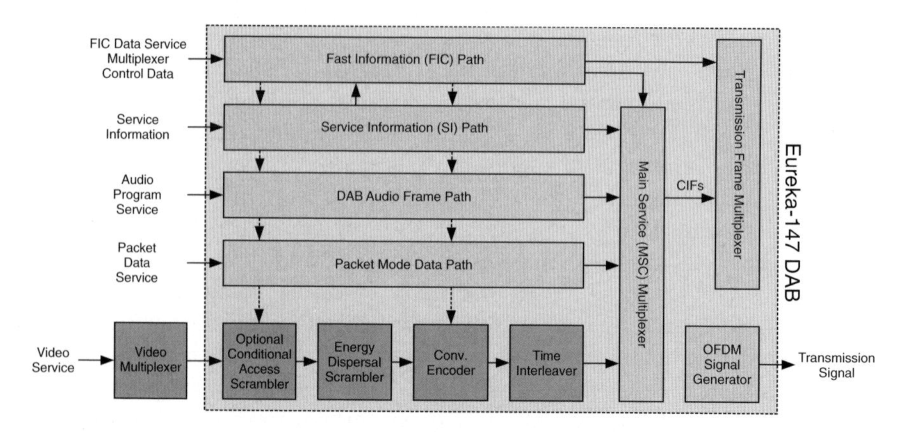


**Fig. 8.2.** Conceptual transmission architecture for the video service. The video data is conveyed in the DAB stream mode, which is specifically designed for the delivery of streaming contents

application, i.e., the multiplex configuration information (MCI) and the service information (SI), are transmitted in parallel via the forward information (FIC) and SI paths. The MCI provides multiplex information for the various services within a single ensemble. It addresses subchannel organization, service organization, ensemble information, and re-configuration information of the multiplex. The SI consists of information about services, both for audio programs and data. It deals with service related features, language features, and program related features. Service related features include announcements, the service component trigger and frequency information. The program related features include program number and type.

### 8.2.2 The T-DMB Protocol

The protocol stack utilized by T-DMB is illustrated in Fig 8.3. It is composed of four distinct paths: audio, data, video, and control.

The audio path is used for the delivery of audio frames for the original MUSICAM (MPEG-1 Audio Layer 2) digital radio service [15]. Recently, World DMB [35] also adopted the newer, more efficient HE-AAC v2 [20] audio codec in order to provide the users with an enhanced audio service dubbed DAB+ [11]. With this new codec, DAB/DMB can offer about three times as many audio channels with the same perceptual quality than with the original MUSICAM technology. In order to enhance the robustness of the transmissions a stronger error correction coding based on a Reed-Solomon coder and a virtual interleaver has also been specified.

Concerning the data path, there are two types of data services: program associated data and non-program associated data. Most of the data services are independent from the audio program, but some of them are also closely

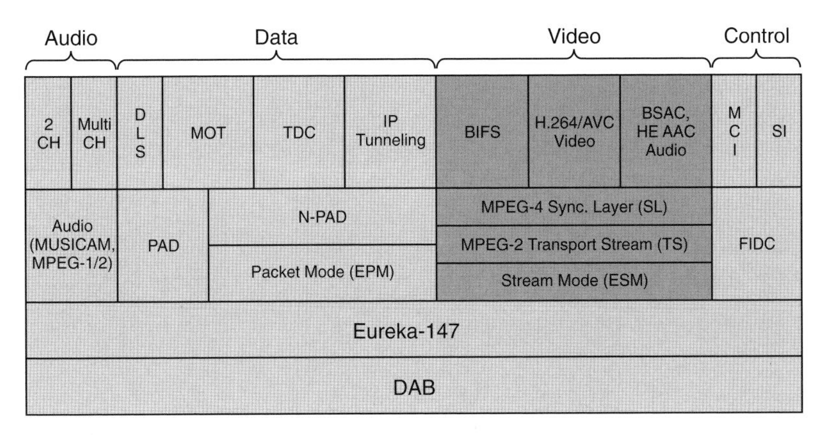


**Fig. 8.3.** Protocol stack utilized by T-DMB. It is composed of four distinct parts: Audio, data, video, and control

related to the audio signal. The program associated data (PAD) is directly attached to the compressed audio streams. Simple text applications like the dynamic label service (DLS), file objects, and enhanced text services are examples of PAD. On the other hand, N-PAD is a protocol for audio independent data services. As can be seen from the block diagram, N-PAD is related to the enhanced packet mode (EPM) and upper layer stacks such as multimedia object transfer (MOT) [9] for the delivery of multimedia objects, transparent data channel (TDC) for the transparent delivery of data, and IP tunneling for IP encapsulation. Through the N-PAD route, applications such as the MOT slideshow [8], the MOT broadcast websites (BWS) [4], and various object-based data services are available.

The video path exhibits four layers ranging from media generation to error protection. Starting from the top, the media layer consists of BIFS, video, and audio blocks, which create an interactive scene description and carry out the encoding of the audio/visual data. The underlying MPEG-4 synchronization layer (SL) handles the synchronization of the compressed elementary streams (ESs) from the upper layer. Each ES is encapsulated into a stream of MPEG-4 SL packets. These are refered to the MPEG-2 TS layer, where they are converted into MPEG-2 TS packets. Finally the TS converted data is forward error protected using the enhanced stream mode (ESM) scheme. In addition to the main video service, this part of the protocol stack also enables advanced, BIFS-based interactive audio/visual data services (see also Section 8.2.3).

The conceptual architecture of the T-DMB video service is further outlined in Fig 8.4. In detail, it involves the following media processing steps:

1. In accordance with the MPEG-4 Systems [18] standard, an Initial Object Descriptor (IOD) as well as further Object Descriptors (ODs) and BIFS data are created in the IOD and OD/BIFS generators. The IOD provides

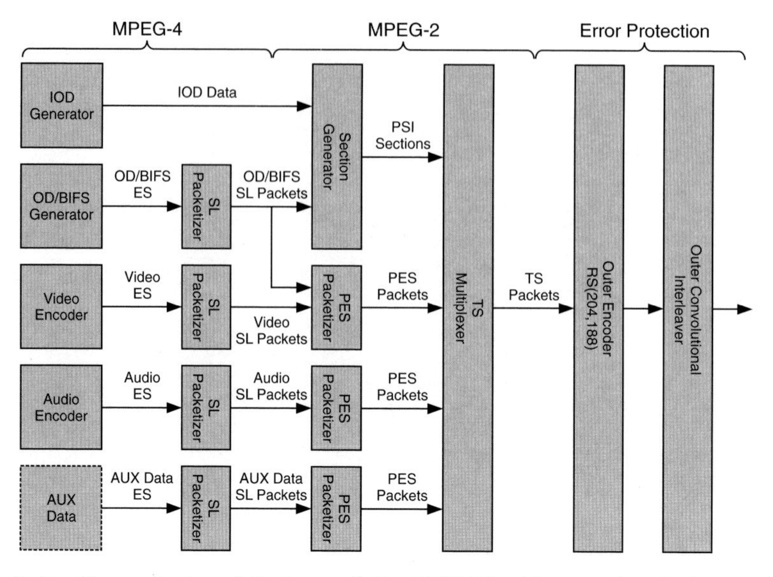


**Fig. 8.4.** Conceptual architecture of the T-DMB video service. It incorporates latest high-performance MPEG-4 audio/visual coding technologies, MPEG-2 and MPEG-4 Systems tools for synchronization and multiplexing as well as an efficient error protection

initial access to the Program Specific Information (PSI) sections while the ODs inform the receiver about the properties (codec, profile, etc.) of the individual audio/visual elementary streams.

2. The video encoder compresses the video according to the H.264/AVC [19] DMB profile [5]; the audio signal is compressed using either the BSAC or the HEAAC v2 [20] audio codec. Any auxiliary data, such as JPEG still pictures, can be provided as an auxiliary data ES.
3. The OD/BIFS streams as well as all other elementary streams (audio, video, and auxiliary data) are encapsulated into SL packets within the SL packetizers in order to attach MPEG-4 Systems [18] compliant timing information.
4. Each SL packet is then wrapped into a single PES (packetized elementary stream) packet compliant with the MPEG-2 Systems [16] specification. The IOD/OD/BIFS streams are converted into MPEG-4 PSI sections in a way that is also compliant with this standard.
5. The TS multiplexer integrates all (PESs) and PSI sections into a single MPEG-2 Transport Stream. The TS packets are then further augmented with error correction codes from a Reed-Solomon coder and interleaved within the outer convolutional interleaver.

In the control path, both MCI and SI as well as a traffic message channel (TMC) and an emergency warning service (EWS) can be transmitted. In order to deliver such time-critical information without any latency the protocol stack provides a suitable Fast Information Data Channel (FIDC) that deals with the multiplexing and signaling of these highly important service elements.

### 8.2.3 Applications

The T-DMB system provides a vast number of media services ranging from the original broadcast of digital radio (MUSICAM [15] and DAB+ [11]) over electronic program guides (EPGs) [6, 7], web services such as the MOT broadcast websites (BWS) [4], the MOT slideshow [8], travel and traffic information as well as text and voice applications, up to an advanced Java-based middleware engine that allows for the execution of applications downloaded either via the broadcast network or over a bidirectional network.

In addition to these valuable media services, World DMB [35] also approved the world's first mobile TV application: the DMB video service. The specification allows for the transmission of videos with up to $352 \times 288$ pel (CIF) resolution and a framerate of up to 30 frames per second (FPS) [5]. Depending on the chosen video format typical distribution bitrates are in the area of around 200–800 kbps. The DMB video service is already used commercially in countries such as the Republic of Korea, Germany, etc., and a large variety of fixed and mobile receivers are available to the market.

Finally, DMB also provides dynamic and interactive audio/visual data services based on MPEG-4 BIFS [18]. BIFS is a flexible scene description language that allows for the synchronous representation of presentations comprising 2-D and 3-D graphics, images, text, audio, and video. The specification of such a presentation can include the description of the spatio-temporal organization of the different scene components as well as user interactions and animations. The BIFS standard is based in large part on the Virtual Reality Modeling Language (VRML), but defines a more efficient binary distribution format.

## 8.3 The 3-D DMB System

The flexible design of the T-DMB system, as well as its rapidly increasing availability, provide us with the highly attractive possibility to introduce new 3-D video and data services into the mobile TV market. How this can be done in an efficient manner and what requirements such services should fulfill to be commercially successful is described in detail in the following sections.

### 8.3.1 Basic Concepts and Requirements

Mobile devices, such as cell phones or personal media players (PMPs) are inherently single user devices. It follows that a mobile TV service, be it 2-D or 3-D, can also be designed as a single viewer application. For our envisaged 3-D DMB scenario, this has two very important implications: [1] The 3-D display need support a single viewer and must provide a good 3-D effect only from a single, central viewpoint. [2] Such a 3-D display can be driven efficiently with conventional stereoscopic imagery, i.e., with two synchronized video streams one for the left eye and one for the right eye.

Based on these two crucial observations, we developed the concept of a 3-D DMB system that extends the capabilities of the available 2-D T-DMB video and data services. This mobile 3-D video and data technology is supposed to fulfill the following requirements:

- **Forward and backward compatibility:** For the commercial success of a future 3-D video service, it is important that it also supports the existing functionalities of the 2-D legacy system [3, 12]. This is known as forward compatibility and it implies that a new 3-D DMB receiver must be able to decode and display (in 2-D mode) a regular 2-D T-DMB video stream. In addition, the 3-D system should be backward compatible in the sense that a conventional 2-D T-DMB terminal is able to decode and display one of the two views of a stereoscopic 3-D DMB video stream.
- **Switchability between 2-D and 3-D mode:** In order to fulfill the forward and backward compatibility requirement, the 3-D DMB receiver must have the ability to switch between 2-D and 3-D display mode – either automatically, depending on the type of content, or manually. This feature is also important for providing the user with the highest possible 2-D viewing quality when he/she works with day-to-day applications like spreadsheets, the calender, or an e-mail client.
- **Autostereoscopic 3-D display:** The 3-D effect must be perceptible without the need to wear special glasses or other viewing aids. This calls for the use of so-called autostereoscopic 3-D displays, where the technology that separates the left and right-eye views is integrated into the display itself. Fortunately, a number of high quality autostereoscopic 3-D display solutions (see for example [14, 23]) are already available on the market today. These 2-D/3-D switchable displays are able to provide a 3-D experience that seems perfectly acceptable for a mobile video application.
- **Reduction of eye fatigue:** For a 3-D video service to be acceptable to the users, the viewing experience must be comfortable – even over a longer period of time. This necessitates the use of high quality autostereoscopic 3-D displays on the other hand, it also requires the production of 3-D content that does not lead to eye fatique due to excessive parallax and unnatural depth effects.
- **Low overhead:** A stereoscopic video stream consists of two synchronized, monoscopic videos and, thus, comprises twice as much raw pixel data as conventional 2-D moving imagery. To be commercially viable, a 3-D video service must, therefore, be designed to achieve very high bitrate reductions while, at the same time, maintaining a good 2-D and 3-D video quality. In order to achieve this goal, the use of video coding algorithms that are highly optimized to this specific type of content is required.

### 8.3.2 The 3-D DMB Service Framework

In order to meet the basic requirements defined above, we designed our 3-D DMB system as a forward and backward compatible extension to the original 2-D T-DMB system. From the DAB/DMB service framework (see again Section 8.2) we transparently utilize the underlying protocol layers (i.e., the DAB transmission architecture, the MPEG-2/4 Systems tools, and the ESM error protection mechanisms). Changes and upgrades are mainly required in the media layers (i.e., the audio/visual coding tools and the BIFS scene description). Our concept provides two new types of 3-D services: (a) a 3-D video service, and; (b) a 3-D data service. These are described in more detail in the following sections.

### 8.3.3 3-D Video Service

The 3-D video service enhances the original 2-D T-DMB video service by supplementing the basic monoscopic video (base layer) with an additional 2-D video stream (enhancement layer) that was captured from a slightly different viewing position. When displayed jointly on an autostereoscopic display, the perspective differences between the two views provide the user with a binocular 3-D depth effect. The major functional building blocks of our system are described in more detail in the following sections.

#### 3-D DMB Sender

In order to provide the required forward and backward compatibility with legacy 2-D T-DMB receivers, the 3-D DMB sender uses the H.264/AVC [19] coding tools to compress the base layer video according to the DMB profile [5]. For the encoding of the additional enhancement layer video, we have, in principle, a number of thinkable options:

1. **Simulcast:** The most straightforward approach would be to encode both videos independently with the same quality using the basic H.264/AVC coding technology. This so-called *Simulcast* solution is very easy to implement, however, it is not very efficient as it requires twice the bitrate for achieving the same video quality. To say it the other way around, for staying within the limited bit budget of the T-DMB system the compression ratio has to be chosen twice as high than for the transmission of 2-D video. This, of course, results in stronger coding distortions and, therewith, a reduction in perceived visual quality in the 2-D viewing case.
2. **Disparity-compensated prediction (DCP):** Compared to Simulcast a higher compression efficiency can be reached if we are able to also exploit the statistical dependencies that exist between the two stereoscopic views. This can be achieved by combining the conventional motion-compensated prediction (MCP) with a so-called disparity-compensated

prediction (DCP), where the individual pictures of the extra enhancement layer video are also predicted from the temporally coinciding pictures of the base layer video.
3. **Asymmetric coding:** Another well-known approach for reducing the amount of data required to encode a stereoscopic video sequence is to spend more bits on the base layer video – which must also look good on conventional 2-D displays – and less bits on the extra enhancement layer stream. Such *asymmetric* coding can be implemented by either using different compression ratios (e.g., by choosing different quantization parameters) for the two views or, preferably, by reducing the resolution of the second view before encoding – the decoded images then have to be interpolated at the receiver side to their original size before display.
4. **Asymmetric coding with DCP:** Ideally, it is preferable combine both asymmetric coding and DCP in order to exploit the benefits of both coding methodologies and, therewith, achieve an even higher compression efficiency.

Our current 3-D DMB real-time demonstrator platform (see also below) works with an asymmetric coding approach that is based on the spatial downsampling of the additional enhancement layer video, but we are also working on combining both asymmetric coding and DCP into a highly efficient, integrated stereoscopic video coding scheme (see also Section 8.1).

In addition to the stereoscopic imagery our system also supports multichannel stereophonic sound. The five input channels (left, center, right, left surround, right surround) are pre-processed and mixed according to the ITU-R BS.775-2 specification [21]. In order to achieve the necessary forward and backward compatibility the resulting L0 and R0 channels are encoded with the MPEG-4 Part 3 BSAC audio codec [20]. The remaining three channels (T, Q1, Q2) are encoded seperately as additional enhancement layers.

In the MPEG-4 Systems layer, the encoded 3-D audio/visual streams – as well as the generated OD/BIFS data – are packetized into SL packets as already described in Section 8.2. The IOD data, on the other hand, is directly embedded into MPEG-2 PSI sections. The remaining part of the processing chain (i.e., PES packetization and TS multiplexing in the MPEG-2 layer, error correction coding, etc.) is identical to the regular 2-D T-DMB case.

### 3-D DMB Receiver

The 3-D DMB receiver basically performs the "inverse" processing steps of the sender. The data received by the DAB module is error corrected and demultiplexed/depacketized in the respective MPEG-2 and MPEG-4 Systems layers. The compressed audio/visual streams as well as the IOD/OD/BIFS data are decoded in the media layer and the 2-D or 3-D scene, consisting of either monoscopic video with stereo sound or stereoscopic video with multichannel audio, is rendered to the display (in either 2-D or 3-D viewing mode) according to the BIFS scene description.

## Implementation and Evaluation

To evaluate our 3-D video service concept and to verify the backward compatibility with existing 2-D T-DMB receivers, we implemented a complete 3-D DMB transmission chain in a PC-based demonstrator setup (see Fig. 8.5).

Our prototype 3-D DMB receiver has the capability to not only support regular 2-D T-DMB video services, but also to receive and display 3-D video streams. The current version of the setup consists of a laptop that implements the basic DAB processing (incl. OFDM signal reconstruction, error correction, etc.), the required MPEG-2/4 Systems operations (including TS demultiplexing, PES and SL depacketization, decoding of the IOD/OD/BIFS data, etc.), the decoding of the compressed audio/visual streams, and, finally, the BIFS-based compositing of the 2-D or 3-D imagery. The 3-D DMB receiver laptop is connected to an autostereoscopic 3-D LCD display with a native resolution of 640×480 pel (VGA). This single-user device is based on an active parallax barrier that can be turned on and off in order to switch between 2-D and 3-D viewing. In the basic 2-D mode, only the base layer video is rendered to the full VGA resolution of the LCD panel; in the advanced 3-D mode, both the base layer video and the enhancement layer video are each enlarged to a resolution of 320×480 pel and displayed on the screen in a column-interleaved spatial multiplex. The opaque columns formed by the active parallax barrier then shield the left and the right eye from seeing the unwanted image information.

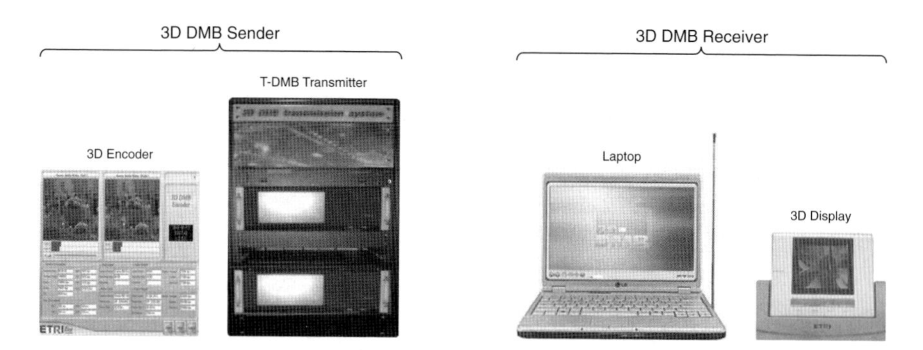


**Fig. 8.5.** The implemented 3-D DMB real-time demonstrator platform. It is mainly composed of PC components and an autostereoscopic 3-D display based on an active parallax barrier that can be switched between 2-D and 3-D viewing mode

Our experimental 3-D DMB sender (a) is comprised of a PC application that performs in real-time both the encoding of the stereoscopic video and multichannel audio inputs and the generation of the necessary IOD/OD/BIFS streams compliant to T-DMB. The 3-D DMB transmitter implements the DAB/DMB protocol stack as described in Section 8.2. In our first tests, we distributed 3-D video services using video sequences with a resolution of 352×288 pel (CIF format) and a framerate of 30 fps The total video bitrate was set to 740 kbps from which 512 kbps were assigned to the 2-D base layer video. The remaining bits were spend on the spatially down-sampled enhancement layer video. While this simple configuration seems to provide a reasonably good viewing quality we want to achieve a higher compression performance by implementing the more efficient stereoscopic video coding method described in Section 8.4.

### 8.3.4 3-D Data Service

The 3-D data service provides the user with an enhanced viewing experience by overlaying a conventional 2-D video program with 3-D data. This could be used, for example, for advertisement scenarios, where certain elements of the video could be highlighted in order to achieve a stronger effect (see Fig. 8.6).

Technically, our approach is based on the complementary distribution of stereoscopic still pictures, which are synchronized to the basic 2-D audio/visual information. The spatio-temporal positioning of these 3-D images on top of the 2-D background video, as well as any information about user interactions and animations, is described by means of BIFS commands.

**Fig. 8.6.** A simple example of the 3-D data service. The butterflies are rendered stereoscopically in front of a conventional 2-D video that is played in the background

Because the current version of the T-DMB specification only supports a limited 2-D subset of BIFS (Core2-D@Level1) we had to extend it by adding appropriate 3-D BIFS nodes for representing stereoscopic still imagery. An important goal for the future would, therefore, be to upgrade today's 2-D T-DMB BIFS profile in a backward and forward compatible fashion based on our proprietary solution.

## 8.4 Efficient Coding of Stereoscopic Video

The distribution of stereoscopic 3-D sequences over the existing T-DMB network requires efficient video coding algorithms that are optimized for the new type of content, and that are able to compress the larger amount of visual information to a bitrate that is compatible with the transmission capacity of the network. In addition, it will also be important for the successful introduction of future 3-D video services that they are backward compatible to today's 2-D T-DMB system. Therefore, it must be ensured that: (a) a regular 2-D T-DMB decoder is able to parse a compressed stereoscopic bitstream and extract/decompress one of the two views, and that; (b) this view can be reproduced on a conventional, monocular (2-D) display with an acceptable image quality (see again Section 8.3.1).

In order to fulfill these essential requirements, we investigated, how the use of DCP could be combined with the concept of *asymmetric coding*, i.e., with a compression method where the two views of a stereoscopic video sequence are represented with different image qualities. This is a very promising approach that is motivated by the *suppression theory* of binocular vision [22], which indicates that the perceived sharpness and depth effect of a mixed-resolution image pair is dominated by the higher quality component [26, 29, 34].[1] By exploiting the benefits of both an efficient combination of motion- and disparity-compensated prediction (MCP, resp. DCP) and a "subjectively acceptable" reduction of the total amount of data to be encoded (asymmetric coding), we expect our 3-D DMB codec to achieve very high compression gains without reducing the overall visual quality of the resulting three-dimensional (3-D) percept.

### 8.4.1 Related Work

The concept of mixed-resolution coding was first introduced by Perkins in 1992 [30]. He applied a simple low-pass filtering as a compression algorithm resulting in high- and low-resolution views of a stereo image pair. In his seminal

[1] Interestingly, this observation only holds true for stereoscopic imagery degraded by blur. If, instead, one of the views is impaired by blockiness, the perceived visual quality seems to be an average of the image quality of the left and right-view images [26, 34].

paper, he also presented a second algorithm based on DCP. Since then, quite a number of works have been published on both asymmetric coding using standard video codecs such as MPEG-2 [17] and H.264/AVC [19] (see for example [13,26, 34]), and DCP-based stereo image coding (i.e., the work on the MPEG-2 Multi-view profile [28, 31] or the latest work on H.264/AVC-based Multi-view Video Coding (MVC) within the Joint Video Team (JVT) of MPEG [27, 33]). However, to our knowledge, only a few researchers have tried to combine both coding methodologies into a joint, integrated video compression system (see [1, 12, 24]).

### 8.4.2 Structure of 3-D DMB Codec

The spatio-temporal prediction structure of our 3-D DMB codec is shown in Fig. 8.7. In order to achieve backward compatibility with conventional 2-D DMB devices, the left view images $V_l$ (base layer) are compressed independently in a typical *IPPP...* GOP (Group of Pictures) structure using MCP only. Thus, they can be decoded without any references to the additional right-view imagery. To be compliant with the specifications for video services over DAB [5] only forward-predictive coding (*P*-pictures) with a maximum number of three temporal reference frames and a single *I*-picture per second is utilized. In contrast, the right-view images $V_r$ (enhancement layer) are encoded using both MCP with up to two reference frames as well as DCP from the temporally coinciding lefteye views.

Making use of DCP already provides two potential sources of a coding gain compared to the basic Simulcast case, i.e., the case where both the left and the righteye views are compressed independently using MCP only:

1. No intra-coded pictures (*I*-pictures) have to be used in the righteye views. Instead, the first right-view image in a GOP is also predicted (DCP) from the temporally associated leftview image. Because *P*-pictures usually require a smaller number of bits than *I*-pictures, this will almost always result in a coding gain compared to Simulcast.

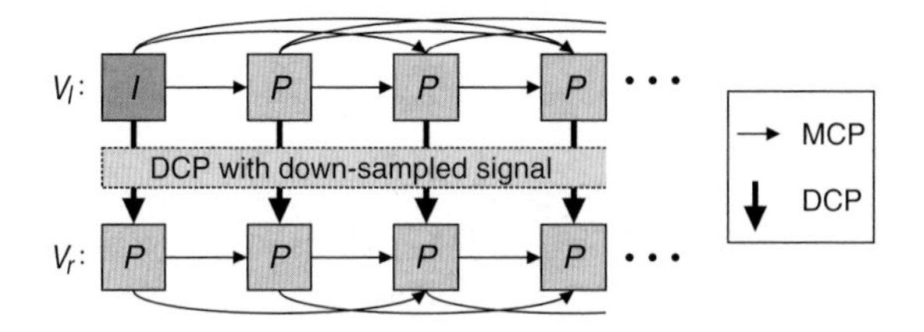


**Fig. 8.7.** Spatio-temporal prediction structure. The left view images (base layer) are compressed independently in a typical *IPPP...* GOP structure using MCP only. The right-view images $V_r$ (enhancement layer) are compressed at full, half, or quarter resolution using both MCP and DCP from the lefteye views

2. The temporally associated lefteye view used for DCP might, in some cases, provide a better reference than the temporally preceding right-view images utilized for MCP. The size of the gain, however, is highly dependent on the characteristics of the stereoscopic video sequence, i.e., on the motion content as well as on the difference in perspective between the two views.

The so-far described stereoscopic video compression concept does not yet support asymmetric coding and simply equals the H.264/AVC-based MVC codec described in [27]. In order to extend it to mixed-resolution imagery, two additional down-sampling blocks have to be integrated into the encoder structure. This is illustrated in the block diagram shown in Fig. 8.8. When used, the first of these blocks is responsible for down-sampling all right-view images $V_r$ to half or quarter resolution while leaving the lefteye views $V_l$ unaltered. The second block, attached to the reference picture buffer, is required for down-sampling the reconstructed leftview images $\hat{V}_l$ DCP of the right-eye views (see again Fig. 8.7). In the associated 3-D DMB decoder, the exact same down-sampling block is required for DCP while an appropriate up-sampling block is used in order to resize the reconstructed lower resolution right-view images to the original (lefteye view) dimensions.

### 8.4.3 Implementation

We have integrated the changes required to support our stereoscopic video coding approach into the H.264/AVC-based Joint Multi-view Video Model (JMVM) codec, which is maintained by MVC group within MPEG. The JMVM implements DCP by simply re-ordering the input pictures according

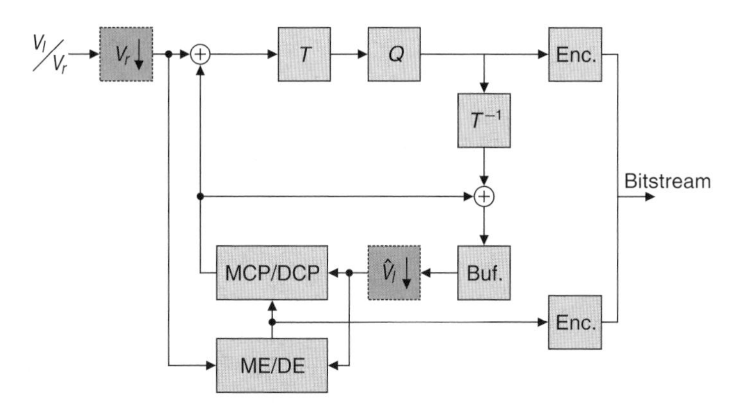


**Fig. 8.8.** Block diagram of our 3-D DMB encoder. Conceptually, it can be understood as consisting of an H.264/AVC-based MVC encoder as described in [27] with two additional down-sampling blocks (shown in darker gray)

**Table 8.1.** Filters that were used for down-, resp. up-sampling. The filter taps are provided for an integer implementation, i.e., the filtered pixel values finally have to be divided by the divisors given in the third column of the table (from [32])

| Type | Filter taps | / |
|---|---|---|
| Down | 2, 0, −4, −3, 5, 19, 26, 19, 5, −3, −4, 0, 2 | 64 |
| Up | 1, 0, −5, 0, 20, 32, 20, 0, −5, 0, 1 | 32 |

(a) (b) (c) (d)

**Fig. 8.9.** Original stereo image pair and associated, down-sampled righteye views (**a,b**) left and righteye views at full resolution; (**c,b**) Righteye view at half and quarter resolution

to the specified prediction structure (see again Section 8.4.2) and by reusing the regular MCP tools [27]. For the down-, resp. the up-sampling of the stereoscopic imagery we used the filters specified in Table 8.1.

An exemplary original stereo image pair at full resolution and the associated, down-sampled righteye views at half and quarter resolution are shown in Fig. 8.9.

### 8.4.4 Experimental Results

We have evaluated the compression performance of our 3-D DMB codec in comparison to regular asymmetric coding on different stereoscopic video sequences in QVGA format (320×240 pel, 30 fps). The results for two of these tests are shown in Fig 8.10. As can be seen from the rate-distortion (R-D) curves for the righteye views (b,e), the use of DCP can lead to noticeable coding gains, or, equivalently, reductions in required bitrate, compared to an independent encoding of the base and enhancement layer streams.[2] This

[2] Note that the coding distortions (PSNR values) of the righteye views were calculated on the down-sampled imagery.

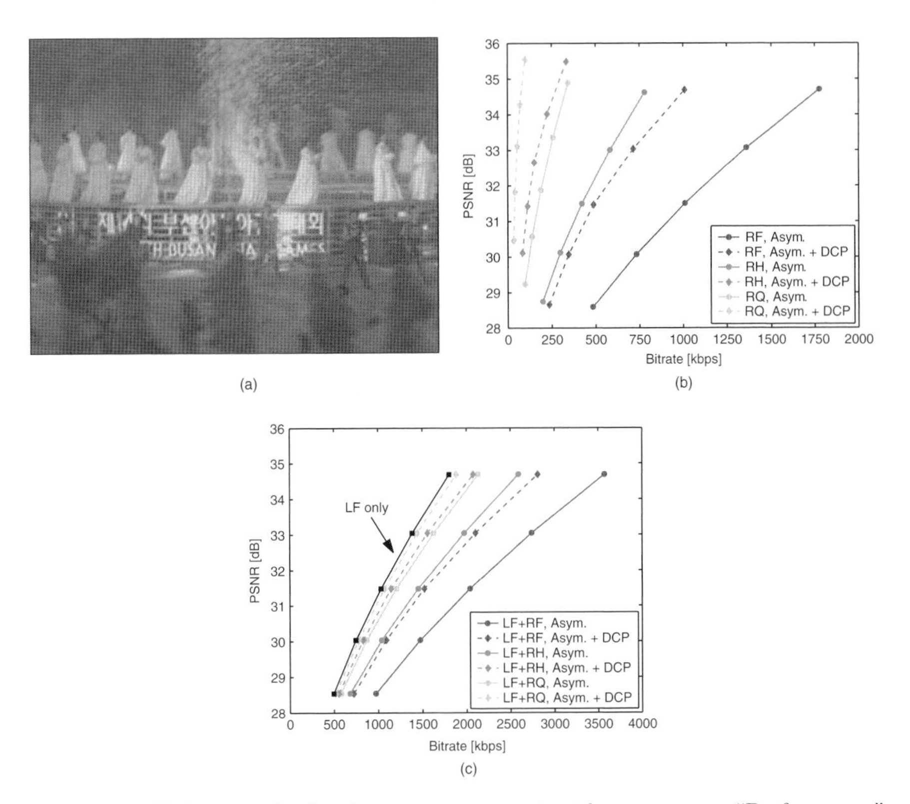


**Fig. 8.10.** Coding results for the two stereoscopic video sequences "Performance" and "Soccer2". (**a,d**) Overlaid left and righteye views; (**b,e**) R-D curves for the righteye views (R) encoded at full (F), half (H), and quarter (Q) resolution using both regular asymmetric coding and our 3-D DMB codec; (**c,f**) R-D curves for the combined left and righteye views (L+R). The R-D curves for encoding only the lefteye views (L) at full (F) resolution provide a reference for the case of monoscopic (2-D) distribution

effect is seen regardless of the resolution (full, half, or quarter) used for the righteye views. Furthermore, we find for both our 3-D DMB codec and the regular asymmetric coding that the lower resolution versions of the right-view images can be compressed at good quality (in terms of PSNR) to significantly lower bitrates than the correspond full resolution variants. The R-D curves for the combined left and righteye views (c,f) signify that by making use of both DCP and asymmetric coding we are able to reduce the overhead necessary for transmitting the additional right-view imagery down to a point where the resulting joint bitrate in only slightly larger than what is required for encoding a single monoscopic (2-D) video stream alone.

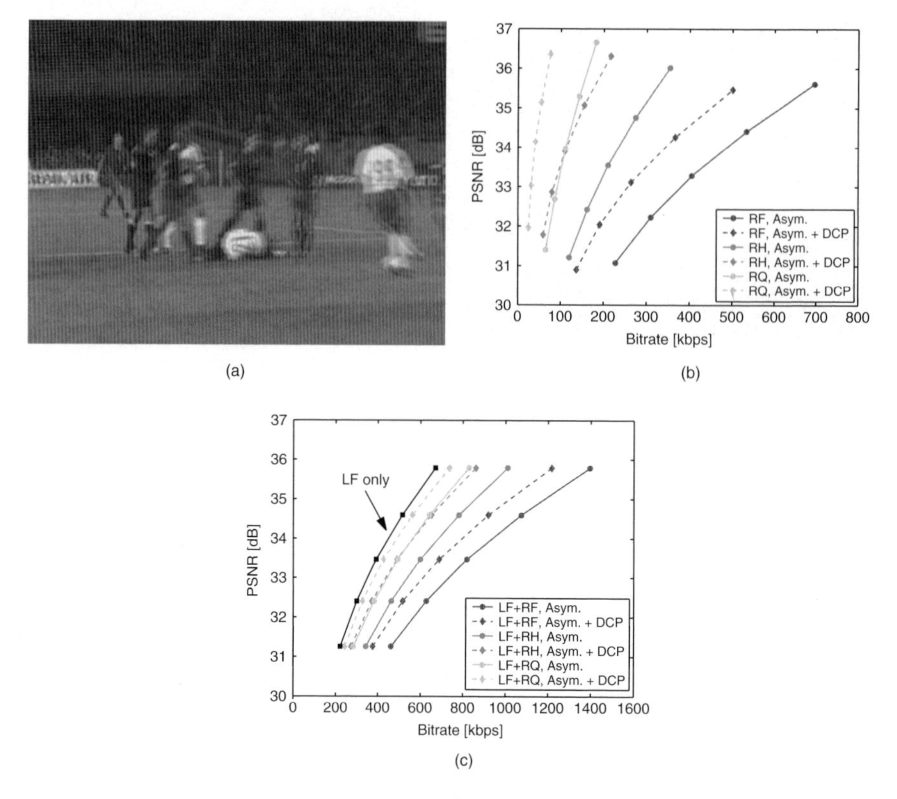


**Fig. 8.10.** (continued)

## 8.5 Conclusions and Future Work

We presented a novel approach for mobile 3-D video and data services that extend the regular 2-D video service of the Terrestrial Digital Multimedia Broadcasting (T-DMB) system. The system concept fulfills important technical and commercial requirements such as: (1) forward and backward compatibility to 2-D T-DMB receivers; (2) switchability between 2-D and 3-D viewing mode; (3) low overhead compared to a conventional 2-D video transmission.

In order to evaluate our approach we have implemented a complete 3-D DMB chain, including content creation, coding, transmission, decoding, and 3-D display, in a PC-based demonstrator setup. Our current platform supports two new services: (1) A 3-D video service that enhances the conventional 2-D video broadcast by supplementing the basic monoscopic video (base layer) with an additional 2-D video stream (enhancement layer) that was taken from a slightly different viewing position. When displayed jointly on an autostereoscopic display, the perspective differences between the two views provide the user with a binocular 3-D depth effect. (2) A 3-D data service that provides the user with an enhanced viewing experience by overlaying a monoscopic video with 3-D data (e.g., stereoscopic still pictures).

Based on our experiences with the system so far we are confident that 3-D DMB could become an interesting, well perceived, and commercially successful future service enhancement. The underlying hardware components (i.e., powerful 2-D T-DMB receivers, 2-D/3-D switchable autostereoscopic 3-D displays, etc.) are already available to the market and the step from 2-D to 3-D DMB will mainly require an upgrade in audio/visual compression technology as well as an appropriate extension of the DMB audio, video, and BIFS profiles.

Our future work will focus on bringing our 3-D DMB system to a state that is ready for marketing, and on finding the best possible solution for the compression of stereoscopic video. Based on our preliminary research on combining asymmetric coding with disparity-compensated prediction (DCP) we plan to develop an efficient, integrated coding scheme and to propose this solution as the future standard for 3-D mobile video services. In addition, we will also implement a complete, end-to-end 3-D data service chain that will provide an attractive short-term solution for future 3-D DMB services.

**Acknowledgment** This work was supported by the IT R&D program of MIC/IITA [2007-S004-01, Development of glassless single-user 3-D broadcasting technologies].

## References

[1] A. Aksay, C. Bilen, E. Kurutepe, T. Ozcelebi, G. Bozdagi Akar, M. R. Civanlar, and A. M. Tekalp. Temporal and Spatial Scaling For Stereoscopic Video Compression. In *Proceedings of 14th European Signal Processing Conference*, Florence, Italy, September 2006.

[2] P. Barr. Flying False Flags. The Scotsman, June 2002.

[3] S. Cho, N. Hur, J. Kim, K. Yun, and S.-I. Lee. Carriage of 3-D Audio-Visual Services by T-DMB. In *Proceedings of International Conference on Multimedia & Expo*, pages 2165–2168, Toronto, Canada, July 2006.

[4] ETSI. Digital Audio Broadcasting (DAB); Broadcast Website; Part 1: User Application Specification. ETSI Technical Specification TS 101 498-1.

[5] ETSI. Digital Video Broadcasting (DVB); DMB Video Service; User Application Specification.ETSI Technical Specification TS 102 428 V1.1.1, June 2005.

[6] ETSI. Digital Audio Broadcasting (DAB); Digital Radio Mondiale (DRM); Transportation and Binary Encoding Specification for Electronic Programme Guide (EPG). ETSI Technical Specification TS 102 371 V1.2.1, February 2006.

[7] ETSI. Digital Audio Broadcasting (DAB); Digital Radio Mondiale (DRM); XML Specification for DAB Electronic Programme Guide (EPG) . ETSI Technical Specification TS 102 818 V1.3.1, February 2006.

[8] ETSI. Digital Audio Broadcasting (DAB); MOT Slide Show; User Application Specification. ETSI Technical Specification TS 101 499 V2.1.1, January 2006.

[9] ETSI. Digital Audio Broadcasting (DAB); Multimedia Object Transfer (MOT) Protocol. ETSI European Telecommunication Standard EN 301 234 V1.4.1, June 2006.

[10] ETSI. Radio Broadcasting Systems; Digital Audio Broadcasting (DAB) to Mobile, Portable and Fixed Receivers. ETSI European Telecommunication Standard EN 300 401 V1.4.1, June 2006.

[11] ETSI. Digital Audio Broadcasting (DAB); Transport of Advanced Audio Coding (AAC) Audio. ETSI Technical Specification TS 102 563 V1.1.1, February 2007.

[12] C. Fehn, P. Kauff, S. Cho, N. Hur, and J. Kim. Asymmetric Coding of Stereoscopic Video for Transmission over T-DMB. In *Proceedings of 3-DTV-CON – Capture, Transmission, and Display of 3-D Video*, Kos Island, Greece, May 2007.

[13] K. Hari, L. Mayron, L. Christodoulou, O. Marques, and B. Furht. Design and Evaluation of 3-D Video System Based on H.264 View Coding. In *Proceedings of International Workshop on Network and Operating Systems Support for Digital Audio and Video*, Newport, RI, USA, May 2006.

[14] J. Harrold and G. J. Woodgate. Autostereoscopic Display Technology for Mobile 3-DTV Applications. In *Proceedings of SPIE Stereoscopic Displays and Applications XVIII*, San Jose, CA, USA, January 2007.

[15] W. Hoeg and T. Lauterbach. *Digital Audio Broadcasting: Principles and Applications*. John Wiley & Sons, Chichester, UK, 2003.

[16] ISO/IEC JTC 1/SC 29/WG 11. Information Technology – Generic Coding of Moving Pictures and Audio: Systems.ISO/IEC 13818-1:2000, November 2000.

[17] ISO/IEC JTC 1/SC 29/WG 11. Information Technology – Generic Coding of Moving Pictures and Audio: Video. ISO/IEC 13818-2:2000, April 2000.

[18] ISO/IEC JTC 1/SC 29/WG 11. Information Technology – Coding of Audio-Visual Objects – Part 1: Systems. ISO/IEC 14496-1:2005, December 2005.

[19] ISO/IEC JTC 1/SC 29/WG 11. Information Technology – Coding of Audio-Visual Objects – Part 10: Advanced Video Coding. ISO/IEC 14496-10:2005, December 2005.

[20] ISO/IEC JTC 1/SC 29/WG 11. Information Technology – Coding of Audio-Visual Objects – Part 2: Audio. ISO/IEC 14496-3:2005, December 2005.

[21] ITU-R. Multichannel Stereophonic Sound System With and Without Accompanying Picture. BS Series Recommendation 775-2, July 2006.

[22] B. Julesz. *Foundations of Cyclopean Perception*. University of Chicago Press, Chicago, IL, USA, 1971.

[23] D. Kim, H. Kang, and C. Ahn. A Stereoscopic Image Rendering Method for Autostereoscopic Mobile Devices.In *Proceedings of 6th International Symposium on Mobile Human-Computer Interaction*, pages 441–445, Glasgow, UK, September 2004.

[24] H. Kwon, S. Cho, N. Hur, J. Kim, and S.-I. Lee. AVC Based Stereoscopic Video Codec for 3-D DMB. In *Proceedings of 3-DTV-CON – Capture, Transmission, and Display of 3-D Video*, Kos Island, Greece, May 2007.

[25] S. Lee and D. K. Kwak. TV in Your Cell Phone: The Introduction of Digital Multimedia Broadcasting (DMB) in Korea. In *Proceedings of Annual Telecommunications Policy Research Conference*, Arlington, VA, USA, September 2005.

[26] L. M. J. Meesters, W. A. IJsselsteijn, and P. J. H. Seuntiëns. A Survey of Perceptual Evaluations and Requirements of Three-Dimensional TV. *IEEE*

*Transactions on Circuits and Systems for Video Technology*, 14(3):381–391, March 2004.

[27] P. Merkle, K. Müuller, A. Smolic, and T. Wiegand. Efficient Compression of Multi-view Video Exploiting Inter-view Dependencies Based on H.264/MPEG4-AVC. In *Proceedings of International Conference on Multimedia & Expo*, pages 1717–1720, Toronto, Canada, July 2006.

[28] J.-R. Ohm. Stereo/Multiview Video Encoding Using the MPEG Family of Standards. In *Proceedings of Stereoscopic Displays and Virtual Reality Systems VI*, pages 242–253, San Jose, CA, USA, January 1999.

[29] S. Pastoor. 3-D-Television: A Survey of Recent Research Results on Subjective Requirements.*Signal Processing: Image Communication*, 4(1):21–32, November 1991.

[30] M. G. Perkins. Data Compression of Stereo Pairs. *IEEE Transactions on Communications*, 40(4):684–696, April 1992.

[31] A. Puri, R. V. Kollarits, and G. B. Haskell. Stereoscopic Video Compression Using Temporal Scalability. In *Proceedings of Visual Communications and Image Processing*, pages 745–756, Taipei, Taiwan, May 1995.

[32] A. Segall. Study of Upsampling/Downsampling for Spatial Scalability. JVT-Q083, Nice, France, October 2005.

[33] A. Smolic, K. Müller, P. Merkle, C. Fehn, P. Kauff, P. Eisert, and T. Wiegand. 3-D Video and Free Viewpoint Video – Technologies, Applications and MPEG Standards. In *Proceedings of International Conference on Multimedia & Expo*, pages 2161–2164, Toronto, Canada, July 2006.

[34] L. B. Stelmach, W. J. Tam, D. Meegan, and A. Vincent. Stereo Image Quality: Effects of Mixed Spatio-Temporal Resolution. *IEEE Transactions on Circuits and Systems for Video Technology*, 10(2):188–193, March 2000.

[35] WorldDMB [Online], 2007. http://www.worlddab.org.

# 9

# Depth Map Generation for 3-D TV: Importance of Edge and Boundary Information

Wa James Tam, Filippo Speranza, and Liang Zhang

## 9.1 Introduction

As an ongoing and natural development, it is reasonable to regard stereoscopic three-dimensional television (3-D TV) as the next desirable milepost beyond high-definition television (HDTV). This direction of development is evidenced by the increasing number of scientific and engineering conferences held each year on 3-D TV and related technologies. At the same time, we see that industry, academia, and standard organizations have launched major research and standardization initiatives, such as the 3-D Consortium in Japan [1], the European IST research project ATTEST [2, 3] and the JVT/MVC (MPEG-3-D) standardization effort [4, 5].

Three-dimensional television is desirable not only because it can enhance perceived depth, but also because it promises to increase the feeling of presence and realism in television viewing [6, 7]. These attributes can heighten the entertainment value of the television medium, thereby attracting larger audiences and increasing revenues. Despite these potential benefits, the broadcasting industry has not really paid much attention to 3-D TV. One reason for such neglect is that 3-D technology is simply not mature enough to satisfy the critical conditions that are required to ensure the success of 3-D TV broadcasting.

W.J. Tam
Communications Research Centre Canada, 3701 Carling Avenue, Ottawa, Ontario, Canada K2H 8S2
e-mail: james.tam@crc.ca

B. Javidi et al. (eds.), *Three-Dimensional Imaging, Visualization, and Display*,
DOI 10.1007/978-0-387-79335-1_9, © Springer Science+Business Media, LLC 2009

For 3-D TV to be practical and attractive to broadcasters and users alike, several critical requirements and conditions must be met. First and foremost, broadcasters need a 3-D TV approach and technology that can be easily integrated within the existing infrastructure. This condition would be best addressed by 3-D capture, transmission and display systems that are, as much as possible, backward compatible with the existing 2-D technology. It is easy to predict that the level of compatibility will be a major determinant to the adoption rate of 3-D TV services, at least in the short-term. The full implementation of these services will take several years and considerable financial resources. Therefore, it would be unreasonable to expect the broadcasting industry to quickly switch to 3-D TV unless the cost of such change is reasonably low, particularly because HDTV services are just being introduced. In fact, it would be economically beneficial if the same coding and transmission standards, as well as newly developed television sets, could be used for both 2-D and 3-D video content.

For users, the main requirement is the quality of the 3-D TV service offered. In addition to providing enhanced depth sensation, realism, and sense of presence, the picture quality of the 3-D images must be as good as or better than that offered by HDTV images. On top of all these requirements, consumers need to enjoy 3-D TV that is unobtrusive and easy to use. This implies no cumbersome glasses, no limitations on viewing position, no restrictive room light conditions, and no visual discomfort. Last, but not least, there must be a rich supply of 3-D program material. In particular, at the rollout stage, it is critical that broadcasters have a sufficient amount of 3-D program content to attract viewers.

## 9.2 Primed for 3-D TV

Despite the obstacles mentioned above, the opportunity to establish 3-D TV has never been better. Importantly, one key ingredient is in place. The backbone for knowledge representation and communication is now mainly digital, instead of analog, and digital technologies are rapidly being deployed in television broadcasting. The digital format for communication not only allows for media integrity, but also interoperability, interchangeability, and flexibility. More concretely, a digital framework provides a way to integrate format, transmission, and reception of 3-D and 2-D signals.

In conjunction with this digital revolution, there have been significant advances in autostereoscopic display technology that have paved the way for bringing 3-D TV to the home. Although, digital 3-D cinema has been making great inroads (with encouraging box office receipts of full-length, commercial 3-D movies), it is still based on display technology that uses only two views (left eye and right eye), polarized or shutter eyewear, as well as large and expensive projection systems. These solutions are not appropriate for the home environment. Fortunately for home viewers, the last few years have been marked by the rapid development of autostereoscopic display systems that are capable of displaying multiple views of the same scene. Some of these systems are already available commercially. Autostereoscopic systems eliminate the need to wear special glasses to experience the stereoscopic effect by combining flat panel display technology with view separation techniques (i.e., lenticular and barrier technology). Importantly, in the last few years multi-view autostereoscopic displays have steadily increased in display size, picture resolution, as well as range of viewing angles and distances. Currently, autostereoscopic displays have become practical with a display size that can be as large as 42" (in diagonal length) and, significantly, with the capability to display both 2-D and 3-D images with decent picture quality and good freedom of movement for multiple viewers. Given the fast pace of technological improvements in autostereoscopic 3-D displays, it is reasonable to expect their performance and price to reach acceptable levels for broadcasting within a decade.

## 9.3 Depth Image-based Rendering

Although 3-D TV was initially envisioned within a framework in which two streams of images – one for the left eye and the other for the right eye – have to be compressed and then transmitted for stereoscopic viewing, the direction in which autostereoscopic displays is advancing has altered this framework. Current autostereoscopic displays that enable good freedom of movement without any need to wear cumbersome glasses require more than two views. Existing autostereoscopic displays use as many as nine views, and this number is probably going to increase in the future (see [8]). Thus, the capture and transmission of such a large number of views clearly pose unique technical challenges. In particular, the need to capture, to encode, and to transmit multiple views of the same scene could severely limit the successful adoption of 3-D TV because of the extra bandwidth that will be required. For this reason, as well as for targeting display systems that allow virtual navigation and free point viewing, there has been a flurry of research and standardization efforts to find efficient and practical ways for coding images from multiple camera configurations for both storage and transmission purposes [4, 5, 9].

A promising solution, that has recently received much attention, is to render novel views directly from input images consisting of standard full-color images and a corresponding set of images containing depth information [10, 11, 12, 13, 14]. In this approach, adopted from a field in computer graphics known as depth image-based rendering (DIBR) [15, 16], the three-dimensional structure of a visual scene is represented not as a classical stereo pair with a left eye view plus a right eye view, but rather as a 2-D image and its associated depth map.[1] The depth map consists of a grey scale image whose pixel values represent the depth information of corresponding pixels in the 2-D image, with a low value (darker intensity) indicating a farther distance and a higher value (lighter intensity) indicating a closer distance. Using geometrical projections, the data contained in the 2-D-plus-depth representation are processed to generate new perspective views of the scene, as if the scene had been captured from different vantage points.

At the center of the DIBR approach is the concept of 3-D image warping. Consider a parallel camera configuration with three cameras as depicted in Fig. 9.1. The central camera ($c_c$) is associated with a known center image and its corresponding depth map. The two virtual cameras located on the left and right ($c_l$ and $c_r$) are associated with two virtual images (stereoscopic pair) that must be rendered.

For this camera configuration, a point $p$ with depth $Z$ in the 3-D world (of dimensions $X$, $Y$, $Z$) is projected onto the image plane of the three cameras at pixel ($x_l$, $y$), ($x_c$, $y$) and ($x_r$, $y$), respectively. Note that the vertical coordinate ($y$) of the projection of any 3-D point on each image plane of the three cameras is the same because the cameras are parallel, so only $x_l$ and $x_r$ need to be

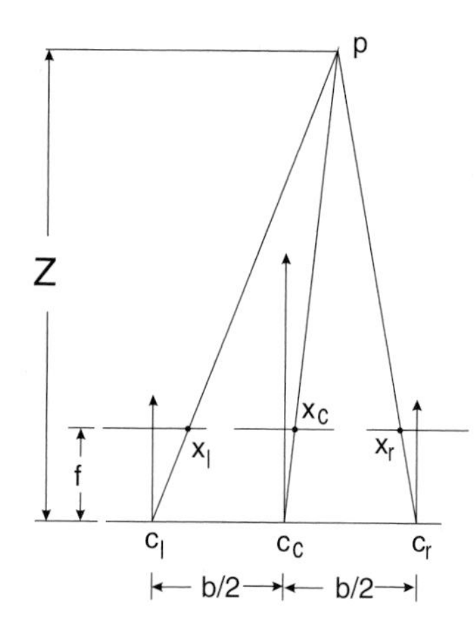


**Fig. 9.1.** Top-view of camera configuration for generation of virtual stereoscopic images from a single image and a depth map. $p$ is a point of interest in the actual scene. Given $x_c$, the position of point $p$ in an image taken from a camera located at $c_c$ and the distance (depth) $Z$, the position of $p$ ($x_l$ and $x_r$) in the virtual images viewed from cameras located at $c_l$ and $c_r$ can be calculated. The camera focal length, $f$, and the camera baseline, $b$, are required for the calculation

[1] In this chapter we do not distinguish between depth maps containing relative depth information of how far apart objects are from one another ("disparity maps") and those containing absolute depth information with respect to the camera ("range maps"). Theoretically, absolute depth information can be derived from the relative depth information if sufficient camera and capture information is available.

computed. Thus, if $f$ is the focal length of the cameras and $b$ is the baseline distance between the two virtual cameras then, from the geometry shown in Fig. 9.1, we have

$$x_l = x_c + \frac{b}{2}\frac{f}{Z}, \quad x_r = x_c - \frac{b}{2}\frac{f}{Z}, \tag{9.1}$$

where information about $x_c$ and $Z$ is available from the center image and its associated depth map, respectively. In short, given formulation (9.1) for 3-D image warping, the virtual left-view and right-view images can be generated from the original center image and its depth map by providing the value of the baseline distance $b$ and the focal length $f$.

For 3-D TV, the advantages of the 2-D-plus-depth format are many. The format is backward compatible with the current television broadcasting framework, requiring no major changes in either the coding or the transmission protocols that are currently used for standard 2-D images. Compared to multiplexing or multi-view encoding, 2-D-plus-depth is not as demanding on bandwidth. In general, a set of color images and their associated depth maps are more efficient to compress than two or more streams of color images [11]. An additional advantage is that the number of new camera views that can be rendered is not fixed (even though the rendered image quality can increasingly suffer as the viewpoint is shifted away from the original camera position). This makes the method flexible, especially given that the number of views used in multi-view autostereoscopic systems might vary between manufacturers and might increase in the future. Furthermore, this format would allow users to control rendering parameters. By varying the rendering parameters, the user could increase or decrease the perceived three-dimensionality of the scene to accommodate his/her individual preferences of depth quality and viewing comfort. Given these advantages, it is not surprising that use of the 2-D-plus-depth format for rendering images for 3-D TV is being advocated by an increasing number of researchers [10, 11, 12, 13, 14]; furthermore, this framework for 2-D-plus-depth has recently been adopted as part of the MPEG standard [4].

Admittedly, DIBR is not without its weaknesses. The most serious problem is that of disoccluded regions. In the original image these areas are hidden from view by objects in the foreground. When a different view is generated from a new viewpoint, these areas should become visible because of the new camera position. Unfortunately, there is no information either in the original image or in its associated depth map explaining how to render these disoccluded regions; if not filled, the disoccluded areas would look like empty black outlines around the foreground object. How best to fill in these regions is an active research topic [17, 18]. A possible solution is applying a Gaussian smoothing (blurring) to the depth maps before DIBR processing. Depth map smoothing can reduce and even completely eliminate the size of the disoccluded areas in DIBR, thereby circumventing the issue of filling in disoccluded areas. We

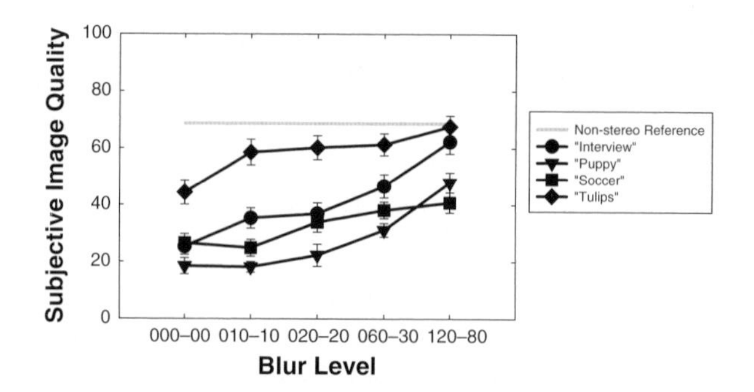


**Fig. 9.2.** Effect of depth map smoothing (*blur*) on image quality for four different stereoscopic image sequences. The x-axis shows the extent of Gaussian smoothing. The first number in each pair refers to the standard deviation and the second number refers to the size (*in pixels*) of the kernel of the filter. The larger the value the greater the level and the extent of blur. The symbols represent the mean and the error bars are standard errors of the ratings of image quality obtained from 23 viewers

have collected empirical evidence to show that depth map smoothing before DIBR can significantly improve image quality (see Fig. 9.2) [19, 20].

## 9.4 Depth Map Generation

The 2-D-plus-depth format as an underlying framework for 3-D TV will work out only if simple and effective methods for the capture or creation of depth maps are made available. The efficacy of this new aspect of content production will have a strong influence in determining the success of 3-D TV. A brief survey of the available technology and commercial solutions for producing depth maps for video indicates that there are several possible methods, with different advantages and disadvantages.

Direct methods involve the use of range finding cameras such as the Z-Cam™ developed by 3-DV Systems, Ltd. [21] and the HDTV Axi-Vision Camera from NHK Science & Technical Research Laboratories, Japan [22]. These cameras not only capture standard 2-D video of a scene, but also produce gray-level images containing the depth of each pixel or block of pixels in the standard 2-D images. The depth associated with each pixel or block of pixels in the 2-D image is obtained with a range finder mechanism that measures the time of flight (TOF) of an infrared light source. The major limitation of the current generation of range cameras is that they are not "general purpose" and are mainly suitable for indoor capture in a broadcasting studio.

An alternative method for obtaining depth maps of a visual scene is through disparity estimation based on images that have been captured using two or more camera viewpoints. For disparity estimation based on stereoscopic image pairs, many algorithms have been developed and a good overview of them can be found in [23]. Fundamentally, to estimate disparity this image processing approach computes the lateral spatial separation of corresponding features in stereoscopic image pairs. If both internal and external camera parameters, such as focal length and camera separation, are available, absolute distances of features and objects with respect to the camera position can also be derived. Current research on disparity estimation has been extended to the generation of depth maps based on multiple images captured from more than two camera positions [24]. The information gathered from more than two images could provide, at least theoretically, more accurate estimates of disparity.

Another approach for generating depth maps is inspired by the fact that the human visual system can utilize monocular visual cues to recover the depth information of a scene. Depth can be recovered based on the information provided by many monocular cues, such as familiar size, interposition, shading and shadows, linear perspective, atmospheric perspective, texture gradient, and motion parallax [25]. Thus, in principle, depth information can be recovered from almost any 2-D video image of a natural scene because the latter contains information about the 3-D geometry in the form of the abovementioned cues to depth. As a matter of fact, artists have used these so-called 2-D depth cues, alone or in combination, to generate an effective impression of depth in otherwise two-dimensional paintings and murals. In this vein, one interesting approach combines analysis of gradient and linear perspective cues with a process of image classification and "object" segmentation to extract the depth information [26, 27]. For video sequences, motion information might be used to extract the depth information of a scene [28]. There are some commercial applications (such as from Digital Dynamic Depth) which use motion information and semi-automated processes to convert existing video material into 2-D-plus-depth format [11]. However, this approach frequently requires manual intervention to optimize results, and much research is required in order to make this method more attractive.

In general, researchers have focused on achieving accurate and complete depth information about the scene. In our laboratory, we have also been investigating extraction of qualitative depth information, which might be sufficient for 3-D TV; qualitative depth should be distinguished from veridical depth that is required for critical applications such as remote manipulation.

Our approach for generating depth maps is based on the analysis of information at edges and object boundaries in the image. It originated from our search for efficient methods to extract depth information from 2-D images, which initially led us to a technique referred to in the literature as "depth from focus" [29, 30]. This technique is founded on the relationship between the depth of an object, i.e., its distance from the camera, and the amount of blur of that object in the image [31]. An object placed at a specific position that produces a sharp picture of the object in the image, on the plane of focus, will appear blurred as the object is displaced away from that position (image blur is also dependent on aperture setting and focal length of camera lens). Note, however, that the same blur extent could be observed for an object in front or behind the focus plane. Thus, the extent or level of blur in local regions of an image can be an indication, albeit an ambiguous one, of the position of the object in the scene. In extracting depth information from the blur contained in images, we noticed that the depth information is mainly located at edges and at object boundaries. This finding motivated us to investigate different methods for generating depth maps based on the processing of information at edges and object boundaries. Interestingly, the tenuous depth information contained in the resulting depth maps produces an effective depth effect, compared to the original 2-D images. The effectiveness of depth information located at edges and object boundaries might stem from the known ability of the human visual system to interpolate depth information [32, 33]. To emphasize the incomplete nature of these depth maps we have termed them "surrogate" depth maps. The rest of the chapter describes our methods and experimental findings using this approach.

## 9.5 Effect of Edge and Boundary Information

We conducted two studies to assess the viability of generating effective depth maps using information found at edges and object boundaries in 2-D source images. In the first study, we investigated different methods for generating the depth maps. One set of depth maps was generated based on the assumption that the extent of blur at local edges and object boundaries is related to the distance of the edges and boundaries from the camera. In this case, blur was estimated by implementating the Zucker-Elder technique for edge detection [34]. Since a textured region tends to be more homogenous when it is blurred than if not, a second set of depth maps was generated based on the assumption that the blur level would be proportional to the variance in intensity levels in local regions of an image. As a first approximation, we used a measure of the standard variance within a given block of pixels to obtain an estimate of the extent of "blur" in local regions within the 2-D image. Finally, we generated a third set of depth maps by applying a standard Sobel kernel to the 2-D image to localize the position of edges and object boundaries. This last set of "depth maps" was originally intended to be a reference set of depth maps because it contained information about the location of edges, but not about blur level, i.e., depth. Further details of the three methods for depth map generation are provided in the next subsections.

### 9.5.1 Local Scale Method

To construct our blur based depth maps, we used the approach proposed by Elder and Zucker to determine the best scale to utilize for correct edge detection and blur estimation [34]. In that approach, the optimal size of a filter (and, therefore, the best scale) for detecting an edge is the one that best matches the width of the edge. Consequently, the optimal size of a filter for detecting a given edge or line is also an index of the magnitude of blur for that edge or line. In the Local Scale method, there are several steps involved in generating images containing the blur information that are then used as "depth maps" for stereoscopic DIBR.

In the first step, blur is estimated by analyzing the intensity gradient at local regions of an image, looking for details and fine edges. Blurred edges manifest a shallow gradient and sharp edges a steep gradient. The optimal scale for the detection of an edge is estimated as a luminance function using a steerable Gaussian first derivative basis filter:

$$g_1^x(x, y, \sigma) = \frac{-x}{2\pi\sigma^4} e^{-\frac{(x^2+y^2)}{2\sigma^2}}$$

The minimum reliable scale, $\sigma$, to estimate the gradient for each pixel of the image is determined from a finite set of scales so as to reduce the number of computations. Once the minimum scale is found, the magnitude of the gradient estimate is assumed to be an estimate of the blur at the pixel location and is recorded as a depth value in the depth map. Specifically, the sequence of processing is:

(i) Construct the Gaussian first derivative basis filters for a set of scales $\sigma$: {16, 8, 4, 2, 1, 0.5}.

(ii) Compute the convolutions for the Gaussians and the image pixels using the first scale in the $\sigma$ set. If the magnitude of the convolution is larger than a critical value, set the depth value to that magnitude. If not, set the magnitude to 0 and repeat step (i) with the next $\sigma$ scale value. The critical value is selected such that the output contains visibly clear outlines of object boundaries, and this was generally not too difficult to determine.

(iii) Adjust the range of depth values to lie within the range {0, 255}. The resulting depth map contains depth values that mainly correspond to the location of detected edges and lines.

(iv) Expand the depth information to neighboring regions where there are no values by calculating the local maximum value within a $9 \times 9$ window and replacing pixels in the image with no depth information with this local maximum value. This is repeated for the next $9 \times 9$ window and so on until the whole image is processed.

### 9.5.2 Standard Deviation Method

A second type of depth map was created by calculating the standard deviation of luminance values of local regions in the image. We speculated that this method would provide depth maps functionally similar to those obtained with the Local Scale method. Indeed, the variability of luminance values within a local region, measured in terms of standard deviation from the mean value, is related to the extent of blur within that region. We reasoned that the intensity values of pixels in a local region would exhibit more similar luminance values if the region were blurred than if it were not blurred. Admittedly, highly textured regions can complicate matters because these regions can be blurred and still give rise to large standard deviations. Thus, like the depth obtained with the Local Scale method, these depth maps also contain potentially ambiguous information. With the Standard Deviation method, surrogate depth maps were generated by first calculating the standard deviation of the luminance values of the pixels within a $9 \times 9$ block window and then replacing all pixels within the block with the standard deviation value calculated. This was repeated for the next $9 \times 9$ block and so on until the whole image was processed. The resulting image was used as a "depth map."

### 9.5.3 Sobel Method

In the third method we created "depth maps" by applying a Sobel edge detection algorithm on the image. The Sobel method produces binary depth maps whose values do not carry any depth information, but simply signal the presence or absence of an edge (mainly object boundaries). In contrast, values obtained from blur estimation using the Local Scale method reflect potentially actual depth values at object boundaries. Thus, one can compare ratings of depth quality and image quality for the rendered images obtained with the Sobel method (edge localization) against those obtained with the Local Scale method (depth from blur plus edge localization) to better assess the roles of depth information and depth localization at object boundaries.

With the Sobel method the output image greatly depends on the threshold selected as input to the Sobel operator. In this case, we chose the "best" threshold that would provide an output that was neither too bare nor too finely textured. The thresholds for the set of original images that we used ranged from 0.04 to 0.10; the larger the value the greater the suppression of spurious edges and lines. The operation resulted in a binary value of 1 where a line is detected and 0 elsewhere. This binary image, showing object outlines, was then transformed to values of 255 (corresponding to 1) and 0 (corresponding to 0).

### 9.5.4 Subjective Assessment

We tested the validity of these methods for depth extraction from 2-D images by conducting a subjective assessment test. We asked a panel of 18 non-expert viewers to assess the image quality and depth quality of stereoscopic images that were rendered using the surrogate depth maps obtained with the three methods. Seven still images were used in the test: "Student," "Group,"

**Fig. 9.3.** Test images used in the first study. Note that the original images were in full color, but are shown as gray-level pictures here. From the top, and left to right, "Group," "Student," "Tulips," "Meal," "Cheerleaders," and "Interview"

"Interview," "Tulips," "Meal," "Cheerleaders," and "Hand." These are shown in Fig. 9.3, except for "Hand" because of copyright issues.

The rendered stereoscopic images had the original images for the left eye view and a rendered image for the right eye view, following the approach of asymmetrical coding as described in [35, 36]. To create the rendered right-view images, we first generated the depth map of the image using the three depth map creation methods (Local Scale, Standard Deviation, Sobel) described above. Next, we applied a DIBR algorithm that consisted of three steps: preprocessing the depth maps, warping the test images based on information in the depth maps, and filling disoccluded regions around object boundaries that tend to arise because of new camera viewpoints [14, 17].

The first step involved smoothing the depth maps using an asymmetrical two-dimensional Gaussian filter. The depth maps are shown in Fig. 9.4(a)–(c). Smoothing in the vertical direction was three times larger than that in the horizontal direction and was manipulated by changing the standard deviation and the window size of the filter. This pre-processing was done because it has been demonstrated that smoothing depth maps before DIBR can improve image quality of the rendered stereoscopic images by reducing the area of the disoccluded regions [19, 20]. Interestingly, after smoothing with the Gaussian blur filter the output of all three methods resulted in surrogate depth maps

(a)

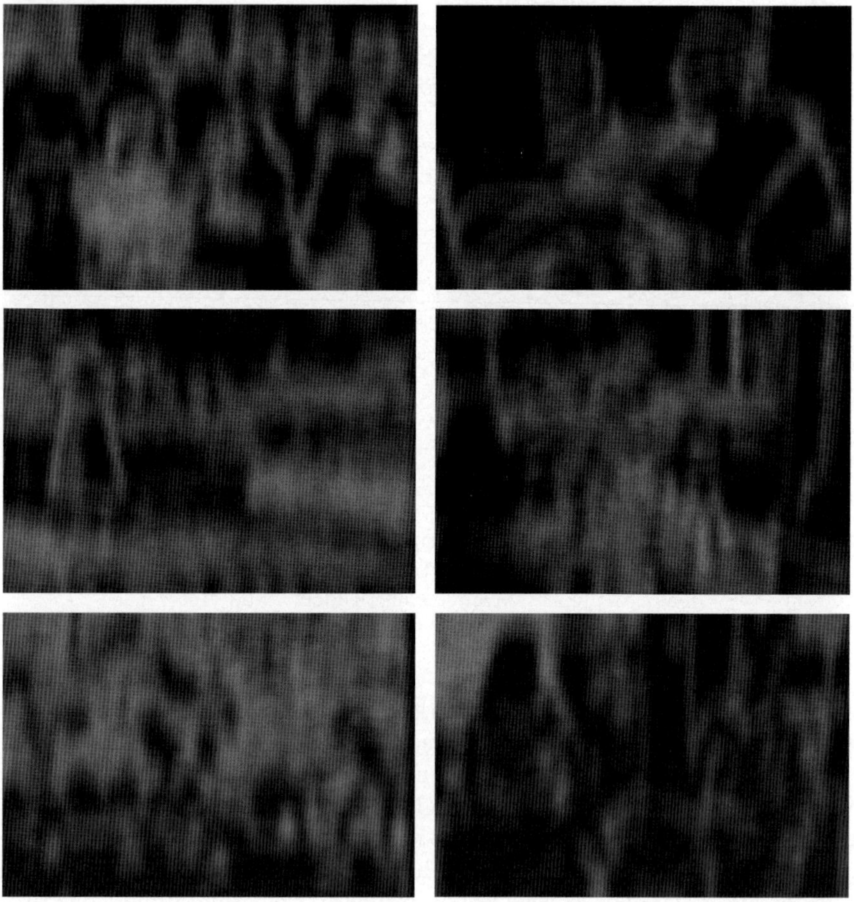

**Fig. 9.4.** Surrogate depth maps generated from the test images shown in Fig. 9.3. They were created using (**a**) the Local Scale method; (**b**) the Standard Deviation method; (**c**) the Sobel method followed by asymmetrical blurring. Please see main text for details

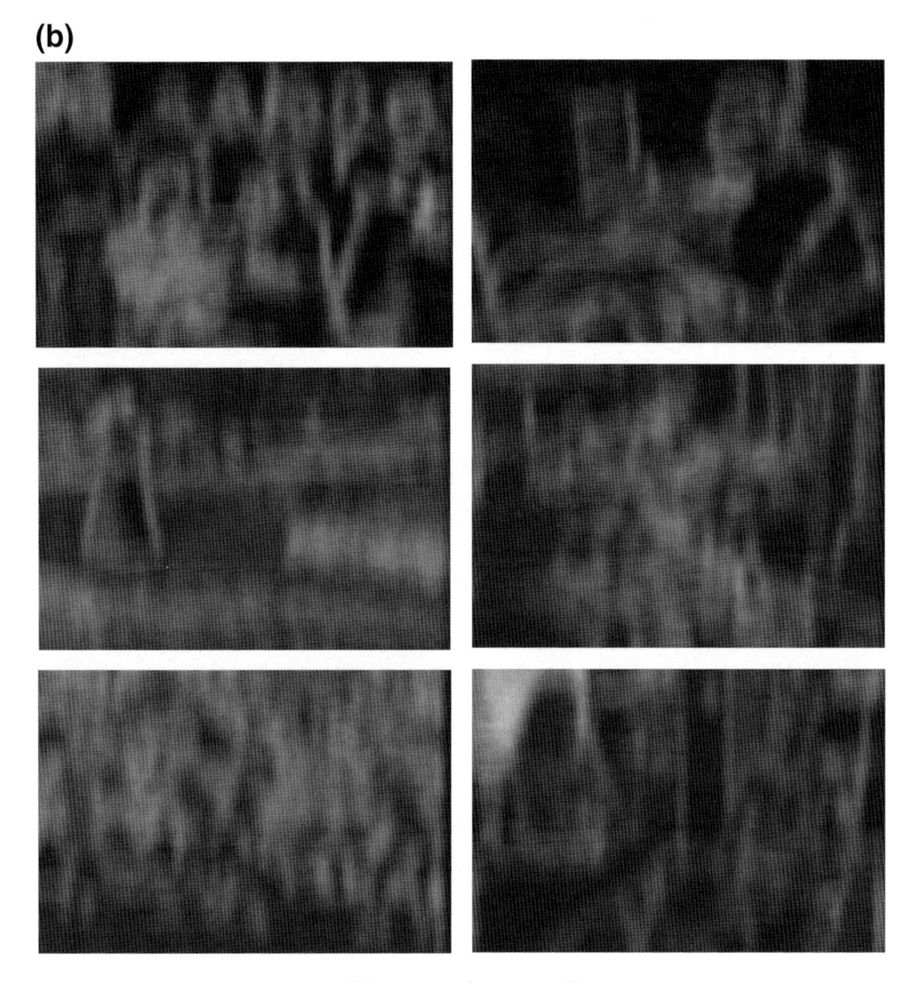


**Fig. 9.4.** (continued)

that were remarkably similar, suggesting that they might produce stereoscopic images with comparable image and depth quality. The similarity can be observed by comparing the depth maps obtained with the three methods as shown in Fig. 9.4(a)–(c). In the second step, the depth maps were used in a 3-D warping algorithm to obtain the rendered stereoscopic images. The process of 3-D warping was described in Section 9.3. The third step was carried out to eliminate disoccluded areas that have not been remedied with the smoothing process. Specifically, pixels without an assigned depth value were provided with a depth value based on the average depth of neighboring pixels.

For comparison purposes, stereoscopic images that had been captured with an actual stereoscopic camera were also assessed, but only "Tulips" and "Meal" were available for testing (Real Stereo Camera Method). As a baseline

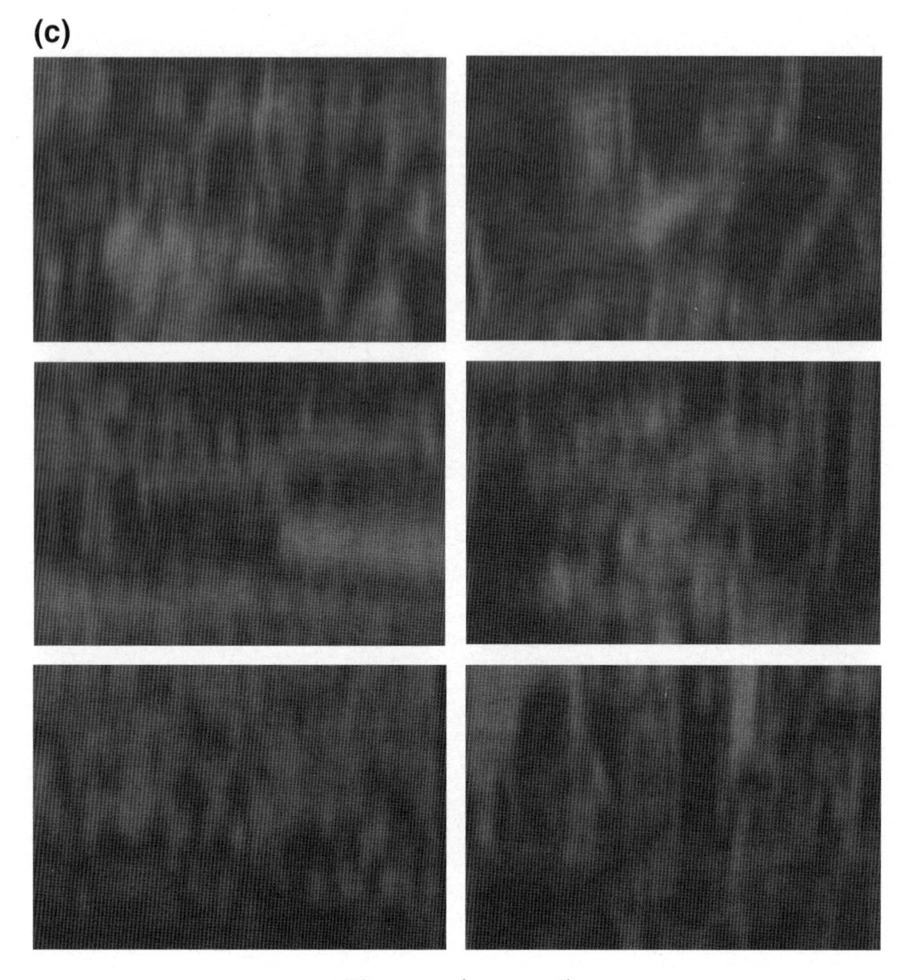


**Fig. 9.4.** (continued)

reference, viewers were asked to rate 2-D versions of the rendered stereoscopic images: in this case, the original image was presented to both the left and right eyes (non-stereoscopic view).

The image quality and depth quality of the rendered stereoscopic images were assessed in separate sessions. Each session consisted of a series of trials. The structure of a trial was identical in both sessions. Within each trial, viewers were presented with two versions of the same image. One version, the "Test" version, was a stereoscopic image, either actual or rendered, using one of the methods described above. The other version, labeled the "Reference," was a non-stereoscopic image consisting of the original test image presented to both the left and right eyes (non-stereoscopic view). In both cases, the images were displayed using a field sequential method and viewed with liquid crystal shutter eyewear (see [37] for details of the apparatus).

Viewers rated the Test and Reference images using the Double-Stimulus Continuous-Quality Scale (DSCQS) method described in ITU-R Recommendation 500 [38]. In this method, the Test and Reference images are presented successively, twice, and in a random order, i.e., Test, Reference, Test, Reference. Viewers are asked to rate both versions separately using a continuous quality scale. The scale provides a continuous rating system in the range 0 to 100, but it is divided into five equal segments, as shown in Fig. 9.5. In correspondence of these segments, five quality labels: Bad (range 00–20), Poor (range 20–40), Fair (range 40–60), Good (range 60–80), and Excellent (range 80–100), are included for general guidance. This scale was used for both image and depth quality. As is customary with the DSCQS method, the data were expressed as a difference mean opinion score (DMOS), which is simply the mean of the arithmetic differences between the ratings of the Test sequence and that of the Reference sequence obtained on each trial.

*Image Quality.* The mean ratings of image quality for the Local Scale, the Standard Deviation, and the Sobel methods were 54.5, 56.4, and 57.9, respectively. These values corresponded to the upper end of the "Fair" response category on the rating scale. The mean rating of the two actual stereoscopic images, Real Stereo Camera method, was higher at 67.3. The corresponding non-stereoscopic images received a mean rating of 62.2.

The results, obtained by averaging ratings across all images and viewers, are shown in Fig. 9.6 as DMOSs. In the figure, the DMOS represents the difference in subjective video quality between the stereoscopic Test images and their corresponding non-stereoscopic Reference versions. A difference score of 0 on the y-axis indicates no difference in subjective image quality of the

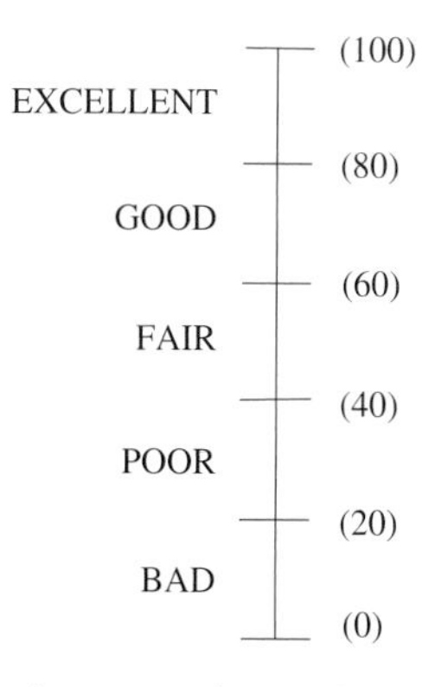


**Fig. 9.5.** Rating scale used for image quality and overall depth quality assessments. The numbers in parentheses were not printed on the actual scale, but are indicated here to suggest how the ratings were digitized to range between 0 and 100 for data analysis

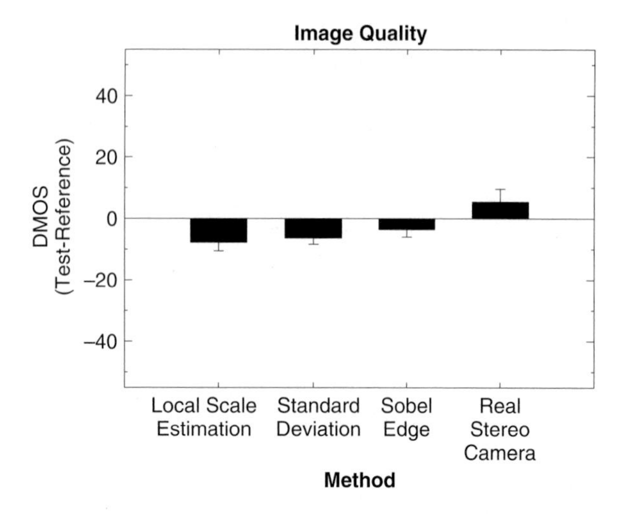


**Fig. 9.6.** Mean ratings of image quality for the three depth map generation methods: Local Scale, Standard Deviation, and Sobel. The mean rating for the Real Stereo Camera on the far right was obtained using only the camera-captured versions of the two images, "Tulips" and "Meal;" stereoscopic versions were not available for the other test images. The error bars represent the standard errors

stereoscopic images as compared to that of the corresponding non-stereoscopic images. A negative value indicates that the stereoscopic images had lower image quality than the non-stereoscopic images, whereas a positive difference indicates the opposite.

In the figure, the image quality of the actual stereoscopic images, labeled as "Real Stereo Camera," was slightly higher than that of the non-stereoscopic sequences. On the contrary, the image quality of the stereoscopic sequences generated with the three surrogate depth map methods was lower than that of the non-stereoscopic images. However, this difference was very small, less than eight units. The other important aspect of Fig. 9.6 is that it shows no practical difference between the different methods used in generating the surrogate depth maps.

Overall, the results show that even though the surrogate depth maps contained quite limited and tenuous depth information, which was extracted from blur and edge information, the image quality of the rendered stereoscopic images did not suffer as much as one might have anticipated.

*Depth quality.* The mean rating of the actual stereoscopic images (Real Stereo Camera method) was 81.3. The non-stereoscopic images received a lower mean rating of 44.3. The mean ratings for the Local Scale, Standard Deviation, and Sobel methods were similar and all located in the low "Good" category; the actual values were 62.0, 63.5, and 61.0, respectively.

The corresponding DMOS are shown in Fig. 9.7. As indicated in the figure, the depth quality of the actual stereoscopic images (see the data for the Real

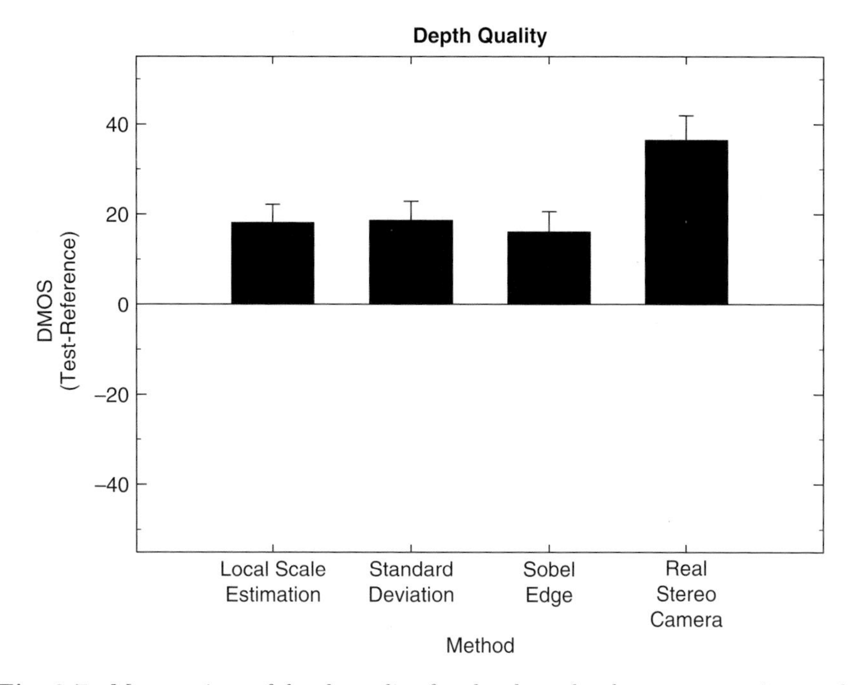


**Fig. 9.7.** Mean ratings of depth quality for the three depth map generation methods: Local Scale, Standard Deviation, and Sobel. The mean rating for the Real Stereo Camera on the far right was obtained using only the camera-captured versions of the two images, "Tulips" and "Meal;" stereoscopic versions were not available for the other test images. The error bars represent the standard errors

Stereo Camera) was substantially higher than that of the non-stereoscopic images. This result simply reflects the usual observation that stereoscopic sequences provide a better depth experience than conventional 2-D images. More important for the purposes of this investigation were the results for the stereoscopic images generated with the three surrogate depth map methods. These sequences were also judged as having better depth quality than the non-stereoscopic images, although to a lesser extent than for the actual stereoscopic sequences. The reduction in depth quality could be due to a number of possible reasons, such as inconsistent depth among different regions of a given object, inaccurate stereoscopic depth in detailed 2-D areas, etc. Nevertheless, the main finding of this study was that the limited and tenuous depth information contained in the surrogate depth maps was sufficient to generate stereoscopic sequences whose depth quality was overall "Good," and higher than that of the corresponding non-stereoscopic images.

Another interesting finding of this study was that the ratings for the stereoscopic sequences generated with the three surrogate depth map methods were statistically similar. This finding was both interesting and somewhat puzzling. The Local Scale method and the Standard Deviation method both attempt to obtain a measure of local depth by estimating blur throughout the image, but vary with respect to the complexity of the processes employed for this estimate. The depth maps obtained with these two methods provide complete, but ambiguous depth information. In contrast, the Sobel method is a very simple method based on edge localization that provides limited binary depth information.

The similarity of performance across methods indicates that surrogate depth maps can still provide an enhanced and stable sensation of depth even though the depth information they contain is neither accurate nor complete. Admittedly, the quality of depth that can be achieved with these surrogate depth maps is limited; nonetheless, these maps can be used in applications where depth accuracy is not critical.

## 9.6 Role of Edge and Boundary Information

To better understand the role of depth information at edges and object boundaries, we conducted a second study in which we varied the extent and accuracy of information available at those locations. A panel of viewers assessed the image and the depth quality of six stereoscopic still images: "Aqua," "Cones," "Meal," "Red Leaves," "Tulips," and "Xmas." The left eye views are shown in Fig. 9.8. In these stereoscopic images, the left eye view was always the original image and the right eye view was an image rendered with a depth map created using one of four different methods: Full-Depth, Boundary-Depth, Edge-Depth, and Uniform-Depth methods.

The first three experimental methods, which constituted the focus of this experiment, were characterized by depth maps that differed in the degree and/or accuracy of depth information that they contained. The first method (Full-Depth) was used to generate "full" depth maps. For all images but "Cones," the depth information was obtained through disparity estimation based on a stereoscopic pair of images that were captured with an actual camera. The parameters of the disparity estimator had been chosen to maximize accuracy of estimation. The black areas in these estimated depth maps were either occluded areas or unresolved estimation areas without available disparity values. The depth map for the image "Cones" was obtained from a resource website and was created using a structured lighting technique [39]. The second method (Boundary-Depth) was used to generate depth maps whose depth information was obtained from the previous "full" depth maps, but restricted to edges and boundaries of objects that were in the source images. Thus, these depth maps contained the same depth information as in the Full-Depth maps, but only at object boundaries and edges. Pixels not at the boundaries and edges were given a value of 0. The third method (Edge-Depth map) was used to generate depth maps based on the Sobel method of edge localization described in Section 9.5.3; as noted, the depth generated by this method contained only binary depth information at the location of object outlines and

**Fig. 9.8.** The left eye image from the six stereoscopic images that were used in the second study. From the top, and from left to right, the names for the images are "Aqua," "Cones," "Meal," "Red Leaves," "Tulips," and "Xmas." The images are shown as gray-level pictures here, but the originals that were used in the study were in full color

edges. The depth maps for these three methods are shown in Figs. 9.9, 9.10 and 9.11. The fourth method (Uniform-Depth) was used to generate depth maps with a constant depth value. With these Uniform-Depth maps, the rendered images of the right eye view were identical to those of the left eye view except for a horizontal shift of the image. When viewed stereoscopically, the whole scene appeared to shift in depth behind the display screen. The extent of the shift in depth was comparable to that obtained for the other rendered images.

**Fig. 9.9.** "Full-Depth" maps of the images shown in Fig. 9.8. Except for "Cones," the depth maps were created using a disparity estimation technique based on the original stereoscopic images. The depth map for "Cones" was obtained using a structured lighting technique. The lightness of the pixels indicates the relative distance from the camera, with the lightest shade corresponding to the closest distance. Note that these maps were smoothed before depth image-based rendering

The rendered stereoscopic images were obtained using the three step DIBR algorithm: Gaussian smoothing of the depth maps, 3-D warping, and disocclusion post-processing, which was described in Section 9.5.4.

For comparison, viewers were also asked to rate the image and the depth quality of a version of the stereoscopic images in which the left eye and right eye views were captured using an actual stereoscopic camera (Real Stereo Camera method). Finally, as a baseline reference, viewers were also asked to rate 2-D versions of the stereoscopic images; this non-stereoscopic image was created by presenting the left eye view to both eyes.

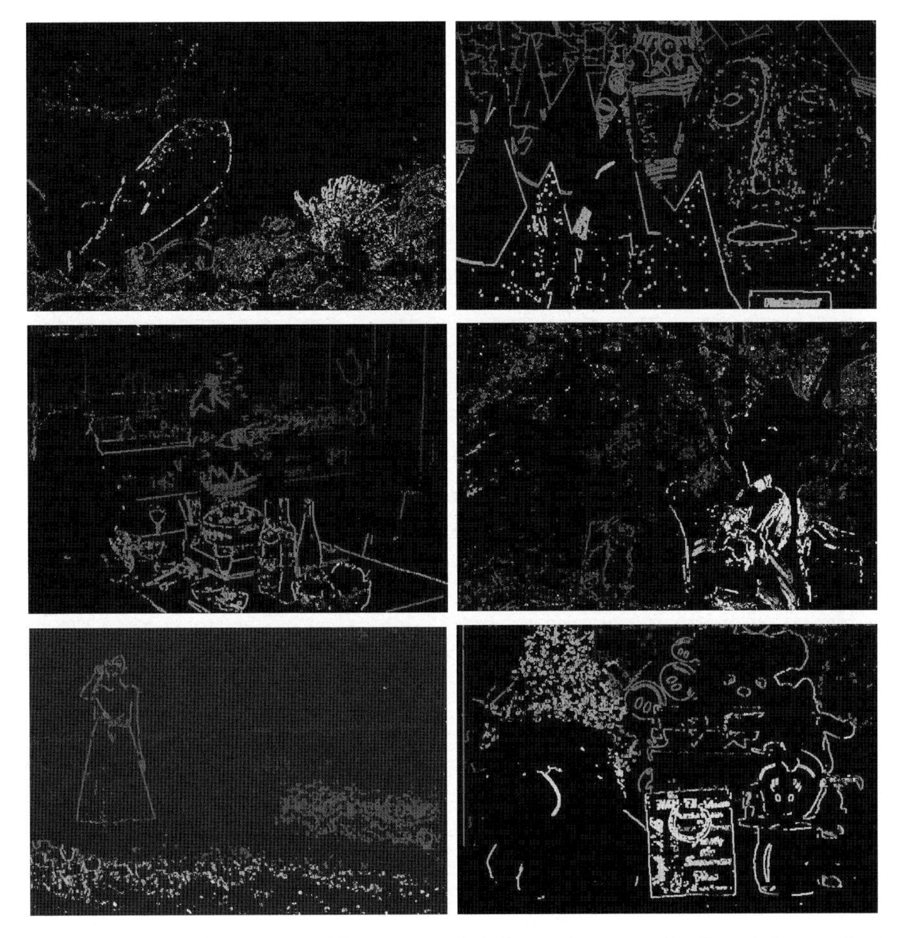

**Fig. 9.10.** "Boundary-Depth" maps with information contained mainly at edges and object boundaries. These Boundary-Depth maps were created by removing all depth information contained in the depth maps shown in Fig. 9.9 except at edges and object boundaries. Note that these maps were smoothed before depth image-based rendering

### 9.6.1 Subjective Assessment

Seventeen non-expert viewers assessed the image and depth quality of the stereoscopic images with the subjective assessment methodology described in Section 9.5.4 above. Thus, within the same trial, the viewers were presented with two versions of the same image: the "Test" version was a stereoscopic image obtained with one of the described methods, whereas the "Reference" version was a non-stereoscopic image. The images were displayed using a field sequential method and viewed with liquid crystal shutter eyewear, as described in [40].

**Fig. 9.11.** "Edge-Depth" maps obtained with the Sobel method. Depth information is limited to the edges and object boundaries. Note that these maps were smoothed before actual depth image-based rendering

*Image quality.* The mean rating of the non-stereoscopic Reference was 62.6; the mean rating of the Real Stereo Camera method was 63.3. These ratings corresponded to "Good" on the rating scale. Mean ratings for the Full-Depth, Boundary-Depth, Edge-Depth, and Uniform-Depth method were slightly lower at 57.7, 56.8, 58.3, and 57.7, respectively, and corresponded to the higher end of "Fair" on the rating scale.

Figure 9.12 shows the DMOS. For the Real Stereo Camera method, the DMOS is very close to the zero line indicating that the actual stereoscopic images were rated as having virtually the same image quality as the non-stereoscopic images. In the figure, the DMOS corresponding to the four depth map creation methods of Full-Depth, Edge-Depth, Boundary-Depth

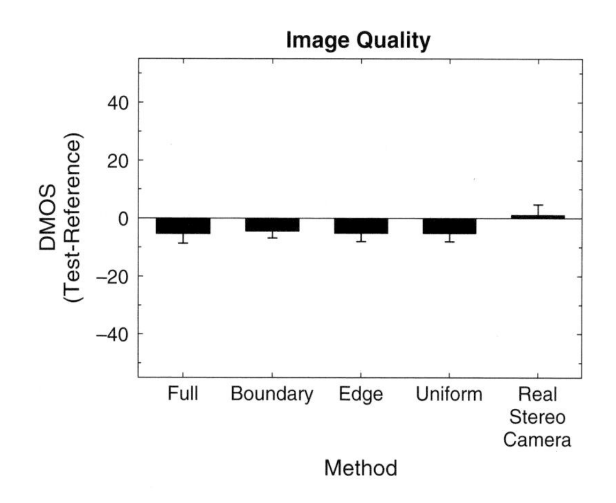


**Fig. 9.12.** Mean ratings of image quality for the four depth map generation methods: Full, Boundary, Edge, and Uniform. The mean rating for the Real Stereo Camera on the far right was obtained using the camera-captured versions of the images. The error bars represent the standard errors

and Uniform-Depth, are negative. This indicates that the image quality ratings for the rendered stereoscopic images were marginally lower than the ratings for the non-stereoscopic images, with a difference score of under five units. Interestingly, the image quality rating for the Uniform-Depth method was also slightly lower than that of the non-stereoscopic condition. This was unexpected since most of the original 2-D image would have been intact except for a narrow column at the left border of the image that had to be "filled" because of the horizontal shift in virtual camera perspective (in the right eye view). This finding suggests that the DIBR algorithm used in this study introduced a degradation in image quality in the rendered right eye view. This degradation might have occurred because of interpolation errors that could have been introduced when the required shift was not an integer pixel value.

*Depth quality.* For the non-stereoscopic Reference, the mean rating of depth quality was 42.0, corresponding to the low end of "Fair" on the rating scale. The mean rating for the actual stereoscopic images, i.e., the Real Stereo Camera method, was 81.5, corresponding to "Excellent." The mean ratings for the Full-Depth, Boundary-Depth, Edge-Depth, and Uniform-Depth methods were 61.8, 57.7, 55.0, and 52.4, respectively. These ratings corresponded to the high end of "Fair" on the rating scale.

The DMOS results for depth quality in Fig. 9.13 show that the DMOSs were all positive, indicating that depth quality was higher for the stereoscopic Test images than for the Reference non-stereoscopic images in all experimental conditions. The highest positive difference in ratings, i.e., the largest DMOS, was observed for the images in the Real Stereo Camera method. This was also expected because these images were genuine stereoscopic images.

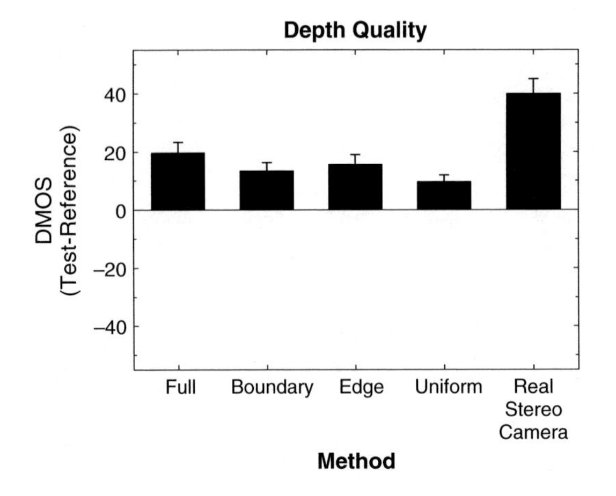


**Fig. 9.13.** Mean ratings of depth quality for the four depth map generation methods: Full, Boundary, Edge, and Uniform. The mean rating for the Real Stereo Camera on the far right was obtained using the camera-captured versions of the images. The error bars represent the standard errors

Of the four experimental methods used to generate rendered images, the lowest difference in ratings, i.e., the smallest DMOS, was observed for the "Uniform-Depth" method. The difference is small, about 10 units, but statistically different from 0. This difference indicates that the images processed with the "Uniform-Depth" method still provided a depth quality higher than that of the non-stereoscopic images. This might be expected because, although the images were inherently flat (i.e., 2-D) they were recessed behind the screen plane. In other words, these images had some form of depth and the depth ratings for this condition probably reflected the presence of such depth.

The depth maps that produced the highest DMOS difference in depth rating were from the Full-Depth method. Recall that the Full-Depth map contained all the depth information acquired through disparity estimation of an actual stereoscopic image pair. As one might expect, images that were generated with a Boundary-Depth map did not produce as high a difference in rating; however, this value is only seven units lower than that in the Full-Depth map condition. Importantly, the condition in which the images were rendered with an Edge-Depth map produced the second highest difference in depth ratings for the depth map based images. Recall that with the Edge-Depth method, actual depth information was not available because the depth maps contained only binary information at object boundaries and edges. Notwithstanding this limited depth information, the difference between this condition and the Full-Depth condition is relatively small at five units.

In summary, the results of the present study are consistent with those reported in Section 9.5 and confirm that enhanced depth can be attained with very limited depth information at object boundaries and edges.

## 9.7 Conclusions and Future Research

Our findings indicate that a robust stereoscopic depth effect can be achieved simply by locating edges and object boundaries in an image and using the resulting surrogate depth map for stereoscopic DIBR. Furthermore, results from both studies suggest that this effect can be obtained even when the depth information contained in the surrogate depth maps is largely inaccurate and/or incomplete. This observation suggests that the human visual system can somehow still use very limited and tenuous stereoscopic depth information to arrive at a relatively stable and enhanced perception of depth. It is possible that stereoscopic cues are integrated with other depth cues, i.e., occlusion that is typically available in natural images, such that any inaccuracy is suppressed so as to arrive at an overall stable and comprehensible view of the scene. This interpretation is consistent with studies pertaining to cue combination, depth interpolation, and surface or boundary completion in the perception literature [32, 33, 41, 42, 43, 44, 45]. Recently we have begun an investigation to examine whether this process of integrating deficient information is an effortful one, and whether it might negatively affect visual comfort.

The results reported here indicate that there is still much room for improvement in the generation of surrogate depth maps. The depth effect that can be achieved with surrogate depth maps is relatively modest, and it is less than half of that obtained with an actual stereoscopic camera setup. Although one would not expect the depth quality obtained with a surrogate depth map and DIBR to match that of an authentic stereoscopic image, a 75 percent target level would be a desirable and practical goal. More research is needed to determine how degradation in depth quality originates for stereoscopic images rendered using surrogate depth maps. In part some of this loss could be due to the rendering process itself. Notice that in the second study the Full-Depth maps produced reduced depth ratings, even though they contained more accurate and complete depth information than the surrogate depth maps. This pattern of results suggests that the DIBR process requires improvement to further enhance perceived depth obtained from depth maps.

The reported findings were obtained with still images. However, we have conducted pilot studies whose results indicate that the method for the generation of surrogate depth maps is amenable for video. In particular, we have found that the asymmetrical Gaussian smoothing of depth maps, a pre-process incorporated into our DIBR method, eliminates potential inconsistencies of local depth between frames. Thus, depending on the strength of smoothing, there is no noticeable depth instability when a rendered stereoscopic video clip is played out.

Our approach is intended for the extraction of qualitative depth. This term is used to refer to a depth effect arising from depth information that is not necessarily veridical, quantitative or complete. We have shown that despite its inaccuracy, this type of depth information can still enhance the depth quality of 3-D images. Although much improvement is required, this approach could potentially be used to quickly and efficiently convert 2-D material to 3-D material. An efficient 2-D to 3-D conversion technology could play a relevant role in the future of 3-D TV, particularly in the initial stage of 3-D TV rollout when 3-D program material will be relatively scarce and in great demand.

**Acknowledgements** The authors wish to thank Ron Renaud for his technical assistance in setting up the experimental apparatus and for helping with the data collection for the studies. We also would like to thank co-op students Guillaume Alain, Jorge Ferreira, Sharjeel Tariq, Anthony Soung Yee and Jason Chan Wai for developing software tools and algorithms for the experiments. Results of the first study were presented at the conference on Three-Dimensional Displays and Applications, San Jose, CA., USA, January 2005, and those for the second were presented at the conference on Three-Dimensional Television, Video and Display, Philadelphia, PA., USA, October 2005. Thanks are also due to Carlos Vázquez, Xianbin Wang, and André Vincent for their suggestions and comments on an earlier version of the manuscript. Finally, "Interview" was provided by HHI (Germany) while the images, "Aqua," "Meal," "Red Leaves," and "Tulips," were produced by NHK (Japan). "Cones" is a cropped image of an original that was obtained from the Stereo Vision Research Page maintained by D. Scharstein and R. Szeliski at http://http://cat.middlebury.edu/stereo (accessed June 2007).

## References

[1] 3D Consortium. [June 20, 2007 available on-line http://www.3-Dc.gr.jp/english/]
[2] European IST-2001-34396 project: ATTEST, 2002–2004. [June 20, 2007 available on-line http://www.extra.research.philips.com/euprojects/attest/]
[3] C. Fehn, "A 3D-TV system based on video plus depth information," 37th Asilomar Conference on Signals, Systems, and Computers, Vol. 2, pp. 1529–1533, 2003.
[4] ISO/IEC JTC1/SC29/WG11, "Information technology – MPEG Video Technologies – Part 3: Representation of auxiliary video and supplemental information," ISO/IEC FDIS 23002-3, January 19, 2007.
[5] A. Smolic & H. Kimata, "Report on 3DAV exploration," ISO/IEC JTC1/SC29/WG11 Doc N5878, July 2003.
[6] S. Yano & I. Yuyama, "Stereoscopic HDTV: Experimental system and psychological effects," Journal of the SMPTE, Vol. 100, pp.14–18, 1991.
[7] W. A. IJsselsteijn, H. de Ridder, R. Hamberg, D. Bouwhuis, & J. Freeman, "Perceived depth and the feeling of presence in 3DTV," Displays, Vol. 18, pp. 207–214, 1998.
[8] J. Ouchi, H. Kamei, K. Kikuta, & Y. Takaki, "Development of 128-directional 3D display system," Three-Dimensional TV, Video, and Display V, Vol. 6392, pp. 63920I, 2006.
[9] P. Merkle, K. Müller, A. Smolic, & T. Wiegand, "Efficient compression of multi-view video exploiting inter-view dependencies based on H.264/MPEG4-AVC,"

IEEE International Conference on Multimedia and Exposition (ICME2006), pp. 1717–1720, 2006.
[10] K. T. Kim, M. Siegel, & J. Y. Son, "Synthesis of a high-resolution 3D stereoscopic image pair from a high-resolution monoscopic image and a low-resolution depth map," Stereoscopic Displays and Applications IX, Vol. 3295A, pp. 76–86, 1998.
[11] J. Flack, P. Harman, & S. Fox, "Low bandwidth stereoscopic image encoding and transmission, " Stereoscopic Displays and Virtual Reality Systems X, Vol. 5006, pp. 206–214, 2003.
[12] C. Fehn, "A 3D-TV approach using depth-image-based rendering (DIBR)," Visualization, Imaging, and Image Processing (VIIP'03), pp. 482–487, 2003.
[13] C. Fehn, "Depth-image-based rendering (DIBR), compression and transmission for a new approach on 3D-TV", Stereoscopic Displays and Virtual Reality Systems XI, Vol. 5291, pp. 93–104, 2004.
[14] L. Zhang & W. J. Tam, "Stereoscopic image generation based on depth images for 3D TV," IEEE Transactions on Broadcasting, Vol. 51, pp. 191–199, 2005.
[15] L. McMillan, "An image-based approach to three-dimensional computer graphics," Ph.D. thesis, University of North Carolina at Chapel Hill, 1997. [June 20, 2007 available on-line http://ftp://ftp.cs.unc.edu/pub/publications/techreports/97-013.pdf]
[16] W. R. Mark, "Post-rendering 3D image warping: Visibility, reconstruction and performance for depth-image warping," Ph.D. thesis, University of North Carolina at Chapel Hill, 1999. [June 20, 2007 available on-line: http://www-csl.csres.utexas.edu/users/billmark/papers/dissertation/TR99-022.pdf]
[17] L. Zhang, W. J. Tam, & D. Wang, "Stereoscopic image generation based on depth images," IEEE Conference on Image Processing, Vol. 5, pp. 2993–2996, 2004.
[18] C. Vázquez, W. J. Tam, & F. Speranza, "Stereoscopic imaging: Filling disoccluded areas in image-based rendering," Three-Dimensional TV, Video, and Display (ITCOM2006), Vol. 6392, pp. 0D1–0D12, 2006.
[19] W. J. Tam, G. Alain, L. Zhang, T. Martin, & R. Renaud, "Smoothing depth maps for improved stereoscopic image quality," Three-Dimensional TV, Video and Display III (ITCOM'04), Vol. 5599, pp.162–172, 2004.
[20] W. J. Tam & L. Zhang, "Non-uniform smoothing of depth maps before image-based rendering," Three-Dimensional TV, Video and Display III (ITCOM'04), Vol. 5599, pp. 173–183, 2004.
[21] G. Iddan & G. Yahav, "3D imaging in the studio," Videometrics and Optical Methods for 3-D Shape Measurement, Vol. 4298, pp. 48–55, 2001.
[22] M. Kawakita, K. Iizuka, H. Nakamura, I. Mizuno, T. Kurita, T. Aida, Y. Yamanouchi, H. Mitsumine, T. Fukaya, H.Kikuchi, & F. Sato, "High-definition real-time depth-mapping TV camera: HDTV Axi-Vision camera," Optics Express, Vol. 12, pp. 2781–2794, 2004.
[23] D. Scharstein & R. Szeliski, "A taxonomy and evaluation of dense two-frame stereo correspondence algorithms," International Journal of Computer Vision, Vol. 47, pp.7–42, 2002.
[24] L. Zhang, W. J. Tam, G. Um, F. Speranza, N. Hur, & A. Vincent, "Virtual view generation based on multiple images," IEEE Conference on Multimedia & Expo (ICME), Beijing, July, 2007.

[25] M. W. Eysenck, "Visual perception." In *Psychology: An international perspective*, Chapter 7, pp.218–259, London: Psychology Press, 2004.
[26] S. Battiato, S. Curti, E. Scordato, M. Tortora, & M. La Cascia, "Depth map generation by image classification," Three-Dimensional Image Capture and Applications VI, Vol. 5302, pp. 95–104, 2004.
[27] S. Battiato, A. Capra, S. Curti, & M. La Cascia, "3-D stereoscopic image pairs by depth-map generation," Second International Symposium on 3-D Data Processing, Visualization and Transmission, pp. 124–131, 2004.
[28] L. Zhang, B. Lawrence, D. Wang, & A. Vincent, "Comparison study on feature matching and block matching for automatic 2-D to 3-D video conversion," The 2nd IEE European Conference on Visual Media Production, pp. 122–129, 2005.
[29] J. Ens & P. Lawrence, "An investigation of methods for determining depth from focus," IEEE Transactions on Pattern Analysis and Machine Intelligence Vol. 15, pp. 97–108, 1993.
[30] N. Asada, H. Fujiwara, & T. Matsuyama, "Edge and depth from focus," The 1st Asian Conference on Computer Vision, pp. 83–86, 1993.
[31] A. P. Pentland, "Depth of scene from depth of field," The Image Understanding Workshop, pp. 253–259, 1982.
[32] G. J. Mitchison & S. P. McKee, "Interpolation in stereoscopic matching," Nature, Vol. 315, pp. 402–404, 1985.
[33] S. M. Wurger & M. S. Landy, "Depth interpolation with sparse disparity cues," Perception, Vol.18, pp. 39–54, 1989.
[34] J. H. Elder & S. W. Zucker, "Local scale control for edge detection and blur estimation," IEEE Transactions on Pattern Analysis and Machine Intelligence, Vol. 20, pp. 699–716, 1998.
[35] L. B. Stelmach &W. J. Tam, "Stereoscopic image coding: Effect of disparate image-quality in left- and right-eye views," Signal Processing: Image Communication, Special Issue on 3-D Video Technology, Vol.14, pp.111–117, 1998.
[36] W. J. Tam, L. B. Stelmach, D. Meegan, & A.Vincent, "Bandwidth reduction for stereoscopic video signals," Proceedings of the SPIE: Stereoscopic Displays and Virtual Reality Systems VII, Vol. 3957, pp. 33–40, 2000.
[37] W. J. Tam, A. Soung Yee, J. Ferreira, S. Tariq, & F. Speranza, "Stereoscopic image rendering based on depth maps created from blur and edge information," Stereoscopic Displays and Applications XII, Vol. 5664, pp.104–115, 2005.
[38] ITU-R Recommendation BT.500-11, "Methodology for the subjective assessment of the quality of television pictures," 2005.
[39] D. Scharstein & R. Szeliski, "High-accuracy stereo depth maps using structured light," IEEE Computer Society Conference on Computer Vision and Pattern Recognition (CVPR'03), Vol. 1, pp. 195–202, 2003.
[40] W. J. Tam, F. Speranza, L. Zhang, R. Renaud, J. Chan, & C. Vazquez, "Depth image based rendering for multiview stereoscopic displays: Role of information at object boundaries," Three-Dimensional TV, Video, and Display IV, Vol. 6016, p. 97–107, 2005.
[41] M. Ichikawa, S. Saida, A. Osa, & K. Munechika, "Integration of binocular disparity and monocular cues at near threshold level," Vision Research, Vol. 43, pp. 2439–2449, 2003.
[42] E. B. Johnston, B. G. Cumming, & A. J. Parker, "Integration of depth modules: stereopsis and texture," Vision Research, Vol. 33, pp. 813–826, 1993.

[43] S. Grossberg & E. Mingolla, "Neural dynamics of form perception: Boundary completion, illusory figures, and neon color spreading," Psychological Review, Vol. 92, pp. 173–211, 1985.
[44] P. J. Kellman & T. F. Shipley, "A theory of visual interpolation in object perception," Cognitive Psychology, Vol. 23, pp. 141–221, 1991.
[45] C. Yin, P. J. Kellman, & T. F. Shipley, "Surface completion complements boundary interpolation in the visual integration of partly occluded objects," Perception, Vol. 26, pp. 1459–1479, 1997.

# 10

# Large Stereoscopic LED Display by Use of a Parallax Barrier

Hirotsugu Yamamoto, Yoshio Hayasaki, and Nobuo Nishida

**Abstract** Since the development of high-brightness blue and green LEDs, the use of outdoor commercial LED displays has been increasing. Because of their high-brightness, good visibility, and long-term durability, LED displays are a preferred technology for outdoor installations such as stadiums, street advertising, and billboards. This chapter deals with a large stereoscopic full-color LED display using a parallax barrier. We discuss optimization of viewing areas, which depend on LED arrangements. Enlarged viewing areas have been demonstrated by using a three-in-one chip LED panel that has wider black regions than ordinary LED lamp cluster panels. We have developed a real-time system to measure a viewer's position and investigated the movements of viewers who watch different designs of stereoscopic LED displays, including conventional designs and designs to eliminate pseudoscopic viewing areas. The design of parallax barrier for plural viewers was utilized for a 140-inch stereoscopic LED display.

**Keywords:** LED (Light Emitting Diode), stereoscopic display, parallax barrier, viewing area

---

H. Yamamoto
Department of Optical Science and Technology, The University of Tokushima, 2-1 Minamijosanjima, Tokushima 770-8506, Japan
e-mail: yamamoto@opt.tokushima-u.ac.jp

B. Javidi et al. (eds.), *Three-Dimensional Imaging, Visualization, and Display*,
DOI 10.1007/978-0-387-79335-1_10, 

## 10.1 Introduction

Recently, many viewers enjoyed watching images of big events such as the Olympic Games, World Cup soccer, and the Super Bowl on large LED screen televisions. The number of large screen televisions that use full-color LED panels have been increasing after the development of high-brightness blue LED in 1993 and green LED in 1995 by NICHIA Corporation [1]. Because of LED's high brightness, good visibility, and long-term durability, LED displays are a preferred technology to realize the dream of the large 3-D display. This chapter deals with the large stereoscopic LED display for outdoor use by the general public. In the past large projection-type stereoscopic displays have only been used in indoor venues at exhibitions and theaters. The use of LED panels enables extra large outdoor stereoscopic displays for the general public. Stereoscopic displays using full-color LED panels need image separating, i.e., a parallax barrier, a lenticular sheet, polarizing eyewear, or eyewear consisting of liquid crystal shutters synchronized to the display field rate. Experiments using polarizing eyewear have shown the feasibility of large stereoscopic LED displays [2].

For use by the general public it is preferable that stereoscopic images be viewed without any special glasses. A parallax barrier is composed of slits that separates interleaved stereoscopic images [3, 4, 5]. This composition has scalability and is considered to be suitable for implementing a large stereoscopic display using a full-color LED panel. In order to increase the number of viewers, it is necessary to enlarge the viewing areas. The viewing areas of stereoscopic displays using parallax barriers are restricted by cross talk and pixels disappearing when the viewer's eye position deviates from the designed position. Using black regions between LED pixels allows us to enlarge the viewing areas. We have analyzed the relationships between the viewing areas and the pixel and parallax barrier arrangement, and we have discussed optimization to enlarge the viewing area [6, 7]. The possibility of many viewers seeing such a stereoscopic display has been shown in numerical examples [6]. The analysis suggests that the arrangement of the LED panel determines the viewing area. Experiments using different types of full-color LED panels are necessary to confirm the analysis of the viewing area of stereoscopic full-color LED displays using parallax barriers [7].

In practice, a viewer's interpupillary distance is not always equal to the designed distance. The difference of interpupillary distances may also restrict the viewing area. However, enlarging this area can provide viewers with different interpupillary distances with a stereoscopic image. We have investigated a stereoscopic full-color LED display using parallax barriers for different interpupillary distances [8]. Parallax barriers are determined by interpupillary distances and their viewing areas are formulated. Experiments on distance perception of stereoscopic targets have been conducted utilizing the developed stereoscopic LED displays which employ parallax barriers. For comparison, distance perception is examined using a full-color LED display and polarizing eyewear. Enlargement of the viewing area is shown to allow several viewers whose interpupillary distances differ to be able to see with stereoscopic perception.

Although the developed stereoscopic display enables viewers to view the stereoscopic images without any special glasses, viewers need to stand within the viewing areas. In order to guide the viewers to the designed viewing areas by use of viewed images, pseudoscopic viewing areas are elminiated by introducing viewing areas of a black image [9]. These designs should help the viewers to move to the viewing areas. For the sake of evaluating the different viewing areas that are realized by use of different designs of parallax barriers and LED arrangements, it is important to record the movements of the viewer who comes to see a stereoscopic image. We have developed a real-time system to measure a viewer's position and investigated the movements of viewers who watch different designs of stereoscopic LED displays. We have evaluated the performances of different viewing areas that have different designs of parallax barriers, including conventional designs to provide multiple perspective images and designs to eliminate pseudoscopic viewing areas [9, 10].

In order to show real world images on the developed stereoscopic display, it is necessary to capture stereo images and process them in real-time. We have developed an active binocular camera and demonstrated real-time display of real world images on a stereoscopic full-color LED display by using a parallax barrier. Convergence of the binocular camera has been controlled so that both cameras tracked the target object [11, 12].

The purpose of this chapter is to describe the large stereoscopic LED display by using a parallax barrier. Enlargement of the viewing area and dependence of perceived distances for different parallax barrier designs are shown in Sections 10.2 and 10.3. Elimination of pseudoscopic viewing areas is illustrated in Section 10.4 and is experimentally evaluated based on traces of viewers' positions in Section 10.5. Finally, we describe future works and summary.

## 10.2 Enlargement of Viewing Areas of Stereoscopic LED Display by Use of a Parallax Barrier

For use by the general public, it is preferable that stereoscopic images be viewed without any special glasses. A parallax barrier is considered to be suitable for implementing a large stereoscopic display using a full-color LED panel. The principle of stereoscopic LED display by using a parallax barrier is shown in Fig. 10.1. An array of slits, called parallax barrier, is located in front of an LED panel. Through the apertures of the parallax barrier, both eyes view slightly different areas on the LED panel. When viewed at the designed viewing points, the column sequentially displayed stereoscopic

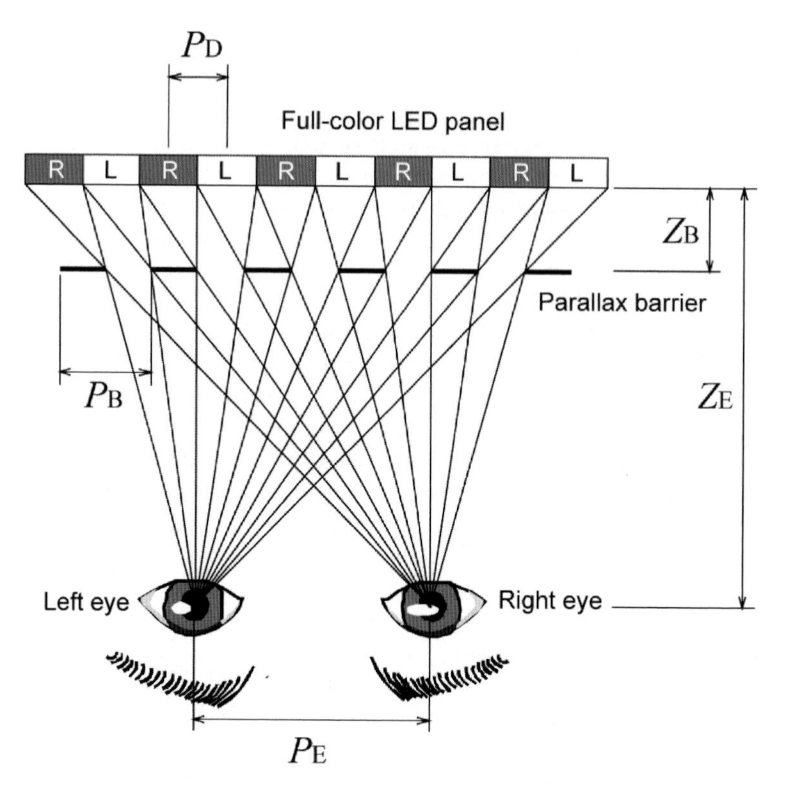


**Fig. 10.1.** Design of parallax barrier

images are separated into the corresponding eyes without any special eyewear. The viewing distance $Z_E$ is given by:

$$Z_E = \frac{2P_D Z_B}{2P_D - P_B}, \tag{10.1}$$

where $Z_{\mathrm{B}}$ denotes the distance of the parallax barrier from the LED panel. $P_{\mathrm{D}}$ and $P_{\mathrm{B}}$ are the pitch of the LED panel and the parallax barrier, respectively. The distance between the right and left viewing points must be equal to the interpupillary distance, which is indicated by $P_E$. Then, the relation of

$$P_B = \frac{2P_E P_D}{P_E + P_D} \tag{10.2}$$

is obtained. The pitch of the parallax barrier is determined by the designed interpupillary distance and the pitch of LED display.

The viewing area of stereoscopic displays using a parallax barrier is restricted by cross talk or pixel disappearance according to deviations of the viewer's eye position from the position designed as shown in Fig. 10.2(a)

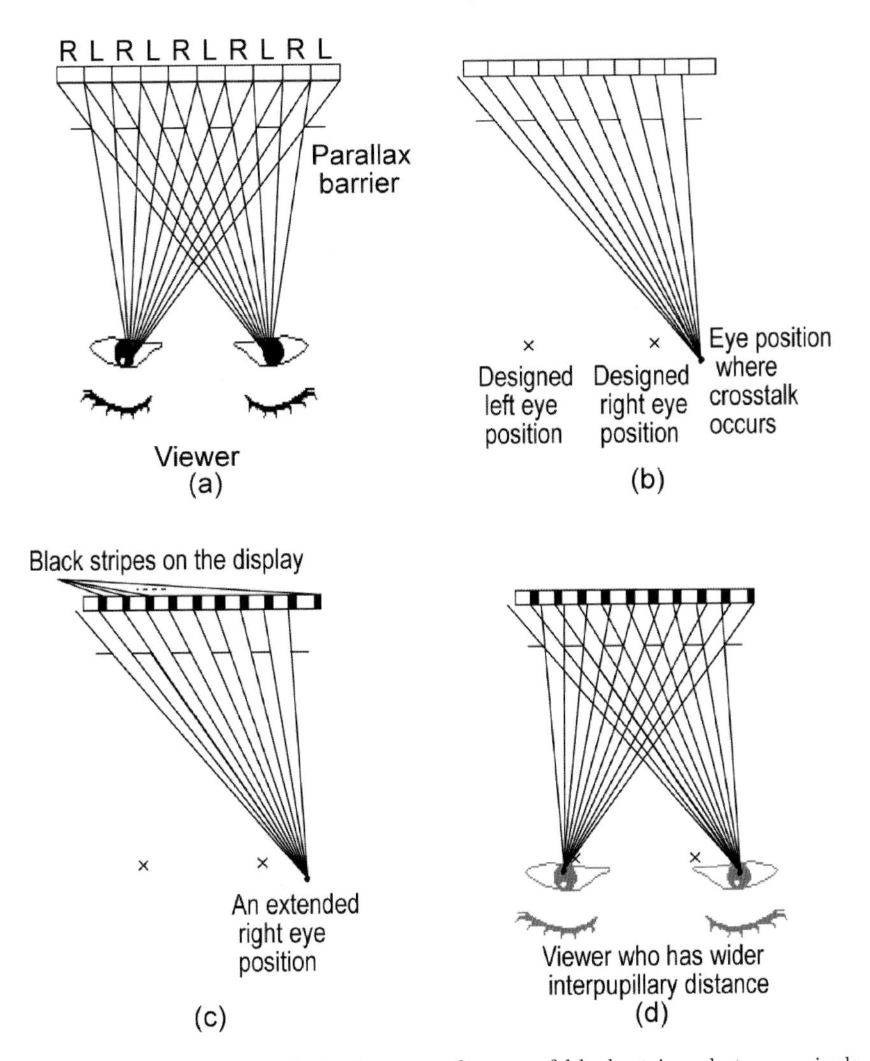


**Fig. 10.2.** Enlargement of viewing areas by use of black stripes between pixels. (**a**) Basic setup of stereoscopic display by use of a parallax barrier, (**b**) cross talk of stereoscopic images according to a deviation of an eye from the designed position, (**c**) enlargement of the viewing area by introducing black regions between pixels, and (**d**) a viewing position for another viewer with a wider interpupillary distance

and (b). Optimization to enlarge the viewing area of stereoscopic full-color LED displays using a parallax barrier is analyzed; such enlargement permits a certain amount of deviation in viewing position, as shown in Fig. 10.2 (c). The possibility of a stereoscopic LED display for crowds of viewers has been shown in numerical examples based on the model of a viewer whose interpupillary distance was the same as the 65 mm interpupillary distance designed in the stereoscopic display [6]. In practice, a viewer's interpupillary distance is not always equal to the designed distance. The difference of interpupillary distances can also restrict the viewing area. However, enlargement of this area can provide viewers with different interpupillary distances with a stereoscopic image, as shown in Fig. 10.2 (d). In this section, we describe the analytical formulas and experimental results of viewing areas of stereoscopic LED display that use a parallax barrier.

Viewing areas of the stereoscopic display that use a parallax barrier are limited by cross talk of both perspective images and the disappearance of pixels due to deviations of eye positions from the designed viewing points. One method is the use of a parallax barrier with a narrow aperture. However, a narrow emitting width on the display or a narrow aperture in the parallax barrier causes pixels to disappear. Another method to reduce the cross talk is to use an emitting width narrower than the pixel pitch on the display, as shown in Fig. 10.3. The ratio $a_{\mathrm{D}}$ of the emitting width to the pixel width, which is referred to as "the pixel aperture ratio" hereafter in this chapter, is decreased by increasing the black spacing between individual LEDs of the LED panel. Viewing areas where only the left or right image is seen without the emitting areas disappearing are determined under so-called "no cross talk" and "no disappearance" conditions. The relationship between the viewing area and the pixel aperture ratio and the aperture ratio $a_{\mathrm{B}}$ of the parallax barrier has

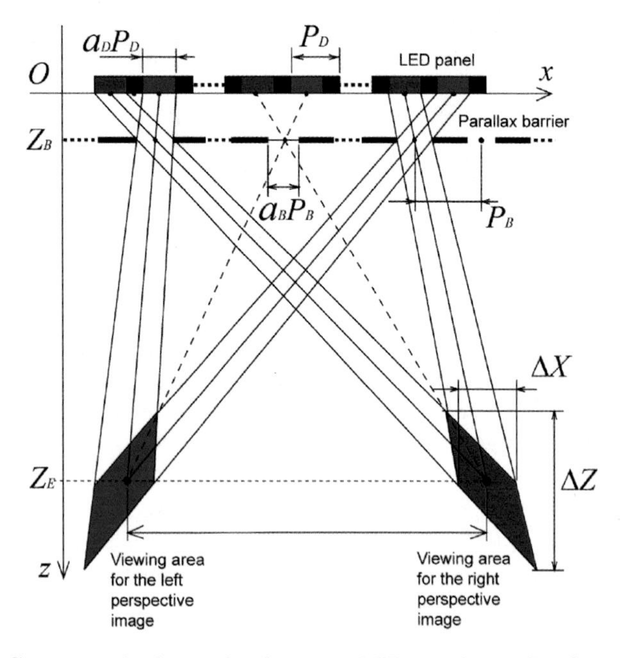


**Fig. 10.3.** Geometrical relationship between LED panel, parallax barrier, designed eye positions

been analyzed and optimization of the parallax barrier to enlarge the viewing areas has been discussed [7]. Consequently, LED panels with smaller pixel aperture ratio enlarge the viewing areas. Furthermore, the optimal aperture ratio of a parallax barrier for two perspective images is obtained as 1/2. Under the optimized condition, the width $\Delta X$ of the viewing area is expressed by:

$$\Delta X = P_E \left(1 - a_D\right). \tag{10.3}$$

The width decreases with the pixel aperture ratio and is independent of the number of horizontal pixels. The depth $\Delta Z$ of the viewing area is expressed as:

$$\Delta Z = \frac{Z_E P_E \left(1 - a_D\right)}{P_D \left(N - 1\right)}. \tag{10.4}$$

This is the case when the LED panel contains $2N$ horizontal dots. Thus, the depth linearly increases as the pixel aperture ratio decreases.

For the purpose of experimentally examining the design to enlarge the viewing areas, we have developed a large stereoscopic display using a parallax barrier by employing full-color LED panels. For the experiments, two types of full-color LED panels, each having 160 × 80 (=12,800) pixels, were used. Their size was 1.28 × 0.64 m. The pitch of each LED panel was 8.0 mm. The widths of the black regions of the lamp cluster type and three-in-one type LED panels were 0.6 and 5.4 mm wide, respectively. Therefore, the pixel aperture ratios of the LED panels were 0.925 and 0.325, respectively. The LED lamps in the LED lamp cluster panel had specially designed packages that allowed wide directivity in the horizontal direction. The parallax barrier was made so as to match the 8.0-mm-dot pitch of the full-color LED panel and the viewer's interpupillary distance, which was set to 65 mm. The pitch of the parallax barrier was 14.25 mm and its aperture ratio was 0.5. The number of apertures, which must be at least one-half the number of horizontal pixels, was set to 100 to provide a sufficient number of horizontal pixels for viewers not at the center. The viewing distance was set to 4 m. The effect of enlarging the viewing areas according to the above-described optimization procedure remains the same as the distance increases. The arrangement of LEDs and demonstrated viewing areas are shown in Figs. 10.4 and 10.5. In the case of the LED lamp cluster type, there are small areas where the contrast of stereoscopic images remains at the maxima or minima. Outside those areas, the contrast gradually changes between the maxima and the minima. Enlarged viewing areas can be obtained in the case of the three-in-one chip LED type. In the lateral direction, the contrast rapidly changes between the maxima and the minima. The calculated viewing areas according to the optimization analyses agree with experimental results. The calculated viewing areas in the case of the three-in-one chip LED type were 4.5 cm wide and 28.6 cm deep. The full-color LED panel of the three-in-one chip LED type makes the viewing

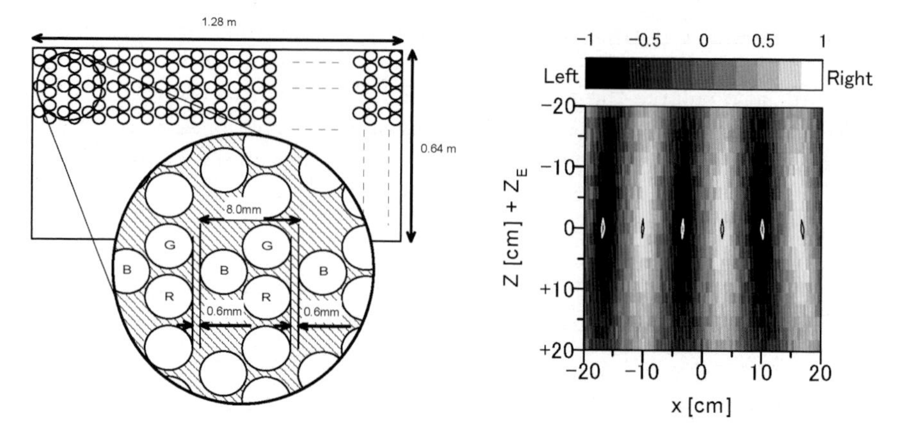


**Fig. 10.4.** Arrangements of pixels and obtained contrast distribution of stereoscopic images in case of a full-color LED panel of LED lamp cluster type

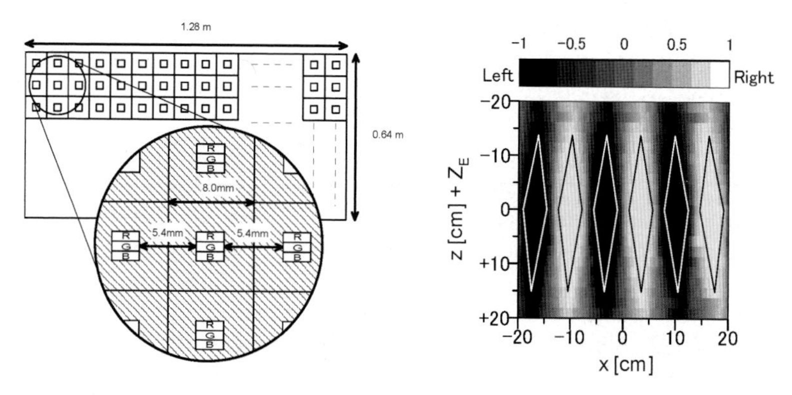


**Fig. 10.5.** Arrangements of pixels and obtained contrast distribution of stereoscopic images in case of a full-color LED panel of three-in-one chip LED type

areas 81 times larger than the viewing areas in the case of the LED lamp cluster type.

Figure 10.6 shows a developed stereoscopic LED display using a parallax barrier. The parallax barrier is placed in front of a full-color LED panel of $1.54 \times 0.77$ m. The parallax barrier is optimized to enlarge the viewing areas. Stereoscopic images and stereoscopic movies can be displayed on the developed stereoscopic full-color LED display using the parallax barrier. Examples of viewed images are shown in Fig. 10.7. Observers whose eyes are located in the viewing areas viewed the stereoscopic images or movies. When the viewing distance was 4 m, observers standing within a 2.6 m wide region (at least four observers) viewed the stereoscopic images at the same time. It is possible to increase the number of viewers, for example, to four people sitting on chairs and three people standing behind and between the sitting people. The width

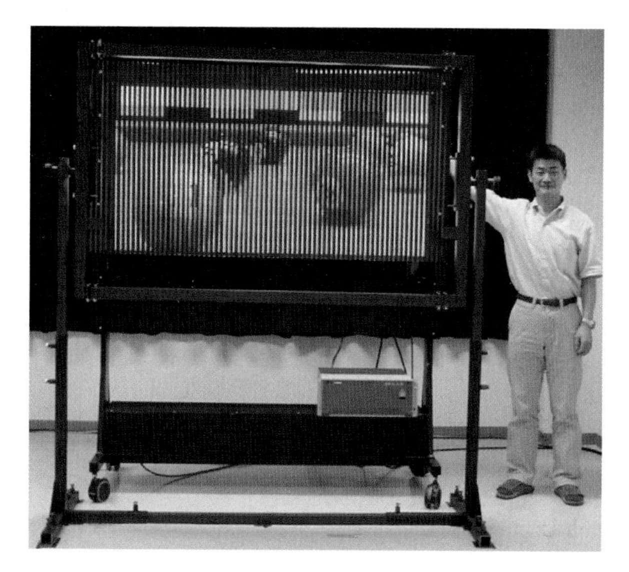

**Fig. 10.6.** Developed stereoscopic LED display by use of a parallax barrier. Parallax barrier of a large scale is placed in front of a full-color LED panel

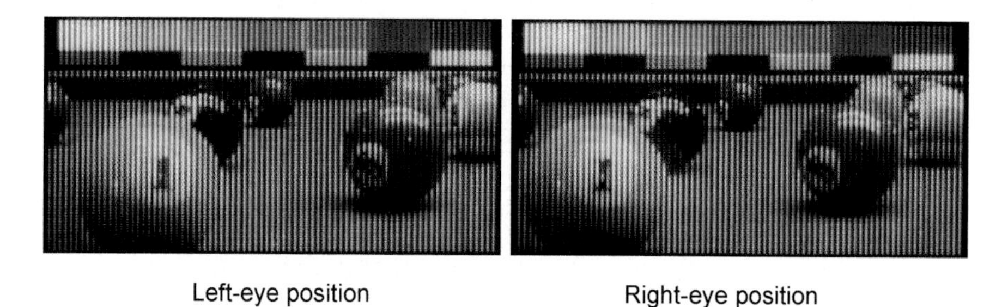


**Fig. 10.7.** Viewed images at a left eye position and a right eye position

was limited by the number of apertures of the parallax barrier. Observers standing outside this region viewed only the central part of the stereoscopic images, and the left and right ends were obscured by the frame of the parallax barrier. It is difficult to determine the range of depth perception because it depends on various factors, such as brightness of the background, viewing distance, size of the target, and the observer's ability to see stereoscopic images. At a viewing distance of 4 m, a dot pattern on a dark background was perceived to float 3.6 m from the LED panel.

The viewing distance was set to 4 m for these experiments. In practice, however, the viewing distance for the 8 mm pitch LED panel is over 18 m. The width of the viewing area is independent of the viewing distance, while the depth of the viewing area is proportional to the viewing distance [7]. For example, when the three-in-one chip type full-color LED panel is used while maintaining a viewing distance of 20 m, the width and the depth of each viewing area is 4.5 cm and 1.4 m, respectively. The depth of 1.4 m allows viewers to stand in front and behind. Viewing areas with the same width and depth are located side by side at an interval twice the designed interpupillary distance. Viewers standing off the horizontal center of the LED panel may turn towards the center of the LED panel. According to the analysis based on a viewer's model [3], in the case of the three-in-one chip type LED panel, cross talk and the disappearance of pixels are caused by turning through 70 degrees. In this case, the critical turning angle is wider than the horizontal directivity of the LED panel. When the parallax barrier is wide enough, the total width of the viewing areas at a distance of 20 m is almost 70 m, which is determined by the horizontal directivity of the LED panel. Therefore, a crowd of viewers can view the stereoscopic full color LED panel at the same time.

## 10.3 Design of Parallax Barrier for Plural Viewers

The viewing areas depend on the pixel aperture ratio of the LED panel and the designed interpupillary distance and are independent of the number of viewing zones. When the pixel aperture ratio is set to 0.4875, including the following experimental cases, the viewing areas for both perspective images are plotted in Fig. 10.8. The pitch of the LED panel is 8 mm and it has 160 dots horizontally. A ratio of 0.4875 can be achieved by using the LED panel of LED lamp cluster type in a single elemental color. In the case of parallax barriers for wider interpupillary distance than those of viewers, only slight deviation is allowed. The viewers must locate their eyes exactly in the corresponding viewing areas. Only a little shift of eye position is allowed even though the viewing areas are wider than the allowable shift. In this case, the effective viewing areas for those viewers become small. Even tilting of their heads let their eyes outside the viewing areas. On the contrary, in the case of parallax barriers which offer narrower interpupillary distance, tilting their heads will allow their eyes to go inside the viewing areas.

A full-color LED panel 128 × 64 cm in size with 160 × 80 pixels was used for experiments. A parallax barrier was placed in front of the full-color LED panel to maintain a 27 m viewing distance. Three types of parallax barriers were used for different interpupillary distances; the designed distances were 56, 60, and 67 mm. The number of apertures in the parallax barrier was 100, which allowed 20 extra viewing areas.

For experiments with the display using the polarizing eyewear, the full-color LED panel mentioned above was covered with polarizing film strips; these were arranged in columns on the panel. The columns of the film had orthogonal axes, and each alternating column covered a vertical line of image with its own perspective view. When viewed with the polarizing eyewear, the interlaced stereoscopic images were separated into images for the left and the right eye.

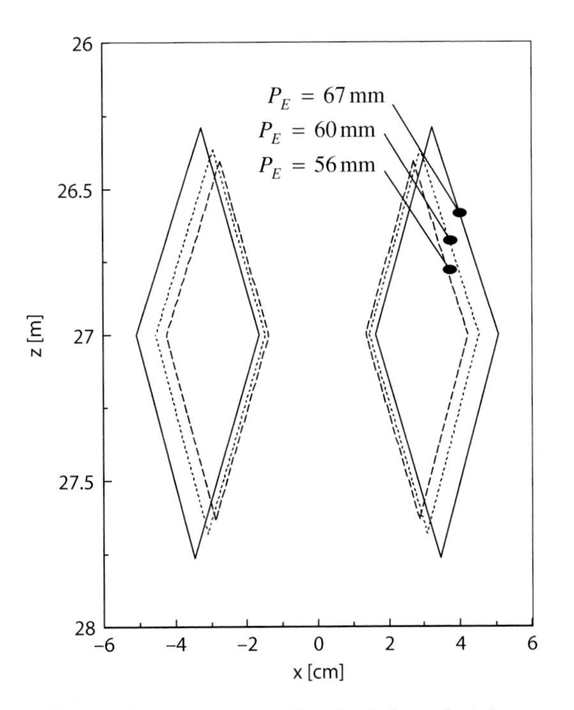


**Fig. 10.8.** Dependence of viewing areas for the left and right perspective images upon the designed interpupillary distance. LED panel has 160 horizontal dots. The LED pitch is 8 mm. The viewing distance is set at 27 m

Three students 23- or 24-years-old served as subjects. All had normal vision and good stereopsis, and the interpupillary distances of subjects A, B, and C were 58, 63, and 72 mm, respectively. For the stereoscopic LED display using polarizing eyewear, each subject wore this eyewear, and for the display using parallax barriers, they viewed stereoscopic images without any eyewear. Subjects were required to move the reference LEDs via a remote controller at the same distance as the virtual image of the target pattern, as shown in Fig. 10.9. The distance between the LED panel and the reference LEDs was recorded as the perceived distance. The distance between the stereoscopic dot targets was changed from three pixels (24 mm) to 29 pixels (232 mm) until the distance between targets became too wide for a subject to fuse the targets.

The perceived distance depended upon the distance between the stereoscopic targets on the stereoscopic full-color LED displays using parallax barriers as shown in Fig. 10.10, where solid circles, open circles, and crosses indicate the results for subjects A, B, and C, respectively. Figures 10.10(a)–(c) are the results in the cases of parallax barriers designed for 56 mm, 60 mm, and 67 mm interpupillary distances, respectively. The curves are geometrical calculations of the perceived distance according to the designed interpupillary

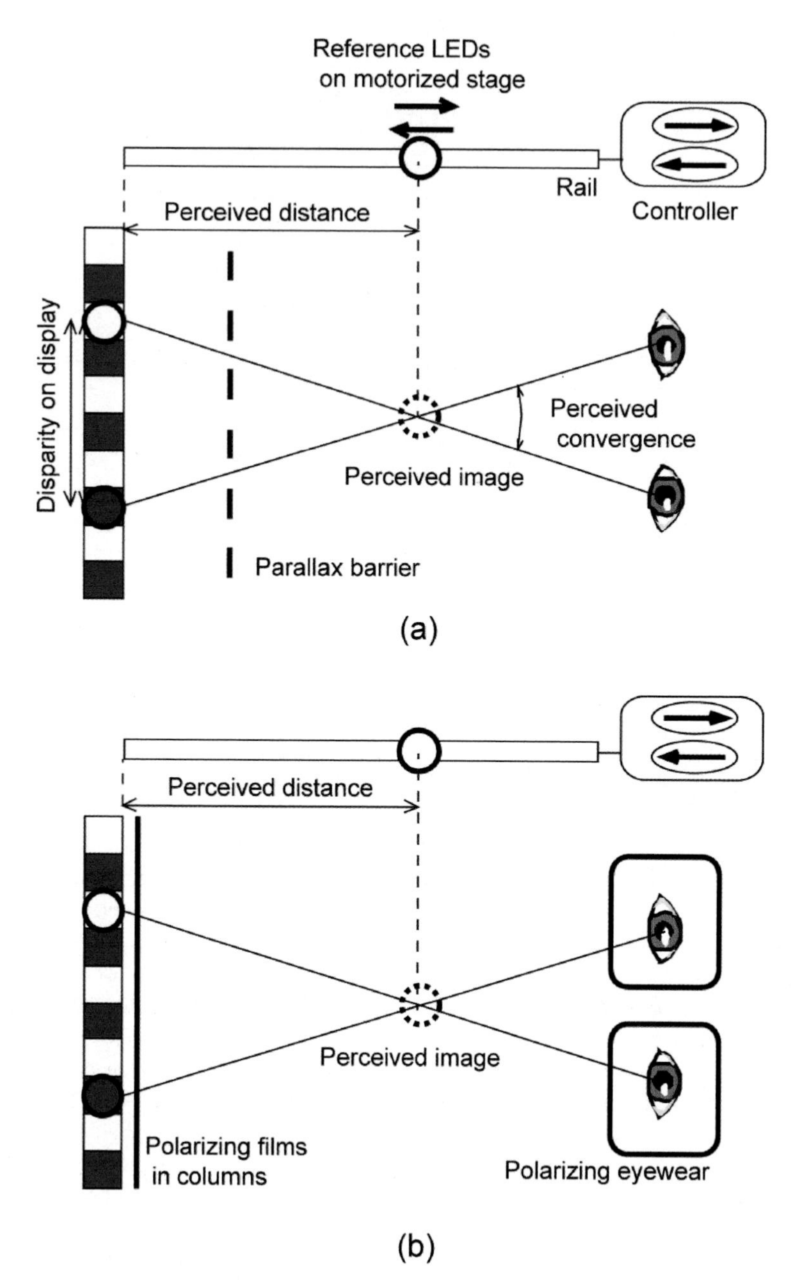


**Fig. 10.9.** Schematic illustrations of experimental setup. The viewing distance was set at 27 m. The subject was required to put his chin on a support. Reference LEDs were moved at the perceived distance by the subject. (**a**) Experiments using a parallax barrier and (**b**) experiments using polarizing eyewear

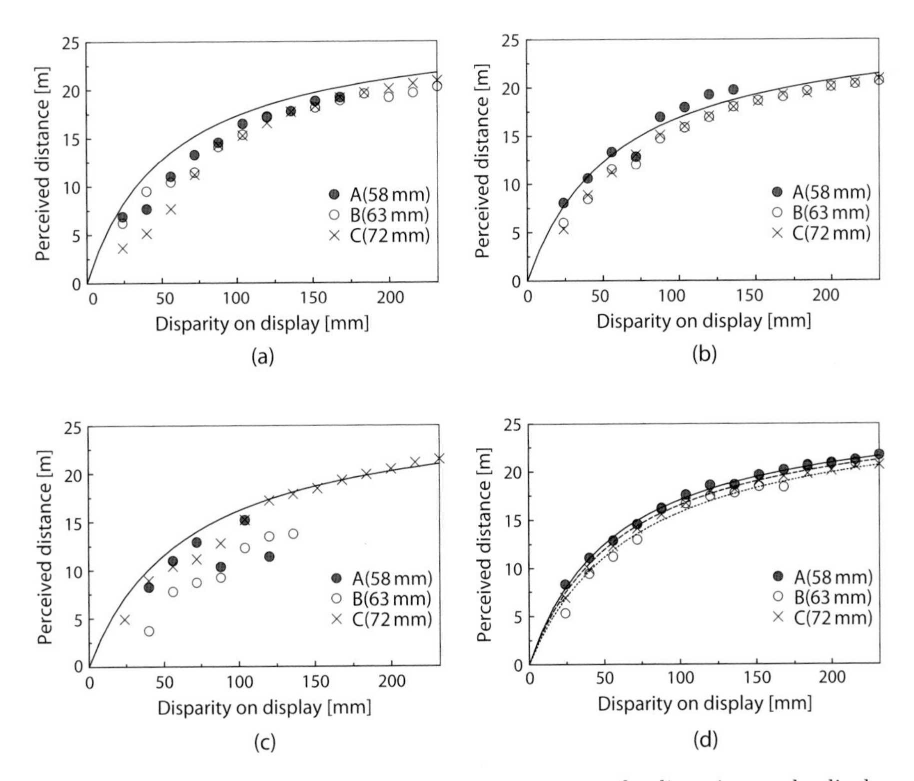


**Fig. 10.10.** Dependence of the perceived distance upon the disparity on the display using (**a**) the parallax barrier for 56 mm interpupillary distance, (**b**) the parallax barrier for 60 mm interpupillary distance, (**c**) the parallax barrier for 67 mm interpupillary distance, and (**d**) the polarizing films and polarizing eyewear. Filled circles, open circles, and crosses indicate the results for subject A whose interpupillary distance was 58 mm, subject B whose interpupillary distance was 63 mm, and subject C whose interpupillary distance was 72 mm

distances. Dependence of the perceived distance upon the distance between the stereoscopic targets in the stereoscopic full-color LED display using polarizing eyewear is shown in Fig. 10.10(d); the curves are geometrical calculations of the perceived distance according to subjects' interpupillary distances.

Using polarizing eyewear, the calculated curves of perceived distances agree with the experimental results. Thus, subjects A, B, and C moved the reference LED at the perceived distance of the stereoscopic images. Using parallax barriers, the results are not exactly consistent with the curves, but the trends are similar. This is because the curves were calculated according to different interpupillary distances from those of the subjects. In the cases of parallax barriers for 60 mm and 67 mm interpupillary distance, subject A fused the stereoscopic targets on the display with a certain amount of disparity. Subject B fused all the stereoscopic targets, except the 67 mm designed interpupillary distance, in which case the subject fused the targets 128 mm apart. Subject C fused all the stereoscopic targets in the experiments. This showed that enlarging the viewing area provided the subjects with stereoscopic images with a certain disparity.

The results suggest that viewers can fuse stereoscopic targets when the parallax barrier is designed for narrower interpupillary distance than the subject's own interpupillary distance. This tendency can be explained based on the viewing areas. The viewing areas are narrow (about 3 cm) according to the above-mentioned viewing area formula. When a viewer's interpupillary distance is narrower than the designed interpupillary distance of the parallax barrier, the viewer must locate his/her eyes exactly in the corresponding viewing area. Small deviations of eye positions prevent the subject from fusing the stereoscopic images. When a viewer's interpupillary distance is wider than the designed distance, however, tilting his head allows his eyes to go inside the viewing area. Consequently, for a crowd of simultaneous viewers, including children and adults, it is suggested that the parallax barrier be designed to accommodate children whose interpupillary distance is narrower.

## 10.4 Elimination of Pseudoscopic Viewing Area

In order to eliminate pseudoscopic viewing areas, a parallax barrier for multiple perspective images is utilized. As shown in Fig. 10.11, viewing positions for different perspective images, numbered from one to three, locate side by side. In the arrangement shown in Fig. 10.11, pseudoscopic viewing is caused by positioning the right eye in the viewing area for the image numbered one. The arrangements to eliminate the pseudoscopic positioning of eyes are shown in Fig. 10.12. The perspective images of images two and three are the left and the right perspective images, respectively. The pixels for image number one are assigned black. Therefore, the pseudoscopic viewing area is eliminated.

When both eyes of an observer locate within orthoscopic viewing areas, the observer is provided with both stereoscopic images. When one of the eyes locates within the viewing area for the black image, the observer cannot view both of the stereoscopic images and notice that one of the images is black. Thus, observers recognize that they are not standing within the orthoscopic viewing areas. Then, the observers will move right and left to position the eyes within the orthoscopic viewing areas.

In order to implement the above design to eliminate pseudoscopic viewing areas, a parallax barrier is fabricated to provide viewers with three perspective images. The parallax barrier with pitch $P_{3\mathrm{B}}$ is placed at a distance of $Z_{3\mathrm{B}}$ from the LED panel with horizontal pitch $P_{\mathrm{D}}$. To maintain a viewing distance of $Z_{\mathrm{E}}$ and a distance of $P_{\mathrm{E}}$ between the viewing positions, the parallax barrier is determined by the following relations: $P_{3\mathrm{B}} = 3P_{\mathrm{D}}\ P_{\mathrm{E}}\ /\ (P_{\mathrm{E}} + P_{\mathrm{E}})$ and

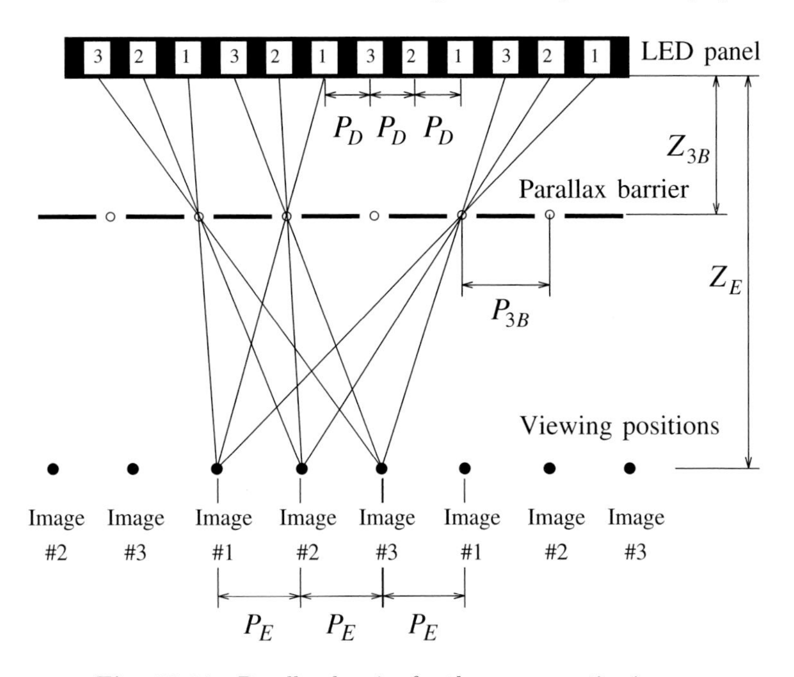


**Fig. 10.11.** Parallax barrier for three perspective images

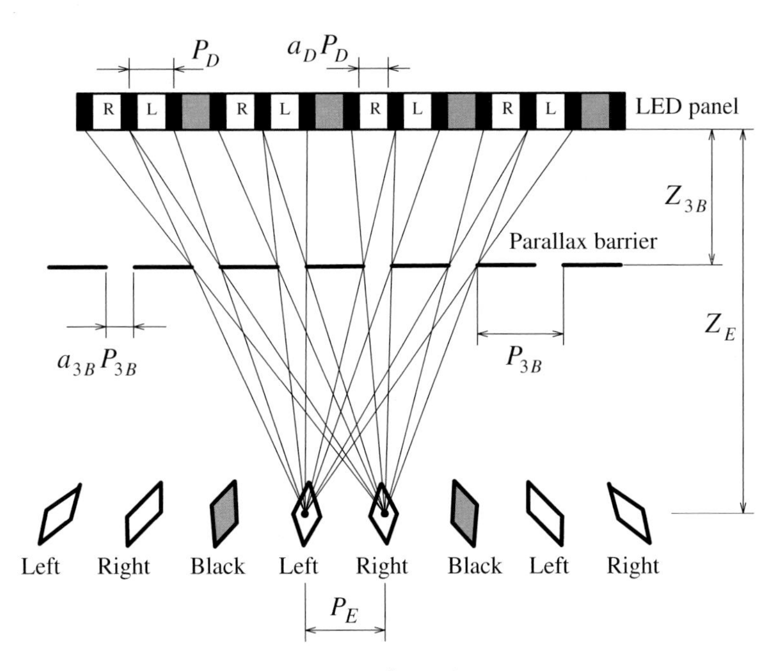


**Fig. 10.12.** Elimination of pseudoscopic viewing areas

$Z_{3B} = P_D Z_E / (P_E + P_E)$. The viewing positions depend on a right and left shift of LED panel and parallax barrier. The interval of positions where the same perspective image is seen is $3P_E$, which is independent of right and left shift of the parallax barrier. The viewing areas where only the right or left image is seen without the emitting area disappearing are determined based on the no disappearance and no cross talk conditions. The viewing areas depend on the aperture ratio $a_{3B}$ of the parallax barrier and the pixel aperture ratio $a_D$ that is defined by the proportion of the light emitting width in each pixel to the display pitch. When a parallax barrier with aperture ratio 1/3 is used for display of three perspective images, the equality condition that is necessary to maximize the width and depth of the viewing areas is satisfied [9]. Therefore, the optimal aperture ratio of the parallax barrier is 1/3.

Figure 10.13 shows the intensity distributions of the three perspective images obtained at the designed viewing distance (4 m). Viewing areas for the perspective images locate side by side with by having a sufficiently wide tolerance. Contrast distributions of stereoscopic images in the arrangement to eliminate the pseudoscopic viewing areas are shown in Fig. 10.14. Viewing areas for the right and left perspective images are indicated in white and black, respectively. The calculated viewing areas, surrounded by solid lines, agree with the experimental results. There are pairs of white and black regions separated by gray regions. When an observer stood in the appropriate viewing locations, both eyes of the observer viewed the corresponding perspective images. When the observer stood at the viewing distance and outside the viewing area, only one of the eyes viewed a perspective image and the other eye viewed a black pattern. Thus, the observer realized that he stood at a wrong position.

Since the pitch of the LED panel used for the experiments was uniform, the pseudoscopic viewing area was eliminated by interleaving a black image between both perspective images. However, the elimination is realized by

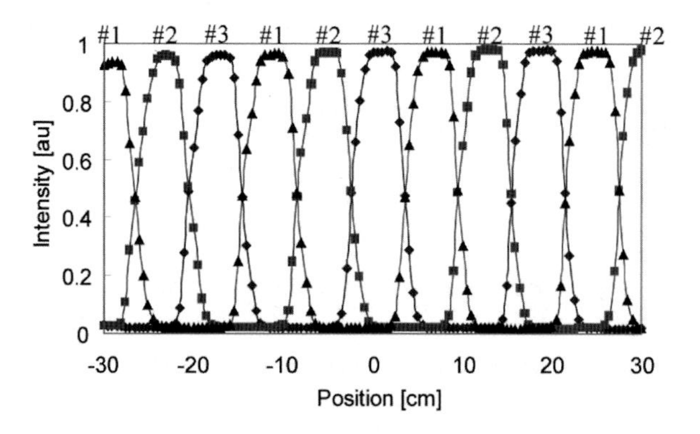


**Fig. 10.13.** Intensity distributions of test patterns for three perspective images at the viewing distance

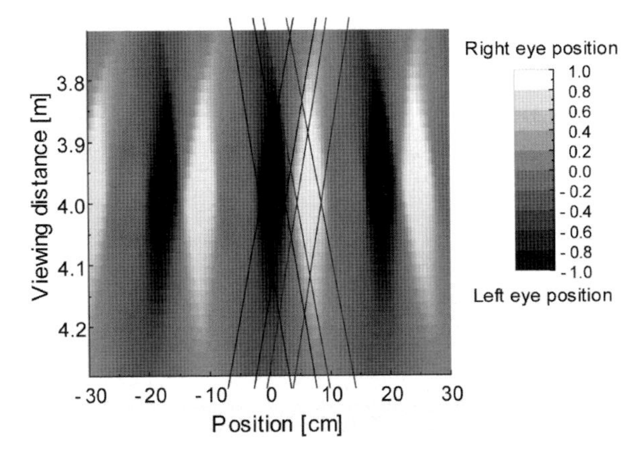


**Fig. 10.14.** Contrast distributions of stereoscopic images in the arrangement to eliminate the pseudoscopic viewing areas. Designed viewing distance was set to 4 m

widening the spacing between the pairs of LED columns in the LED panel. Furthermore, even when conventional LED panels are used, rapid switching of the order of the black image and two of the successive three perspective images may eliminate the pseudoscopic viewing area and increase the orthoscopic viewing areas.

## 10.5 Trace of Viewer's Movements

Schematic diagram of the developed real-time measurement system of a viewer's position is shown in Fig. 10.15. The system consists of a CCD (charge coupled device) camera fixed on the ceiling of the experimental room, a frame grabber, and a computer for image processing. The height of the camera from the floor was 2.6 m. An example of the detected images of a viewer is also shown in Fig. 10.15. The camera detects a viewer who put on a headset with an LED so that the LED locates at the middle of both eyes. The images detected by the CCD camera are captured with a frame memory and processed with a PC. The LED position is extracted with an image processing method by use of self windowing, which was originally proposed for image segmentation and matching with high-speed vision. That is, the image processing employs renewing a self window (a reduced scanning area around the detected LED position) in order to reduce the scanning time. The central position of the LED is calculated inside the self window. The LED position is detected and stored in real-time (30 Hz). Scaling of the camera coordinate is converted to the real coordinate.

Figure 10.16 shows the experimental setup for recording the path of a viewer. The experiments are conducted with eight subjects who watched the stereoscopic LED display with a parallax barrier. The starting position in

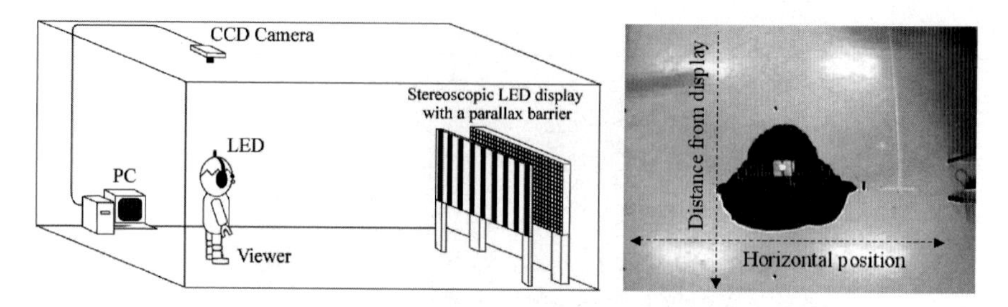


**Fig. 10.15.** Schematic diagram of real-time measurement system of the position of a viewer. The left image is obtained with the CCD camera fixed on the ceiling

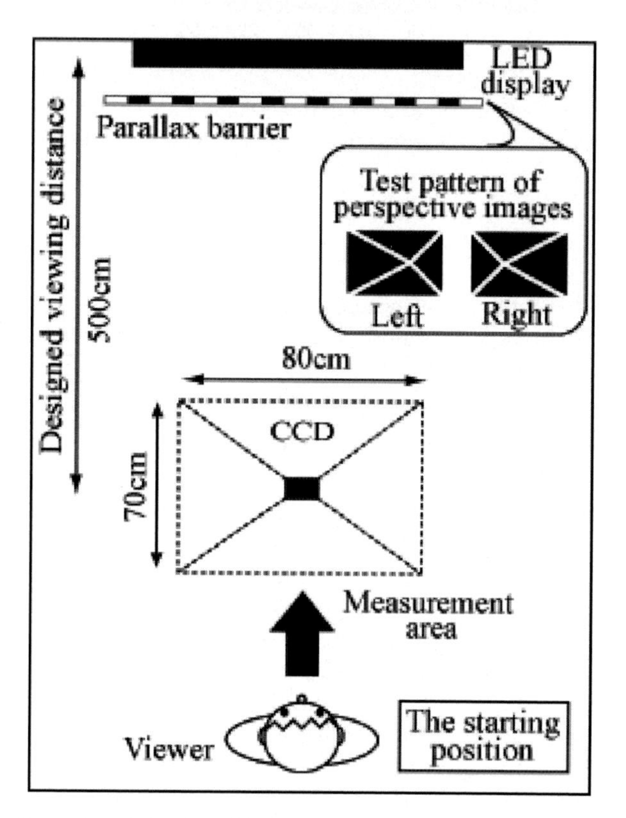


**Fig. 10.16.** Experimental setup. The starting position in experiments was located outside the viewing areas

the experiments was located outside the viewing areas and away from the designed viewing distance. Then, Moiré fringes appear at the starting position. Thus, the viewer has to move to find a viewing position. The position of the viewer is detected and recorded with the developed real-time measurement system using the camera fixed on the ceiling. During the experiments, the

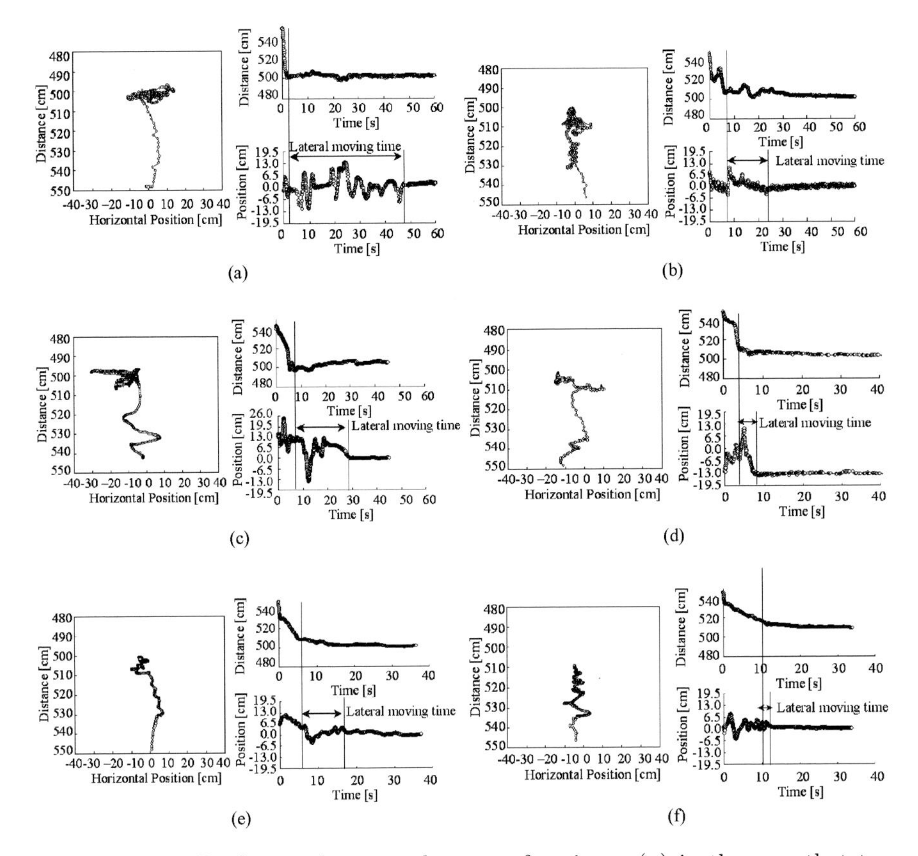


**Fig. 10.17.** 2-D plots and temporal traces of a viewer (**a**) in the case that two perspectives were displayed in the conventional design, (**b**) in the case that two perspectives were displayed in design to eliminate pseudoscopic areas, (**c**) in the case that three perspectives were displayed in the conventional design, (**d**) in the case that three perspectives were displayed in design to eliminate pseudoscopic areas, (**e**) in the case that four perspectives were displayed in the conventional design, and (**f**) in the case that two perspectives were displayed with a parallax barrier for four perspectives, that is, when the width of the black stripes is doubled

room lights are turned off. In each experiment, the subjects are not informed about the design of the stereoscopic LED display. The order of the types of the stereoscopic LED displays is determined at random. Before the experiments were conducted, the subjects were given the only directions of "Please move to see a stereoscopic image. When the stereoscopic image is perceived, please stand still at the position."

Firstly, Figs. 10.17(a) and (b) show the traces of the viewer position in the cases that two perspective images were displayed in the conventional design and in the design without pseudoscopic viewing areas, respectively. In both designs, the viewer moved toward the display at the start and then moved laterally. In the conventional design, the viewer moved right and left repeatedly in order to find an orthoscopic viewing position at the same distance from the display. The distance of lateral movements was 6.5 cm, which agreed with the designed interpupillary distance and the distance between the orthoscopic and pseudoscopic viewing positions. All the viewers tended to move toward the display at the start and then moved laterally. This is because the Moiré fringes appear when a viewer stands at a shorter and a further distance than the designed viewing distance. At the starting position, the viewer perceived Moiré fringes. Then, the viewer moved forward. If the viewer moved backward, Moiré fringes increased and the viewer moved forward. There is a significant difference in the lateral movements. In order to compare the lateral movements of the viewer, we have calculated lateral moving time, which is defined by the time required to move right and left after the viewer found the viewing distance. As the lateral moving times are also shown in Fig. 10.17, the lateral moving time was reduced in the design without pseudoscopic viewing areas.

Secondly, Fig. 10.17(c) and (d) show the traces of the viewer position in the cases that three perspective images were displayed in the conventional design and in the design without pseudoscopic viewing areas, respectively. In both designs, the viewer moved forward at the start and then moved right and left. Without pseudoscopic viewing areas, the lateral moving time was reduced.

Thirdly, Fig. 10.17(e) and (f) show the traces of the viewer position in the cases that four perspective images were displayed in the conventional design, and that two perspective images and two black images were displayed in the design without pseudoscopic viewing areas, respectively. In the conventional design, the viewer moved forward at the start and then moved right and left. In this design without pseudoscopic viewing areas, the viewer moved toward the display while moving right and left. This is because the widened spacing between pairs of viewing areas helped the viewer maintain a lateral position to view the stereoscopic images. Thus, the lateral moving time, which is defined as the time for lateral movements at the viewing distance, is much reduced in the design without pseudoscopic viewing areas.

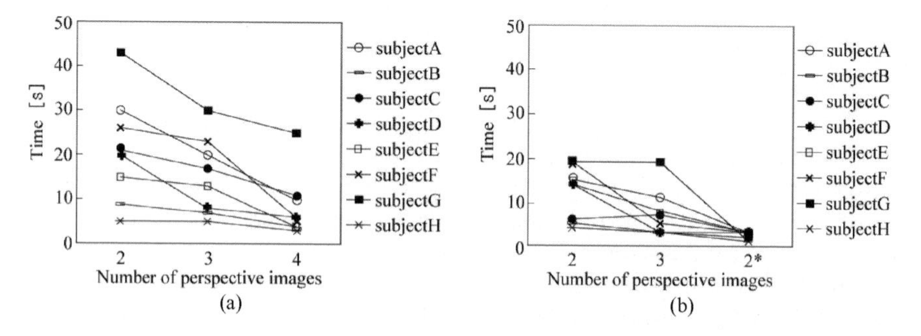


**Fig. 10.18.** Lateral moving times (**a**) in the conventional designs and (**b**) in the designs without pseudoscopic viewing areas, where 2* indicates the design to provide two perspective images with two black images

Calculated lateral moving times for eight viewers are shown in Fig. 10.18 (a) and (b). In the conventional designs, the average and the maximum of the lateral moving times decreased as the number of perspective images increased. This is because the number of orthoscopic viewing areas increased with the number of perspective images. There was no difference observed between the minimums of the lateral moving times in these experiments. The other lateral moving times in the design without pseudoscopic viewing areas were shorter than in the conventional designs. In particular, the maximum lateral moving time was significantly reduced. Thus, eliminating pseudoscopic viewing areas effectively reduces the lateral moving time, especially of novice viewers.

## 10.6 Recent Developments

In order to show real world images, it is necessary to capture, process and show stereo images in real-time. We have constructed a stereoscopic television system that consists of binocular cameras, a PC, and a stereoscopic LED display. We have developed active binocular cameras and software for a stereoscopic LED display [11]. The binocular cameras consist of a pair of CCD cameras, each of which is mounted on a stepping motor. The images detected by the active binocular cameras are captured with frame grabbers and processed to show on the stereoscopic LED display with a parallax barrier. The camera separation and convergence angle have been analyzed based on the reproduced area of the field and the range of reproduced depth [12].

**Fig. 10.19.** Developed 140-inch stereoscopic display with a parallax barrier

**Fig. 10.20.** Plural viewers within the range of 16 m in the width at the same time

We have developed a large parallax barrier using an aperture grille. 140-inch LED panel with 512×288 pixels has been utilized for stereoscopic display. The pitch of the LED panel was 6.0 mm. The pitch of the parallax barrier was 11 mm. The size of the parallax barrier was 4 × 2 m. As shown in Fig. 10.19, the large stereoscopic display was installed in a lecture hall. The viewing distance is 15 m. It was confirmed that plural viewers within the range of 16 m in the width simultaneously enjoyed stereoscopic movies on the large LED screen, as shown in Fig. 10.20. Interleaved stereoscopic images are separated into both perspective images at viewing positions.

## 10.7 Summay

We have developed stereoscopic full-color LED displays by using parallax barriers that provide different viewing areas. Optimization of the viewing areas has been discussed and confirmed by using a three-in-one chip LED panel that has wider black regions than ordinary LED lamp cluster panels. We have developed a real-time measurement system of a viewer's position and utilized the measurement system to evaluate the performance of different stereoscopic LED display designs, including conventional designs, to provide multiple perspective images and designs to eliminate pseudoscopic viewing areas. The developed stereoscopic LED display provides a number of side lobes, which allow plural viewers to view the stereoscopic image at the same time.

## References

[1] K. Bando, "Application of high-brightness InGaN LED for large size full color display," in Proceedings of The Sixth International Display Workshops (IDW'99), pp. 997–1000 (1999).

[2] H. Yamamoto, S. Muguruma, T. Sato, Y. Hayasaki, Y. Nagai, Y. Shimizu, and N. Nishida, "Stereoscopic large display using full color LED panel," Proceedings of The Fifth International Display Workshops (IDW'98), pp.725–728 (1998).

[3] F. E. Ives, "Parallax stereogram and process of making same," U. S. Patent 725567 (1903).
[4] S. H. Kaplan, "Theory of Parallax Barriers," J. SMPTE, Vol. 59, pp. 11–21 (1952).
[5] G. B. Kirby Meacham, "Autostereoscopic displays – past and future," in Advances in Display Technology VI, Proc. SPIE, Vol. 624, pp. 90–101(1986).
[6] H. Yamamoto, S. Muguruma, T. Sato, K. Ono, Y. Hayasaki, Y. Nagai, Y. Shimizu, and N. Nishida, "Optimum Parameters and Viewing Areas of Stereoscopic Full-Color LED Display Using Parallax Barrier," IEICE Trans. on Electronics, Vol. E83-C, No. 10, pp. 1632–1639 (2000).
[7] H. Yamamoto, M. Kouno, S. Muguruma, Y. Hayasaki, Y. Nagai, Y. Shimizu, and N. Nishida, "Enlargement of viewing area of stereoscopic full-color LED display by use of a parallax barrier," Applied Optics, Vol. 41, No. 32, pp. 6907–6919 (2002).
[8] H. Yamamoto, T. Sato, S. Muguruma, Y. Hayasaki, Y. Nagai, Y. Shimizu and N. Nishida, "Stereoscopic Full-Color Light Emitting Diode Display Using Parallax Barrier for Different Interpupillary Distances," Optical Review, Vol. 9, No. 6, pp. 244–250 (2002).
[9] H. Yamamoto, M. Kouno, Y. Hayasaki, S. Muguruma, Y. Nagai, Y. Shimizu, and N. Nishida, "Elimination of Pseudoscopic Viewing Area of Stereoscopic Full-color LED Display Using Parallax Barrier," Proceedings of The Ninth International Display Workshops (IDW '02), pp. 1249–1252 (2002).
[10] S. Matsumoto, H. Yamamoto, Y. Hayasaki, and N. Nishida, "Real-time Measurement of a Viewer's Position to Evaluate a Stereoscopic LED Display with a Parallax Barrier," IEICE Trans. on Electronics, Vol. E87-C, No. 11, pp. 1982–1988 (2004).
[11] Y. Yamada, H. Yamamoto, Y. Hayasaki, S. Muguruma, Y. Nagai, Y. Shimizu, and N. Nishida, "Real-time display of stereoscopic images by use of an active binocular camera," Proceedings of The 10th International Display Workshops (IDW '03), pp. 1445–1448 (2003).
[12] H. Noto, H. Yamamoto, Y. Hayasaki, S. Muguruma, Y. Nagai, Y. Shimizu, and N. Nishida, "Analysis of Reproduced 3-D Space by Stereoscopic Large LED Display," IEICE Trans. on Electronics, Vol. E89-C, No. 10, pp. 1427–1434 (2006).

# 11

# Synthesizing 3-D Images with Voxels

Min-Chul Park, Sang Ju Park, Vladmir V. Saveljev, and Shin Hwan Kim

## 11.1 Introduction

3-D images provide viewers with more accurate and realistic information than 2-D images. They also bring immersive feeling to the viewers with depth sense, on the other hand, often causing dizziness and serious eye fatigue. The main demands of 3-D images occur in the movies, broadcasting, medical applications, advertisement, telepresence, education and entertainment, and so on.

Generally 3-D images adopt "voxel" representation, which is analogous to the concept of "pixel" in 2-D images. The voxels, basic elements of 3-D images, are used to describe virtual points. Any desired 3-D image can be displayed by synthesizing it with voxels of pre-defined coordinate values because 3-D images are formed by voxels. Voxels can be visible if a group of pixels in the display panel, which is responsible for making each voxel visible at the viewing zone, is defined because voxels are virtual points in a pre-defined space. The viewing zone is a spatial location where viewers can see entire images displayed on the screen.

The multi-view (MV) [1, 2, 3, 4, 5, 6, 7, 8, 9, 10, 11, 12, 13, 14] and IP (Integral Photography) [15, 16, 17, 18, 19, 20, 21, 22, 23] are the typical methods of displaying a full parallax 3-D image on a flat panel display. As autostereoscopic image displays these methods have been a matter of great concern since 1990. MV and IP have the same optical structure composed of a viewing zone forming optics and a display panel located at the focal plane

M.-C. Park
Intelligent System Research Division, Korea Institute of Science and Technology, Seoul, South Korea
e-mail: minchul@kist.re.kr

B. Javidi et al. (eds.), *Three-Dimensional Imaging, Visualization, and Display*,
DOI 10.1007/978-0-387-79335-1_11, © Springer Science+Business Media, LLC 2009

of the optics. The images projected to viewers' eyes in MV and IP have a conjugate relationship between them [16]. In this chapter, synthesizing 3-D images with voxels in a contact-type imaging system for these methods will be discussed.

MV images for generating full parallax images can be easily obtained by using a two-dimensional (2-D) camera array. The images can also be synthesized with a computer by considering the relative viewing direction of each camera in the array for a given object [5, 6, 7, 12, 13, 14]. In order to configure these MV images, the display panels of MV 3-D imaging systems must be divided into a number of segments, called pixel cells. A 3-D imaging system can provide full parallax 3-D images.

Pixel cells (the number of segments divided) corresponds either to (1) the number of pixels in the image from each camera in the array [8, 9] or (2) to the number of cameras in the array [10, 11, 15, 16]. The number of pixels in each pixel cell is equal to the number of cameras in the array. Each pixel in the cell represents a pixel from each camera for (1). For (2), each pixel cell in the display panel presents the whole image of each camera. The each camera position in the array corresponds to that of the cell in the panel. The typical shape of a pixel cell is either rectangular or square. Since these shapes are vulnerable to the Moire effect, rhomb shaped pixel cells can also be used [24]. In these configuration methods, we typically scale the proper resolution of each camera image to fit into the resolution of the display panels available. This scaling process is somewhat cumbersome and time consuming.

3-D images are formed by voxels, so we can synthesize the 3-D image with voxels of pre-defined coordinate values [16, 25]. Finally, we can display any desired 3-D image. The voxels are used to describe virtual points. Any desired 3-D image can be displayed by synthesizing it with voxels of pre-defined coordinate values because 3-D images are formed by voxels. Voxels can be visible if a group of pixels in the display panel, which is responsible for making each voxel visible at the viewing zone, is defined because voxels are virtual points in a pre-defined space. The scaling step can be eliminated and the computational time for preparing MV images can be minimized. To obtain these effects a set of voxels is defined in the optical configuration of a full parallax MV 3-D imaging system based on a 2-D PLS (Point Light Source), and the set is used to display 3-D images.

In this configuration, the group of pixels provides passage for rays from PLSs such that the voxel is visible at the viewing zone′s cross section, where the viewing zone is centered. The pixel pattern formed by the group of pixels has a unique pattern to represent a voxel in a certain location. Finding pixel patterns in the display panel is required so that it will be able to display 3-D images. The configuration provides two voxel types; one is seen at the entire viewing zone′s cross section (complete voxel) and the other is only partially seen (incomplete voxel). The spatial volume is where 3-D images appear, and it depends on the spatial distribution of voxels. Similarly, the resolution of the images depends on their available number. As the number of complete voxels is limited and the voxels occupy only a small space, the image space and the voxel resolution of the displayable images will be extremely limited as well. Using incomplete voxels will effectively increase the volume and the resolution of the images. The incomplete voxels will successfully increase the image volume in the MV 3-D imaging systems, since most of these systems [9, 10, 11] are based on the optical configuration [26].

The pixel patterns for these voxels strongly depend on the shape of the pixel cells. The pixel pattern for the complete voxel has the same pixel arrangement in both vertical and horizontal directions. For this reason, we are easily able to determine the 2-D pixel pattern from the vertical or horizontal direction pattern for a rectangular or square pixel cell. However, for the rhomb shaped pixel cell, the difference in pixel numbers in both directions changes the arrangements of directions from each other.

This chapter outlines synthesizing 3-D images with voxels in a contact-type imaging system by describing voxels in the PLS array and defining the incomplete voxels and their corresponding pixel patterns to increase the volume. Also, we mathematically identify their positions. In addition, we extend the pixel pattern to pixel cells with a rhombus shape for a Moire-free image display.

## 11.2 Description of Voxels in the Point Light Source Array

The central viewing zone forms the geometry of a 3-D imaging system.The 3-D image system is based on a PLS array shown in Fig. 11.1 that has the same number of PLSs in both horizontal and vertical directions. The PLSs are all in the same plane with an equal separation in all directions. A display panel is located in front of the PLS plane. The panel is divided into segments called pixel cells. Each segment that consists of pixel cells has an equal area and shape. These cells are arranged for full parallax image generation. They are the units of the MV images. Each cell is illuminated by its corresponding PLS. Therefore, the total number of cells in the panel is equal to the number of PLSs. The general shape of a cell is rectangular or square. Since each PLS illuminates the image display panel in front of it, all PLSs in the array should be shown at the viewing zone′s cross section through the panel without any mismatch. All the magnified pixel cell images are superposed in the viewing zone. The viewing zone is the spatial region where viewers can observe 3-D images from the MV images on the panel.

Figure 11.2 shows the voxels and the voxel space defined in Fig. 11.1. The voxels are defined as the crossover points. These crossover points make the connection lines from each PLS to the four corners of the viewing zone′s cross section.

This means: (1) each voxel is created by rays from four distinct PLSs in order to form the same shape as the viewing zone′s cross section; (2) each

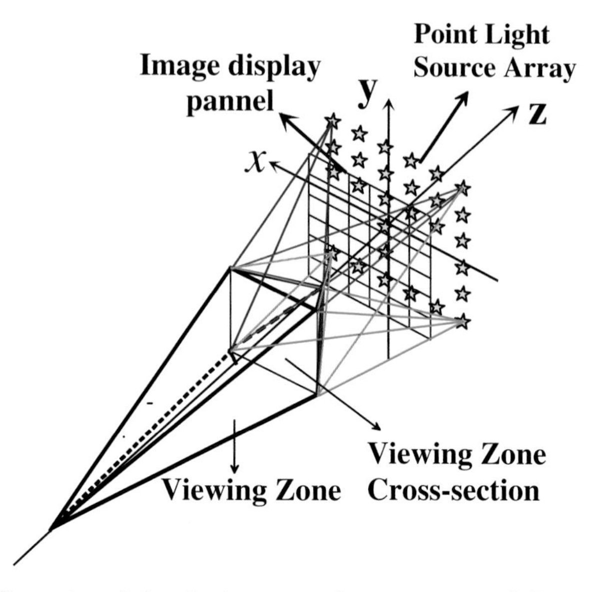


**Fig. 11.1.** Geometry of the viewing zone of a contact-type full parallax imaging system based on a PLS array

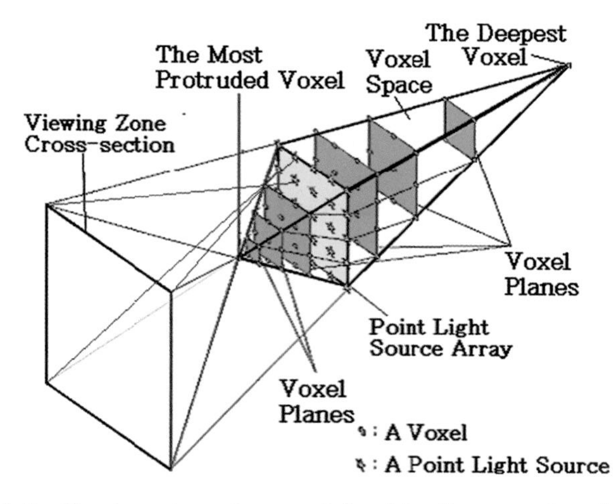


**Fig. 11.2.** Voxels and voxel space defined by the geometry of Fig. 11.1

voxel can be displayed at any location in the viewing zone's cross section, and (3) each voxel can be placed in a specific plane.

We define the pixel pattern for each voxel in the capacity of a portion of pixel cell through the PLS. The PLS corresponding to the cell can be seen at the cross section. The voxel space where the voxels are located is defined as a volume. The volume has a diamond shape because the synthesizable 3-D images with these voxels can exist only in a diamond shaped voxel space.

$M_S$ and $N_S$ represent the numbers of PLSs in horizontal and vertical directions, respectively. $N_T$ is the total number of voxels in the voxel space. When $N_S > M_S$ as shown in Fig. 11.2, $N_T$ is calculated as

$$N_T = \frac{M_S(3N_S M_S - M_S^2 + 1)}{3}. \tag{11.1}$$

When $N_S < M_S$, $N_S$ and $M_S$ in Eq. (11.1) will be transposed. For the case when $N_S = M_S$, Eq. (11.1) can be simplified as,

$$N_T = \frac{M_S(2M_S^2 + 1)}{3}. \tag{11.2}$$

Figure 11.3 shows the top (horizontal plane) and side (vertical plane) views of the geometry in Fig. 11.2. This figure specifies the definition of complete and incomplete voxels for 6 X 6 PLS array. In this figure, voxels are described as the points dotted from PLSs to two edges of the viewing zone's cross section. This series of points (voxels) comprises the crossover lines in-between (between PLSs and two edges). Voxels are represented as squares for the first type and circles for the second type. The area with the incomplete voxels is much greater

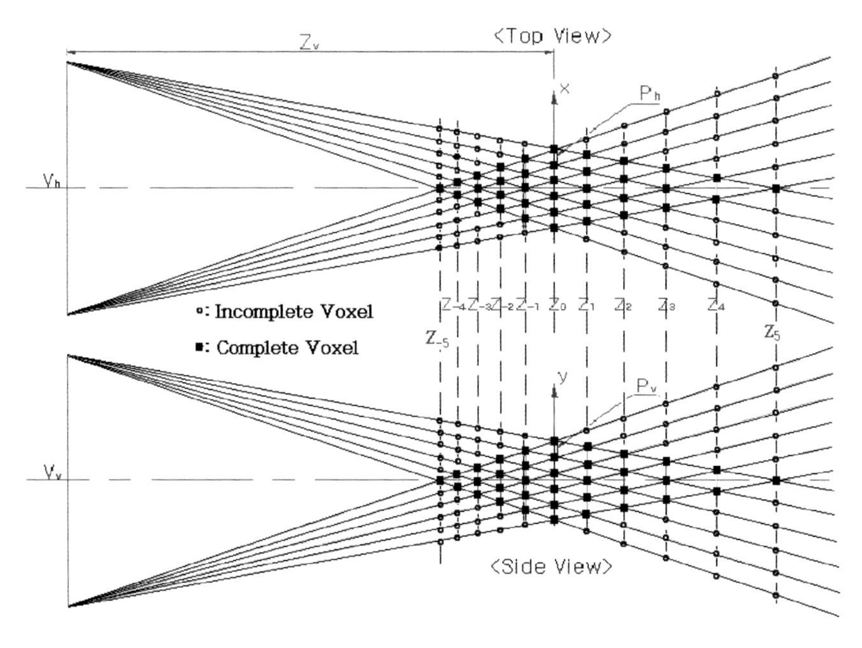


**Fig. 11.3.** Plane view of Fig. 11.2 defining complete and incomplete voxels

than that with complete voxels. This means that the voxel volume created by the incomplete voxels is much greater than the voxel space in Fig. 11.2.

From Fig. 11.3, the positions of the voxel planes in reference to the PLS array plane, which is located at $z = 0$, are calculated as,

$$z_k = |z_V| \frac{kP_h}{V_h - kP_h}, \quad z_k = |z_V| \frac{kP_v}{V_v - kP_v}. \tag{11.3}$$

$z_k$ and $z_V$ represent the locations of $k_{th}$ voxel plane and the viewing zone cross section, respectively. $P$ is the end of PLSs and it can be expressed as $P_h$ in the horizontal direction and $P_v$ in the vertical direction. $V$is the size of the viewing zone's cross section, and it also can be expressed as $V_h$ and $V_v$ in horizontal and vertical directions, respectively. $k$ is the order of voxel planes and has values $k = 0, \pm 1, \pm 2, \pm 3, \pm 4, \cdots\cdots, \pm M_S - 1$ and $k = 0, \pm 1, \pm 2, \pm 3, \pm 4, \cdots\cdots, \pm N_S - 1$, where $M_S$ and $N_S$represent the numbers of PLSs in horizontal and vertical directions, respectively. One can obtain the maximum value of k for the full parallax voxels taking the smaller value of |Ms| and |Ns|. In Eq. 11.3, $z_k$ is not defined when $V_h - kP_h \leq 0$ or $V_v - kP_v \leq 0$. In the coordinate defined in Fig. 11.3, $k = 0$ and –1 represents the plane of PLS array and the image display panel, respectively. Zv has a negative value. If $V_h/P_h = V_v/P_v = s$, then $z_k = k\,|z_V|\,/(s-k)$, i.e., $z_k$ is just a function of $k$. This means that $z_k$has the same values in both horizontal and vertical directions.

The distance between voxels in the $k_{th}$ plane, $G_h^k$ and $G_v^k$ ($h$ and$v$represent horizontal and vertical directions, respectively) is calculated as,

$$G_h^k = \frac{V_h P_h}{V_h - kP_h}, \quad G_v^k = \frac{V_v P_v}{V_v - kP_v}. \tag{11.4}$$

The number of voxels in $k_{th}$ plane is $(M_S - |k|)(N_S - |k|)$ and the relative positions of voxels in $k_{th}$ and $(k+1)_{st}$ planes are different. They are different about one-half voxel distance to each other in both horizontal and vertical directions. The difference is approximately half of the distance between two neighboring voxels in each plane in both the horizontal and vertical directions.

From Eq. (11.4), if $i$ and $j$ represent the order of voxels in horizontal and vertical directions from the $z$axis, respectively, the coordinate of each voxel in the $k_{th}$plane, $X_k^{ij}(C_k^i, C_k^j, z_k)$ , are defined as,

$$C_k^i = iG_h^k \quad \text{and} \quad C_k^j = jG_v^k. \tag{11.5}$$

For

$$j = 0, \pm 1, \cdots, \pm\left(\left[\frac{N_S + 1 - |k|}{2}\right] - 1\right)$$
$$i = 0, \pm 1, \cdots, \pm\left(\left[\frac{M_S + 1 - |k|}{2}\right] - 1\right) \tag{11.5a}$$

when

$$\left(\frac{N_S + 1 - |k|}{2} - \left[\frac{N_S + 1 - |k|}{2}\right]\right) = \left(\frac{M_S + 1 - |k|}{2} - \left[\frac{M_S + 1 - |k|}{2}\right]\right) = 0. \tag{11.5b}$$

Also,

$$C_k^j = \frac{j}{2\,|j|} G_v^k + (j \mp 1) G_v^k (- : j > 0, + : j < 0),$$
$$C_k^i = \frac{i}{2\,|i|} G_h^k + (i \mp 1) G_h^k (- : i > 0, + : i < 0), \tag{11.6}$$

$$j = \pm 1, \cdots, \pm\left[\frac{N_S - |k|}{2}\right],$$
$$j(i) = \pm 1, \cdots, \pm\left[\frac{M_S - |k|}{2}\right] \tag{11.6a}$$

when

$$\left(\frac{N_S + 1 - |k|}{2} - \left[\frac{N_S + 1 - |k|}{2}\right]\right) = \left(\frac{M_S + 1 - |k|}{2} - \left[\frac{M_S + 1 - |k|}{2}\right]\right) = \frac{1}{2}. \tag{11.6b}$$

In Eqs. (11.5) and (11.6), $i$ and $j$ are 0 when voxels are on the horizontal and vertical planes, respectively. Both $i$ and $j$ have positive values if the voxels are in the first quadratic plane when they are seen from the VZCS (Viewing Zone's Cross Section), and [] sign denotes integer values. The conditions specified in Eqs. (11.5a) and (11.6b) are for odd and even values of $N_S - |k|$ and $M_S - |k|$, respectively.

The voxels defined in Eqs. (11.1) and (11.2) represent those that have the pixel patterns shown in Fig. 11.4 . Figure 11.4(a) comprises ray diagrams that define pixel patterns for voxels specified as squares in Fig. 11.3. The voxels are identified as $X_k^i$ where i and k are the values defined above. In the instance of voxels $X_{-2}^1$ and $X_2^1$, light sources 2, 3, and 4 can make these voxels visible at the viewing zone's cross section. Voxel $X_{-2}^1(X_2^1)$ is visible at the bottom (top), middle and top (bottom) thirds of the viewing zone's cross section by the rays explained below. From light source 2, rays pass through the bottom (top) third of pixel cell 2 (from the top); from light source 3, rays pass through

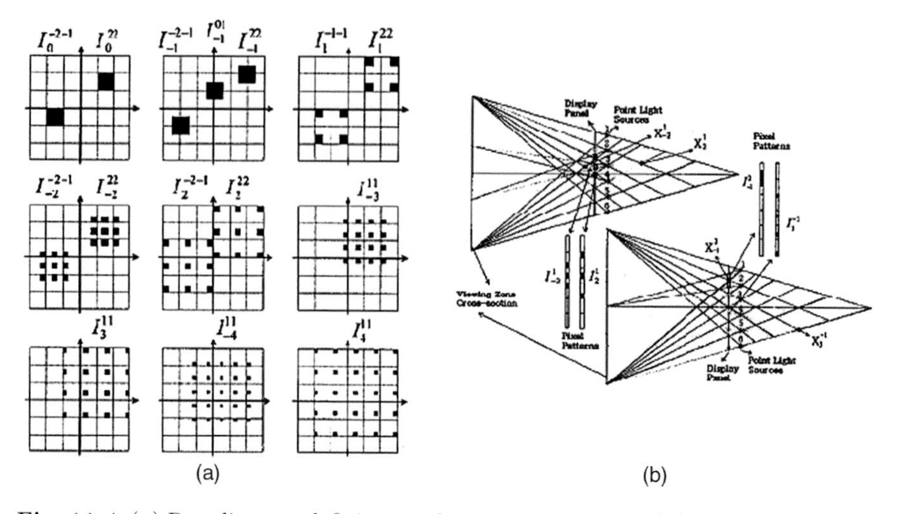

Fig. 11.4 (a) Ray diagram defining pixel patterns for voxels. (b) 2-D pixel patterns for the complete voxels defined in Fig. 11.3

the mid third of pixel cell 3 and, finally, from light source 4, rays pass through the top (bottom) third of pixel cell 4. Voxel $X^{1}_{-2}$ is visible at the bottom and top halves of the viewing zone's cross section by the rays explained below. From light source 1, rays pass through the bottom half of pixel cell 1, and from light source 2, rays pass through the top half of pixel cell 2. In the same way, voxel $X_{3}^{-1}$ is visible at the top fourth, the second fourth, the third fourth, and the bottom fourth of the viewing zone's cross section, respectively, by the rays. These rays from light sources 3, 4, 5 and 6 pass through the top fourth of pixel cell 3, the second fourth of pixel cell 4, the third fourth of pixel cell 5, and the bottom fourth of pixel cell 6, respectively.

These rays eventually make the voxel visible at the top fourth, the second fourth, the third fourth, and the bottom fourth of the viewing zone's cross section, respectively. The one-dimensional pixel patterns for voxels $X^{2}_{-1}$, $X^{1}_{-2}$, $X^{1}_{2}$, and $X_{3}^{-1}$ and for voxels $I^{2}_{-1}$, $X^{1}_{-2}$, $I^{1}_{2}$, and $I_{3}^{-1}$ are shown in Fig. 11.4(a).

The pixel patterns obtained here reveal one serious problem related to this voxel method. We have made two assumptions about obtaining the patterns that have been explained above. The first is that the voxels in different $k$ planes divide the viewing zone's cross section into $k + 1$ different segments. A second assumption is that each segment is illuminated by its corresponding PLS. Based on these two assumptions of the patterns, the viewer's eyes should be positioned in the two different segments to understand the sense of depth with a specific voxel. Thus, the viewer would not be able to perceive the voxel to have a certain depth, if the widths or the segments are bigger than one's interocular distance. This problem applies to voxels in smaller $|k|$ values. The solution to this problem is to reduce the width of the viewing zone's cross section and to watch the images at a short distance from the display panel. A solution is to create 3-D images without using the voxels in the smaller $|k|$ value plane; but in this case, $|k|$ should be large enough. Figure 11.4(b) shows the 2-D extension of the patterns in Fig. 11.4(a). In this case the voxels and their pixel patterns are identified as $X_k^{ij}$ and $I_k^{ij}$, respectively. It is shown that a pixel cell is divided into $(|k|+1)^2$, and $(|k|+1)^2$ is equal segments for voxels in a $k$th plane. $I_k^{ij}$ is composed of a segment from each $(|k|+1)^2$ neighboring to pixel cells. $I_k^{ij}$ is a pattern with two-fold symmetry. The total area is occupied by these $(|k|+1)^2$ segments. This area is equal to the area of a pixel cell. Each pixel cell pattern corresponds to a complete voxel and the area occupied by the pattern (each pixel cell pattern) is equal to that of a pixel cell. The distance between the segments that compose the pixel pattern, in the unit of number of segments, is $|k| \pm 1$. The minus sign is for negative values of $k$. The voxels in the $k_{th}$ plane are seen by each one of $(|k|+1)^2$ PLSs. Therefore, only two PLSs at a time can be seen by the viewer's two eyes.

## 11.3 Pixel Patterns of Incomplete Voxels

As Figs. 11.2, 11.3 and 11.4 show, each voxel in the voxel space has its corresponding pixel pattern in the display panel and, through the pixel pattern, it can be seen in any place in the viewing zone's cross section. If the patterns in the boundary of the panel are shifted one pixel cell at a time in each of four directions, (i.e., up-down and left-right), they become incomplete patterns and their corresponding voxels will be seen only at a specific part of the cross section. In Fig. 11.5, the patterns created shift the boundary pattern toward the right or the upper direction or both. Voxels that correspond to these patterns are called incomplete voxels and are marked as circles in Fig. 11.3.

These voxels are located at the crossover points of the extended row and column lines of the voxels in the $k_{th}$ plane, and also on the lines drawn from each PLS to the four corners of the viewing zone's cross section. When these incomplete voxels are included, the voxel space has the shape of two trapezoidal pyramids sharing the PLS array's plane as their top sides. The height of the shape can be defined from the most protruding (outermost) voxel to the most submerged voxel.

The distances between incomplete voxels in the $k_{th}$ plane are the same as those in the voxel space. As the pixel patterns for the voxels in the $k_{th}$ plane are composed of $|k|+1$ pixel cells in each direction, they can be shifted $k$ times in each direction. This means that, in the $k_{th}$ plane, there are $2|k|$ incomplete voxels in both the horizontal and vertical directions. The positions of the incomplete voxels can also be defined by Eqs. (11.5) and (11.6), but, in this case, the $j$ and $i$ values should be redefined as

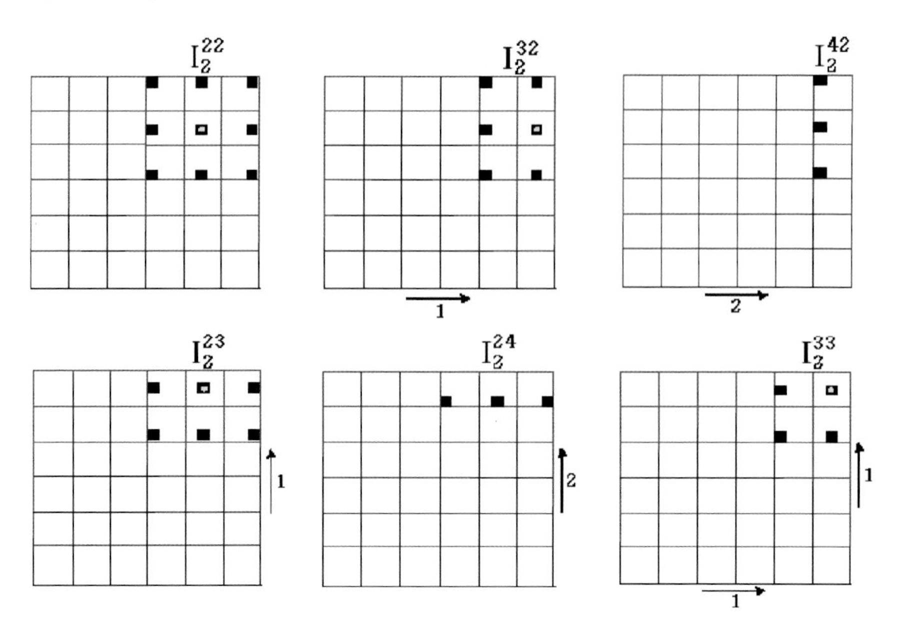


**Fig. 11.5.** Incomplete pixel patterns created by shifting the complete pixel pattern toward the right or upper direction or both

$$j = 0, \pm 1 \ldots, \pm \left( \left\lceil \frac{N_S + 1 + |k|}{2} \right\rceil - 1 \right), \qquad i = 0, \pm 1, \ldots, \pm \left( \left\lceil \frac{M_S + 1 + |k|}{2} \right\rceil - 1 \right) \tag{11.7a}$$

for

$$\left( \frac{N_S + 1 - |k|}{2} - \left\lceil \frac{N_S + 1 - |k|}{2} \right\rceil \right) = \left( \frac{M_S + 1 - |k|}{2} - \left\lceil \frac{M_S + 1 - |k|}{2} \right\rceil \right) = 0,$$

$$j = \pm 1, \cdots, \pm \left( \left\lceil \frac{N_S + |k|}{2} \right\rceil \right), \qquad i = \pm 1, \cdots, \pm \left( \left\lceil \frac{M_S + |k|}{2} \right\rceil \right) \tag{11.7b}$$

for

$$\left( \frac{N_S + 1 - |k|}{2} - \left\lceil \frac{N_S + 1 - |k|}{2} \right\rceil \right) = \left( \frac{M_S + 1 - |k|}{2} - \left\lceil \frac{M_S + 1 - |k|}{2} \right\rceil \right) = \frac{1}{2}.$$

In Eq. (11.7), the incomplete voxels are represented by j and i values bigger than those defined in Eqs. (11.5a) and (11.6a). With Eqs. (11.7) the total number of voxels, including the incomplete voxels, $N_I$, are calculated as

$$
\begin{aligned}
N_I &= 2\sum_{K=1}^{M_s-1} (N_S + K)(M_S + K) + N_S M_S \\
&= 2\sum_{K=1}^{M_s-1} \{N_S M_S + K(M_S + N_S) + K^2\} + N_S M_S, \\
N_I &= \frac{M_S(5M_S^2 + 9N_S M_S - 6M_S - 6N_S + 1)}{3}.
\end{aligned}
\tag{11.8}
$$

For Eqs. (11.8) it is assumed that $N_S > M_S$. Because of the total number of incomplete voxels, $N_{IV}$ equals to $N_I - N_T$, it is calculated from Eqs. (11.1) and (11.8) as,

$$N_{IV} = 2M_S(N_S + M_S)(M_S - 1) \tag{11.9}$$

When $N_S = M_S$, Eq. (11.9) can be rewritten as,

$$N_{IV} = 4M_S^2(M_S - 1) \tag{11.10}$$

A comparison of Eqs. (11.2) and (11.10) verifies that the number of incomplete voxels is ~6 times greater than that of complete voxels. This means that the image space and voxel resolution will be increased ~7 times greater than those with complete voxels only. The viewers can see the incomplete voxels in the parts of the viewing zone's cross section shown in Fig. 11.6.

There are 15 incomplete voxels: 5, 4, 3, 2 and 1 voxels on the lines connecting PLSs a, b, c, d, and e to the upper end of the viewing zone's cross section, respectively. The outermost incomplete voxels, numbered 1–5, are seen only at section A by PLS a. A is the uppermost section in six and it is equally divided viewing zone cross sections. Among the next four voxels numbered 6–9, voxels 6–9 are seen at sections D and F and at sections and E and F by PLSs f and e, respectively. These voxels are seen at two different segments. Accordingly, the inner voxels numbered 10–12, 13 and 14, and 15 are seen at three, four, and five sections, respectively. As the voxel arrangements in the upper and lower parts of Fig. 11.6 are symmetric along the z axis, the above analysis is valid for the voxels in both parts. There are another 15 incomplete voxels in the back side of the PLS array plane. These voxels can also be seen at parts of the viewing zone's cross section as their counterpart voxels in the front side of the array plane.

The incomplete pixel patterns look more like the pixel patterns in the voxel planes with smaller $|k|$ values because the number of segments in the horizontal and vertical directions can be equal to each other. However, the differences in the relative position and size of each segment in each pixel cell, and the distances between neighboring segments, still distinguish the patterns. Using the incomplete voxels, we can remarkably increase the 3-D image space, compared with the complete-voxels-only case.

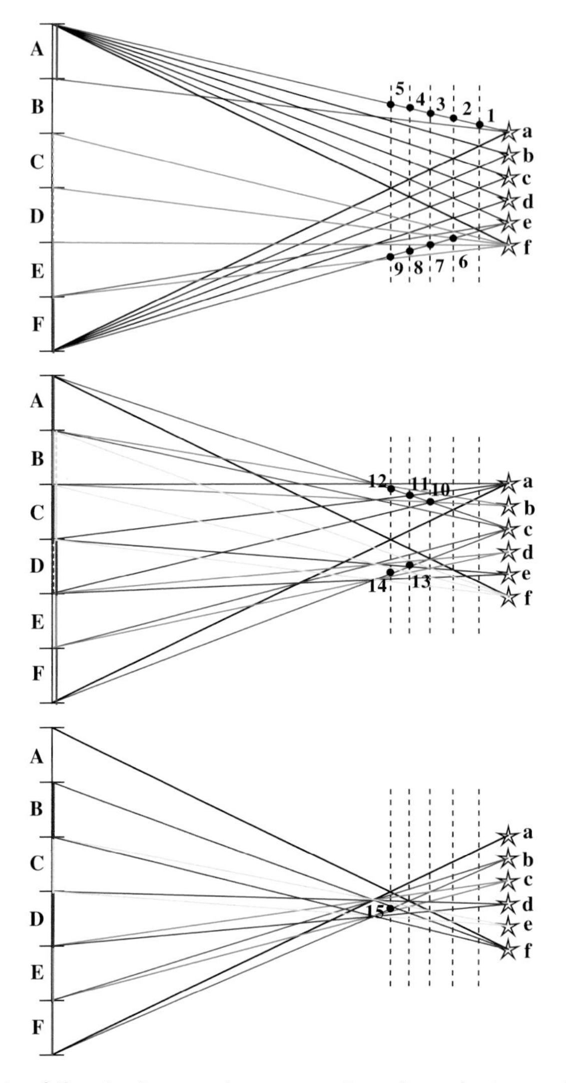


**Fig. 11.6** Parts of the viewing zone's cross section where the incomplete voxels can be seen

## 11.4 Pixel Patterns of Pixel Cells with Rhomb Shapes

To minimize the Moire effect rhomb shaped pixel cells have been used. The Moire effect occurs when overlaying optical plates to form a viewing zone on the display panel in a full parallax 3-D imaging system. There are two steps to create the rhomb cell. The first step is to overlay a rhomboidal net. A rhomboidal net consists of rhombs with a proper vertex angle onto a display panel. The second step is to approximate the sides of each component rhomb. This process is necessary to discrete lines composed of boundaries of pixels along the sides. The net result of this operation is the same as rotating a square or rectangular pixel cell 45 degrees, and then squeezing or stretching the cell in either the horizontal or vertical direction to produce an appropriate vertex angle. Therefore, the pixel pattern for the rhomb shaped pixel cells will have the same pattern as that for the square or rectangular-shaped pixel cells rotated by 45 degrees. Figure 11.7 shows an image display panel with an arrangement of six rhomb shaped pixel cells in both the horizontal and vertical directions in Fig. 11.7(a). Its corresponding PLS array is shown in Fig. 11.7(b). The surface area of the display panel is divided into 72 equal rhombs. However, the total number of PLSs required to illuminate the panel is 84 because the edges of the panel are comprised of isosceles triangles with rhombic half-areas. Figure 11.8shows voxel arrangements in voxel planes of $|k| = 1$ in Fig. 11.8(a) and $|k| = 2$ in Fig. 11.8(b). The voxel positions correspond to the four corner points of the pixel cells in the display panel in Fig. 11.7.

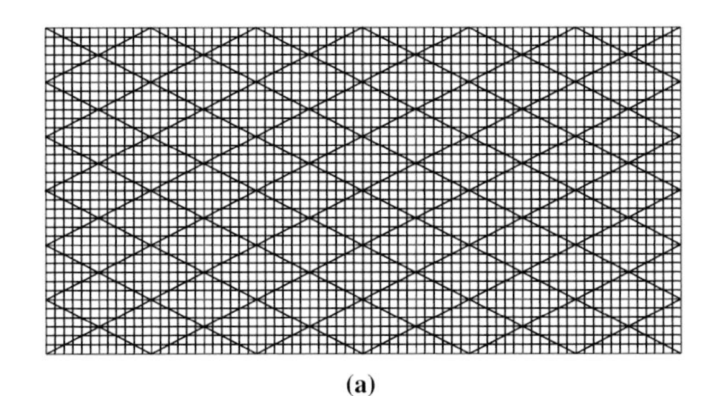

(a)

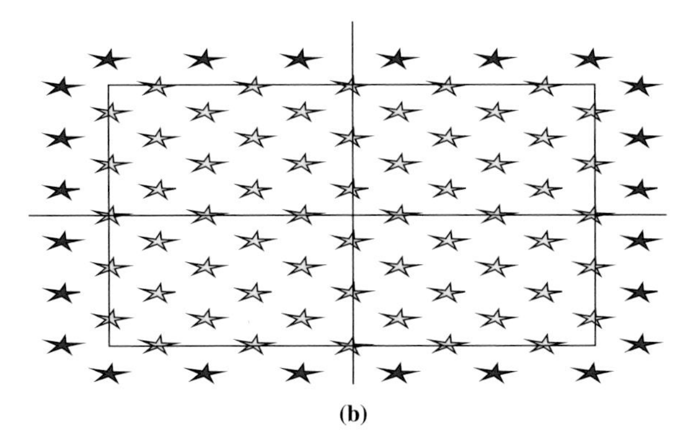

(b)

**Fig. 11.7.** Image display panel with an arrangement of six rhomb shaped pixel cells (**a**) in both horizontal and vertical directions, and (**b**) its corresponding PLS array

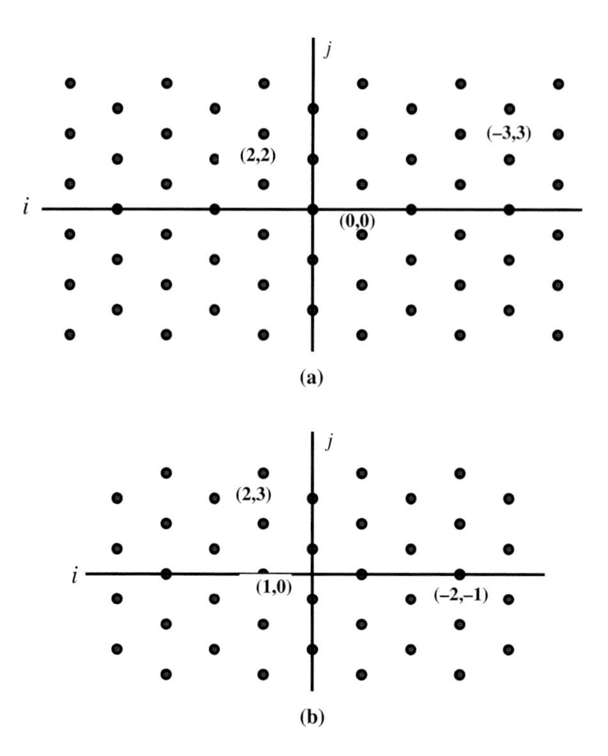


**Fig. 11.8.** Voxel arrangement in voxel planes of (**a**) $k=1$ and (**b**) $k=2$

In this arrangement, the coordinate of each voxel in the $k_{th}$ plane, $X_k^{ij}(C_k^i, C_k^j, z_k)$ , are defined as,

$$C_k^i = \frac{i}{2}G_h^k \quad \text{and} \quad C_k^j = \frac{j}{2}G_v^k \tag{11.11}$$

However, Eq. (11.11) is valid only for the following sets of $i$ and $j$ values: for $i = 0, \pm 2, \pm 4, \cdots\cdots, \pm(N_S - |k|)$, $j = \pm 1, \pm 3, \pm 5 \cdots\cdots, \pm(N_S - 1 - |k|)$ and for $i = \pm 1, \pm 3, \pm 5 \cdots\cdots, \pm(N_S - 1 - |k|)$, $j = 0, \pm 2, \pm 4, \cdots\cdots, \pm(M_S - |k|)$, when $N_S - |k|$ and $M_S - |k|$ are even numbers. These sets of $i$ and $j$ values indicate that there are no voxels with $i = \pm 1, \pm 3, \cdots$ and $j = \pm 1, \pm 3, \cdots$. When $N_S - |k|$ or $M_S - |k|$ are odd numbers, Eq. (11.11) will be satisfied for the following sets of $i$ and $j$ values: $i = 0, \pm 2, \cdots, \pm(N_S - 1 - |k|)$ and $i = \pm 1, \pm 3, \cdots, \pm(M_S - |k|)$. These sets indicate that no voxels with even $i$ and odd $j$ values, or vice versa, exist.

A problem occurs with the pixel pattern for the rhomb shape. There is no 1:1 matching condition of pixels with rhomb shaped cells or segments, since the pixels are either square or rectangular. One can come up with two solutions to work out this problem: (1) any pixel in the rhomb is considered to

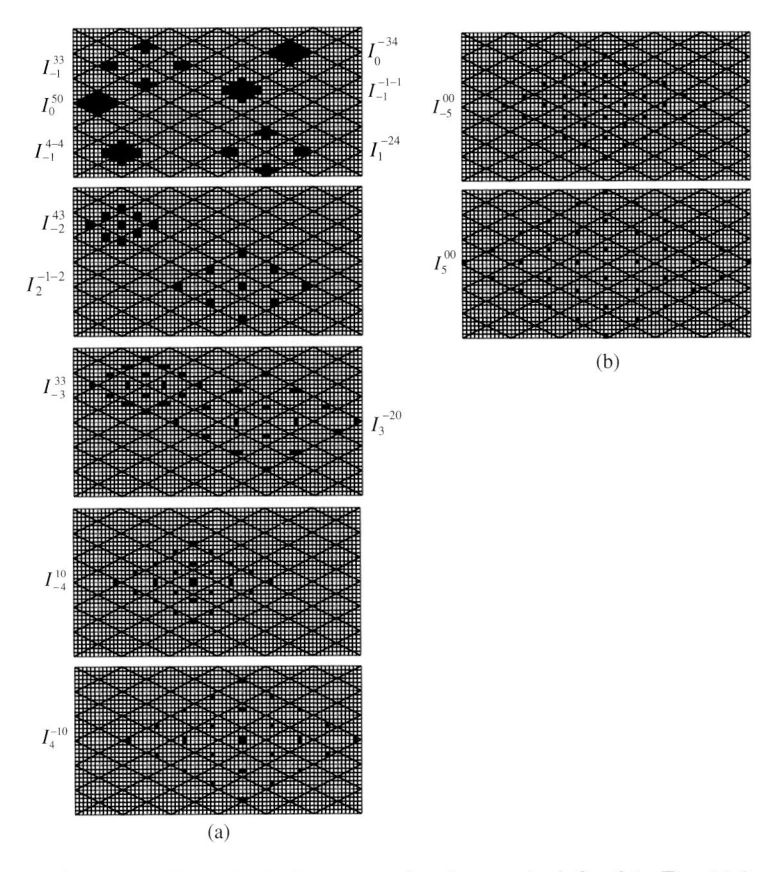


**Fig. 11.9.** Several pixel patterns for the voxels defined in Fig. 11.8

be in a rhomb or rhomb shaped segment if at least one-half of its area belongs to the rhomb, or (2) the total number of pixels should be close to the number in the rhomb. Figure 11.9 shows several pixel patterns for the voxels displayed in Fig. 11.8. The number of pixels in the patterns is 36. This is equal to the number of pixels in a rhomb cell.

Making use of the pixel patterns shown in Fig. 11.9, we are required to show the validity of the pixel patterns (Fig. 11.10). In order to show the validity, 3-D images of five different Platonic solids are generated on an LCD monitor with 300 $\mu$m pixel size in both directions. Each of the solids is composed of pixel patterns. The pixel patterns correspond to $\sim$150 voxels scattered in nine different voxel planes ($|k| = 4$). The solids reveal a good depth. But the images look discrete and change abruptly as the viewing direction changes. This is most likely so because the distances between voxel planes are longer than those between voxels in each voxel plane. Also some voxels in $k = \pm 1$ and $\pm 2$ planes do not appear in the 3-D sense because the designed width of

**Fig. 11.10.** 3-D images of five Platonic solids generated on a LCD monitor with 300 $\mu$m pixel in both directions

the viewing zone's cross section is 20 cm. In any case, Fig. 11.10 shows that the pixel patterns are valid. A magnified image of the combined pixel pattern on the LCD panel is shown in Fig. 11.11. This pattern is the pinnacle of an octahedron in the center of Fig. 11.10. It is encircled in the inset of Fig. 11.11. The rhomb shaped pixel pattern can be traceable from the pattern.

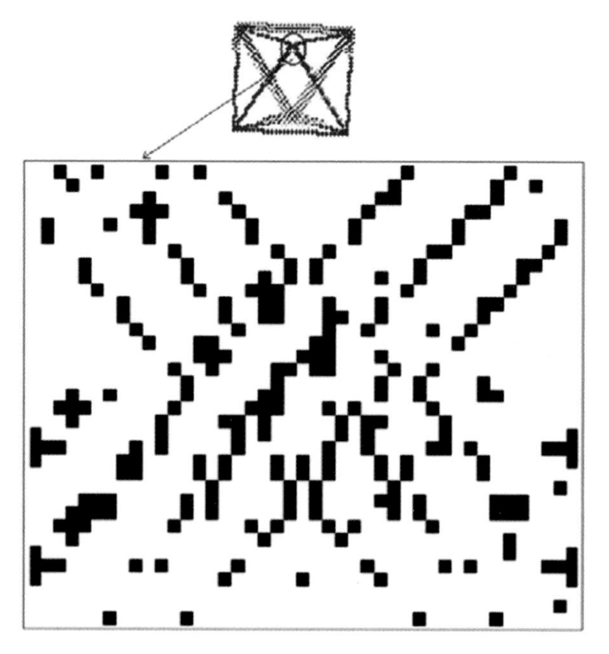

**Fig. 11.11.** Magnified image of the combined pixel pattern on the LCD display panel.

## 11.5 Comparison of 3-D Image Synthesis Between MV and IP Imaging Systems

The IP imaging system acquires and displays light rays passing through a plane. It has been regarded as one of the ideal 3-D photographic technologies. Integral imaging has the advantages of high 3-D luminance and full parallax, but has disadvantages of low 3-D resolution, Moire pattern, and color dispersion.

The MV and IP based on 3-D imaging methods practically have the same optical composition and structure, although the arranging methods of their MV image are different. The same optical structure causes the depth sense mechanism of both methods to be binocular parallax. On the other hand, the difference between them leads the projected images to be different view images [16]. These different view images are in the MV and are virtually synthesized images from the same number of pixels from all elemental images in the display panel.

In MV and IP, the image display panel is divided into pixel cells of an equal size and shape [4]. The pixel cell is located where MV images are arranged. It is the primary unit of display panel that presents MV images together. The viewing zone's cross section represents that all the magnified images of pixel cells by the corresponding lenslets are superposed. The viewing zone's cross section is parallel to the display panel. Pixel pattern in each pixel cell perfectly matches with those in other cells to a pixel unit.

When the viewing zone forms optics, it is said to be a special optical plate. The special optical plate is almost the same size as the panel. It consists of a 2-D array of elemental optics for full parallax image display. The array dimension is the same as the pixel cells in the panel. Various types of plates have been used as a role of optics, such as two superposed lenticular plates, a 2-D pinhole plate, a parallax barrier plate and a lenticular plate. The lenticular plate must be used with either a cross angle between them or 2-D microlens array plate. The lenslet is created by two superposed plates. The shape and size of each lenslet are the same as those of the pixel cells in the center of the panel. A 2-D PLS array can also be used as the optics. In this case it is located at the back side of the panel. We denote the horizontal length of the viewing zone's cross section, the distance between the viewing zone's cross section, and the microlens plate, as $V_W$.

Viewing distance and pitch of each lenslet (Pinhole, PLS) are denoted as $a$ and $P_L$, respectively. Immediately the width of each pixel cell, $C_W$ and focal length of the lenslet, $f$can be defined as,

$$C_W = \frac{P_L V_W}{V_W - P_L}\left(\frac{P_L V_W}{V_W + P_L} \text{ for PLS}\right) \quad \text{and} \quad f = \frac{a P_L}{V_W - P_L}, \tag{11.12}$$

respectively.

It is remarkable that the image depth can be maximized when these two conditions are satisfied. The parallax between left and right most viewing images has to be smaller than interocular distance when the width of the viewing zone's cross section is made as close as the viewer's interocular distance.

Though current IP systems have a greater number of different view images than the MV, IP systems still provide a lower quality image than the MV. This is because the IP system, compared to MV, has the lower quality of resolutions in both recorded and synthesized images, and the effective size of its common viewing zone cross section is smaller. Therefore, it will be good to design IP's optical structure to satisfy Eq. (11.12) and to reduce the resolution of the recorded images for an improved image quality. This process can be done by minimizing the field of view angle of each lenslet. This improvement will accompany the increase in the total number of different view images, i.e., the total number of lenslets in the microlens plate [16].

## 11.6 Conclusion

Voxels are defined in the optical configuration of a full parallax MV imaging system based on a 2-D PLS array. The 3-D images synthesized by the voxels that have been described above, will be more spacious and refined with incomplete voxels. The voxels increase the image space and the image resolution up to $\sim 7$ times greater than those with the complete voxels. We can assume that it is practicable to rotate a square shaped pixel cell 90 degrees. According to the procedure explained in this chapter, we can derive the pixel patterns for rhombic shaped pixel cells by rotating the pixel patterns for a square pixel cell 90 degrees, either clockwise or counterclockwise. The patterns permit images to be displayed with good 3-D quality, but the image quality can be improved. To improve the image quality, one should use the incomplete voxels effectively and use voxels in the higher $|k|$ value planes. The MV and IP based on 3-D imaging methods practically have the same optical composition and structure, although the methods to arrange their MV image are different. To obtain an improved image quality in the IP system its optical structure should be designed to meet specific conditions described in Chapter 5, and reduce the resolution of the recorded images while increasing the total number of lenslets in the microlens plate.

## References

1. R. Borner: FKT. 45 (1991) 453
2. H. Isono, M. Yasudsa and H. Sasazawa: 12th Int. Display Research Conf. Japan, Display'92 (1992) 303
3. T. Honda, Y. Kajiki, K. Susami, T. Hamaguchi, T. Endo, T. Hatada and T. Fuji: Three-Dimensional Television, Video, and Display Technologies. Ch. 19 (2002) 461

4. J. Y. Son, Ch. 2, in Three-Dimensional Television, Video, and Display Technologies, B. Javidi and F. Okano (eds.), Springer (2002)
5. L. F. Hodges and D. F. McAllister: "Computing Stereoscopic Views," in Stereo Computer Graphics and Other True 3D Technologies, D. F. McAllister (ed.), New Jersey, U.S.A.: Princeton University Press, Ch. 5 (1993) 71–89
6. K. Ji and S. Tsuji: 3 Dimensional Vision, Tokyo, Japan, Kyoritz Publishing Company, Ch. 7 (1998) 95–110
7. Y. N. Gruts, J. Y. Son and D. H. Kang: "Stereoscopic Operators and Their Application," J. Opt. Soc. Korea, 5 (2002) pp 90–92
8. T. Izumi: Fundamental of 3-D Imaging Technique, NHK Science and Technology Lab., Tokyo, Japan: Ohmsa (1995)
9. J. Y. Son, V. V. Saveljev, Y. J. Choi, J. E. B. and H. H. Choi: "Parameters for Designing Autostereoscopic Imaging Systems Based on Lenticular, Parallax Barrier and IP Plates," Opt. Eng., 42 (2003) 3326–3333
10. O. Matoba, T. Naughton, Y. Frauel, N. Bertaux and B. Javidi: "Real-time Three-dimensional Object Reconstruction Using a Phase-encoded Digital Hhologram," Appl. Opt., 41 (2002) 6187–6192
11. J. Ren, A. Aggoun and M. McCormick: "Computer Generation of Integral 3D Images with Maximum Effective Viewing," SPIE Proc. 5006 (2003) 65–73
12. J.-Y. Son, V. V. Saveljev, B. Javidi, D.-S. Kim and M.-C. Park: "Pixel Patterns for Voxels in a Contact-type 3-D Imaging System for Full-parallax Image Display," Appl. Opt., 45(18) (2006) 4325–4333
13. J. Y. Son, B. Javidi and K. D. Kwack: Proceedings of the IEEE, Special Issue on: 3-D Technologies for Imaging & Display. 94(3), (2006) 502
14. J-Y. Son, and B. Javidi: "3-Dimensional Imaging Systems based on Multiview Images," IEEE/OSA J. Display Technol. 1 (2005) 125–140
15. F. Okano, H. Hosino, J. Arai, M. Yamada and I. Yuyama: "Three Dimensional Television System Based on Integral Photography," in Three-Dimensional Television, Video, and Display Technique, B. Javidi and F. Okano (eds.) Berlin, Germany: Springer (2002)
16. J.-Y. Son, V. V. Saveljev, K.-T. Kim, M.-C. Park and S.-K. Kim: "Comparison of Perceived Images in Multiview and IP based 3-D Imaging Systems," Jpn. J. Appl. Phys., 46(3A) (2007) 1057
17. F. Okano, H. Hoshino and I. Yuyama: Appl. Opt. 36 (1997) 1598
18. J.-S. Jang and B. Javidi: Opt. Let. 28 (2003) 324
19. H. Liao, M. Iwahara, N. Hata, T. Dohi: Opt. Exp. 12 (2004) 1067
20. L. Erdmann and K. J. Gabriel: Appl. Opt. 40 (2001) 5592
21. H.E. Ives: J. Opt. Soc. Am. 21 (1931) 171
22. C. B. Burckhardt: J. Opt. Soc. Am. 58 (1967) 71
23. T. Okoshi: 3 Dimensional Imaging Techniques. Ch 2 (1976) 21
24. I. Amidror: The Theory of the Moire Phenomenon, Kluwer Academic Publishers (2000)
25. J.-Y. Son and B. Javidi: "Synthesizing 3 Dimensional Images Based on Voxels," SPIE Proc. V5202 (2003) 1–11
26. A. Watt: 3D Computer Graphics, 3rd Edition, Addison-Wesley, Harlow, England, Chap. 13 (2000) 370–391

12

# Multi-view Image Acquisition and Display

Sung Kyu Kim, Chaewook Lee, and Kyung Tae Kim

**Abstract** This chapter discusses distortion analysis in parallel and radial (toe-in) type stereo image capture and display and parallel type multi-view image capture and display. The distortions in the perceived image from a stereoscopic image pair displayed on a screen are analyzed for different conditions of photographing, projecting, and viewing when using either a stereo camera or stereo projector with a parallel configuration. The conditions used for the analysis are positions and stereo bases of the viewer and the camera and magnification of the displayed image. A closed form solution of describing perceived images in the stereoscopic imaging systems with radial recording and projecting geometry for arbitrary viewer positions is presented. This solution is derived by making the heights of homologue points in both left and right images projected on the screen in the geometry to be equal. The solution has the same equation form as that of the parallel geometry, except a constant shifting term in the horizontal direction. This is a primary source of distortions in the perceived image. The condition of eliminating the term makes the solution the same as that for stereoscopic imaging systems with parallel recording and projecting geometry. And a solution for describing both the multi-view image set obtained with a parallel camera layout and the perceived image in a projection-type full parallax multi-view imaging system with a parallel projector layout, is derived by using $4 \times 4$ homogenous matrices to quantitatively analyze the image quality in the system. The solution provides a mean of finding properties and/or behavior of the perceived image changes depending on the viewer's position in the system. The solution can analytically describe the appearance of three-dimensional images in the space generated by the multi-view image set displayed on a projection screen.

S.K. Kim
Imaging media center, Korea Institute of Science and Technology, Seoul, Korea
e-mail: kkk@kist.re.kr

B. Javidi et al. (eds.), *Three-Dimensional Imaging, Visualization, and Display*, 
DOI 10.1007/978-0-387-79335-1_12, 

**Keywords:** Distortions, stereo base, perceived image, geometrical distortion, stereo image pair, projection screen, radial configuration, keystone distortion, multi-view image acquisition, parallel camera layouts, full parallax, homogeneous matrix, perceived image

## 12.1 Introduction

The basic image unit in the autostereoscopic imaging systems is a stereo image pair that can be fused into a stereoscopic image to our eyes. This image pair provides binocular parallax to our eyes. However, the stereoscopic images are more subjected to distortions, compared with 2-D ones, because of their extra dimension in the depth direction. The distortions are the results of the differences in viewing and photographing conditions, i.e., position and stereo base [1]. The distortions increase as the differences increase. To better present stereoscopic images, the distortions should be minimized. Distortions involved with the stereoscopic images are both introduced by psychological and geometrical reasons. Puppet theater and cardboard effects are examples of psychologically generated distortions by the human perception mechanism [2]; however, they are also related to geometry. Keystone and nonlinearity are among the geometrical distortions in stereo image pairs introduced by the photographing and projection mechanisms [3, 4, 5]. The geometrical distortions are mostly caused by the stereo camera with the toed-in configuration. The distortions can be minimized by preparing stereo image pairs with a stereo camera with a parallel configuration with the stereo base equal to that of our eyes, and making photographing and display conditions such as (1) distances of camera from the convergence plane, and viewer to the image projection screen, and (2) equal positions of the camera and the viewer relative to the plane and the screen, respectively [6]. However,, in practice, it is difficult to make the photographing and display conditions the same lead to the distortions. The geometrical distortions are quantitatively analyzed by several authors [3, 7]; however, they are not fully developed to include distortions in the image perceived by a viewer due to changes in the viewer position relative to the screen. It was shown that the difference in the conditions of viewing and photographing introduces distortions in the perceived image [1]. But, no quantitative analysis relating all those conditions has been done yet. In this chapter, distortions in the perceived stereoscopic images due to the differences in the conditions of photographing, displaying and viewing are quantitatively analyzed and a method of minimizing the distortions is also discussed.

In the parallel type layout, cameras (projectors) are aligned with a constant interval to have their optical axes in parallel to each other in a camera (a projector) array, but in the radial type layout, they are aligned with a constant angle interval on an arc with a certain radius of curvature. Between these two, in aligning point of view, the radial type layout is relatively easier than the parallel type layout because it has a reference point as its center of the radius curvature. In the distortion point of view, both types are suffering from nonlinearity distortions caused by geometrical mismatches between left and right eye images in the stereo image pair [6, 8, 9, 10]. For the the radial type layout, it is also suffering from a perspective distortion – keystone – which is originated by size difference between left and right side images projected on the screen [3, 11, 12]. Furthermore, mismatches in the geometrical parameters of recording, projecting and viewing relative to the screen/panel also cause distortions in the perceived image which is defined as the 3-D image fused as a 3-D image by the viewers' eyes from the stereoscopic image pair projected to their eyes. The distortion in the perceived image is investigated for the parallel type by deriving a closed form expression for the image [13] in this chapter. But for the radial type, no closed form expression is derived and the investigation was limited to when the stereo image pair from the radial type is displayed on a flat display panel/screen [3] by assuming that viewer eyes are located along the horizontal line passing the center of the panel/screen. This probably due to the height difference between left and right images in a stereo image pair introduced by the radial type camera array making the geometrical analysis of the perceived image's behaviors practically impossible. Since radial recording and projection is a typical practice in projection type stereoscopic imaging systems, image projection by a radial type stereo projector will make the analysis impossible, unless the height difference between the projected images of left and right eyes is eliminated. Therefore a mathematical expression of the perceived image in the projection type 3-D imaging systems with a radial type recording and projecting geometry is derived by finding a condition to make the height difference between left and right eye images projected on the screen the same [14].

With full parallax multi-view 3-D imaging systems, there is no appropriate solution for the perceived image, i.e., the 3-D image which the viewer actually perceives from a stereoscopic image pair in the multi-view images displayed on the screen. In the multi-view imaging systems, the viewing zone where viewers can perceive depth sense is pre-defined at the space in front of the image display screen as a design parameter [15, 16], and divided into many subzones to provide viewing regions for each view image displayed on the screen. The viewer's two eyes should be located at two different viewing regions to perceive a stereoscopic image. Hence, the distance between the neighboring viewing regions should be either a submultiple of or the same as the viewer's interocular distance. It is also noticed that the viewing zone geometries are not much different from those of the camera layouts. A generalized solution which can predict distortions in the perceived image in projection-type multi-view 3-D imaging systems are derived by introducing a multi-view image transform method based on $4 \times 4$ homogeneous matrices [17]. The solution can be applied for many parallel cameras and projectors with different configurations and parameters. The homogeneous matrix method was originally developed to find the screen coordinate of 3-D graphic images. This method is extended to map the world space into the perceived image space by going through two transforms: (1) transforming the world space of a target object to the screen images and (2) the images to the perceived image space. The homogeneous representation allows composing the resulting transform as a sequence of basic transformations. This approach can be applied for quantitatively evaluating perceived image quality in various multi-view 3-D imaging systems with different viewing zone configurations [18, 19, 20, 21].

The multi-view image is assumed here as the images taken from a camera array consisting of N X M cameras arranged in a matrix form, and a target object is located near the origin [22].

## 12.2 Stereoscopic Image Distortion Analysis in Parallel Type Camera Configuration

In the Cartesian coordinate centered at the upper left corner of an image projection and/or display screen, as shown in Fig. 12.1, when a stereo camera with a stereo base (distance between two entrance pupils of the stereo camera) $2a$, positioned at $C(x_C, y_C, z_C)$ is photographing a point object $O_O(x_O, y_O, z_O)$, the stereo image pair recorded on film can be considered the same as that projected on the screen surface which is normal to the camera axis and located behind the object because of the similar triangle relationship. The screen can

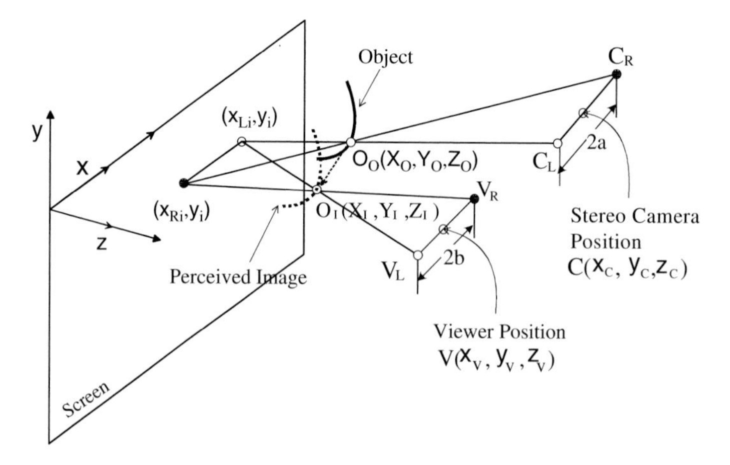


**Fig. 12.1.** A coordinate for stereo image calculation

be considered a monitor surface. The stereo image pair on the screen surface can be represented as a column matrix $\hat{S}_i$ if the camera is parallel to the $x$ axis [23]. $\hat{S}_i$ is represented as,

$$\hat{S}_i = P_a \bullet \hat{O}_O, \tag{12.1}$$

where $P_a$ is a matrix relating $\hat{S}_i$ and a column matrix defining the object point, $\hat{O}_O.\hat{S}_i$, $\hat{O}_O$and $P_a$ are represented as,

$$\hat{S}_i = \begin{pmatrix} x_{L_i} \\ x_{R_1} \\ y_i \end{pmatrix}, \quad \hat{O}_O = \begin{pmatrix} X_O \\ Y_O \\ Z_O \end{pmatrix} \quad \text{and} \quad P_a = \frac{1}{(z_C - Z_O)} \begin{pmatrix} z_C & 0 & a - x_C \\ z_C & 0 & -a - x_C \\ 0 & z_C & -y_C \end{pmatrix}, \tag{12.2}$$

where $x_{Li}$ and $x_{Ri}$, and $y_i$ are $xy$- axis values of i-th point in the stereo image pair. From Eqs. 12.1 and 12.2, $x_{Li}, x_{Ri}$ and $y_i$ are found as,

$$\begin{aligned} x_{Li} &= \frac{z_C X_O - Z_O(x_C - a)}{z_C - Z_O} \\ x_{Ri} &= \frac{z_C X_O - Z_O(x_C + a)}{z_C - Z_O} \\ y_i &= \frac{z_C Y_O - y_C Z_O}{z_C - Z_O} \end{aligned} \tag{12.3}$$

The object location corresponding to a given stereo image pair can be found by reversing the Eq. 12.1. Then $\hat{O}_O$ is expressed as,

$$\hat{O}_O = P_a^{-1} \bullet \hat{S}_i, \tag{12.4}$$

where $P_a^{-1}$ is the inverse matrix of $P_a$. When a viewer whose position $V(x_V, y_V, z_V)$, is defined by the midpoint between his/her two eyes specified by $V_L$ and $V_R$, respectively, and whose eye line, i.e., a line segment $V_L V_R$ is parallel with the screen and the $x$axis, is watching the screen where a stereo image pair represented by $\hat{S}_i$ is displayed, the perceived image position $O_I(X_I, Y_I, Z_I)$ can be found by Eq. 12.4 by replacing $P_a^{-1}$ to $P_b$, which defines the position of a viewer. Hence, a column matrix $\hat{O}_I$ that defines the perceived image position is expressed as,

$$\hat{O}_I = P_b \bullet \hat{S}_i \tag{12.5}$$

Where,

$$P_b = \frac{1}{(2b + x_{Li} - x_{Ri})} \begin{pmatrix} b + x_i & b - x_i & 0 \\ y_V & -y_V & 2b \\ Z_V & -z_V & 0 \end{pmatrix}, \tag{12.6}$$

and b is a half distance between our two eyes. By combining Eqs. 12.5 and 12.6, $\hat{O}_I$ can be rewritten as,

$$\hat{O}_I = P_b \bullet P_a \bullet \hat{O}_O \tag{12.7}$$

By combining Eqs. 12.2 and 12.6, $P_b \bullet P_a$ in Eq. 12.7, is calculated as,

$$P_b \bullet P_a = \frac{2}{(2b + x_{Li} - x_{Ri})(z_C - Z_O)} \begin{pmatrix} b_{zC} & 0 & ax_V - bx_C \\ 0 & bz_C & ay_V - by_C \\ 0 & 0 & az_V \end{pmatrix} \tag{12.8}$$

By substituting Eq. 12.8 to Eq. 12.7, $O_I(X_I,\ Y_I,\ Z_I\ )$ can be found. $X_I$, $Y_I$ and $Z_I$ are calculated as,

$$\begin{aligned} X_I &= \frac{bX_O z_C + Z_O(ax_V - bx_C)}{bz_C + Z_O(a-b)} \\ Y_I &= \frac{bY_O z_C + Z_O(ay_V - by_C)}{bz_C + Z_O(a-b)} \\ Z_I &= \frac{aZ_O z_V}{bz_C + Z_O(a-b)} \end{aligned} \tag{12.9}$$

With Eq. 12.9, the distortion in the perceived image, i.e., the difference between the object and the image, can be found. Equation 12.9 offers the following facts:

(1) Since the stereo camera and object positions are fixed, terms $Z_O(ax_V - bx_C)$ and $Z_O(ay_V - by_C)$ indicate that $X_I$ and $Y_I$ are functions of the viewer position. This means that the perceived image will be shifted to the viewer direction and the shift amount will increase as the $Z_O$ value increases. Hence, the perceived image will follow the viewer. The consequence is that the image will be distorted more for bigger $Z_O$ values and bigger deviations of the viewer position from the $z$-axis. Making $x_C$(and $y_C$) and $x_V$ (and $y_V$) zero will minimize the distortion.

(2) $X_I$ and $Y_I$ have the scaled values of $X_O$ and $Y_O$ with a scale factor $bz_C / \{bz_C + Z_O(a-b)\}$, respectively, because terms $Z_O(ax_V - bx_C)$ and $Z_O(ay_V - by_C)$ are simply making the perceived image shift to the viewer direction. The scale factor is a constant for a given $Z_O$ value. This means that the shapes of an $xy$ plane cross section of the perceived image and of its corresponding object cross section are the same, but their sizes are different by the scale factor. When $a > b$, as the $Z_O$ value increases, the scale factor becomes smaller and, consequently, the perceived image cross section decreases as well, compared with its corresponding object cross section. For the opposite case, the perceived image cross section becomes larger as the $Z_O$ value increases. Hence, the $xy$ plane cross section of the perceived image can be smaller or bigger than its corresponding object cross section, i.e., the image

will be distorted due to the differences between photographing and viewing conditions. In $z$-axis direction, $Z_I$ is scaled by $az_V/\{bz_C + Z_O(a-b)\}$. Hence, the perceived image will be distorted more in the $z$ -axis direction.

(3) When $a = b$, Eq. 12.9 is simplified as $Z_I = Z_O z_V / z_C$ and $W_I = W_O + (\omega_V - \omega_C) Z_O / z_C$, where $W$ and $\omega$ represent $X$(or $Y$) and $x$(and $y$), respectively. These relationships dictate that the perceived image size in the $z$-axis direction is different from $Z_O$ by the factor $z_V/z_C$, but in $x$ and $y$ -axis directions, the image has the same size as the object, but it is shifted by the amount $(\omega_V - \omega_C)Z_O/z_C$, to the viewer direction. The shift will be more for larger $Z_O$ values and smaller for larger $z_C$ values. When the viewer and the camera positions are also the same, the positions of the image and the object become the same, i.e., the perceived image and the object are spatially matched.

(4) When $a \neq b$, the $Z_O$ term in the denominator of Eq. 12.9 makes the sizes of the image and the object differ from each other. For the object points having the same $z$-axis value, the corresponding image points are shifting the same amount in both $x$ and $y$-axis directions. The amount of shifting becomes more as the $z$-axis value increases. Hence, the object with a cube shape will be perceived as a slanted trapezoidal pillar. This is shown in Fig. 12.2 which depicts the perceived image of a cube when the photographic and viewing conditions are different. Figure 12.2 clearly shows that the perceived image is shifted to the viewer direction, its front surface is bigger than its inner surface and it is extended in the $z$ -axis direction. The image is not a cube, but has the shape of a slanted trapezoidal prism.

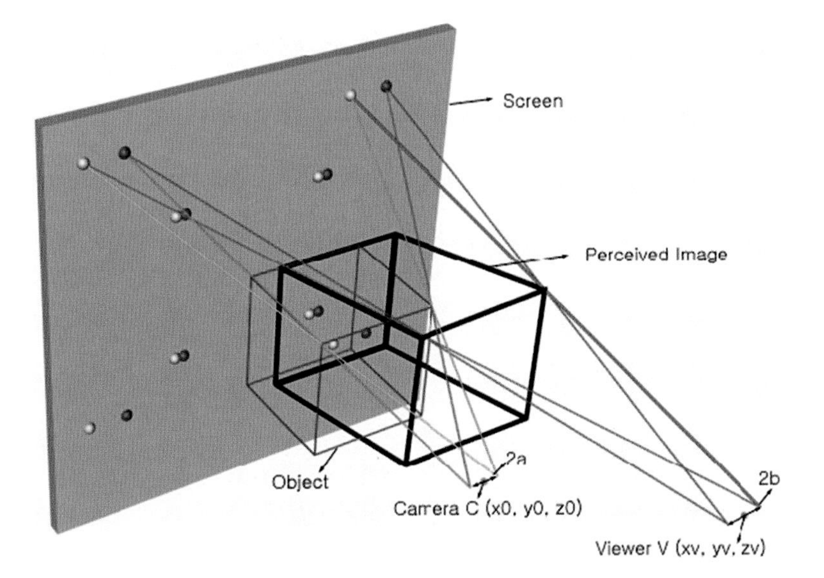


**Fig. 12.2.** The perceived image of a cube

(5) When $b > a$, the denominator of Eq. 12.9 always has a positive value because $z_C > Z_O$, i.e., the object is located in front of the camera. In this case, if$Z_O$ is in $-\propto$, $X_I$, $Y_I$ and $Z_I$ are simplified as $(ax_V - bx_C)/(a - b)$, $(ay_V - by_C)/(a - b)$ and $-az_V/(b - a)$, respectively. These relationships determine that the space behind the screen is compressed to a point because $X_I$ and $Y_I$ are independent to $X_O$ and $Y_O$, respectively, and the maximum depth the perceived image can have is $-az_V/(b - a)$.

(6) When $a > b$, i.e., the camera stereo base is larger than the viewer's eye separation, if $Z_O = -bz_C/(a - b)$, the denominator of Eq. 12.9 will be zero and, hence, $X_I$, $Y_I$ and $Z_I$ become infinity, respectively. This is the case when the parallax difference between the stereo image pair of an object in $Z_O$ is the same as our eye separation. It occurs only when $Z_O$ has a negative value, i.e., the object point is located behind the screen. The object points more distanced than the $Z_O$ value will produce stereo image pairs with the parallax difference more than our eye separation and, consequently, they cannot be fused.

(7) The simplified way to recognize distortions in the perceived image is to calculate the difference between positions in the perceived image and the original object. It is given as,

$$\begin{aligned} X_O - X_I &= \frac{(b-a)X_O + (ax_V - bx_C)}{(bz_C/Z_O) + (a-b)} \\ Y_O - Y_I &= \frac{(b-a)Y_O + (ay_V - by_C)}{(bz_C/Z_O) + (a-b)} \\ Z_O - Z_I &= \frac{(b-a)Z_O + (az_V - bz_C)}{(bz_C/Z_O) + (a-b)} \end{aligned} \tag{12.10}$$

Equation 12.10 shows that the difference in each coordinate direction has the same equation form; however, it depends on the object positions in the same coordinate direction and $z$ -axis. This means that the difference is different for different object positions, i.e., the perceived image will be distorted. For the case of $a = b$, i.e., the stereo bases of both camera and viewer's eye are equal, the difference in each axis direction becomes proportional to the difference between the positions of the camera and the viewer, scaled by $Z_O/z_C$. If the positions of the camera and the viewer are also the same, the difference will be zero, i.e., no distortion in the perceived image.

## 12.3 Stereoscopic Image Distortion Analysis in Radial Type Camera Configuration

For the analysis the following symbols are used to represent geometric parameters related to cameras, projectors and viewer:

$C_L(C_R)$ – left (right) camera; $P_L(P_R)$– left (right) projector; $V_L(V_R)$ – left (right) eye of viewer; 2c(2p,2v) – distance between left and right cameras (projectors, eyes) of stereo camera (stereo projector, viewer); $2\varphi_C(2\varphi_P)$ – convergence angle of stereo camera (stereo projector), and $M_C(M_P, M_V)$ – center position of the distances between left and right cameras (projectors, eyes) of the stereo camera (stereo projector, viewer). The position is defined by $(x_C, y_C, z_C)\{(x_P, y_P, z_P), (x_V, y_V, z_V)\}$. It is also assumed that the coordinate origin is the upper left corner of the screen and the screen has a rectangular shape with size $l \times h$, and the stereo camera (stereo projector and viewer's eyes) is located along the normal line originated from the center of the screen. Hence, $x$ and $y$ axis coordinate values of camera and projector, $x_C$ and $x_P$, and $y_C$ and $y_P$ when camera image is not magnified in the process of projecting, are represented as,

$$\begin{aligned} x_C &= x_P = l/2, \\ y_C &= y_P = h/2 \end{aligned} \tag{12.11}$$

When $m$ times magnification is introduced, they are represented as,

$$\begin{aligned} x_C &= l/2 \quad \text{and} \quad x_P = m(l/2) \\ y_C &= h/2 \quad \text{and} \quad y_P = m(h/2) \end{aligned} \tag{12.12}$$

The 3-D view of a stereo image pair recording geometry for both parallel and radial type layouts is shown in Fig. 12.3. In Fig. 12.3, $C_L$ and $C_R$ represents left and right cameras in a stereo camera, respectively. Two cameras are $2c$ distance apart, and the lines connecting $C_L$ to $C_R$, and the center of the screen, F to the center of the stereo camera, $M_C$ are parallel and normal to the screen, respectively. The screen in this geometry is the same as the film plane of the stereo camera. When the optical axes of left and right cameras are in parallel, i.e., in parallel type layout, the stereo image pair of an object point I recorded on the left and right camera will appear at points $I_{S_R}$ and $I_{S_L}$ on the screen plane, respectively. Hence, $y$-axis values of $I_{S_R}$ and $I_{S_L}$ are the same, but the two cameras are directed to F. The stereo image pair, $I_{C_R}$ and $I_{C_L}$, is no longer appearing on the screen, but on the plane normal to the optical axes of their corresponding cameras, though $I_{S_R}$ and $I_{C_R}$ and $I_{S_L}$ and $I_{C_L}$ are in the same lines originated from their corresponding cameras due to the fact that $C_L$ and $C_R$ represent the centers of left and right camera objectives, respectively. Since the distances from $I_{C_R}$ and $I_{C_L}$ to their corresponding cameras are different, $y$-axis values of $I_{C_R}$and $I_{C_L}$ will not be the same to each other.

When a stereo camera is in parallel type layout, the stereo image pair of an object point I is obtained by the following relationships [13]:

$$s_I = A_C \cdot V_I / (z_C - Z_I), \tag{12.13}$$

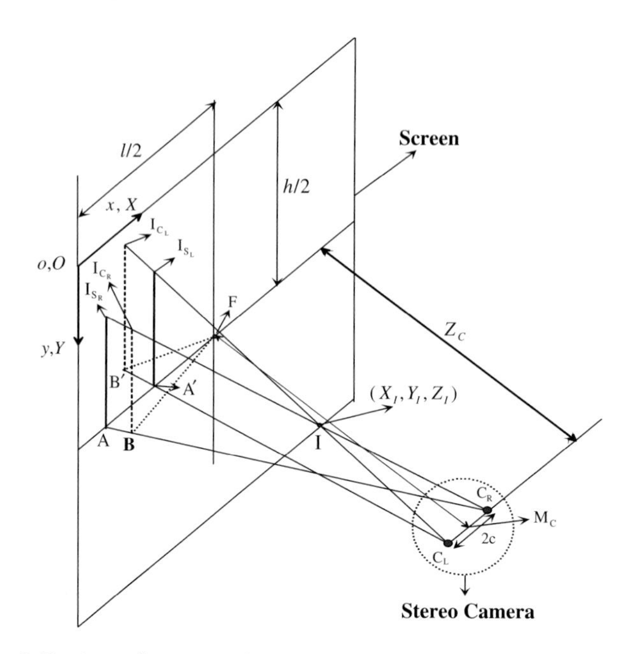


**Fig. 12.3.** 3-D view of a stereo image pair recording geometry for both parallel and radial type layouts

where

$$s_I = \begin{bmatrix} x^i_{C_L} \\ x^i_{C_R} \\ y^i_{C_{LR}} \end{bmatrix}, \quad V_I = \begin{bmatrix} X_I \\ Y_I \\ Z_I \end{bmatrix} \quad \text{and} \quad A_C = \begin{bmatrix} z_C & 0 & C - l/2 \\ z_C & 0 & -C - l/2 \\ 0 & -z_C & -h/2 \end{bmatrix}$$

represent the positions of the stereo image pair and object, and a $3 \times 3$ matrix defining stereo camera base, respectively. $\mathrm{I_{S_L}}(x^i_{C_L}, y^i_{C_{LR}}, 0)$ and $\mathrm{I_{S_R}}(x^i_{C_R}, y^i_{C_{LR}}, 0)$ represent positions of left and right images, respectively.

From Eq. 12.13, $x^i_{C_L}$, $x^i_{C_R}$ and $y^i_{C_{LR}}$ are calculated as,

$$\begin{aligned} x^i_{C_R} &= \frac{X_I z_C - Z_I\,(l/2 + c)}{z_C - Z_I} \\ x^i_{C_L} &= \frac{X_I z_C - Z_I\,(l/2 - c)}{z_C - Z_I} \\ y^i_{C_{LR}} &= \frac{Y_I z_C - Z_I\,(h/2)}{z_C - Z_I} \end{aligned} \tag{12.14}$$

In Fig. 12.3 coordinate $z_C$ is always greater than $Z_I$, i.e., $z_C > Z_I$ because the camera should be away from the object and $x^i_{C_L} \geq x^i_{C_R}$. When the axes of left and right cameras in the stereo camera are rotated $90° - \alpha = \varphi_c$ in both CW and CCW directions, respectively, by pivoting the center of their

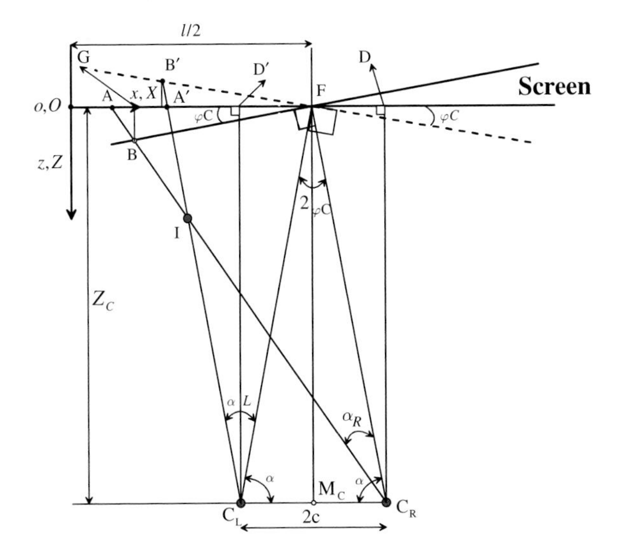


**Fig. 12.4.** Top view of Fig. 12.1 geometry

objective lenses, i.e., the stereo camera is reconfigured as the radial type layout, the stereo image pair $I_{C_R}$ and $I_{C_L}$ is no longer focused at the screen. The coordinates of $I_{C_R}$ and $I_{C_L}$ can be found from Fig. 12.4 which is the top view of Fig. 12.3 geometry.

In Fig. 12.4, the positions of $I_{S_R}$ and $I_{S_L}$ are specified as A and A′, respectively, and those of $I_{C_R}$ and $I_{C_L}$ as B and B′, respectively. The positions of points B and B′ will be determined by $\alpha_R$ and $\alpha_L$ which are angles defining the positions of points A and A′ from optical axes of their corresponding cameras, respectively. Since $\alpha_R$ and $\alpha_L$ have different values depending on the relative positions of A and A′ along the $x$-axis, from Fig. 12.4, $\alpha_R$ and $\alpha_L$ are represented as:

$$\begin{aligned} \alpha_R &= \tan^{-1}\left(\frac{z_C}{c}\right) - \tan^{-1}\left[\frac{z_C - Z_I}{l/2 + c - X_I}\right] \\ \alpha_L &= \tan^{-1}\left(\frac{z_C}{c}\right) + \tan^{-1}\left[\frac{z_C - Z_I}{(l/2 - c) - X_I}\right] \end{aligned} \tag{12.15}$$

Since the distances are defined as $\bar{B}\bar{G}(= \bar{B}\bar{F}\sin\varphi_C)$ and $\bar{B}'\bar{G}'(= \bar{B}'\bar{F}\sin\varphi_C)$ from triangles $\Delta FAB$ and $\Delta FA'B'$, respectively, $y^i_{C_R}$ and $y^i_{C_L}$ are calculated as,

$$\begin{aligned} y^i_{C_R} &= \frac{h}{2} - (\frac{h}{2} - y^i_{C_{LR}})\frac{z_C - \bar{B}\bar{F}\sin\varphi_C}{z_C} = y^i_{C_{LR}} + (\frac{h}{2} - y^i_{C_{LR}})\frac{\bar{B}\bar{F}\sin\varphi_C}{z_C} \\ y^i_{C_L} &= \frac{h}{2} - (\frac{h}{2} - y^i_{C_{LR}})\frac{z_C + \bar{B}'\bar{F}\sin\varphi_C}{z_C} = y^i_{C_{LR}} - (\frac{h}{2} - y^i_{C_{LR}})\frac{\bar{B}'\bar{F}\sin\varphi_C}{z_C} \end{aligned} \tag{12.16}$$

When I is in the light side of the line connecting F to $M_c$, the sign of the terms containing $\bar{B}\bar{F}\sin\varphi_C$ and $\bar{B}'\bar{F}\sin\varphi_C$ will be reversed in Eq. 12.16. Equation 12.16 clearly shows that $y^i_{C_R}$ and $y^i_{C_L}$ are different from each other. From these results, the coordinates of points $I_{C_R}$ and $I_{C_L}$ in Fig. 12.1 are defined as $(l/2 - \bar{B}\bar{F}\cos\varphi_C, y^i_{C_R}, \bar{B}\bar{F}\sin\varphi_C)$ and $(l/2 - \bar{B}'\bar{F}\cos\varphi_C, y^i_{C_L}, -\bar{B}'\bar{F}\sin\varphi_C)$, respectively. When $I_{C_R}$ and $I_{C_L}$ are displayed on the image display panel, they may not be fused as a spatial point because lines connecting $I_{C_R}$ and $I_{C_L}$ to their corresponding eyes will not meet together due to the height difference in the $y$-axis direction. However, since the radial type projector array is compensating the distortions in the stereo image pair obtained by the radial type camera array, a fusing condition can be found.

The image projection geometry in the $xz$ plane is shown in Fig. 12.5. In this geometry, if the projectors introduce magnification in the image, the screen should also be magnified equally because it is the magnified image of the film plane of the stereo camera.

In Fig. 12.5, the left and right projectors in the stereo projector are directed $90° - \beta = \varphi_p$ to the right and left directions, respectively, and converged to a point F which is in the center of the screen plane. The $x$- axis values of the stereo image pair projected to the screen, $x^i_{P_L}$ and $x^i_{P_R}$, are obtained as,

$$x^i_{P_R} = m(l/2) \mp \overline{SF} \text{ and } x^i_{P_L} = m(l/2) \mp \bar{S}'\bar{F} \tag{12.17}$$

$y^i_{P_R}$ becomes greater than $y^i_{C_R}$ and $y^i_{P_L}$ less than $y^i_{C_L}$ due to the radial projection when there is no magnification. Since the magnification factor $m$ is

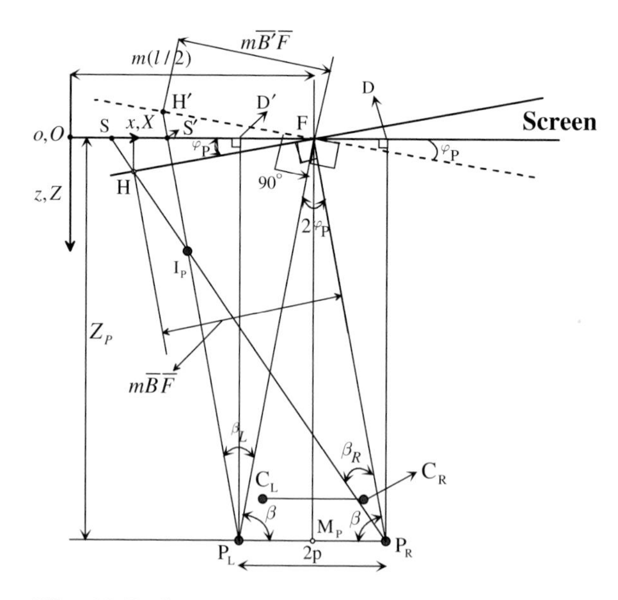


**Fig. 12.5.** Image projection geometry in the $xz$ plane

also valid for the $y$-axis direction, $y^i_{P_R}$ and $y^i_{P_L}$ are calculated with the help of Fig. 12.5 as,

$$\begin{aligned} y^i_{P_R} &= \frac{mh}{2} - \frac{z_P}{z_P - m\bar{B}\bar{F}\sin\varphi_P} m\left(\frac{h}{2} - y^i_{C_R}\right) \\ y^i_{P_L} &= \frac{mh}{2} - \frac{z_P}{z_P + m\bar{B}\bar{F}\sin\varphi_P} m\left(\frac{h}{2} - y^i_{C_L}\right) \end{aligned} \tag{12.18}$$

In Eq. 12.18, $\varphi_P$ should be replaced by $-\varphi_P$ when the corresponding $x^i_{C_R}$ or $x^i_{C_L}$ to each relationship is greater than $l/2$. The stereo image pair on the screen consists of two image points $\mathrm{I_{P_R}}(x^i_{P_R}, y^i_{P_R}, 0)$ and $\mathrm{I_{P_L}}(x^i_{P_L}, y^i_{P_L}, 0)$. These two image points will not be fused as an image point to viewers' eyes unless $y^i_{P_R} = y^i_{P_L}$ on the geometrical point of view, i.e., the lines connecting a viewer's two eyes to their corresponding image points on the screen will never meet. But in practice, fusing the left and right images a stereo image pair as a stereoscopic image is still possible with geometrical differences between the images, such as size, image center and image direction unless they exceed certain values [24]. For this reason, $y^i_{P_R}$ and $y^i_{P_L}$ could be represented by their average value $(y^i_{P_R} + y^i_{P_L})/2$ [3]. But this value can be applied only when $y^i_{P_R} - y^i_{P_L}$ is small. When either $\varphi_C$ or $\varphi_P$, or both $\varphi_C$ and $\varphi_P$ are big, the difference $y^i_{P_R} - y^i_{P_L}$ on both sides of the stereo image pair can exceed the fusing limit. Hence, the condition of making $y^i_{P_R} = y^i_{P_L}$ is found from Eqs. 12.16 and 12.18. This condition works for the entire screen without regard to $\varphi_C$ and $\varphi_P$ values. It is calculated as,

$$\frac{\sin\varphi_C}{\sin\varphi_P} = m\frac{z_C}{z_P} \tag{12.19}$$

Equation 12.19 indicates that the ratio of sine value of the camera orientation angle to that of the projector orientation angle should be equal to the ratio of the camera distance to the projector distance from the screen, multiplied by the magnification/demagnification factor. Since this condition makes the vertical size of both right and left eye images the same, it is the condition of eliminating keystone distortion. With Eq. 12.19, Eq. 12.18 is reduced as,

$$\frac{mh}{2} - y^i_{P_R} = \frac{mh}{2} - y^i_{P_R} = m\left(\frac{h}{2} - y^i_{C_{LR}}\right) \tag{12.20}$$

Since $\sin\varphi_C = c/\sqrt{c^2 + z_C^2}$ and $\sin\varphi_P = p/\sqrt{p^2 + z_P^2}$, Eq. 12.18 is written as,

$$z_P c\sqrt{p^2 + z_P^2} = m z_C p\sqrt{c^2 + z_C^2} \tag{12.21}$$

In parallel type, the size of the projected stereoscopic image pair is given as $m(l/2 - x^i_{C_L})$ and $m(l/2 - x^i_{C_R})$. The ratio is obtained as,

$$\frac{m(\frac{l}{2} - X^i_{CR})}{\overline{SF}} = \frac{c^2 + z_C(z_C - Z_I) - c^2 Z_I/z_C}{\sqrt{p^2 + z_P^2}\sqrt{c^2 + z_C^2}(z_C - Z_I)/z_P} = \frac{\sqrt{c^2 + z_C^2}/z_C}{\sqrt{p^2 + z_P^2}/z_P} = \frac{\cos\varphi_P}{\cos\varphi_C}$$
$$\frac{m(\frac{l}{2} - X^i_{CL})}{\overline{S'F}} = \frac{c^2 + z_C(z_C - Z_I) - c^2 Z_I/z_C}{\sqrt{p^2 + z_P^2}\sqrt{c^2 + z_C^2}(z_C - Z_I)/z_P} = \frac{\cos\varphi_P}{\cos\varphi_C} \tag{12.22}$$

The results are the same for I in other places specified above. When $\cos\varphi_C = \cos\varphi_P$, i.e., $\varphi_C = \varphi_P$, the projected image size in radial type becomes the same as that in parallel type. This means that no extra distortion in the perceived image will be introduced by the radial type, compared with the parallel type. By combining this relationship with Eq. 12.17, the following is obtained:

$$z_P^2 = mpz_C^2/c \tag{12.23}$$

In this condition no keystone distortion appears and the projected images in both radial and projection types are exactly the same. When $\varphi_C > \varphi_P$, the image size of the radial type becomes bigger than that of parallel type. This result will introduce a somewhat extended image depth and image size in the horizontal direction compared with that of the parallel type. But when $\varphi_C < \varphi_P$, they are opposite.

## 12.4 Multi-view Image Acquisition and Display

Basically, the two possible geometries to layout cameras for multi-view image acquisition are parallel and radial (toed-in) layouts [25] as shown in Fig. 12.6. For the case of the parallel layout, the cameras with equal characteristics are arranged two-dimensionally on the $xy$ plane with an equal distance for cameras in the same direction, i.e., $\Delta x$ and $\Delta y$ as in Fig. 12.6.

In Fig. 12.6, it is assumed that the cameras in each layout have the same magnification, the screen plane is located at $xy$ plane and centered at $z = 0$, along with the object. In this case, the position of each camera $(x_{ij}, y_{ij}, z_{ij})$ in both layouts can be written as,

$$x_{ij} = \Delta x \cdot \left(-\frac{N}{2} + i - \frac{1}{2}\right), \quad y_{ij} = \Delta y \cdot \left(\frac{M}{2} - j + \frac{1}{2}\right) \text{ and } z_{ij} = z_C \tag{12.24}$$

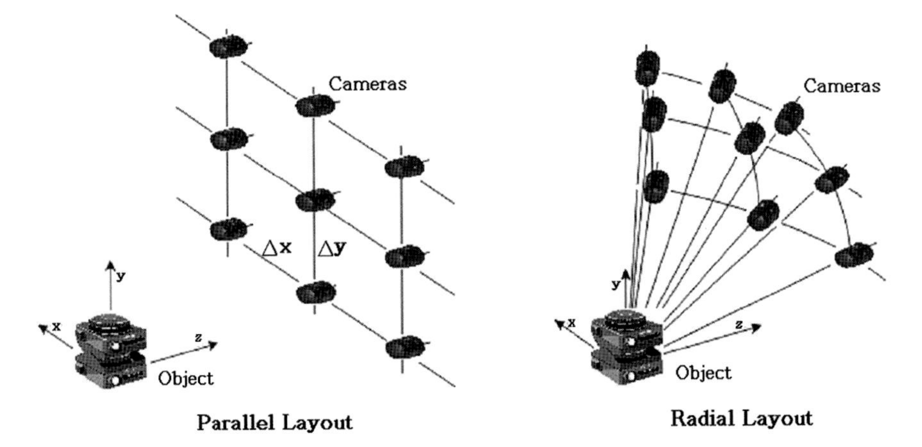


**Fig. 12.6.** Parallel and radial (toed-in) camera layouts

where $i(i = 1, ..., N)$ and $j(j = 1, ...., M)$ represent $(i, j)$th camera position, $N$ and $M$ the total number of cameras in $x$ and $y$ axes directions, respectively. $i$ is counted from left to right and $j$ from top and bottom.

In the form of homogeneous matrix, when a point on an object is photographed with the $(i, j)$th camera and projected to a screen, the image point on the screen and the object point are represented as $I_{ij} = (x^I_{ij}, y^I_{ij}, z^I_{ij}, w^I_{ij})$ and $v = (x, y, z, w)$, respectively. In these representations, $I$ stands for image, and $w^I_{ij}$ and $w$ are dimensionless scaling factors which are working as denominators to define the actual coordinate values from the $x$, $y$ and $z$ values in the matrix. Hence, in these matrices, the actual coordinate values are defined by $(x^I_{ij}/w^I_{ij}, y^I_{ij}/w^I_{ij}, z^I_{ij}/w^I_{ij})$ and $(x/w, y/w, z/w)$. $I_{ij}$ and $v$ can be related by using a $4 \times 4$ homogeneous matrix $M_{ij}$, as follows:

$$I_{ij} = v \cdot M_{ij}. \tag{12.25}$$

In Eq. 12.25, $M_{ij}$ is the matrix defined by $(i, j)$th camera position $C_{ij} = (x_{ij}, y_{ij}, z_{ij}, w_{ij})$. For the case of parallel layout, $M_{ij}$ is determined by three elementary transformations such as forward translation, perspective projection and back translation as shown in Fig. 12.7.

The forward translation translaes the optical axis of $(i, j)$th camera, to coincide with the $z$- axis. The basic 4 x 4 homogeneous matrix of translation, $M_T$ which translates a point $(x_0 + x_{ij},\ y_0 + y_{ij},\ z_0 + z_{ij},\ 1)$ to $(x_0,\ y_0,\ z_0,\ 1)$, is given as,

$$M_T = \begin{bmatrix} 1 & 0 & 0 & 0 \\ 0 & 1 & 0 & 0 \\ 0 & 0 & 1 & 0 \\ -x_{ij} & -y_{ij} & -z_{ij} & 1 \end{bmatrix} \tag{12.26}$$

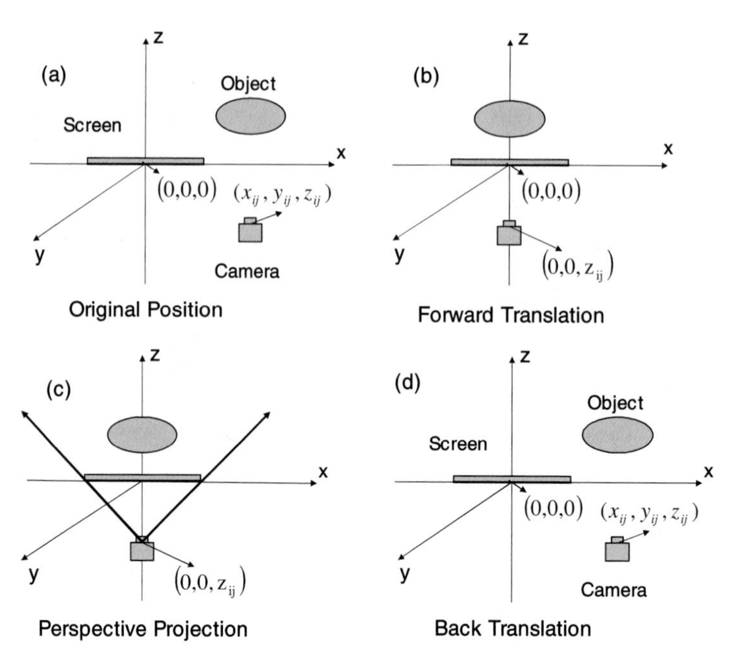


**Fig. 12.7.** Three elementary translations

Hence, in Fig. 12.7(b), since $(x_0,\ y_0,\ z_0,\ 1)$ has the value $(0,\ 0,\ z_{ij},\ 1)$, there is no translation in the $z$- axis direction, i.e., the $z_{ij}$ in $M_T$ should be replaced by zero.

The perspective projection performs photographing an image on the camera. This photographing process will not differ from the projecting image on a screen which is within the field of view of the camera and its direction is normally parallel to the optical axis of the camera. In this analysis, the screen is assumed to be located at the $xy$ plane with its center at the origin. This is shown in Fig. 12.7(a). The basic matrix of the projection $M_P$, when the camera is at $(0, 0, z_0)$ and the screen is centered at $(0, 0, 0)$, is given as,

$$M_P = \begin{bmatrix} 1 & 0 & 0 & 0 \\ 0 & 1 & 0 & 0 \\ 0 & 0 & 1 & -1/z_0 \\ 0 & 0 & -z_0 & 1 \end{bmatrix} \tag{12.27}$$

For the case of Fig. 12.7(c), $z_0$ in $M_P$ should be replaced by $z_{ij}$. The back translation $M_{-T}$ returns the translated point to its original position. The difference between $M_{-T}$ and $M_T$ is the signs of $x_{ij}$, $y_{ij}$ and $z_{ij}$. From these basic matrices, $M_{ij}$ for Fig. 12.2 can be calculated as,

$$M_{ij} = \begin{bmatrix} 1 & 0 & 0 & 0 \\ 0 & 1 & 0 & 0 \\ 0 & 0 & 1 & 0 \\ -x_{ij} & -y_{ij} & 0 & 1 \end{bmatrix} \cdot \begin{bmatrix} 1 & 0 & 0 & 0 \\ 0 & 1 & 0 & 0 \\ 0 & 0 & 1 & -\frac{1}{z_{ij}} \\ 0 & 0 & -z_{ij} & 1 \end{bmatrix} \cdot \begin{bmatrix} 1 & 0 & 0 & 0 \\ 0 & 1 & 0 & 0 \\ 0 & 0 & 1 & 0 \\ x_{ij} & y_{ij} & 0 & 1 \end{bmatrix}$$

$$= \begin{bmatrix} 1 & 0 & 0 & 0 \\ 0 & 1 & 0 & 0 \\ -\frac{x_{ij}}{z_{ij}} & -\frac{y_{ij}}{z_{ij}} & 0 & -\frac{1}{z_{ij}} \\ 0 & 0 & -z_{ij} & 1 \end{bmatrix} \tag{12.28}$$

Equation 12.28 is a transform matrix which converts a point object to multi-view images corresponding to an N X M camera array. In general, the process of projecting an image on the screen involves the magnification and demagnification of the camera image. This magnification/demagnification in disparity can also be included in the matrix by an appropriate scaling. Since the screen is in the $xy$ plane, the process provides the magnification/demagnification of disparity in the image components in $x$- and $y$-axes directions. If the magnifications/demagnifications in $x$- and $y$- axes directions are $p$ and $q$, respectively, then the scaled $M_{ij}$, ${}^{m}M_{ij}$ can be written as,

$${}^{m}M_{ij} = \begin{bmatrix} 1 & 0 & 0 & 0 \\ 0 & 1 & 0 & 0 \\ -\frac{x_{ij}}{z_{ij}} & -\frac{y_{ij}}{z_{ij}} & 0 & -\frac{1}{z_{ij}} \\ 0 & 0 & -z_{ij} & 1 \end{bmatrix} \cdot \begin{bmatrix} p & 0 & 0 & 0 \\ 0 & q & 0 & 0 \\ 0 & 0 & 1 & 0 \\ 0 & 0 & 0 & 1 \end{bmatrix} = \begin{bmatrix} p & 0 & 0 & 0 \\ 0 & q & 0 & 0 \\ -p\frac{x_{ij}}{z_{ij}} & -q\frac{y_{ij}}{z_{ij}} & 0 & -\frac{1}{z_{ij}} \\ 0 & 0 & -z_{ij} & 1 \end{bmatrix} \tag{12.29}$$

With Eq. 12.29, Eq. 12.25 will be rewritten as,

$${}^{m}I_{ij} = v \cdot {}^{m} M_{ij} \tag{12.30}$$

Equation 12.30 represents full parallax multi-view images displayed on the screen. In Fig. 12.8, the projection geometry of full parallax multi-view images in a projection-type 3-D imaging system is depicted. The screen images the exit pupil of each projector objective exit pupil as a viewing sub-zone for the image projected by the projector, such that each viewing sub-zone does not overlap with others. The imaging action of the screen requires that the image order of the projectors be reversed from that of the camera to make viewers who are standing at the viewing zone perceive orthoscopic 3-D images. The viewing zone is defined as the region where all the viewing sub-zones are formed. The distance between the closest neighboring viewing sub-zones should be less than viewers' interocular distance. Hence, each viewer can locate his/her eyes only two of the viewing sub-zones. As a result, among these images, only a pair of them is perceived by the viewer in the

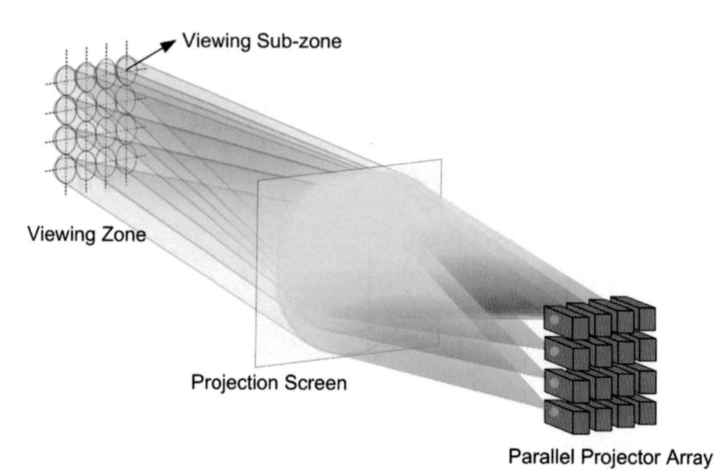


**Fig. 12.8.** Projection geometry of full parallax multi-view images in a projection-type 3-D imaging system

viewing zone at any moment. This pair is called a stereoscopic image pair. If a viewer with eye positions specified as $v_L = (x_L, y_L, z_L, w_L)$ for the left eye and $v_R = (x_R, y_R, z_R, w_R)$ for the right eye, is watching a stereoscopic image pair ${}^mI_{ij}$ and ${}^mI_{kl}$ for the left and right eye, respectively, this image pair will be projected to their corresponding viewer's eyes and perceived as a 3-D image due to their disparity. Hence, when the imaging action of the screen is considered, the perceived 3-D image $V_{ijkl} = (X_{ijkl}, Y_{ijkl}, Z_{ijkl}, W_{ijkl})$ can be expressed as,

$$V_{ijkl} = \frac{1}{x_R - x_L} \left( {}^mI_{ij} \cdot x_R \cdot M_R -{}^m I_{kl} \cdot x_L \cdot M_L \right). \tag{12.31}$$

where $k = 1, \ldots, N$, $l = 1, \ldots, M$ and $i < k$, and $M_L$ and $M_R$ are 4 x 4 matrices defined by the left and right eye positions given above, respectively. They are represented as,

$$M_R = \begin{bmatrix} 1 & y_R/x_R & z_R/x_R & 1/x_R \\ 0 & 1 & 0 & 0 \\ 0 & 0 & 0 & 0 \\ 0 & 0 & 0 & 1 \end{bmatrix}, M_L = \begin{bmatrix} 1 & y_L/x_L & z_L/x_L & 1/x_L \\ 0 & 1 & 0 & 0 \\ 0 & 0 & 0 & 0 \\ 0 & 0 & 0 & 1 \end{bmatrix} \tag{12.32}$$

In Eq. 12.31, $x_R - x_L$ will represent the viewer's interocular distance and $j$ and $l$ should be the same when viewer' eyes are the same distance from the screen and his/her head is straight, i.e., parallel, to the $y$- axis. $k$ should not be less than $i + 1$ when counting view image numbers from left to right. Hence, the maximum number of visible stereoscopic images in the horizontal direction is N – 1.

By substituting Eqs. 12.32 to 12.31,

$$V_{ijkl} = \frac{1}{x_R - x_L} v \cdot M_{ijkl} \quad (12.33)$$

then,

$$M_{ijkl} = {}^m M_{ij} \cdot x_R \cdot M_R - {}^m M_{kl} \cdot x_L \cdot M_L \quad (12.34)$$

From Eqs. 12.30, 12.32, 12.33 and 12.34, $M_{ijkl}$ is calculated as,

$$M_{ijkl} = \begin{bmatrix} p(x_R - x_L) & p(y_R - y_L) & p(z_R - z_L) & 0 \\ 0 & q(x_R - x_L) & 0 & 0 \\ (-\frac{px_{ij}x_R}{z_{ij}} & \{-\left(\frac{px_{ij}y_R + qy_{ij}x_R}{z_{ij}}\right) & (-\frac{px_{ij}z_R}{z_{ij}} & \{p\left(\frac{x_{kl}}{z_{kl}} - \frac{x_{ij}}{z_{ij}}\right) \\ +\frac{px_{kl}x_L}{z_{kl}}) & +\left(\frac{px_{kl}y_L + qy_{kl}x_L}{z_{kl}}\right)\} & +\frac{px_{kl}z_L}{z_{kl}}) & -\left(\frac{x_R}{z_{ij}} - \frac{x_L}{z_{kl}}\right)\} \\ 0 & 0 & 0 & x_R - x_L \end{bmatrix} \quad (12.35)$$

With Eqs. 12.34 and 12.35, each component value of $V_{ijkl} = (X_{ijkl}, Y_{ijkl}, Z_{ijkl}, W_{ijkl})$ for the object point $v = (x, y, z, w)$ is calculated as,

$$\begin{aligned} X_{ijkl} &= \frac{p}{x_R - x_L}\left\{x(x_R - x_L) + z(\frac{x_{kl}x_L}{z_{kl}} - \frac{x_{ij}x_R}{z_{ij}})\right\} \\ Y_{ijkl} &= \frac{1}{x_R - x_L}\left[qy(x_R - x_L) + px(y_R - y_L)\right. \\ &\quad \left.+z\left\{p\left(\frac{x_{kl}y_L}{z_{kl}} - \frac{x_{ij}y_R}{z_{ij}}\right) + q\left(\frac{y_{kl}x_L}{z_{kl}} - \frac{y_{ij}x_R}{z_{ij}}\right)\right\}\right] \\ Z_{ijkl} &= \frac{p}{x_R - x_L}\left\{z(\frac{x_{kl}z_L}{z_{kl}} - \frac{x_{ij}z_R}{z_{ij}}) + x(z_R - z_L)\right\} \\ W_{ijkl} &= \frac{1}{x_R - x_L}\left[w(x_R - x_L) + z\left\{p\left(\frac{x_{kl}}{z_{kl}} - \frac{x_{ij}}{z_{ij}}\right) - \left(\frac{x_R}{z_{ij}} - \frac{x_L}{z_{kl}}\right)\right\}\right] \end{aligned} \quad (12.36)$$

Equation 12.36 is the general solution of a 3-D image perceived by a viewer in a spatial location from a stereoscopic image pair in the multi-view images displayed on the image projection screen for the parallel camera layout case.

From Eq. 12.36, the condition of having no distorted perceived image can be found. To get no distorted perceived image, the different axis component images should not be (1) scaled differently and (2) shifted differently according to different axis values of the object. However, as seen in Eq. 12.36, the scale factor $W_{ijkl}$ is a function of $z$, i.e., depth of the object, $Z_{ijkl}$ is scaled by viewing and camera arrangement conditions and $X_{ijkl}$, $Y_{ijkl}$ and $Z_{ijkl}$ have several extra terms that cause shifts in accordance with other coordinate values of the object. Hence, the perceived image is always distorted unless certain conditions are met. The first condition is that $W_{ijkl}$, which will be the denominator of all three coordinate values, should have a constant value. Since $w = 1$, $z_{ij} = z_{kl}$ and $y_{ij} = y_{kl}$ for parallel camera layout, the z-axis dependent term in $W_{ijkl}$ will be eliminated if $p(x_{kl} - x_{ij}) = x_R - x_L$. This condition leads to $W_{ijkl} = 1$. With this relationship, $Z_{ijkl}$ can be reduced to $z$ only when $z_{ij} = z_{kl} = z_R = z_L$. But these relationships are still not enough to eliminate the terms related to the $z$ axis in $X_{ijkl}$ and $Y_{ijkl}$. Furthermore magnification/demagnification factors $p$ and $q$ in $X_{ijkl}$ and $Y_{ijkl}$ still remain even when the $z$ -axis related terms are completely eliminated. Hence, there no other conditions for a perceived image without distortion, except $p = q = 1$. When $p = q = 1$ if $x_{kl} = x_R, x_{ij} = x_L, z_{ij} = z_{kl} = z_R = z_L$ and $y_R = y_L = y_{ij} = y_{kl}$, $X_{ijkl}$, $Y_{ijkl}$ and $Z_{ijkl}$ are equal to $x$, $y$ and $z$, respectively. These conditions state that the viewer's two eye positions should be exactly matched with those of two cameras in the same row. To achieve the matching, the viewers should stand at the camera distance from the screen, and straighten their heads, making their faces parallel to the screen. In this case, the perceived image becomes a complete replica of the object, i.e., $X_{ijkl} = x, Y_{ijkl} = y, Z_{ijkl} = z$ and $W_{ijkl} = w(= 1)$. When the viewer mistakenly bends his/her neck in either the left or right shoulder direction and matches his/her two eyes to two cameras in a different row, i.e., $y_R \neq y_L, X_{ijkl} = x, Z_{ijkl} = z$ and $W_{ijkl} = w, Y_{ijkl} = y + x(y_R - y_L)/(x_R - x_L)$. This means that the $y$- axis directional component of the perceived image will be shifted in proportion to the object's $x$- axis value. This $x$- axis dependent shifting will cause a distortion and the amount of the distortion will be increased as the dimension of the object's $x$- axis direction increases. It is noted that $y_R \neq y_L$ will be a typical viewing condition for full parallax image display.

When viewing TV or any kind of screen, it is more likely that the viewer's face will not be parallel to them, unless he/she is in or near the z-axis. Hence, $z_R \neq z_L$ is the typical viewing condition. $z_R \neq z_L$ means that only one of the viewer's eyes can be matched to the camera position, i.e., $x_{ij} = x_L$ and $x_{kl} \neq x_R$, or $x_{ij} \neq x_L$ and $x_{kl} = x_R$. In this case, the $z$- axis component in $X_{ijkl}$ and $Y_{ijkl}$ cannot be 0, thus the perceived image will be distorted. The maximum difference in $|z_R - z_L|$ cannot be more than the viewer's interocular distance. As the difference between $z_R$ and $z_L$ increases, the difference between $x_R$ and $x_L$ decreases. As a consequence, the difference between $Z_{ijkl}$ and $z$ value will also be increased. This means that the distortion in the $z$- axis direction is increasing.

Equation 12.36 can be simplified using the actual camera layout condition. For the cameras in the parallel camera layout, the cameras are aligned such that the axis of each camera is parallel to the $z$- axis and their row is parallel to the $x$- axis. This alignment guarantees that $y_{ij} = y_{kl} = y_{jl}$ and $z_{ij} = z_{kl} = z_C$. With these conditions, the real position of the perceived image $V^a_{ijkl} = (X^a_{ijkl}, Y^a_{ijkl}, Z^a_{ijkl})$ is calculated by dividing each coordinate value of $V_{ijkl}$ with $W_{ijkl}$ with $w = 1$, and this can be written as,

$$
\begin{aligned}
X^a_{ijkl} &= p\frac{z_C x(x_R - x_L) + z(x_{kl}x_L - x_{ij}x_R)}{z_C(x_R - x_L) + z\{pC - (x_R - x_L)\}} \\
Y^a_{ijkl} &= \frac{qz_C y(x_R - x_L) + z\left\{p(x_{kl}y_L - x_{ij}y_R) - qy_{jl}(x_R - x_L)\right\}}{z_C(x_R - x_L) + z\{pC - (x_R - x_L)\}} \\
Z^a_{ijkl} &= p\frac{z_C x(z_R - z_L) + z(x_{kl}z_L - x_{ij}z_R)}{z_C(x_R - x_L) + z\{pC - (x_R - x_L)\}}
\end{aligned}
\tag{12.37}
$$

where $C = x_{kl} - x_{ij}$. When viewers are standing with their eyes parallel to the image display screen with their heads straight up, $y_e = y_L = y_R$ and $z_e = z_L = z_R$. In this case, Eq. 12.37 states that the three axes components of the perceived image will be equally magnified/demagnified without any distortion if $pC = x_R - x_L$, i.e., $p(k-i)\Delta x = x_R - x_L$, $Cz_e = (x_R - x_L)z_C$, $py_eC = qy_{ij}(x_R - x_L)$ and $x_{kl}x_L = x_{ij}x_R$. These conditions result in $z_e = pz_C$, $y_e = qy_{ij}$ and $x_L = px_{ij}$. When both $x$ and $y$ axis components are projected with an equal magnification/demagnification, i.e., $p = q$, the result states that if the parameters related with viewers, i.e., interocular distance and viewers' positions in $x$, $y$ and $z$- axis directions, i.e., $x_L$, $y_e$ and $z_e$ are having the ratio equal to the image magnification/demagnification, with their corresponding camera parameters, there will not be any distortion in the perceived image. Since increasing the $z_C$ value results in a wide object scene and requires increasing the $z_e$ value accordingly, the image display screen size should also be increased so that the depth resolution [26] does not deteriorate due to our limited eye resolution. Hence, the relationship $Cz_e = (x_R - x_L)z_C$ can be rewritten as $CS_S = (x_R - x_L)O_S$, where $S_S$ and $O_S$ represent screen and object sizes in transversal direction, respectively.

## 12.5 Conclusions

When the conditions of photographing, projecting, and viewing are different, it was shown that distortions in the stereoscopic images perceived by viewers are usually present. The perceived image will be distorted more as the differences in the conditions increases and will only appear in the space bounded by lines connecting the outsidemost object points and the crossing point of two lines that are created by connecting the pupil positions of left camera and left eye, and right camera and right eye, respectively. The $xy$ plane cross section of the perceived image keeps the same shape as its corresponding object cross section, but it shifts in the viewer direction. The shift amount is proportional to the distance of the cross section from the image projection screen. The distortion amount is more in depth direction than in transverse direction.

The condition of eliminating keystone distortion in the stereoscopic imaging systems configured by radial photographing and projection geometry makes the sine value ratio of the camera orientation angle to that of the projector orientation angle equal to the ratio of the camera distance to the projector distance from the screen, multiplied by the magnification/demagnification factor. This condition causes the projected image on the screen to become the same as that in parallel photographing and projection geometry. The mathematical expression to describe the perceived image in the radial geometry has the same equation form as that in the parallel geometry though the expression has more parameters for the projector and a constant bias term in the horizontal directional component of the image.

In the multi-view case, the perceived image will be distorted unless the parameters related with viewers, i.e., interocular distance and viewers' position, have ratios equal to the image magnification/demagnification, with their corresponding camera parameters. The equation can be used to optimize the design of the full parallax projection-type multi-view imaging systems, and also to generate multi-view images for a given object. It can also quantify all distortions in the perceived images from full parallax multi-view images displayed on a projection-type multi-view imaging system with a parallel projector layout, which projects multi-view images acquired by a parallel camera layout.

## References

[1] J.-Y. Son, Y. Gruts, J.-E. Bahn and Y.-J. Choi, "Distortion Analysis in Stereoscopic Images," IDW'2000, Proceedings of the seventh International Display Workshop, Kobe, Japan (Dec. 2000).
[2] I. Takehiro (ed.), Fundamentals of 3D Images, NHK Broadcasting Technology Research Center, Ohm Sa, Tokyo, Japan (1995).
[3] A. Woods, T. Docherty and R. Koch, "Image Distortions in Stereoscopic Video Systems," Proc. SPIE, V1915, pp 36–48 (1993).
[4] H. Onoda, M. Ohki and S. Hashiguchi, "Methods of Taking Stereo Pictures," Proceeding of 3D Image Conference' 93, Committee of 3D Image Conference, Tokyo, Japan, pp 179–183 (1993).
[5] D. F. McAllister (ed.), Stereo Computer Graphics and Other True 3D Technologies, Princeton, New Jersey (1993).
[6] H. Yamanoue, "The Relation Between Size Distortion and Shooting Conditions for Stereoscopic Images," SMPTE J, pp 225–232 (Apr. 1997).
[7] D. B. Diner and D. H. Fender, Human Engineering in Stereoscopic Viewing Devices, Plenum Press, New York (1993).
[8] J.T. Rule, "The Shape of Stereoscopic Images," J. Opt. Soc. Am., V31, pp 124–129 (1941).
[9] B.G. Saunders, "Stereoscopic Drawing by Computer-Is It Orthoscopic?," Appl. Opt., V7, pp 1459–1504 (1968).
[10] L. F. Hodge, "Time-multiplexed Stereoscopic Computer Graphics," IEEE Comput. Graph. Appl., V12(No. 2), pp 20–30 (1992).
[11] D. L. MacAdam, "Stereoscopic Perception of Size, Shape, Distance and Direction," J. SMPTE V62, pp 271–293 (1954).
[12] H. Dewhurst, "Photography, Viewing and Projection of Pictures with Stereoscopic Effect," U.S. Pat. 2,693,128 (1954).

[13] J.-Y. Son, Y. Gruts, J.-H. Chun, Y.-J. Choi, J.-E. Bahn and V.I. Bobrinev, "Distortion Analysis in Stereoscopic Images," Opt. Eng., V41, pp 680–685 (2002).
[14] J.-Y. Son, Y.N. Gruts and K.-D. Kwack, K.-H. Cha and S.-K. Kim, "Stereoscopic Image Distortion in Radial Camera and Projector Configurations," JOSA-A., V24(No. 3), pp 643–650 (2007).
[15] M. Martinez-Corral, B. Javidi, R. Martinez-Cuenca and G. Saavedra, "Multifacet Structure of Observed Reconstructed Integral Imaging," J. Opt. Soc. Am. A, V22(No.4), pp 597–603 (2005).
[16] J.-Y. Son, V.V. Saveljev, Y.-J. Choi, J.-E. Bahn and H.-H. Choi, "Parameters for Designing Autostereoscopic Imaging Systems Based on Lenticular, Parallax Barrier and IP Plates,", Opt. Eng., V42, pp 3326–3333 (2003).
[17] R. Hartley and A. Zisserman, Multiple View Geometry in Computer Vision, Cambridge University Press, Cambridge, UK, Ch. 5 and 6, pp 139–183 (2000).
[18] S. S. Kim, V. Savaljev, E. F. Pen and J. Y. Son, "Optical Design and Analysis for Super-multiview Three-dimensional Imaging Systems," SPIE Proc., V4297, pp 222–226 (2001).
[19] A. Schwerdtner and H. Heidrich, "The Dresden 3D Display(D4D)," Proc. of SPIE, V3295, pp 203–210 (1998).
[20] R. Boerner, "The Autostereoscopic 1.25 m Diagonal Rear Projection System with Tracking Features," Proceedings of The Fourth International Display Workshops, IDW'97, pp.835–837 (1997).
[21] A. R. L. Travis, S. R. Lang, J. R. Moore and N. A. Dodgson, "Time-manipulated Three-dimensional Video," SID Dig. 26, pp 851–854 (1995).
[22] J.-Y. Son, V.V. Saveljev, J.-S. Kim, K.-D. Kwack and S.-K. Kim, "Multiview Image Acquisition and Projection," IEEE JDT, V 2(No. 4), pp 359–363 (2006).
[23] Y. Gruts, Stereoscopic Computer Graphics, Naukova Dumka, Kiev, Ukrain (1989).
[24] L. Lipton, Foundations of the Stereoscopic Cinema, A Study in Depth, New York, U.S.A., Van Nostrand Reinhold Company, 1982.
[25] I. Takehiro (ed.), Fundamentals of 3D Images, NHK Broadcasting Technology Research Center Published by Ohm-Sa, Tokyo, Japan (1995).
[26] J.-Y. Son, V.I. Bobrinev and K.-T. Kim, "Depth Resolution and Displayable Depth of a Scene in 3 Dimensional Images," J. Opt. Soc. Am. A, V22(No 9), pp 1739–1745 (2005).

# 13

# 3-D Video Processing for 3-D TV

Kwanghoon Sohn, Hansung Kim, and Yongtae Kim

## 13.1 Introduction

One of the most desirable ways of realizing high quality information and telecommunication services has been called "The Sensation of Reality," which can be achieved by visual communication based on 3-D (Three-dimensional) images. These kinds of 3-D imaging systems have revealed potential applications in the fields of education, entertainment, medical surgery, video conferencing, etc. Especially, three-dimensional television (3-D TV) is believed to be the next generation of TV technology. Figure 13.1 shows how TV's display technologies have evolved , and Fig. 13.2 details the evolution of TV broadcasting as forecasted by the ETRI (Electronics and Telecommunications Research Institute). It is clear that 3-D TV broadcasting will be the next development in this field, and realistic broadcasting will soon follow.

The CCIR (International Radio Consultative Committee) established the study program IC-1/11, which outlined the system requirements for 3-D TV broadcasting in 1958 [5]. There has been a consensus among many researchers that the following essential requirements are necessary for 3-D TV broadcasting [6]:

- An orthoscopic 3-D display (the depth of scenes must appear natural and not cause unnecessary viewer discomfort)

K. Sohn
School of Electrical and Electronic Engineering, Yonsei University, Seoul, Republic of Korea
e-mail: khsohn@yonsei.ac.kr

B. Javidi et al. (eds.), *Three-Dimensional Imaging, Visualization, and Display*, 251
DOI 10.1007/978-0-387-79335-1_13, © Springer Science+Business Media, LLC 2009

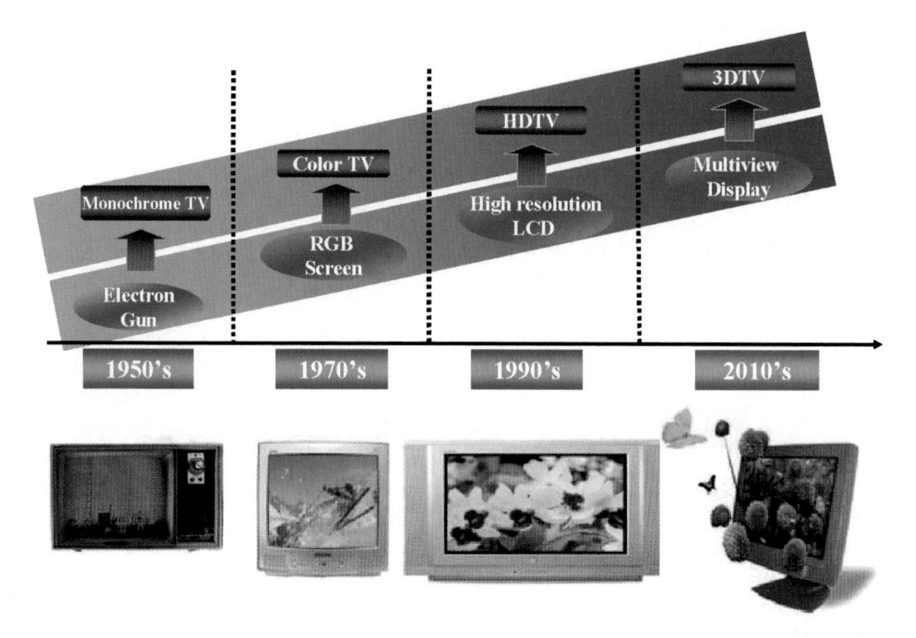


**Fig. 13.1.** Historical evolution of TV technology

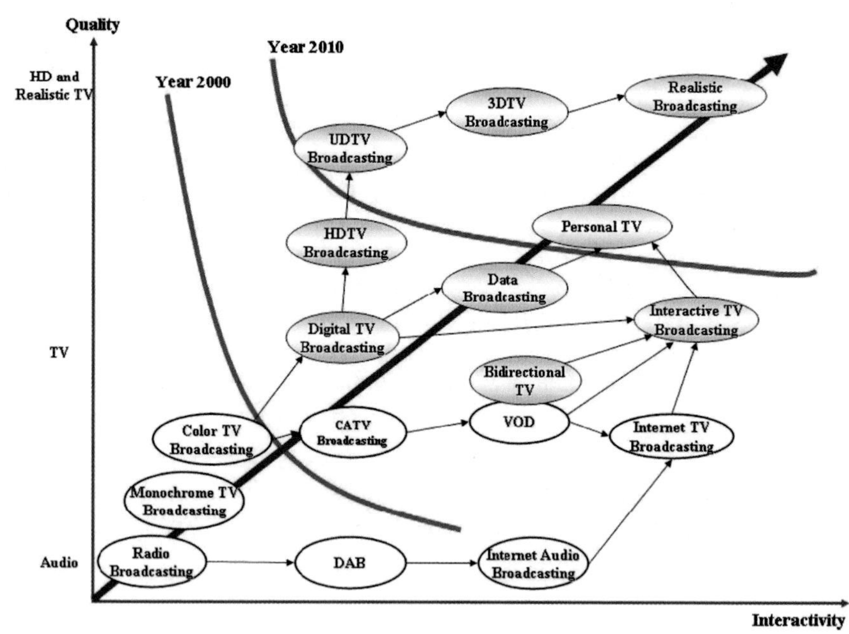


**Fig. 13.2.** Evolution of TV broadcasting

- Group viewing (almost any location in a given room must provide good stereoscopic viewing conditions)
- Compatibility (3-D color receivers must display stereoscopic transmissions in full depth, and two-dimensional (2-D) transmissions operate monoscopically; present 2-D receivers must display stereoscopic transmissions monoscopically)
- Non-degraded pictures (the colorimetry and resolution of 3-D color TV pictures must be comparable to present 2-D color pictures)
- Minimal modification of video standards (industry and government specifications must not require extensive revision)
- Moderate price (the cost and complexity of converting studio and station TV equipment and the cost of stereoscopic TV receivers must not be significantly greater than the cost of the conversion from monochrome images to color images)

A number of experiments have been performed, but no breakthroughs have yet been reached due to some major technical obstacles. There are still many bottleneck problems to overcome, such as the fact that 3-D display technologies usually require special glasses, the prevention of human factors that can lead to eye fatigue problems, lack of 3-D content, etc.

Interest in the use of 3-D images has prompted numerous research groups to report on 3-D image processing and display systems. 3-D TV projects have also been performed in many countries. In Europe, several initiatives researching 3-D TV, such as the DISTIMA (Digital Stereoscopic Imaging and Applications) project, aimed to develop a system for capturing, encoding, transmitting and presenting digital stereoscopic sequences [35]. The goals of the ATTEST (Advanced Three-Dimensional Television System Technologies) project were to develop a flexible, 2-D-compatible and commercially feasible 3-D TV system for broadcast environments [36]. Next, the objective of the NoE (Network of Excellence) was to produce a 3-D TV system with high-end displays that could provide true 3-D views [33]. This system included the recording, processing, interacting, and displaying of visual 3-D information, since capturing visual 3-D information of real-life scenes and creating exact (not to scale) optical duplicates of these scenes at remote sites instantaneously, or at a slightly later time, are the ultimate goals of visual communications. Another noteworthy effort was NHK's 3-D HDTV project in Japan [4]. In addition, 3-D broadcasting with stereo displays was proposed for the FIFA 2002 World Cup in South Korea [14].

Arising from this interest in 3-D multi-view image systems, the MPEG (Moving Picture Experts Group) 3-D AV (Three-Dimensional Audio Video) division was established under the ISO/IEC SC29 WG11 in December 2001, and this division was then transferred to the JVT (Joint Video Team) in July 2006. The JVT-MVC has been finished standardizing the multi-view video coding process in July 2008.

This chapter concentrates on completed research dealing with 3-D video processing technologies for 3-D TV, such as 3-D content generation, 3-D video CODECs and video processing techniques for 3-D displays.

## 13.2 3-D Content Generation

### 13.2.1 Background

To date, one of the most serious problems of 3-D video technology has been a lack of 3-D content. However, new methods of generating 3-D content now include using 3-D computer graphics (CG), capturing images with multiple cameras, and converting 2-D content to 3-D content.

In the field of computer graphics, creating 3-D content is no longer a fresh topic since 3-D modeling and rendering techniques have been researched for a long time, and so many powerful software and hardware applications have been developed. Thus, many kinds of 3-D CG content already exist in animation shows, games, on the internet, and so on. However, 3-D image acquisition is still rare in the field of real videos. When the real world is filmed by a normal camera, all 3-D points are projected onto a 2-D image plane, so the depth information of the 3-D points is lost. In order to recover this information, it is necessary to use more than two images captured from different positions.

Another important fact to consider is that although 3-D content can easily be produced with state-of-the art cameras and computer graphics technologies, there are already massive amounts of existing 2-D content, and there is no way to reproduce this content in 3-D. For this reason, automatic stereoscopic conversion (2-D/3-D conversion) has been used to generate 3-D content from conventional 2-D videos based on computer vision techniques. 2-D videos, CATV (Cable Television) shows and DVDs (Digital Versatile Discs) can now be converted into stereoscopic videos by using 2-D/3-D conversion in real-time. The following section covers a number of topics related to 3-D content generation, from camera setup to synthesizing the 3-D images.

### 13.2.2 3-D Data Acquisition

In all 3-D systems, the depth information of real scenes is an essential component. However, it is theoretically impossible to recover perfect depth information from images since this process represents an inverse problem. Therefore, many active and passive methods have been proposed.

Active range sensors utilize a variety of physical principles, such as ultrasonic factors, radars, lasers, Moiré interference, and so on [12, 39]. These sensors emit short electromagnetic or acoustic waves and detect reflections from objects. Distance is obtained by measuring the time it takes for a wave to hit a given surface and then come back.

Conversely, passive techniques based on computer vision are less sensitive to environmental factors and typically require a simpler and less expensive setup. These techniques estimate depth information from the correspondence of acquired images and camera parameters [18, 22]. In order to recover 3-D information from 2-D image sequences, more than two cameras are usually necessary and all video cameras must be synchronized. Three types of camera setups can be used for 3-D data acquisition: the parallel type, the converging type, and the dorm type.

Figure 13.3 shows the geometries when the 3-D world is captured by two cameras, where a 3-D point $P\ (X,Y,Z)$ is projected into pixels $p_l(x_l,y_l)$ and $p_r(x_r,y_r)$ on each image plane. With parallel camera setups, the cameras are aligned so their optical axes are parallel. In other words, each camera points in the same direction, but there is a distance (the baseline distance) between their focal centers. This is shown in Fig. 13.3(a). In the following equation, $x_l$ and $x_r$ denote the X-coordinates of the projected points on the left and right images, respectively. The level of disparity with respect to the left camera is defined as:

$$d = x_l - x_r = \frac{fX}{Z} - \frac{f(X-B)}{Z} \tag{13.1}$$

The relationship between the level of disparity and the distance is obtained as follows:

$$Z = \frac{fB}{d} \tag{13.2}$$

As a result, with parallel cameras, the depth information from the level of disparity between the left and right image pair can be found. For converging

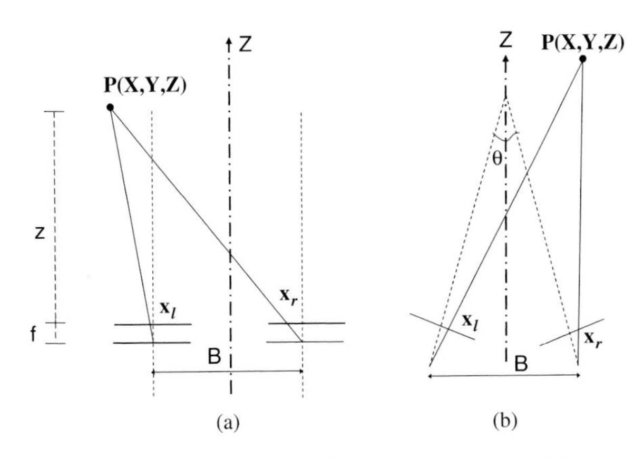


**Fig. 13.3.** Camera geometries: (**a**) parallel camera setup, (**b**) converging camera setup

cameras, as shown in Fig. 13.3(b), the cameras' optical axes went through each optical center $c(c_x, c_y)$ and were perpendicular to each image. This formed an angle ($\theta$). Here, the disparity vector was depicted as the 2-D vector $d(d_x, d_y)$. Equation (13.3) shows the relationship of the camera coordinates for the two cameras.

$$\begin{bmatrix} X_l \\ Y_l \\ Z_l \end{bmatrix} = \begin{bmatrix} \cos\theta & 0 & -\sin\theta \\ 0 & 1 & c_y \\ \sin\theta & 0 & \cos\theta \end{bmatrix} \begin{bmatrix} X_r \\ Y_r \\ Z_r \end{bmatrix} + \begin{bmatrix} B\cos\frac{\theta}{2} \\ 0 \\ B\sin\frac{\theta}{2} \end{bmatrix} \tag{13.3}$$

From the camera geometries, the distance $Z_r$ was extracted, as shown in Eq. (13.4) [10]:

$$\begin{aligned} \overline{A}Z_r &= \overline{B} \\ \overline{A} &= \begin{bmatrix} (x_r + d_x)(x_r \sin\theta + f\cos\theta) - f(x_r\cos\theta - f\sin\theta) \\ (y_r + d_y)(x_r\sin\theta + f\cos\theta) - fy_r \end{bmatrix} \\ \overline{B} &= \begin{bmatrix} Bf(f\cos\frac{\theta}{2} - (x_r + d_x)\sin\frac{\theta}{2} \\ -Bf(y_r + d_y)\sin\frac{\theta}{2} \end{bmatrix} \\ Z_r &= (\overline{A}^T\overline{A})^{-1}\overline{A}^T\overline{B}. \end{aligned} \tag{13.4}$$

The third type of camera setup is the dorm type, in which multiple environmental cameras are set up on walls and ceilings to surround a target object. The "Virtualized Reality" system (Kanade, et al.) consisting of a geodesic dome was equipped with 49 synchronized environmental video cameras [19], and Matsuyama, et al. used 25 pan tilt cameras to track and capture an object [27]. In the dorm type, the 3-D positions of the 2-D points in the images are generally reconstructed by back-projecting each point into a 3-D space by using the camera's parameters and finding the intersection of the projection lines.

### 13.2.3 2-D/3-D Conversion

The principle of 2-D/3-D conversion stems from Ross's psychophysics theory, in which stereoscopic perception can be generated using the Pulfrich effect, which refers to the time delay effect caused by the difference of the amount of light in both eyes [37]. Ross generated a stereo image pair by combining current and delayed images appropriately and then displaying them to both human eyes. Since this research, several 2-D/3-D conversion methods have been developed to allow people to enjoy 3-D experiences with formerly created 2-D content.

The modified time difference (MTD) method shown in Fig. 13.4(a) detects movements of objects and then decides delay directions and times according to the characteristics of the movements. Then, stereoscopic images are selected

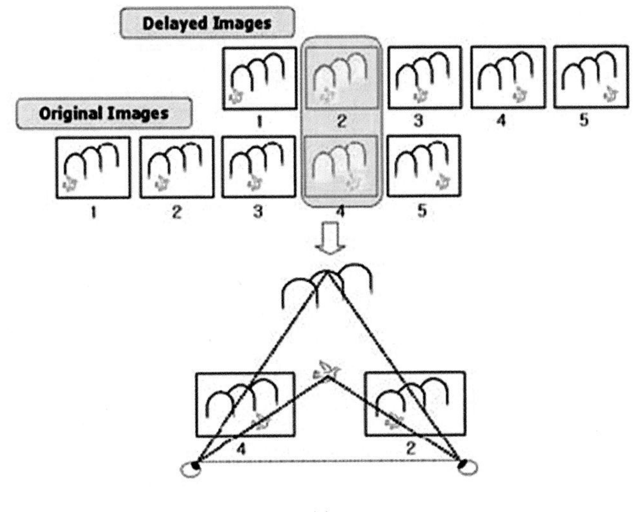

(a)

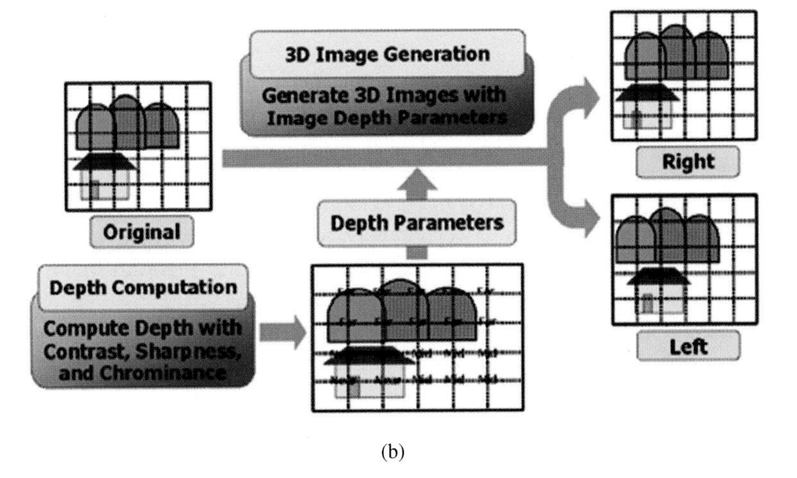


(b)

**Fig. 13.4.** 2-D/3-D conversion methods: (**a**) modified time difference (MTD) method, (**b**) computed image depth (CID) method

according to the time differences in 2-D image sequences [32]. The computed image depth (CID) method, as shown in Fig. 13.4(b), uses the relative position between multiple objects in still images. Image depth is computed by using the contrast, sharpness and chrominance of the input images [29].

The motion-to-disparity (MD) method generates stereoscopic images by converting object motions into horizontal disparity. In general, to eliminate the effect of vertical disparity, the norms of motion vectors are converted

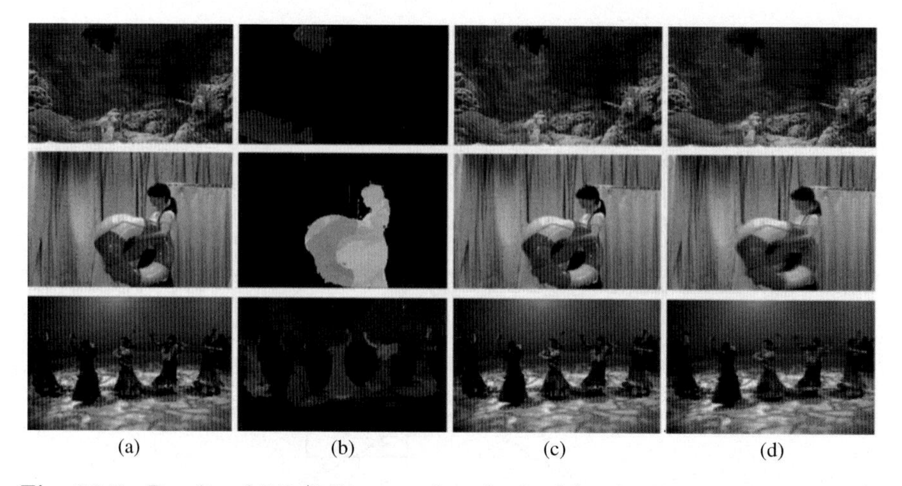


**Fig. 13.5.** Results of 2-D/3-D conversion obtained by the Modified MD method: (**a**) original image, (**b**) motion fields, (**c**) synthesized stereo pair, (**d**) interlaced stereo image

into horizontal disparity [21]. Several structure estimation methods are used to calculate the depth of scenes. Structures can be estimated assuming that camera motions are restricted to translation. Using an extended Kalman filter, camera motions can be estimated in rotation and translation and the structure of scenes can be estimated [7]. Alternatively, a method which detects vanishing lines [3] and the sampling density of spatial temporal interpolation in human visual characteristics [11] has also been proposed.

Recently, a modified motion-to-disparity (MMD) method that considers multiuser conditions and stereoscopic display characteristics has been developed [23]. In this method, the scale factors of motion-to-disparity conversion are determined by considering several cues such as magnitude of motion, camera movement and scene complexity. Figure 13.5 shows some 2-D/3-D conversion results obtained by the MMD method.

### 13.2.4 3-D Mixed Reality Content Generation

Mixed Reality (MR) systems refer to environments in which both virtual and real environments exist [2, 31]. In MR systems, users can immerse themselves and interact in spaces composed of real objects as well as computer-generated objects. Thus, seamless integration and natural interaction of virtual and real worlds are essential components of MR systems.

Chroma keying is an example of the simplest and most typical MR system. Chroma keying uses a particular color vector to distinguish background regions where virtual content can be synthesized. However, chroma keying is limited in terms of the color used in foreground regions and the interaction between real and virtual objects.

In most conventional MR systems, virtual objects are simply overlaid onto images of real objects as if the virtual objects were placed in front of the real ones. When real objects are placed in front of virtual ones, the virtual object images must be pruned before display. Moreover, when virtual objects collide with real objects, the result is the same as usual.

Z-keying is the simplest and most popular way of synthesizing two images when depth information is known [18]. In real 3-D images, disparity vectors can be converted into depth information, as explained in Section 13.2.2. Since virtual objects already contain depth information, they can be combined with real world images by comparing their depth with the estimated depth values of real scenes. Pixels which are closer to the user's viewpoint are displayed and, thus, virtual objects can be placed in any desired positions in real world scenes.

On the other hand, we can also reconstruct 3-D models of scenes by using estimated depth maps. Figure 13.6.(a) shows some scenes of reconstructed

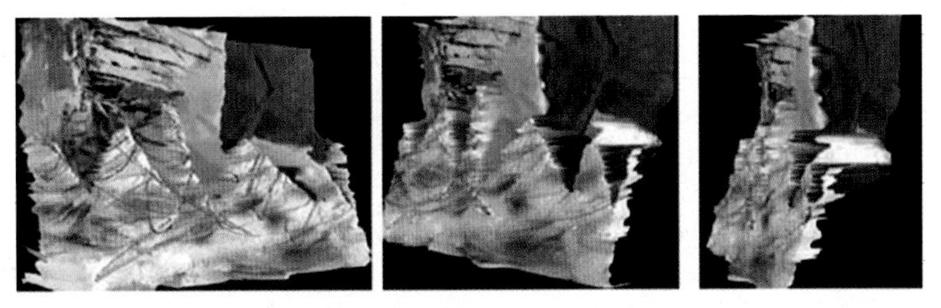

(a)

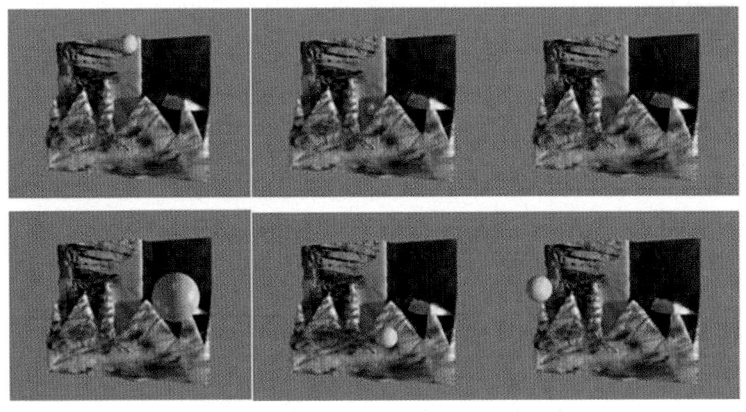

(b)

**Fig. 13.6.** Interaction in MR systems: (**a**) 3-D reconstruction from a stereo pair, (**b**) interaction

models from several viewpoints. With reconstructed 3-D models, we can realize the interaction of virtual objects with real objects. Figure 13.6(b) shows snapshots of interaction with a virtual ball. In the first row, the ball disappears naturally behind the front board when it falls to the ground. The second row shows the active interaction between the real and virtual worlds. The virtual ball is pitched to the front board. When the ball strikes the board, it bounces back in the direction of the mirror. The outgoing angle to the surface normal vector equals the incoming angle.

## 13.3 3-D Video CODECs

### 13.3.1 Background

Stereoscopic and multi-view videos can provide more vivid and accurate information about scene structures than monoscopic videos, since they provide depth information. Transmitting stereo sequences requires twice the amount of bandwidth as that required in conventional 2-D sequences. However, stereoscopic sequences have high correlations with each other. This property can be used to reduce a considerable number of coding bits. Although humans may experience stereoscopic qualities from stereo images, the current progress of autostereoscopic display techniques has been limited to natural 3-D feelings. For this reason, multi-view images have been used for richer experiences because they provide various viewing points. However, the quantity of multi-view data increases according to the number of views. This has been a serious problem when implementing multi-view video systems. Therefore, an efficient compression technique is necessary for multi-view video coding. For this, both the spatial redundancy in the view axis and the temporal redundancy in the time axis have to be efficiently exploited and reduced.

To obtain multi-view video sequences, multiple cameras are necessary. This may cause correspondence problems if there is imbalance among the views. Disparity estimation is one of the most important factors when it comes to ensuring good overall performance. This process uses a cost function that calculates the difference of the intensity values between current and referred frames. Therefore, imbalances among the views decrease performance by decreasing the reliability of the cost function. In ideal disparity estimation, the matching pixels of two different views have the same intensity values. However, imbalances among the views sometimes occur because of differing focus, luminance and chrominance levels. This decreases performance since disparity vectors are obtained by using the difference between the intensity values, such as the MAD (Mean Absolute Difference) or the MSE (Mean Square Error). If this fundamental assumption cannot be satisfied, the reliability of the disparity vectors cannot be guaranteed. To overcome this problem, several balancing algorithms have been proposed.

### 13.3.2 Disparity Estimation

Disparity represents the displacement of corresponding points between left images and right images. From disparity, one can derive the depth information associated with underlying 3-D points and the complete 3-D coordinates in the real world. Hence, disparity estimation is an essential step in any 3-D image or video processing system.

There are several approaches to motion/disparity estimation, including mesh-based, feature-based, pixel-based and block-based approaches [47]. A mesh-based approach (usually used for motion estimation) can similarly be applied to disparity estimation [46]. In this case, a mesh is applied to a referred view (generally the left view) and this mesh tries to find the corresponding nodal positions in the current view (the right view), as shown in Fig. 13.7. Each pair of corresponding 2-D mesh elements can be considered projections of a 3-D surface patch on the left and right images. The nodal disparities can be estimated by minimizing the disparity-compensated prediction error between corresponding elements.

A pixel-based approach matches each pixel by measuring the intensity of a single pixel. The earliest formulations of these methods were based on minimizing the left and right intensity differences, subject to smoothness constraints. This formulation was later refined by including a smoothly varying function to compensate for the view direction dependence of intensity values [13].

Feature-based algorithms first extract pre-defined features, and then match them. The separation of detection and matching represents a restriction on the kind of quality that can be obtained. The most well-known and general feature is edge information.

To allow for convenience and compatibility with conventional 2-D encoders, most multi-view encoders use block-based disparity estimation. This

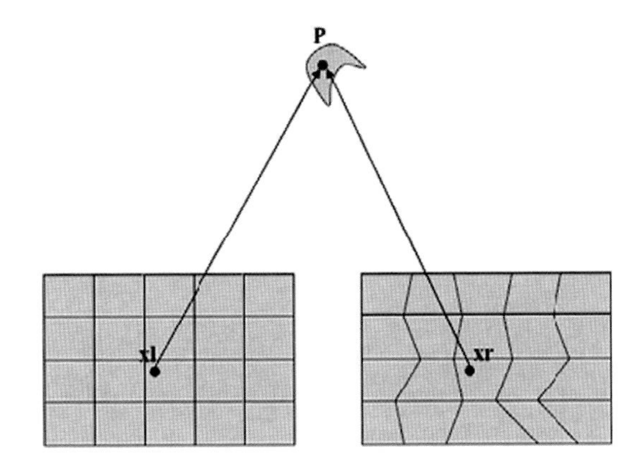


**Fig. 13.7.** Correspondence between 3-D and 2-D meshes

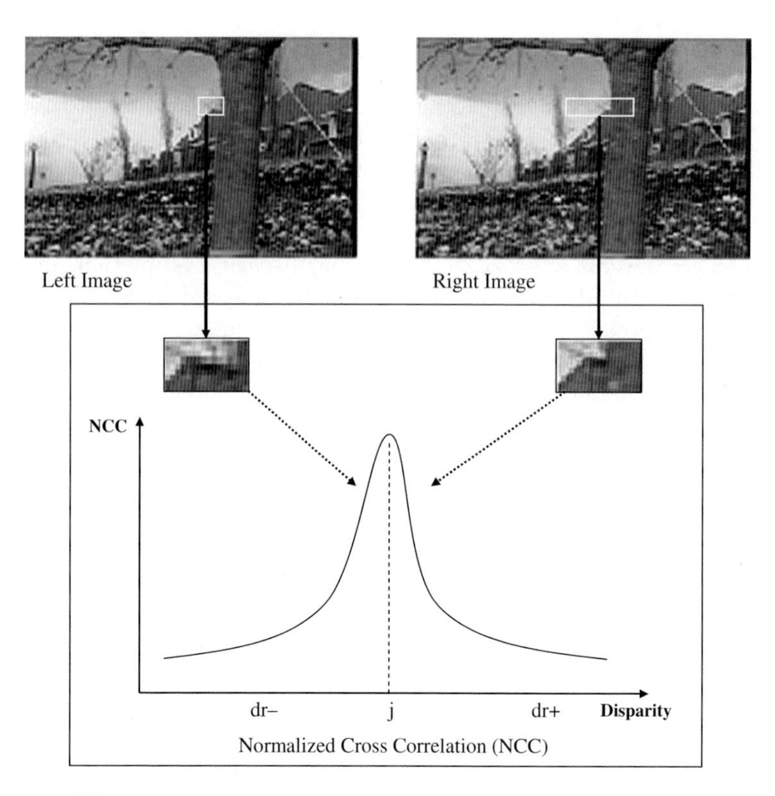


**Fig. 13.8.** Disparity estimation using fixed block matching

method uses rectangular blocks of pixels that can be matched as shown in Fig. 13.8 [1]. Within a given search window, a left image block can be substituted by a rectangular area in the right image which minimizes some criteria, such as the MAE or the MSE.

In block matching algorithms, the resulting vector field assigns the same disparity vector to all the pixels in a given block, resulting in a sparse vector field. However, if cost functions such as the MAE or the MSE consider only the difference of intensity values, it is possible to obtain irregular and incorrect disparity vectors in smooth regions. In addition, they may increase the entropy of the disparity map and decrease coding efficiency. Thus, the regularization technique has been widely used to overcome the above-mentioned problems [25].

Equation (13.5) shows a cost function using the regularization technique. Equation (13.6) shows the MAE term which represents the similarities between the right and left images. The cost function involves the smoothness constraint term which is inversely proportionate to the number of edges, as shown in Eq. (13.7), to preserve edges well. Objects with different depths and distinct boundaries usually have different disparity values. However, a single object usually has similar disparity values. The cost function is as follows:

$$f = \min_{d_k}(f_1 + \lambda f_2) \tag{13.5}$$

$$f_1 = \sum_{i,j \in block} \left| I_r(i,j) - I_l(i+d_k, j) \right| \tag{13.6}$$

$$f_2 = \sum_{i=0}^{3} \frac{\left|\hat{d}_i - d_k\right|}{e_i + e_c + 1} \tag{13.7}$$

where $\lambda$ represents a Lagrange multiplier, $\hat{d}_i$ represents the disparities of the neighboring blocks, and $i$ indicates the direction of the neighboring blocks. $e_c$ represents the magnitude of the edges in a given block and $e_i$ represents the number of edges in the boundary regions of the block, as shown in Fig. 13.9(a) [20]. The magnitudes of the edges can be obtained by the mean values of the edges as extracted by the Sobel operator. Figure 13.9(c) and (d) illustrate the disparity maps with and without regularization for the "piano" images shown in Fig. 13.9(b), respectively. The disparity map in Fig. 13.9(c) shows more irregular disparities in the smooth regions than those of Fig. 13.9(d). Figure 13.9(d) shows regular disparities in the smooth regions and exact matches in the edge regions. It can preserve objects from the reconstructed image and decrease the coding bits of the disparity vectors.

A well-known natural constraint can be derived from the inherent stereo motion redundancy of 3-D structures [15]. A coherence condition between motion and disparity in stereo sequences may be expressed as a linear combination of four vectors (two motion and two disparity vectors) in two successive stereoscopic frame pairs. For a sampling position at time $t$, see Fig. 13.10 and Eq. (13.8).

$$\| \; d_l\,(z,t) + m_r\,[z + d_l\,(z,t)\,,t] - d_l\,[z + m_l\,(z,t)\,,t+1] - m_l\,(z,t) \; \| = \delta \tag{13.8}$$

Setting $\delta = 0$, this relationship can be used to calculate the left motion vector at time $t$ from the right motion vector and the disparity vectors at time $t$ and $t+1$. This kind of joint estimation technique can reduce the effect and the time for stereoscopic sequence coding, but may seriously yield an accumulation of perturbation errors in the vector relationship. For this reason, it has only been used to find initial prediction vectors or as a condition to derive a reliability test for estimated displacements.

### 13.3.3 3-D Video Compression

There are two well-known approaches to encoding 3-D videos: the multi-view image-based approach and the depth-based approach. The first approach

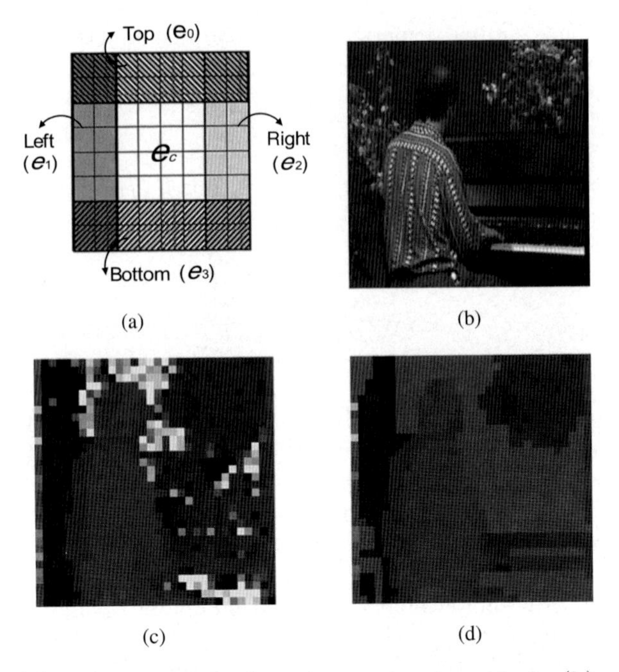


**Fig. 13.9.** (**a**) Definition of the boundary region in a block, (**b**) original right image of "piano," (**c**) disparity map without regularization, (**d**) disparity map with regularization

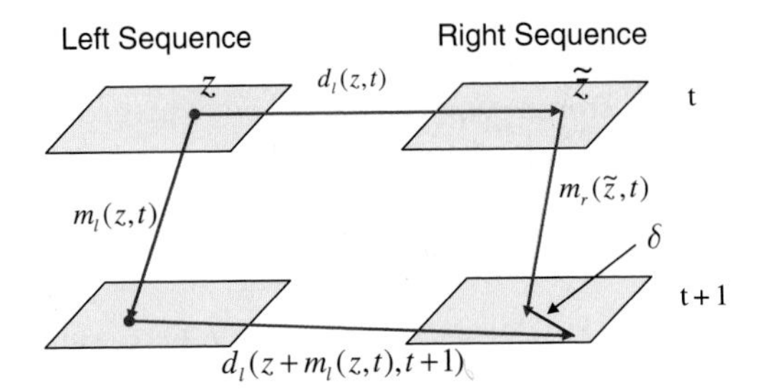


**Fig. 13.10.** Relationship between disparity and motion vectors

directly encodes multi-view images acquired by several independent and temporally synchronized cameras, while the other approach utilizes depth information that can be generated by post-processing algorithms or with special depth cameras.

(1) Multi-view Image-based Approach

A. MPEG-2 Multi-view Profile

The MPEG-2 multi-view profile (MVP) was defined in 1996 as an amendment to the MPEG-2 standard, and the main elements that it introduced were the usage of the temporal scalability mode for multi-camera sequences and the definition of acquisition camera parameters in MPEG-2 syntax [34]. This method made it possible to encode a base layer stream representing a signal with a reduced frame rate, and to define an enhancement layer stream, which has been used to insert additional frames in between to allow reproduction with full frame rates if both streams have been available. In this method, frames from one camera view (usually the left) are defined as the base layer, and frames from the other are defined as the enhancement layer. The enhancement-from-base-layer prediction then turns out to be a disparity-compensated prediction instead of a motion-compensated prediction. Although disparity-compensated predictions can fail, it is still possible to achieve compression by motion-compensated prediction within the same channel. The base layer represents a simultaneous monoscopic sequence. Figure 13.11 is a block diagram of an MPEG-2 MVP encoder/decoder.

B. Disparity-compensated Multi-view Video Coding

This method encodes multi-view sequences by reducing the correlations among the views and using motion/disparity estimation in the time domain as well as in the view domain. Another advantage of this method is that it supports view scalability so that arbitrary views can be chosen. In other words, the desired number of views is only decoded at the receiver side. Given the present state of 3-D displays, decoded multi-view sequences can be properly displayed on various types of displays including 2-D, stereo and 3-D/multi-view monitors or TVs, as shown in Fig. 13.12

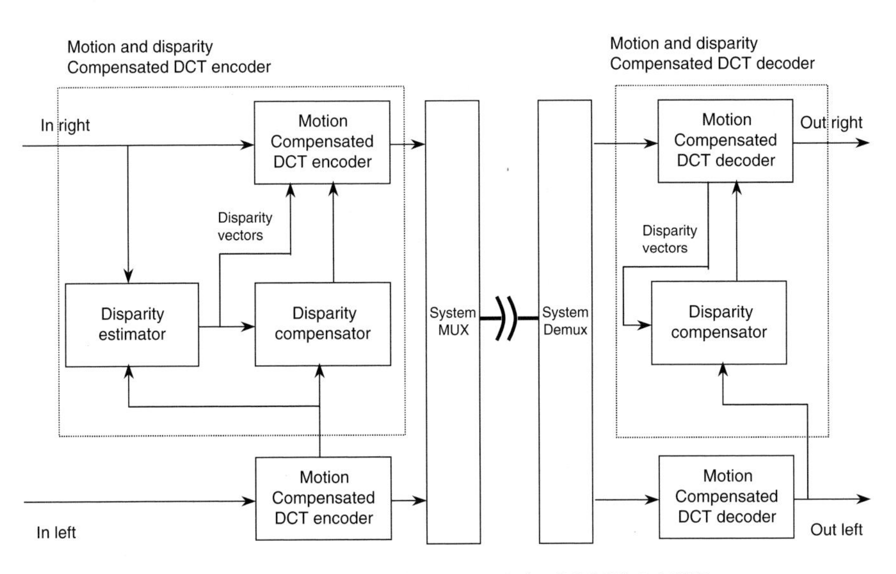


**Fig. 13.11.** Block diagram of the MPEG-2 MVP

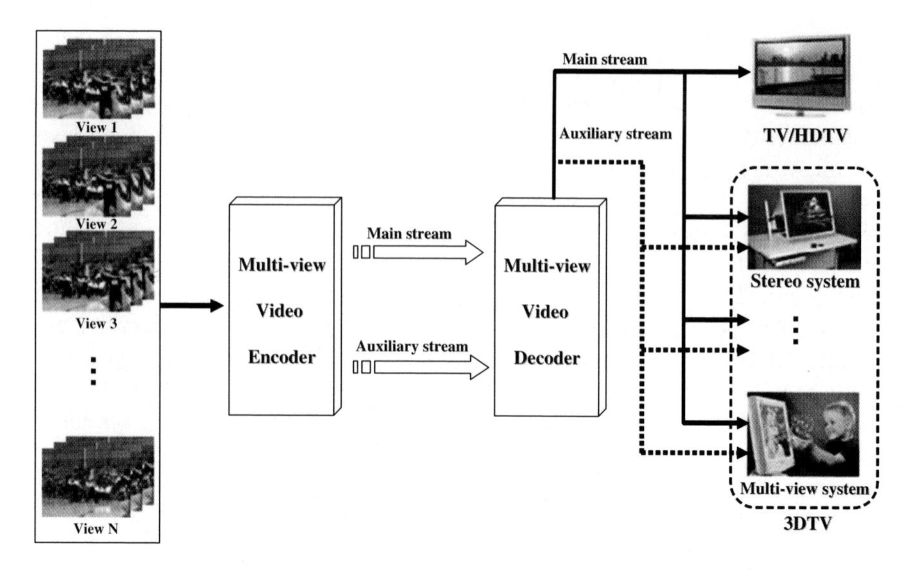


**Fig. 13.12.** Multi-view video CODEC with view scalability

Since conventional 2-D coding only considers temporal predictions, its GOP (Group of Pictures) structure is relatively simple. GGOP (Group of GOP) was defined in order to understand the coding order of multi-view videos [26]. When encoding multi-view sequences, both temporal and spatial predictions must be exploited to obtain higher coding efficiency. The structure of the GGOP in multi-view video coding generally consists of three types of GOPs, shown in Fig. 13.13. First, the GOP including the "I" frames is encoded and refers to the frames of the other views during the estimation stage. The position of this base GOP can change considering the position of the multi-view camera. The GOP including the "P" frame refers to the GOP including the "I" frame or the previously encoded GOP. The other GOPs utilize bidirectional prediction with the previously encoded neighboring GOPs. All the frames of these GOPs are encoded as "B" frames. The temporal coding scheme is similar to the methods generally used in 2-D video coding.

Figure 13.14 shows a block diagram of a multi-view video encoder with the property of view scalability [26]. The CODEC generates two types of bitstreams. The main bitstream (base layer) transmits information from the base GOP to maintain compatibility with the 2-D encoder, and the auxiliary bitstream (enhancement layer) carries the other views. The decoder can adaptively decode these bitstreams based on various types of display modes, as shown in Fig. 13.12. Only selected views can be decoded and appropriately displayed at receivers.

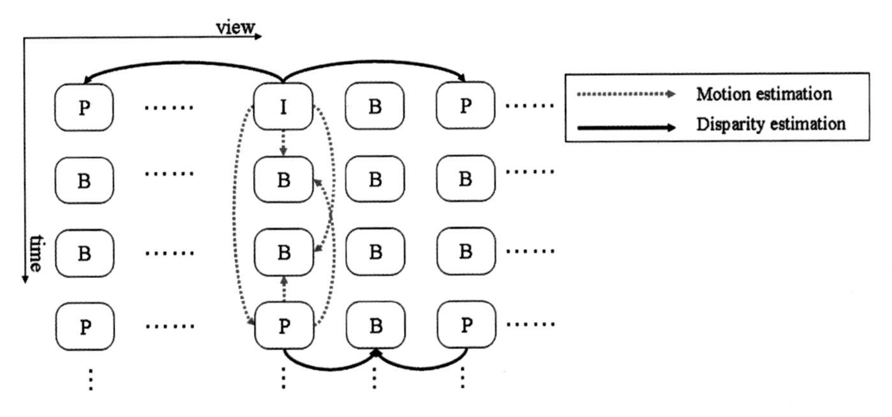


**Fig. 13.13.** General GGOP structure in multi-view sequences

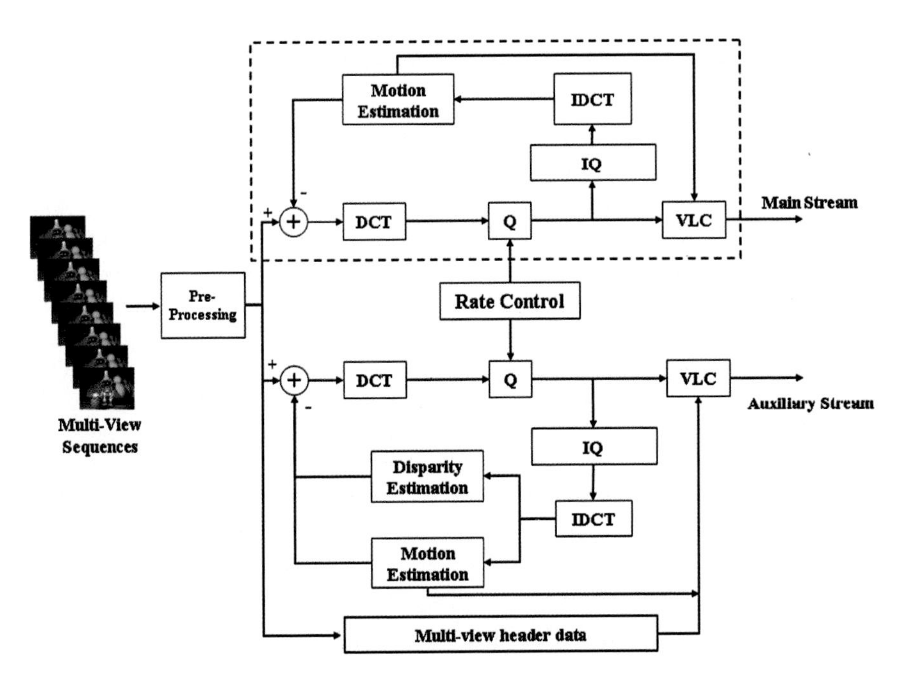


**Fig. 13.14.** Block diagram of the multi-view encoder with view scalability

(2) Depth-based Approaches

Depth-based approaches use depth information as well as texture information for multi-view video coding. Although depth information can be generated by using post-processing techniques such as disparity/depth estimation, a depth camera (Zcam$^{TM}$) similar to that used in the ATTEST project can be more feasible.

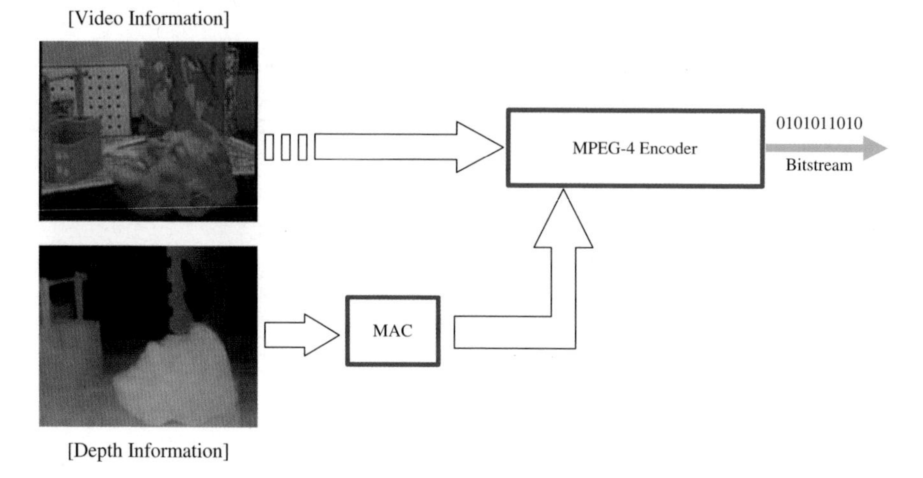


**Fig. 13.15.** Disparity map coding using the MPEG-4 MAC

To encode depth information, a method that uses the MPEG-4 MAC (Multiple Auxiliary Component) has been proposed. The basic idea behind the MAC method is that grayscale shapes are not only used to describe transparency levels of video objects, but can also be defined in more general ways. MACs can be defined for video object planes (VOP) on a pixel-by-pixel basis and contain data related to video objects, such as disparity, depth and additional textures, as shown in Fig. 13.15. Only a limited number of types and combinations have been defined and identified by a 4-bit integer so far, but more applications would be possible by selecting a **User Defined** type or by defining new types. All the auxiliary components can be encoded by the shape coding tools, i.e., the binary shape coding tool and the grayscale shape coding tool which employs a motion-compensated DCT (Discrete Cosine Transform). Therefore, it is possible to encode multi-view videos by aligning pixel-based disparity maps into one layer of MACs.

### 13.3.4 Multi-View Video Coding Standard

The MPEG 3-D AV AHG (Ad-hoc Group) was organized at the 58th MPEG meeting in Pattaya. There had been four exploration experiments (EE) as a response to the call for evidence (CfE) [40]. EE1 had examined the application of mesh objects for omni-directional videos with regard to coding efficiency.

EE2 examined the inter-view prediction scheme, which featured high accessibility of pictures for changing views, based on a simple extension of the MPEG-4 AVC. The Ray-space and model-based scene generation methods were proposed for inter-view prediction. The Ray-space interpolation method was based on adaptive filtering. The model-based scene generation method generated a temporal sequence of topologically consistent 3-D meshes with textures from multi-viewpoint video data. Then, it converted each 3-D video frame into a set of 2-D images and applied them to an existing 2-D image compression method.

The purpose of EE3 was to evaluate MAC for stereoscopic video coding in comparison with several coding methods. After the disparity map was generated from the stereoscopic images, it became an input of one of the MAC layers and was transmitted to the decoder. Since the maximum number of MAC layers was three, the remaining layers contained residual information obtained by disparity–compensated coding in the **User Defined** mode. EE4 presented DIBR (Depth Image-Based Rendering) for 3-D TV and evaluated a number of different MPEG technologies with respect to their depth-image compression performance [41].

In response to a "Call for Comments on 3-D AV" [42], a large number of companies expressed their need for standards. Based on the evidence brought forward for multi-view video coding technology, the call for evidence decided if it was necessary to standardize the MVC in October 2004 [43]. Five organizations proposed their evidence and made calls for proposals (CfP) to determine the reference software in October 2005. In response to these CfP, the coding results from many organizations were compared in both objective and subjective ways at the 75th meeting in Bangkok [44]. The major feature of

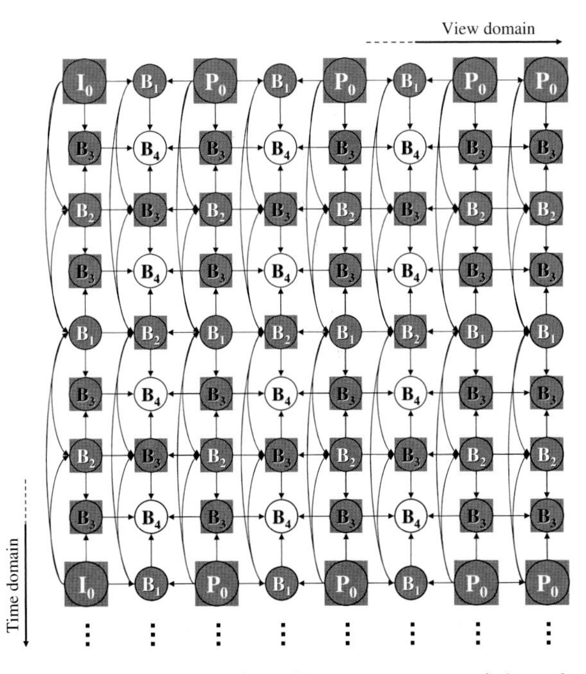


**Fig. 13.16.** Inter-view and temporal prediction structure of the reference software

the reference software was that it used hierarchical B frames and inter-view prediction, as shown in Fig. 13.16.

After determining the reference software, several CEs (Core Experiments) were used to standardize the MVC in detail [45]. In CE1, view-temporal prediction structures were discussed considering flexibility, required syntax changes, application aspects and access issues. The participants of CE2 evaluated coding gains from illumination/color compensation for multi-view coding. View interpolation, vector prediction and macroblock partitioning were discussed in CE3, CE4 and CE5, respectively. The issues of MPEG 3-D AV were moved to the JVT (Joint Video Team) at the 77th meeting in Klagenfurt. The JMVM (Joint Multi-view Video Model) 1.0 was presented, which contained modified high level syntax and reference image management for the MVC. JVT-MVC has been standardized in July 2008.

## 13.4 Video Processing for 3-D Displays

### 13.4.1 Background

The most important goal of 3-D imaging systems is to create realistic images of dynamically changing scenes. Conventionally, these images have been synthesized using geometric models, material properties, and the lighting of scenes. Realistic images can also be synthesized by acquiring geometric descriptions and properties from real scenes.

The ways to obtain arbitrary view generation from multiple images can be subdivided into two classes, Intermediate View Synthesis (IVS) and virtual view rendering, as shown in Fig. 13.17 [16].

IVS performs disparity-compensated interpolation or projection of image textures to synthesize realistic images. This straightforward approach represents the interpolation of a pair of stereo images, where the scaling of disparities allows the selection of a viewpoint on the interocular axis between the cameras. The approach is simple and works quickly, but the main problem is occlusion, which means that no reasonable disparity data is available within areas that are only visible from one of the cameras. The second problem is that the viewpoint is limited to the position of the baseline between the cameras.

On the other hand, the second approach also allows the building up of a 3-D model represented by a 3-D voxel based on the silhouette and disparity fields of objects. For virtual view generation, the 3-D surface model can be rotated towards the desired place and orientation, and the texture data extracted from the original camera can be projected onto this surface. Generated 3-D objects can easily be manipulated, i.e., rotated, to render images from different viewpoints, but this approach's main problem is the slow processing speed for 3-D modeling and rendering. An image-based visual hull technique was developed to accelerate the processing speed [28].

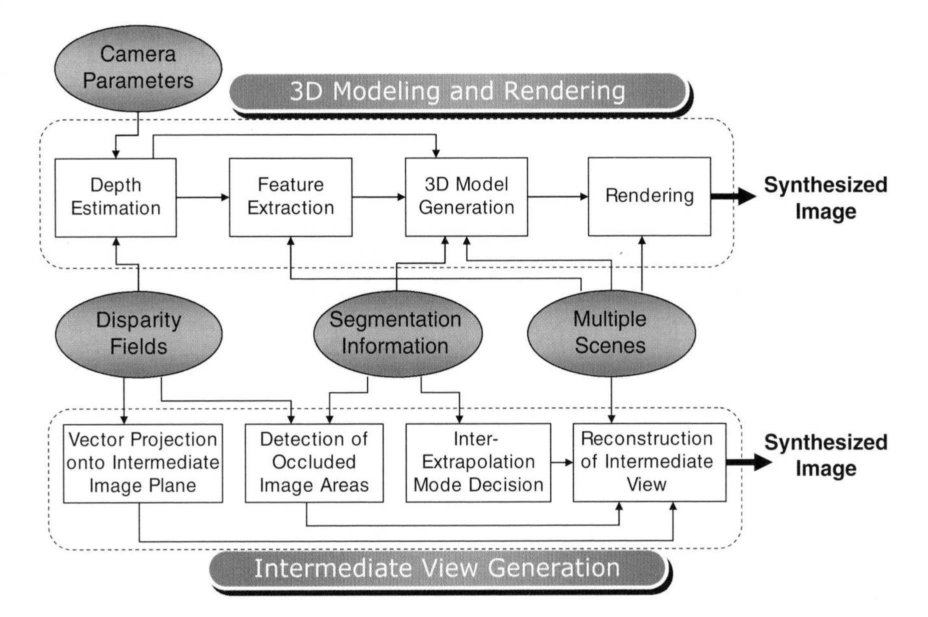


**Fig. 13.17.** Two approaches for arbitrary view generation

### 13.4.2 Intermediate Video Synthesis

Many IVS algorithms based on disparity-compensated interpolation have been developed for the synthesis of images that correspond to the intermediate viewpoints of a pair of images [30]. These algorithms use the estimated disparity vector fields of image pairs to synthesize intermediate views of the given images. Matched regions are interpolated by linearly scaling the disparity vector-related corresponding points in the pair of images, as shown in Eq. (13.9).

$$I(i + \alpha \times d(i,j), j, \alpha) = \alpha \times I_l(i + d(i,j), j) + (1 - \alpha) \times I_r(i,j) \qquad (13.9)$$

$I(\cdot)$ means the pixel value and $\alpha$ represents the relative position between the given views, where $0 \leq \alpha \leq 1$.

Occluded regions are extrapolated by considering the outside disparities of the region. We can assume that occluded regions have constant depth values, which are equal to the depth values of unoccluded background regions adjacent to occluded regions. The smaller the disparity, the farther the object is from the camera pair. Therefore, the unoccluded region with minimum disparity adjacent to the occluded regions represents the unoccluded background region. Figure 13.18 shows the concept of IVS.

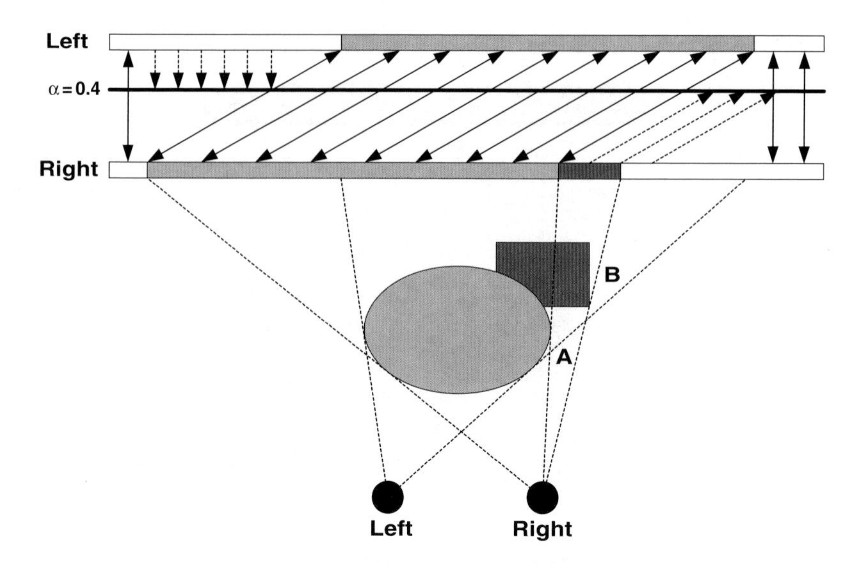


**Fig. 13.18.** Intermediate view synthesis

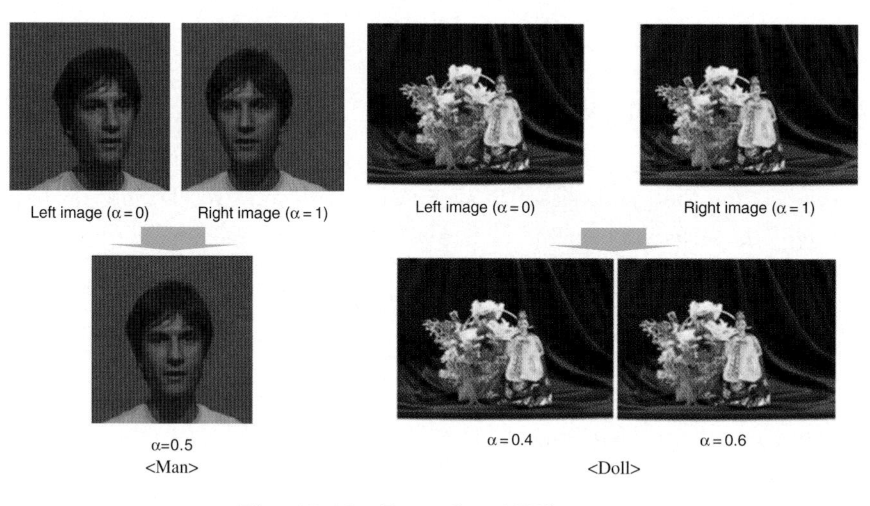


**Fig. 13.19.** Examples of IVS images

Figure 13.19 shows some examples of IVS images obtained from stereo image pairs. The "Man" image pair was captured by a toed-in camera system with an extremely large baseline distance of 80 cm, and the "Doll" image pair was captured by a parallel stereo camera with a 7.5 cm baseline distance.

### 13.4.3 Virtual View Rendering

In computer vision, techniques used to recover shapes are called shape-from-X techniques, where X represents shading, motion, texture, silhouette, stereo, etc. Since the 1980s, much research has been done on computational models for computing surface shapes from different image properties [38].

In recent outcomes of this research, the shape-from-silhouette (SFS) technique has proven to be a very common way of converting silhouette contours into 3-D objects [17, 27]. Silhouettes are readily and easily obtainable and the implementation of SFS methods is generally straightforward. However, serious limitations exist when applying the property to concave surface regions or multiple, disjoint convex objects.

A lot of work has also been done on stereo vision, especially in terms of recovering dense scene structures from multiple images [8]. Figure 13.20 shows an example of the shape-from-stereo technique. Stereo vision needs to solve the correspondence problem, i.e., matching features between the given stereo image pairs. This problem, in general, is under-determined. It is natural to consider integrating stereo and other properties to complement the performance of each.

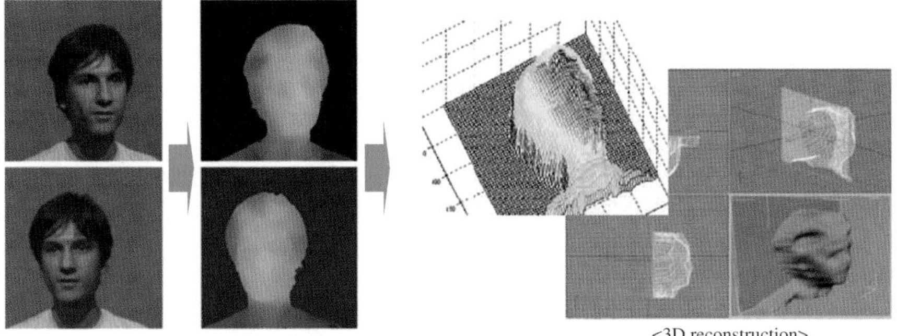


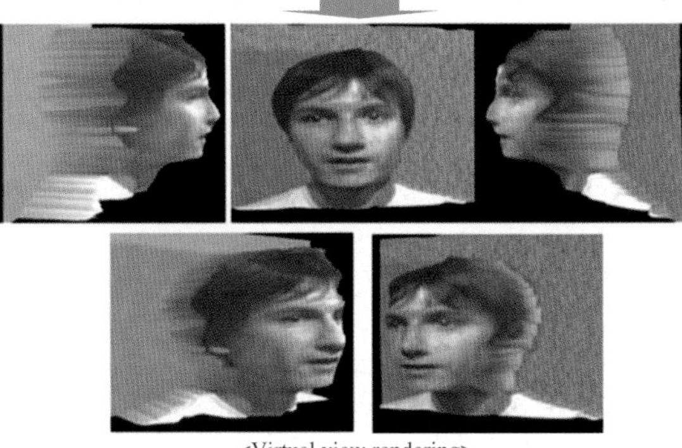


**Fig. 13.20.** Virtual view rendering using shape-from-stereo images

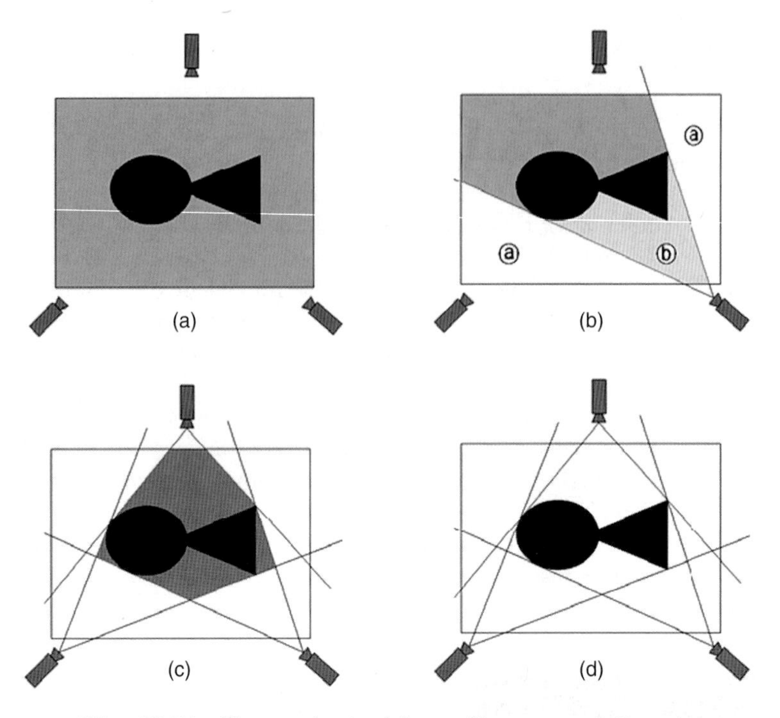


**Fig. 13.21.** Shapes obtained from silhouette and disparity

Several researchers have proposed 3-D modeling and rendering methods that have used both silhouette and disparity information to carve space using stereo cameras [9, 24]. 3-D volumes of these models can be simply carved using silhouette and disparity factors with the camera parameters at the same time. Figure 13.21 shows some simple shape-from-silhouette and disparity (SFSD) techniques on a 2-D plane. Figure 13.21(a) shows a working volume and the object which we tried to reconstruct was only known to be somewhere inside of it. Figure 13.21(b) shows the carving result using the data from the first camera. If we used only silhouette information, region ⓐ as in Fig. 13.21(b) was carved, but region ⓑ remained. By using depth information, it was possible to carve the region ⓑ. Figures 13.21(c) and (d) show the carving results obtained by the SFS and SFSD techniques, respectively.

A 3-D modeling technique that uses SFSD is shown in Fig. 13.22. All voxels $M(X, Y, X)$ in the 3-D spaces were projected onto multiple images $I_n$ using the following equation, where $P_n$ represents the projection matrix of camera $C_n$. If the projected points of $M$ were included in the foreground region of the image, then we checked if the voxel places behind the range of depth were calculated from the disparity. If all projected points of $M$ were included in the foreground region and behind the distance to the surface of multiple images, the point was selected as the inside voxel of the object.

**Fig. 13.22.** 3-D modeling with SFSD

This technique reconstructs 3-D models by merging multiple videos, and it generates 3-D free viewpoint videos by applying CG technology. The system selected the nearest camera from a given viewpoint and projected the texture from it to the model. Users can generate realistic videos of objects in space from an arbitrary viewpoint by controlling a virtual camera.

## 13.5 Conclusion

In summary, this chapter has introduced details of 3-D video processing technologies to 3-D TV, including 3-D content generation, 3-D video CODECs and video processing techniques for 3-D displays.

One of the most serious problems of 3-D video technology has been a lack of 3-D content. For this reason, many recent methods to generate 3-D content have been introduced. These methods include using 3-D computer graphics (CG), capturing images with multiple cameras, and converting 2-D content to 3-D content. The quantity of multi-view data can be increased according to the number of views. This has been a serious problem when implementing multi-view video systems. To efficiently encode multi-view videos, both the spatial redundancy of the view axis and the temporal redundancy of the time axis have to be efficiently exploited and reduced. Current multi-view video coding methods and standard activities have been introduced in this chapter. We also introduced some approaches for arbitrary view generation from multiple images, such as IVS and virtual view rendering.

There is a lot of research actively going on in many countries now. However, human factor problems and 3-D display technologies that do not require special glasses need further study, and additional types of high quality 3-D content need to be developed for 3-D TV to succeed.

## References


[1] M. Accame, F. Natale and D. Giusto, "Hierarchical block matching for disparity estimation in stereo sequences," *Proc. IEEE Int. Conf. Image Processing*, vol. 2, pp. 374–377, 1995.

[2] R. Azuma, Y. Baillot, R. Behringer, S. Feiner, S. Julier and B. MacIntyre, "Recent advances in augmented reality," *IEEE Computer Graphics and Applications*, vol. 25, no. 6, pp. 24–35, 2001.

[3] S. Battiato, "3-D stereoscopic image pairs by depth-map generation," *Proc. 3-DPVT*, pp. 124–131, 2004.

[4] C. V. Berkel and D. W. Parker, "Multiview 3-D-LCD," *Proc. SPIE* 2653, pp. 32–39, 1996.

[5] CCIR 1–1/11 SP1c–1/11, "Constitution of a system of stereoscopic television," 1958–1986.

[6] CCIR Rep. 312–4 "Constitution of a system of stereoscopic television," 1963–1966–1970–1978–1982.

[7] S. Diplaris, "Generation of stereoscopic image sequences using structure and rigid motion estimation by extended kalman filters," *IEEE International Conference on Multimedia and Expo*, pp. 233–236, 2002.

[8] U. Dhond and J. Aggarwal, "Structure from stereo: a review," *IEEE Trans. Syst., Man, Cybern.*, vol. 19, pp. 1489–1510, 1989.

[9] C.H. Esteban and F. Schmitt, "Multi-stereo 3-D object reconstruction," *Proc. 3-DPVT*, pp. 159–167, 2002.

[10] O. Faugeras, *Three-dimensional computer vision: a geometric viewpoint*, The MIT Press, London, 2001.

[11] B. J. Garcia, "Approaches to stereoscopic video based on spatio-temporal interpolation," *SPIE*, vol. 2653, pp. 85–95, 1990.

[12] R. Gvili, A. Kaplan, E. Ofek and G. Yahav, "Depth key," *Proc. SPIE Electronic Imaging*, 2003.

[13] B. Horn, *Robot vision*. Cambridge Mass., MIT Press, 1986.

[14] N. Hur and C. Ahn, "Experimental service of 3-DTV broadcasting relay in Korea," *Proc. SPIE 4864*, pp. 1–13, 2002.

[15] E. M. Izquierdo, "Stereo matching for enhanced telepresence in three dimensional videocommunications," *IEEE Trans. on Circuit and Systems*, vol. 7, no. 4, pp. 629–643, Aug. 1997.

[16] E. M. Izquierdo and S. Kruse, "Image analysis for 3-D modeling, rendering, and virtual view generation," *Computer Vision and Image Understanding*, vol. 71, no. 2, pp. 231–253, 1998.

[17] T. Kanade and P. J. Narayanan, "Historical perspectives on 4D virtualized reality," *Proc. CVPR*, p. 165, 2006.

[18] T. Kanade, A. Yoshida, K. Oda, H. Kano and M. Tanaka, "A stereo machine for video-rate dense depth mapping and its new applications," *Proc. CVPR '96*, pp. 196–202, 1996.

[19] T. Kanade, P. W. Rander, and P. J. Narayanan, "Virtualized reality: constructing virtual worlds from real scenes," *IEEE Multimedia*, vol. 4, no. 1, pp. 34–47, 1997.
[20] S. Kim, M. Kim, J. Lim, S. Son, and K. Sohn, "Forward disparity estimation and intermediate view reconstruction of 3-D images using irregular triangle meshes," *Proc. Conf. 3-D Image*, pp. 51–54, July 2000.
[21] M. Kim and S. Park, "Object-based stereoscopic conversion of MPEG4 encoded data," *PCM*, pp. 491–498, 2004.
[22] H. Kim and K. Sohn, "3-D reconstruction from stereo images for interaction between real and virtual objects," *Signal Processing: Image Communication*, vol. 20, no. 1, pp. 61–75, Jan. 2005.
[23] D. Kim, D. Min and K. Sohn, "A stereoscopic video generation method using stereoscopic display characterization and motion analysis," IEEE Trans. on Broadcasting, vol. 54, no. 2, pp. 188–197, June 2008.
[24] H. Kim, I. Kitahara, K. Kogure and K. Sohn, "A real-time 3-D modeling system using multiple stereo cameras for free-viewpoint video generation," *Proc. ICIAR, LNCS*, vol. 4142, pp. 237–249, Sep. 2006.
[25] M. Kim and K. Sohn, "Edge-preserving directional regularization technique for disparity estimation of stereoscopic images," *IEEE Trans. on Consumer Electronics*, vol. 45, no. 3, pp. 804–811, Aug. 1999.
[26] J. Lim, K. Ngan, W. Yang and K. Sohn, "A multiview sequence CODEC with view scalability," *Signal Processing: Image Communication,* vol. 19, no. 3, pp. 239–256, Jan. 2004.
[27] T. Matsuyama, X. Wu, T. Takai and T. Wada, "Real-time dynamic 3-D object shape reconstruction and high-fidelity texture mapping for 3-D video," *IEEE Trans. CSVT*, vol. 14, no. 3, pp. 357–369, 2004.
[28] W. Matusik, C. Buehler, R. Raskar, S. J. Gortler, and L. McMillan, "Image-based visual hulls," *Proc. ACM SIGGRAPH*, pp. 369–374, 2000.
[29] H. Murata and Y. Mori, "A real-time 2-D to 3-D image conversion technique using computed image depth," *SID*, IGEST, pp. 919–922, 1998.
[30] J. S. McVeigh, *Efficient compression of arbitrary multi-view video signals*, Ph.D. Thesis, Carnegie Mellon Univ., June 1996.
[31] Y. Ohta and H. Tamura, *Mixed reality*, Springer-Verlag, New York Inc., 1999.
[32] T. Okino and H. Murata, "New television with 2-D/3-D image conversion technologies," *SPIE*, vol. 2653, pp. 96–103, 1996.
[33] L. Onural, T. Sikora and A. Smolic, "An overview of a new European consortium integrated three-dimensional television-capture, transmission and display (3-DTV)," *European Workshop on the Integration of Knowledge, Semantics and Digital Media Technology* (EWIMT) Proc., 2004.
[34] R. Puri and B. Haskell, "Stereoscopic video compression using temporal scalability," *Proc. SPIE Visual Communication and Image Processing*, vol. 2501, pp. 745–756, May 1995.
[35] A. Rauol, "State of the art of autostereoscopic displays," RACE DISTIMA deliverable 45/THO/WP4.2/DS/R/57/01, Dec. 1995.
[36] A. Redert, M. Op de Beeck, C. Fehn, W. IJsselsteijn, M. Pollefeys, L. Van Gool, E. Ofek, I. Sexton and P. Surman. "ATTEST – Advanced Three-Dimensional Television System Technologies," *Proc. of 1st International Symposium on 3-D Data Processing, Visualization and Transmission*, pp. 313–319, Padova, Italy, June 2002.

[37] J. Ross and J. H. Hogben, "The Pulfrich effect and short-term memory in stereopsis," *Vision Research*, vol. 15, pp. 1289–1290, 1975.
[38] L. G. Shapiro and G. C. Stockman, *Computer vision*, Chap. 12, New Jersey, Prentice Hall, 2001.
[39] E. Trucco and A. Verri. *Introductory techniques for 3-D computer vision*, New Jersey, Prentice Hall, 1998.
[40] ISO/IEC JTC1/SC29/WG11, Report on Status of 3-DAV Exploration, Doc. W5416, Awaji, Japan, Dec. 2002.
[41] ISO/IEC JTC1/SC29/WG11, Report on 3-D AV Exploration, Doc. W5878, Trondheim, Norway, July 2003.
[42] ISO/IEC JTC1/SC29/WG11, Call for Comments on 3-D AV, Doc. W6051, Gold Coast, Australia, Oct. 2003.
[43] ISO/IEC JTC1/SC29/WG11, Call for Evidence on Multi-View Video Coding, Doc. W6720, Palma, Spain, Oct. 2004.
[44] ISO/IEC JTC1/SC29/WG11, Subjective test results for the CfP on Multi-view Video Coding, Doc. W7779, Bangkok, Thailand, Jan. 2006.
[45] ISO/IEC JTC1/SC29/WG11, Description of Core Experiments in MVC, Doc. W7798, Bangkok, Thiland, Jan. 2006.
[46] R. Wang and Y. Wang, "Multiview video sequence analysis, compression, and virtual viewpoint synthesis," *IEEE Trans. on Circuits and System for Video Technology*, vol. 10, no. 3, pp. 397–410, Apr. 2000.
[47] G. Wei, "Intensity- and gradient-based stereo matching using hierarchical Gaussian basis functions," *IEEE Trans. on Pattern Analysis and Machine Intelligence*, vol. 20, no. 11, pp. 1143–1160, Nov. 1998.

# Part III

# 3-D Image Acquisition, Processing and Display Based on Digital Holography

# 14

# Imaging 3-D Objects by Extending the Depth of Focus in Digital Holography

Pietro Ferraro, Simonetta Grilli, Giuseppe Coppola, and Sergio De Nicola

## 14.1 Introduction

Digital holography is an imaging technique offering both sub-wavelength resolution and real-time capabilities to record 3-D objects using the interference between an object wave and a reference wave captured by an image sensor such as a CCD array. The basic advantage of digital holography is that it can quantitatively extract the three-dimensional (3-D) information of the object from the numerical reconstruction of a single digitally recorded hologram [1,2,3]. Since the information of the optically interfering waves is stored in the form of matrices, the numerical reconstruction process enables full digital processing of the holograms and offers many more possibilities than conventional optical processing. It is possible to numerically focus on any section of the 3-D volume object without mechanical focusing adjustment. This opens the field to a variety applications, such as 3-D microscopic investigations and real-time imaging of biological specimens where wavelength-scanning digital interference holography [4] is used to reconstruct the 3-D volume from a set of scanning tomographic images, hybrid holographic microscopy [5], 3-D microscopy by optical scanning holography [6], phase shifting digital holog-

P. Ferraro
Istituto Nazionale di Ottica Applicata, Consiglio Nazionale delle Ricerche, Via Campi Flegrei 34, 80078 Pozzuoli (Napoli), Italy
e-mail: pietro.ferraro@inoa.it

B. Javidi et al. (eds.), *Three-Dimensional Imaging, Visualization, and Display*,
DOI 10.1007/978-0-387-79335-1_14, 

raphy and particle holography, just to name a few. Accurate reconstruction of the 3-D image volume requires reconstructing the object field at several axial distances from the recording plane. This is a demanding task from the numerical point of view since, for example, if we use the well-known Fresnel transform method [2,3] for hologram reconstructions, reconstructed images at different distances need to be resized to account for the dependence of the reconstruction pixel on the distance [7,8]. The same limitations arise when using multiple wavelength digital holography to record and reconstruct colored objects when reconstructions are being combined at different wavelengths, owing to the linear dependence on the wavelength of the pixel size of the reconstructed image. Furthermore the direct red-green-blue composition of reconstructed amplitude images recorded at different wavelengths is affected mainly by fluctuation of speckle noise that is due to the coherent illumination.

However, coherent imaging based on digital processing of holograms is the great advantage of providing quantitative data bringing, ipso-facto, flexibility in optical data manipulation and simulation of wave propagation. This allows, for example, to devise pure numerical methods for correcting optical component defects, considered mandatory in classical image forming optics, such as lens aberrations, or for compensating the limited depth-of field of a high magnification microscope objective, which is of particular concern when imaging or exploring very small objects that have a 3-D complex shape. Indeed, in this case it is necessary to change the distance between the object and the microscope objective to focus different portions of the object located at different planes and, clearly, if we perform the refocusing numerically by digital holography, the task becomes computationally demanding when many reconstructions along the axial distance must be stuck together and the axial resolution differs from the lateral one.

Below we will describe two methods for reconstructing amplitude images by digital holography. Both methods illustrate the basic potential of digital holography to obtain in-focus images of extended objects. The first method is an angular spectrum-based technique that uses digital holography to reconstruct images on tilted planes (DHT). DHT enables the inspection of an object's characteristics on a plane or a set of plans that tilts with respect to the recording hologram, such as in tomographic application planes or to observe special features of an assembly of objects along selected directions, as in particle holography. Moreover, DHT allows to image in good focus objects in a three-dimensional (3-D) scene tha have extension along the longitudinal direction far behind the depth of the focus of the imaging system [15-17]. The second method is a conceptual novel approach to demonstrate that an extended focused image (EFI) of an object can be obtained by digital holography without any mechanical scanning or any other special optical components. An EFI image is created with a stack of amplitude images and use of a single phase map obtained numerically in the reconstruction process from a digital hologram

The two approaches offer flexible new scenarios for obtaining images of 3-D volume by removing the classic limit of reconstructing by digital hologram single slides perpendicular to the optical axis of the systems. This methods m akes it possible to address a new conceptual step forward in the recording and 3-D display using coherent light.

## 14.2 Angular Spectrum Method for Reconstructing Holograms on Tilted Planes

Digital correction of tilted wavefronts simulates light propagation through calculation of the diffraction between arbitrarily oriented planes. While Leseberg and Frère were the first [9] to address the problem of using the Fresnel approximation, to find the diffraction pattern of a tilted plane, an all-purpose numerical method for analyzing optical systems using full scalar diffraction theory was later proposed [10]. The Fourier transform method (FTM) for numerical reconstruction of digital holograms with changing viewing angles was also described recently [11,12]. Applying digital holography on the tilted plane (DHT) by means of the angular spectrum of plane waves achieved fast calculation of optical diffraction [13]. Fast Fourier transformation is used twice and coordinate rotation of the spectrum enables the hologram to be reconstructed on the tilted plane. Interpolation of the spectral data is effective for correcting the anamorphism of the reconstructed image.

### 14.2.1 Experimental Configuration

The experimental setup for obtaining DHT shown in Fig. 14.1 is a Mach–Zehnder interferometer designed for reflection imaging of microscopic objects. A linearly polarized collimated beam from a diode pumped, frequency doubled Nd:$YVO_4$ laser with $\lambda = 532$ nm is divided by the combination of half-wave plate $\lambda/2$ and a polarizing beam splitter into two beams. The two half-wave plates in combination with the cube act as a variable beam splitter to adjust the fringe contrast of the digital hologram on the CCD plane array. In each arm, spatial filters and beam expanders are introduced to produce plane waves. The object beam is reflected by the test object which consists of a patterned logo "MEMS" covering an area of $(644 \times 1568)\ \mu\text{m}^2$ on a reflective silicon wafer substrate. The object is tilted with respect to the hologram plane $(\xi - 1)$ as shown in Fig. 14.1. The interfering object beam and the reference beam are tilted at small angles with respect to each other to produce off-axis holograms. The digital holograms are recorded by a standard black and white CCD camera and digitized into a square array $N \times N = (1{,}024 \times 1{,}024)$ pixel$^2$,$\Delta\xi \times \Delta 1 = (6.7 \times 6.7)\mu\text{m}^2$ pixel width.

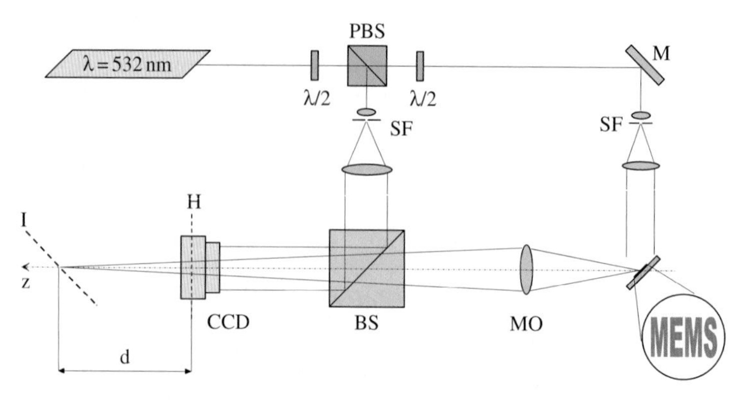


**Fig. 14.1.** Experimental setup to record off-axis digital holograms. PBS polarizing beam splitter; BS beam splitter; SF's spatial filters; MO microscope objective. The hologram is recorded by the CCD camera

The microscope objective produces a magnified image of the object. The hologram plane is located between the MO while the image plane *I* located at a distance *d* from the recording hologram plane is denoted by letter *H*. In digital holographic microscopy we adopted the object wave emerges from the magnified image and not from the object itself [14].

### 14.2.2 Description of the Method and Reconstruction Algorithm

The geometry of the reconstruction algorithm is illustrated in Fig. 14.2. The image plane *I* has coordinates $(\hat{x} - \hat{y})$ and its origin is located along the optical *z*-axis at a distance $d$ from the hologram plane $(\xi - \eta)$. In fact the plane *I* tilts at angle $\theta$with respect to the plane $(x - y)$. The plane $(x - y)$is parallel to the hologram plane *H* and shares the same origin of the tilted plane *I*. Reconstructing the complex wave distribution $o\,(\hat{x},\,\hat{y})$on the tilted image plane using digital holography simulates light propagation from the hologram plane to the image plane by calculating the diffraction of the object wave distribution $o\,(x,\,y)$on the plane *H* (at *z=0*) to the tilted plane.

The angular spectrum-based algorithm (ASA) that we adopted for reconstructing the wave field $o(\hat{x}, \hat{y})$ can be described in two successive steps. Firstly, the wave distribution is reconstructed on the intermediate plane $(\hat{x} - \hat{y})$ at distance $d$. Then the spectrum $\hat{O}(d; u, v)$ where $u = \xi/\lambda d$ and $v = \eta/\lambda d$ denotes the Fourier frequencies in the intermediate plane is calculated. A convenient way of performing the spectrum calculation is by applying the so-called convolution method (CM) for reconstructing the wave field at the translational distance $d$[1]. The formulation uses the convolution theorem to calculate the spectrum as follows:

$$\hat{O}(d;u,v) = O(u,v)\exp\left[i2\pi d\sqrt{\lambda^{-2} - u^2 - v^2}\right] \tag{14.1}$$

Where $O(u,v) = \Im\{o(x,y)\}$ is the spectrum of the object wave distribution $o(x,y)$ at $z = 0$ which is defined by the Fourier transform

$$O(u,v) = \iint o(x,y)\exp\left[-i2\pi(ux+vy)\right]dxdy. \tag{14.2}$$

In CM the field distribution at distance $d$ is derived by taking the inverse Fourier transform of the propagated angular spectrum (14.1), but in our case we must consider the tilting of the reconstruction plane $(\hat{x} - \hat{y})$ with respect to the intermediate plane. The standard transformation matrix is used to rotate the wave vector coordinates $\mathbf{k} = 2\pi\,[u, v, w]$ where $w(u,v) = \left(\lambda^{-2} - u^2 - v^2\right)^{1/2}$, on the $y$-axis with the angle of $\theta$, to express the spatial frequencies $\hat{u}$,$\hat{v}$ and $\hat{w}$ associated to the components of the rotated wave vector $\mathbf{k} = 2\pi\,[\hat{u}, \hat{v}, \hat{w}]$ in terms of the spatial frequencies of the intermediate plane $u$,$v$and $w$. The corresponding transformation law is given by

$$u = \hat{u}\cos\theta + \hat{w}(\hat{u},\hat{v})\sin\theta \tag{14.3a}$$

$$v = \hat{v} \tag{14.3b}$$

$$w = -\hat{u}\sin\theta + \hat{w}(\hat{u},\hat{v})\cos\theta \tag{14.3c}$$

According to the spatial frequencies transformation (14.3a–c), the rotated angular spectrum (RAS) can be written as

$$\hat{O}(d;\hat{u},\hat{v}) = \hat{O}(d;\hat{u}\cos\theta + \hat{w}\sin\theta, \hat{v})\,. \tag{14.4}$$

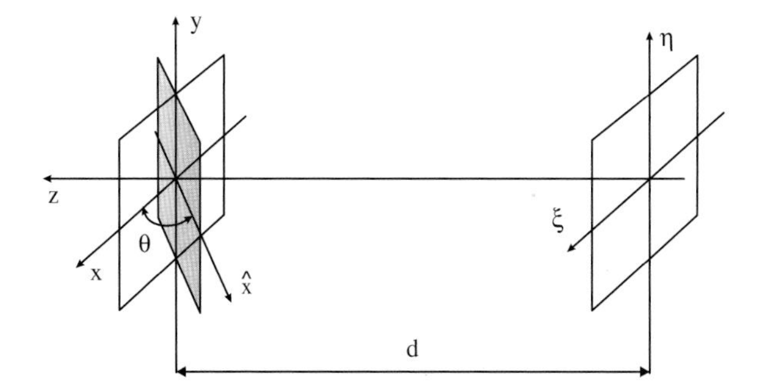


**Fig. 14.2.** Schematic illustration of the ASA algorithm for reconstructing digital holograms on tilted planes

At the second step of ASA, the rotated spectrum is inverse Fourier transformed to calculate the reconstructed wave field $o\,(\hat{x},\,\hat{y})$on the tilted plane,

$$o\,(\hat{x},\,\hat{y}) = \Im^{-1}\left[\hat{O}\,(d;\hat{u}\cos\theta + \hat{w}\sin\theta,\,\hat{v})\right] \tag{14.5}$$

where $\Im^{-1}$ denotes the inverse Fourier transform operation. It is important to note that reconstructing the field according to Eq. (14.5) is valid if we stay within the paraxial approximation, as it applies to our experimental conditions. However, Eq. (14.5) can be generalized to include frequency dependent terms of the Jacobian associated to the RAS. Including these terms does not affect the reconstruction, since they are small in the paraxial case. Eq. (14.5) is a good approximation for the numerical calculation of the wave field on the tilted plane from a digitally recorded hologram as it will be shown in the discussion of the experimental results.

Figure 14.3(a) shows the hologram of the reflective target recorded in off-axis configuration. The plane reference wave interferes with the object waves at small angle ($\leq 0.5$), as required by the sampling theorem. The recording distance $d$ is set to be 265 mm and Figs. 14.3(b), (c) and (d) show a sequence of amplitude images reconstructed at different $z$ planes located at distances 240 mm, 265 mm and 290 mm, respectively, from the hologram plane. The real and not sharp virtual images of the object are separated because of the off-axis geometry. The real amplitude reconstructions, calculated by the FTM with the zero-order term filtered out, can be seen in the bottom-left part of the reconstructed area. Note the three reconstructions by FTM cannot be compared to each other because their size is different in each case, since the reconstruction pixel given by $RP = \lambda d/N\Delta\xi$ depends on the reconstruction distance $d$.

Looking at the reconstructions shown in Fig. 14.3, different features of the amplitude images are more or less highlighted or darkened as the reconstruction distance is changed. At 240 mm the left border of the letter "S" is in-focus; at 265 mm the "E" is in-focus in the middle of the logo script and at 290 mm the "M" letter on the left is in-focus. The result of applying ASA is shown in Fig. 14.3(e). The target is reconstructed on a plane tilted at angle $\theta = 45°$. Now the features of the patterned letters appear to be highlighted uniformly, but the reconstructed image appears shrunk in the horizontal $\hat{x}$ direction, since the pixel size $\Delta\hat{x} \times \Delta\hat{y}$ is different along the $\hat{x}$and $\hat{y}$ directions of the tilted plane.

In fact, the spatial frequency resolution $\Delta\hat{u}$ in the horizontal $\hat{x}$ direction is related to the pixel size $\Delta\hat{x}$ by the relation $\Delta\hat{u} = 1/N\Delta\hat{x}$. If we neglect the $w$ dependent term in the transformation law of $\hat{u}$,$\Delta\hat{u}$ is given to a first approximation by $\Delta\hat{u} = \Delta u\cos\theta$. Since $\Delta u = \Delta\xi/\lambda d$, we readily obtain $\Delta\hat{x} = \lambda d/(N\Delta\xi\,\cos\theta)$. However, the spatial frequency along the $\hat{y}$direction is unchanged, i.e., $\Delta\hat{v} = \Delta v = \Delta\eta/\lambda d$, and the pixel size along the $\hat{y}$direction

**Fig. 14.3.** (**a**) digitized hologram of the logo "MEMS;" (**b**) amplitude image reconstruction at 240 mm; (**c**) amplitude image reconstruction at 265 mm; (**d**) amplitude image reconstruction at 290 mm; (**e**) reconstruction by the angular spectrum-based algorithm; (f)compensation of the anamorphism.

is related to $\Delta\hat{x}$ by the relation $\Delta\hat{y} = \lambda d/N\Delta\eta = \Delta\hat{x}\cos\theta$, where in the last step we have assumed $\Delta\xi = \Delta\eta$.

Clearly, the greater the tilt angle, the larger the size of the reconstruction pixel in the tilted plane and the anamorphism of the reconstructed image.

Let us discuss a simple method for compensating this anamorphism which works straightforwardly in our case. In ASA we are confronted with the numerical problem of applying the FFT algorithm to the RAS (see Eq. (14.5)). The spacing of the RAS's sampling points is not equal, due to the transformation law of the spatial frequencies and a method of interpolation must be unavoidably introduced into the sampled values of the RAS, given that the FFT algorithm works properly for an equidistant sampled grid.

In order to obtain an evenly spaced grid of $N \times N$ interpolated values of the RAS, we use $\Delta\hat{v}$ as a frequency sampling interval for both the $\hat{u}$ and $\hat{v}$ ranges of the RAS. Since the $\hat{u}$ range is less than the corresponding $\hat{v}$ range, a zero is introduced into the spectral array when the corresponding sampled $\hat{u}$ frequency does not fit into the $\hat{u}$ range. Taking the inverse Fourier transform we obtain a reconstructed image with a square pixel of size $\Delta\hat{x} \times \Delta\hat{y} = \lambda d/N\Delta 1 \times \lambda d/N\Delta\eta$ with a corrected anamorphism. This correction procedure works in the Fourier space image on a tilted plane with the anamorphism corrected according to this procedure.

## 14.3 Extended Focus Image (EFI) by Digital Holography

Microscopes allow small objects to be imaged with very large magnifications, but a trade-off exists between magnification and depth-of-focus. In fact, the higher the magnification of the microscope objective (MO), the thinner is the corresponding in-focus imaged volume of the object along the optical axis [16]. In fact, when observing an object with three-dimensional (3-D) shape by miscroscope, only a portion of it appears in good focus since, essentially, a single image plane is imaged. Often it is highly desirable to obtain a single image in which all details of the object under observation are in correct focus. Such an image is an extended focused image (EFI). Two approaches have been developed to get an EFI. Both methods have severe limitations since one requires mechanical scanning while the other requiresspecially designed optics. The use of coherent light can overcome these limitations by obtaining the EFI image of an object via digital holography (DH) in a microscope configuration. The novelty of the proposed approach lies in the fact that it is possible to build an EFI by completely exploiting the unique feature of DH in extracting all the information content stored in the hologram (phase and amplitude).

A DH based method [18–20], leading to a novel concept in optical microscopy, is described here. By using DH it is possible to obtain an EFI of a 3-D object without any mechanical scanning, as occurs in modern microscopy, or as can be done by classical optical holography [21], nor by use of a special phase plate used in the wavefront coding approach [25,26,27]. That is made possible by the unique property of DH, different from film holography, where the phase information of the reconstructed wavefront is available numerically. In DH the buildup of an EFI by a single digital hologram is possible.

### 14.3.1 Approaches Adopted for Extending the Depth-of-Focus in Classical Microscopes

In conventional optical microscopes the approach for constructing of an EFI is based on collecting a stack of images obtained by performing mechanical scanning of the microscope objective on different image planes [22,23]. Another solution that has been developed is based on the use of a specially designed phase plate to use in the optical path of the microscope that allows an extension of the depth–of-focus of the images observable by a microscope [24,25,26].The first solution has already been found through practical application and, in fact, almost all microscopes now offered by manufacturers contain a module that is able to create the EFI image [23].

We will limit our discussion and illustration to the first approach since it is adopted in most optical microscopes. In this approach the EFI is composed by selecting different parts in sharp focus in each image from the stack of a large number of images. Figure 14.4 schematically illustrates the recording process of the stack of images occurring for various distances between the MO and the object.

Usually mechanical translators are actuated by means of a piezoelectric drive element. The MO (or equivalently the plate holder) is displaced and moved along the optical axis with a required and appropriate number of steps in between the highest and lowest points of the object. At each step along the optical axis an image is recorded and stored in a computer and linked to the corresponding depth data. An appropriate software package in the PC controls the microscope and identifies the in-focus portion of each image of the stack. From each image the portion of the object that appears to be, or is numerically recognized as well in-focus, is extracted. Then the different parts are stitched together to give a single image in which all details of the object are in-focus, providing the EFI.

In the EFI all points of an object are, in fact, in-focus independent of their height in the topography of the object. The smaller the stepping increments performed in the mechanical scanning are, the more accurate the result of the EFI. However, more steps require more time to obtain an EFI. The time for single image acquisition essentially depends on the characteristic response time of the piezo-actuator. Generally about 0.10 seconds are needed to acquire a single image. Even if the computing time for a large number of images is not a problem, the length of the acquisition process poses a drawback in obtaining an EFI for dynamic objects.

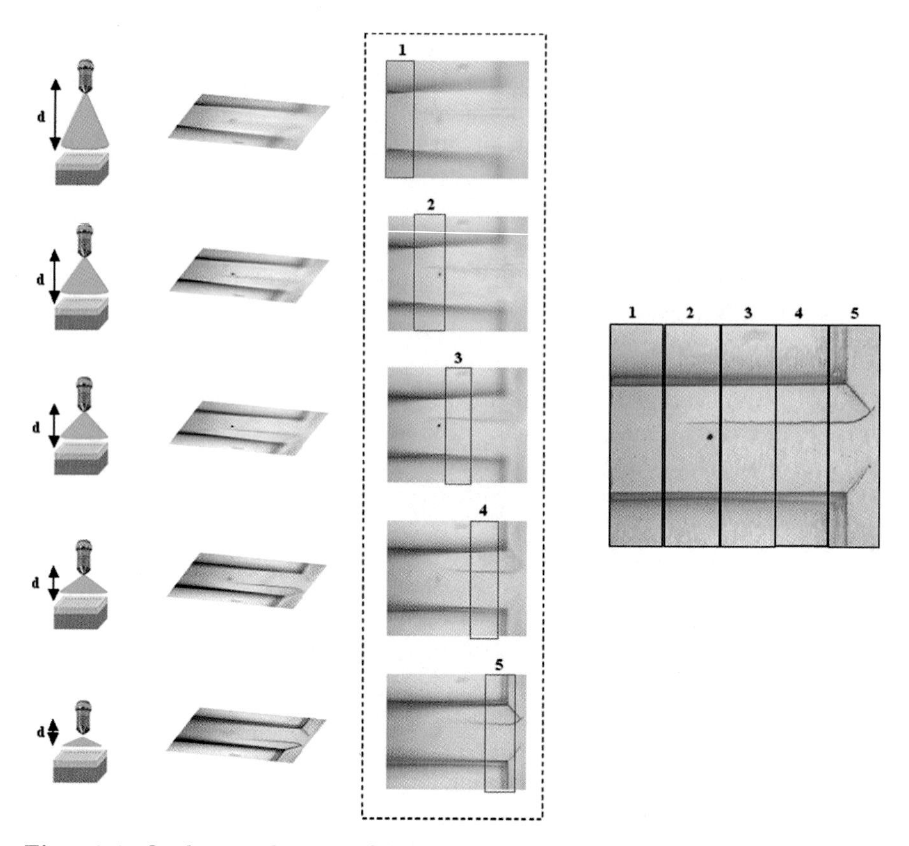


**Fig. 14.4.** Qualitative drawing of the principle of the EFI method. Stack of in-focus images (sequentially numbered) corresponding to different portions of the imaged object are stuck together to get an overall in-focus image (on the *right*)

## 14.4 Construction of an EFI by Means of Digital Holography

As in classical holography, in DH an object is illuminated by collimated, monochromatic, coherent light with the wavelength λ. The object scatters the incoming light forming a complex wavefield (the *object beam*): $O(x, y) = |O(x, y)|e^{-j(x,y)}$ where $|O|$; is the amplitude and φ the phase, $x$ and $y$ denote the Cartesian coordinates in the plane where the wave field is recorded (the *hologram plane*). The phase $\varphi(x, y)$ incorporates information about the topographic profile of the MEMS under investigation because it is related to the optical path difference (OPD):

$$\varphi(x, y) = 4\pi/\lambda * OPD(x, y) \tag{14.6}$$

where a reflection configuration has been considered. Since all light-sensitive sensors respond to intensity only, the phase is encoded in the intensity fringe pattern, adding another coherent background wave $R(x, y) = |R(x, y)|ej\varphi(x,y)$, called the *reference beam*. The two waves interfere at the surface of the recording device. The intensity of this interference pattern is calculated using:

$$I(x, y) = |O(x, y) + R(x, y)|^2 == |R(x, y)|^2 + |O(x, y)|^2 + |O(x, y)R^*(x, y) + R(x, y)O^*(x, y) \quad (14.7)$$

where $*$ denotes the conjugate complex. The *hologram* is proportional to this intensity: $h(x, y) = \alpha I\ (x, y)$.

For hologram reconstruction the amplitude $h(x, y)$ has to be multiplied by the complex amplitude of the reference wave: on

$$R(x, y)h(x, y) == R(x, y)\alpha|R(x, y)|^2 + R(x, y)\alpha|O(x, y)|^2 + \alpha|R(x, y)|^2 O(x, y) + \alpha R^2(x, y)O^*(x, y) \quad (14.8)$$

The first term on the right-hand side of this equation is the attenuated reference wave; the second one is a spatially varying "cloud" surrounding the first term. These two terms constitute the zero-th order of diffraction. The third term is, except for a constant factor, an exact replica of the original wavefront and, for this reason, is called the *virtual image*. The last term is another copy, the *conjugate image*, of the original object wave, but focused on the opposite side of the holographic plane (*real image*). While in classical holography the reconstruction process is performed optically by illuminating the hologram with the reference wave $R$; in DH the reconstruction process is performed numerically by using the Fresnel approximation of the Rayleigh–Sommerfeld diffraction formula [28,29].

Generally, in DH the recording device is a CCD array, i.e., a two-dimensional rectangular raster of $N \times M$ pixels, with pixel pitches $\Delta x$ and $\Delta y$ in the two directions. By DH it is possible to reconstruct and numerically manage not only the amplitude (connected with the intensity of the image of the object), but also the phase that is directly connected to the topography of the object.

As explained above an EFI can, in principle, be obtained by holography. In fact, holography has the unique attribute that allows for recording and reconstructing the amplitude and phase of a coherent wavefront that has been reflectively scattered by an object through an interference process [21]. The reconstruction process allows the object's entire volume to be to imaged. In classical film holography the reconstruction process is performed optically by illuminating the recorded hologram by the very same reference beam. An observer in front of the hologram can view the 3-D scene. Various image planes can be imaged. For example, it is possible to take pictures of different depths during the optical reconstruction process with a photographic camera simply by moving the camera along the longitudinal axis y. Using coherent light, one single hologram obtained using a microscopy setup is sufficient to reconstruct the whole volume of a microscopic object and, by scanning the camera at different depths during the reconstruction process, it is possible to obtain an EFI exactly as modern microscopes do.

It is clear that in the case of holography the scanning process with mechanical movement of the MO must also be performed to image different sections into the imaged volume. However, there is one very important advantage of holography: only one image must be recorded because the scanning is performed not during the recording process, but subsequently during the reconstruction process after the hologram has already been recorded. In this case dynamic events can be investigated. That means the EFI of a dynamic process can be obtained on the basis of using a number of holograms recorded sequentially.

Recently, some advances were achieved in DH in which the recording process of digital holograms was made directly on a solid-state array sensor, such as a CCD camera. In DH the reconstruction process is performed numerically by processing the digital hologram [30]. In fact, the digital hologram is modelled as the interference process between the diffracted field from the object and its interference with a reference beam at the CCD camera (see Fig. 14.5). The use of the Rayleigh-Sommerfield diffraction formula allows the whole wave field in amplitude and phase to be reconstructed backward from the CCD array at one single image plane in the interesting volume. Due to the fact that the reconstruction of the digital hologram is numeric, reconstructions at different image planes can be performed along the longitudinal axis ($z$-axis) by changing the distance of back propagation in the modelled diffraction integral.

For each reconstruction distance $d$, one single image section is reconstructed in good focus. Depending on the optical properties of the employed microscope objective, the depth-of-focus is limited. If the object under investigation has a 3-D shape then, at a fixed reconstruction distance $d$, only some portion of the object will be in focus. Of course it is possible to obtain the entire volume by reconstructing a number of image planes in the volume of interest along the $z$-axis, and with the desired longitudinal resolution. In this way a stack of images of the entire volume can be easily obtained. With those images it is possible to obtain an EFI of the object adopting already developed algorithms for constructing EFI in modern microscopes, provided that some solutions are adopted to control the size of the object independently from the reconstruction distance and centering it by modelling the reference beam appropriately, as it has been demonstrated in recent papers [30,32]. In DH, the reconstruction along the $z$-axis is possible, as is shown in some papers

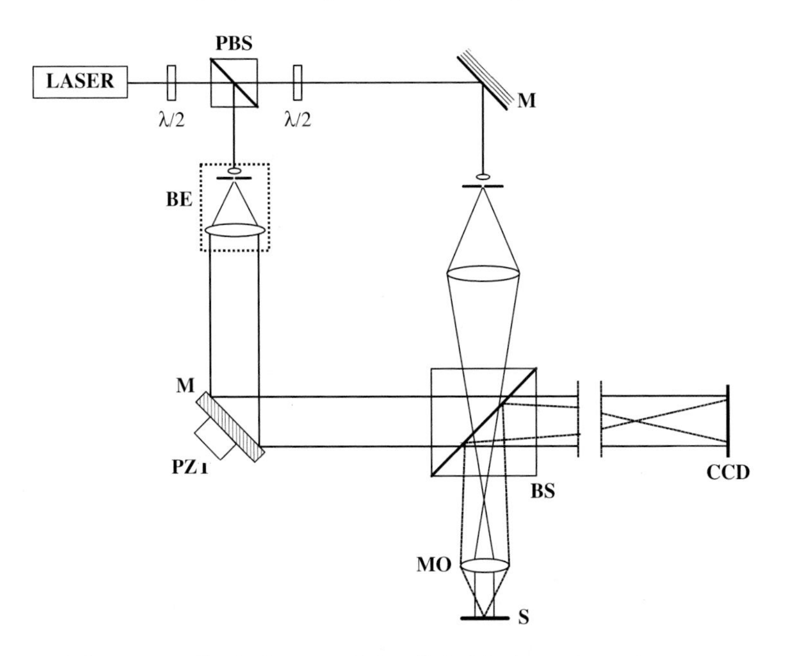


**Fig. 14.5.** Optical setup of the digital holographic microscope

reporting the in-focus reconstruction of a wire positioned in a tilted plane with respect to the $z$-axis, but in a well-known position.

## 14.5 Constructing an EFI by DH Using Amplitude Reconstructions

In DH the EFI is composed of a stack of amplitude images obtained numerically in the reconstruction process. The important advantage of this proposed method is the possibility of obtaining an EFI of a microscopic object without a mechanical scanning operation using only one acquired hologram. We report here the EFI of two different objects to demonstrate the experimental procedure using digital holography, a tilted target on which the logo MEMS is engraved on a silicon wafer and a cantilever MEMS structure made of silicon, respectively.

In this section we report experimental results obtained for the target. The target is tilted at an angle $\varphi = 45$ degrees, and we acquired one hologram shown in Fig. 14.6(a) with the DHM setup shown in Fig. 14.5. The target is $664 \times 1600$ microns. We numerically reconstructed 50 amplitude images at different distances from $d_1$=240 mm up to $d_2 = 290$ mm with a step of $\Delta d = 1$ mm. We adopted FTM to reconstruct amplitude images. [30]. Since

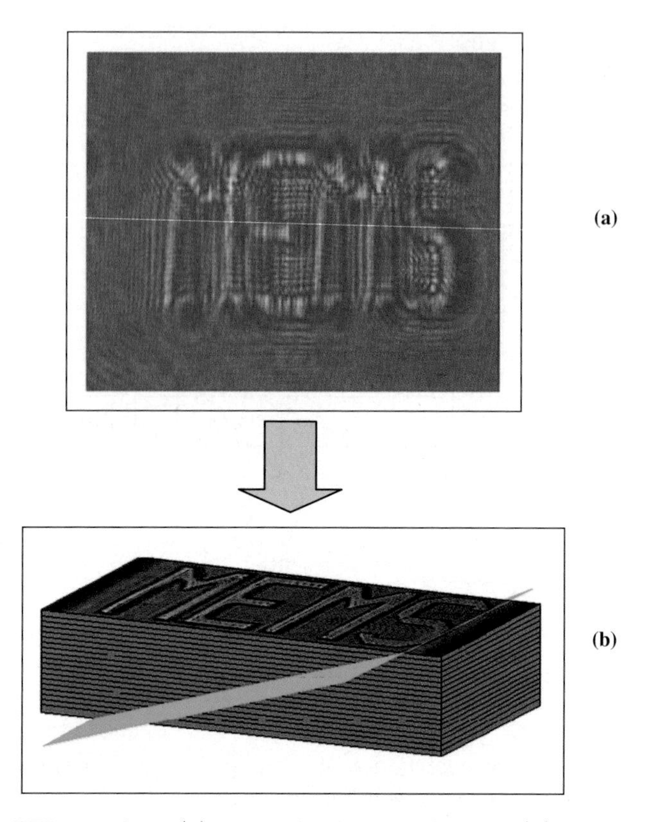


**Fig. 14.6.** EFI procedure: (**a**) acquired hologram of target, (**b**) stack of 50 numerically reconstructed at different distances with the tilted plane along which surface the in-focus pixel is extracted

the reconstruction pixel size changes with the distance when the FTM is applied, we adopted a controlling procedure as described in ref. [31]. In this way all amplitude images of the stack had the same size. We cut the stack of images by a tilted plane with an angle of $\varphi = 45$ degrees, to extract the portion in-focus in each amplitude image, as shown in Fig. 14.6b. In fact that was the original tilting angle of the target, in respect to the optical axis, during the recording of the digital hologram.

Figure 14.7 shows two reconstructed targets at two different distances, $d_1$=240 mm and $d_2$= 290 mm, respectively. In Fig. 14.7a the letter "S" on the right-hand of the logo is clearly in good focus while the remaining part of the logo is out-of-focus; On contrary, in Fig. 14.7b only letter "M" on the left-hand the logo is in good focus. The EFI shown in Fig. 14.8 was obtained by collecting the pixel from each image of the stack that is intersected by the "cutting" tilted plane. It is clearly visible in Fig. 14.8 that the logo "MEMS" is fully in-focus.

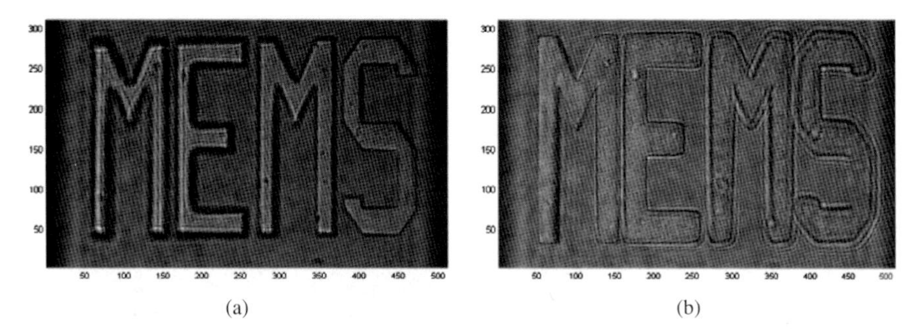


**Fig. 14.7.** Amplitude reconstruction of the 45 degree tilted target at two different distances

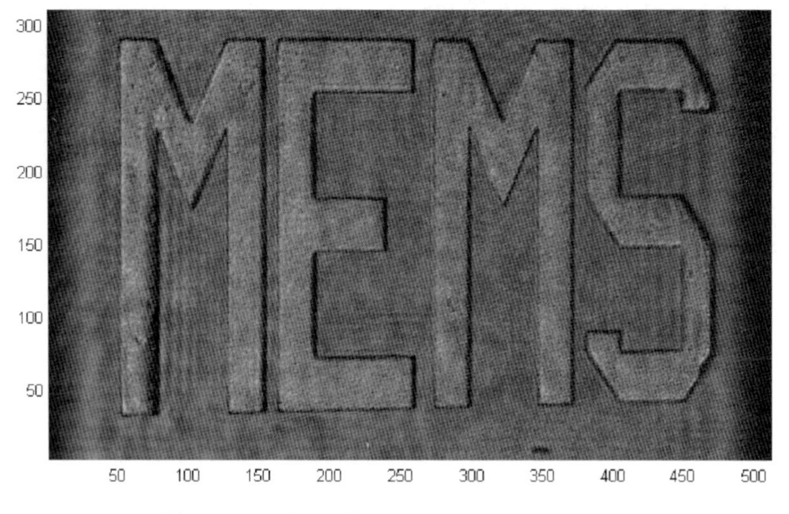


**Fig. 14.8.** EFI of the 45 degree tilted target

## 14.6 Constructing an EFI by DH Using Amplitude and Phase Reconstructions

When using microscopy an important investigative tool is the fabrication of micro-electro-mechanical-systems (MEMS) [33]. Often these structures are made of Silicon, and are the key components for modern microsensors and actuators in many technologically advanced products (accelerometer for air bag systems, gyroscopes, digital display systems devices, etc.) Actually the most reliable techniques for testing such structures are interferometric profilometers, or the scanning electron microscope (SEM), the transmission electron microscope (TEM) that is very important for characterizing the materials. It can be very helpful for an observer to have an EFI of the structure to detect, for example, the presence of cracks or defects as they appear in

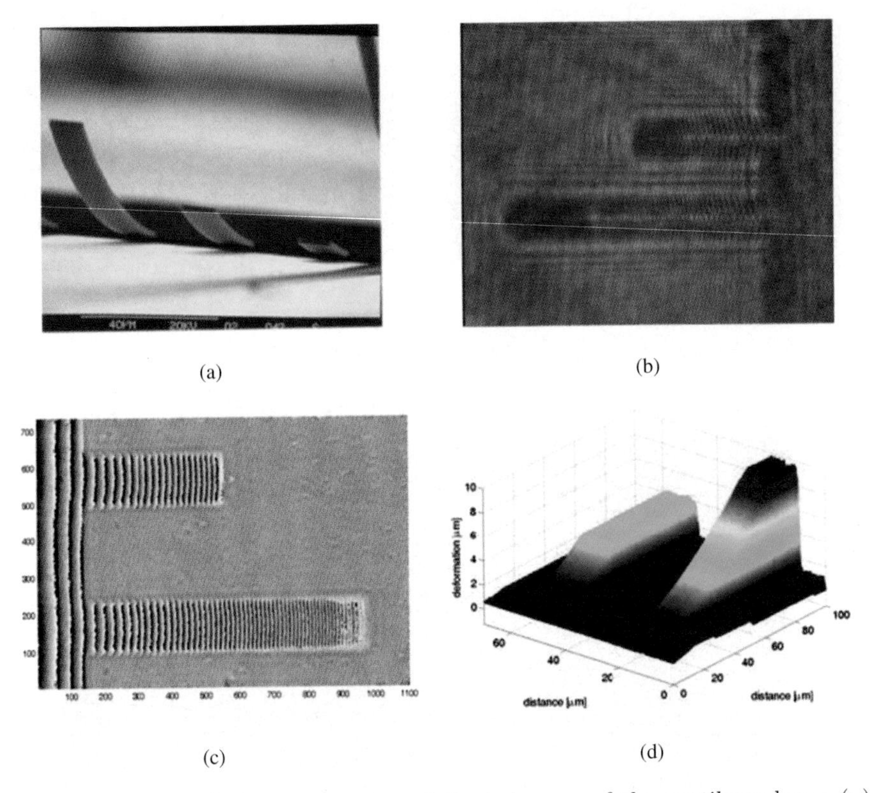


**Fig. 14.9.** Numerical reconstruction of the hologram of the cantilever beam (**a**) SEM image (**b**) digital hologram (**c**) wrapped phase map (**d**) 3-D profile of the cantilever

different locations of the structure under observation. From this the critical start point of the defect and/or the physical mechanism that caused them can be understood and interpreted. Figure 14.9a shows an SEM image of a silicon MEMS structure, a cantilever beam, on which an aluminium layer was deposited. Initially the Silicon cantilever (20 $\mu$m $\times$ 100 $\mu$m) was highly deformed due to the presence of a severe residual stress induced in the fabrication process. The combination of the initial residual stress and the deposition of the aluminium layer caused a progressive breakage of the structure. Figure 14.9b shows the hologram recorded by the DHM setup shown in Fig. 14.4. Figure 14.9c shows the mod.$2\pi$ wrapped phase map that was obtained by reconstructing the digital hologram of Fig. 14.9b. Figure 14.9d shows the profile of the structure obtained by unwrapping it. By observing the figure it is evident that only a part of the beam is in good focus because different parts of the cantilever beam are in-focus at different planes.

We presented the investigations of the MEMS silicon. By observing that MEMS structure with a microscope of very high magnification it is evident

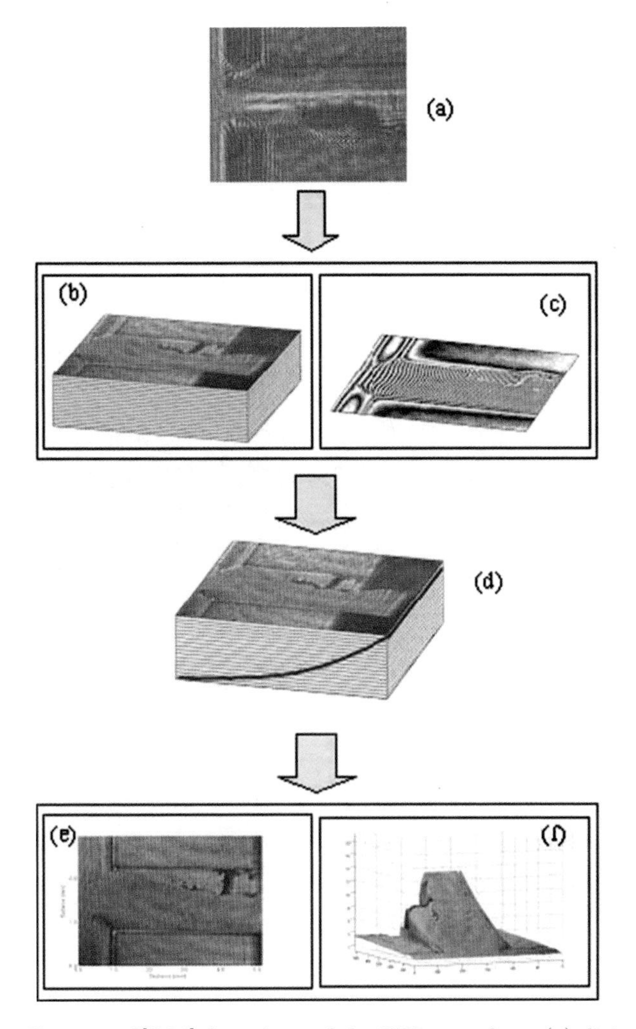


**Fig. 14.10.** Conceptual[Q20] flow chart of the EFI procedure. (a) digital hologram of MEMS; (**b**) stack of the amplitudes images reconstructed at different distances; (**c**) wrapped phase map of the MEMS from which can be extracted its profile; (**d**) the stack of amplitude images is cut by the profile of the MEMS as obtained from the phase map to extract the correct EFI; (**e**) 3-D representation of the MEMS obtained combining the phase map and the stack of the amplitude reconstructed images

that in each focused plane only some portion of the object will be in-focus. In this case, we have acquired a hologram of the object and a reference hologram of a reference plane. We have numerically reconstructed a stack of 34 amplitude images at a distance of $d_1 = 156$ mm up to $d_2=190$ mm with step of $\Delta d = 1$ mm. Then we have reconstructed the phase map at a distance of $\mathrm{d}_1$

**Fig. 14.11.** Comparison of the microscopy and DH reconstruction method (**a**) in-focus image of the base of the cantilever obtained by the microscope (**b**) amplitude reconstruction of the base of the cantilever by DH method (**c**) in–focus image of the tip of the cantilever obtained by the microscope (**d**) amplitude reconstruction of the tip of the cantilever by DH method (**e**) EFI image of the cantilever (**f**) reconstructed amplitude image of the cantilever by DH

to obtain the surface of in-focus for each point of the object. The phase map was used to "cut" the stack of amplitude images for recovering the EFI of the cantilever beam. The conceptual flow chart of the EFI procedure is shown in Fig. 14.10.

Figure 14.11a and c show the images obtained by a classical microscope at two different locations along the optical axis: the base and the tip of the MEMS. Figure 14.11b and d show the corresponding amplitude reconstructions at the base and at the tip of the MEMS structure, obtained by DH, respectively. It is clear in both that by the white light microscope and the DH the tip is severely out-of-focus while the initial part of the crack is visible at the anchor point of the cantilever. On the contrary, at the different plane of focus the tip is in-focus while the base is blurred and out-of-focus. In the amplitude holographic reconstruction of Fig. 14.9, since a coherent light is used, the out-of-focus areas at the sharp edges show highly visible diffraction fringes. Finally, Figs. 14.11e and f- show the EFIs for the microscope and the DHM, respectively.

## 14.7 Conclusions

In this chapter we have described two procedures that use a holographic approach for imaging along inclined planes and for reconstructing extended three-dimensional objects. We have shown that, by means of the angular spectrum-based analysis (ASA), it is possible to image correctly on tilted planes with respect to the optical axis. In-focus images have been reconstructed along planes tilted at angles, up to 45 degrees. In addition the anamorphic distortion can also be removed.

Moreover, we have shown another example in which it is possible to construct an extended focus image (EFI) of a 3-D object in microscopy. Both methods illustrate the potential of DH for 3-D imaging. By means of DH it is possible to obtain and/or reconstruct images along complex surfaces into the volume recorded by a digital hologram. The unique properties of DH that allow such important capability include two important features of DH: (1) numerical focusing and (2) management of complex reconstructed wavefront either in amplitude or phase.

## References

[1] L. Yaroslavsky and M. Eden, Fundamentals of Digital Optics, (BirkhŠuser, Boston, 1996).

[2] O. Schnars and W. Juptner, "Direct recording of holograms by a CCD target and numerical reconstruction," Appl. Opt. **33**, 179–181 (1994).

[3] T. M. Kreis, "Frequency analysis of digital holography," Opt. Eng. **41**, 771–778 (2002).

[4] M. Kim, "Tomographic three-dimensional imaging of a biological specimen using wavelength-scanning digital interference holography," Opt. Express **7**, 305–310 (2000). http://www.opticsexpress.org/abstract.cfm?URI=OPEX-7-9-305

[5] Y. Takaki and H. Ohzu, "Hybrid holographic microscopy: visualization of three-dimensional object information by use of viewing angles," Appl. Opt. **39**, 5302–5308 (2000).

[6] T.C. Poon, K.B. Doh, and B.W. Schilling, "Three-dimensional microscopy by optical scanning holography," Opt. Eng. **34**, 1338–1344 (1995).

[7] P. Ferraro, S. De Nicola, G. Coppola, A. Finizio, D. Alfieri, and G. Pierattini, "Controlling image size as a function of distance and wavelength in Fresnel-transform reconstruction of digital hologram," Opt. Lett. **29**, 854–856 (2004).
[8] T. Zhang and I. Yamaguchi, "Three-dimensional microscopy with phase-shifting digital holography," Opt. Lett. **23**, 1221–1223 (1998).
[9] D. Leserberg and C. Frère, "Computer generated holograms of 3-D objects composed of tilted planar segments," Appl. Opt. **27**, 3020–3024 (1988)
[10] N. Delen and B. Hooker, "Free-space beam propagation between arbitrarily oriented planes based on full diffraction theory: a fast Fourier approach," J. Opt. Soc. A. **15**, 857–867 (1998).
[11] L. Yu, Y. An, and L. Cai, "Numerical reconstruction of digital holograms with variable viewing angles," Opt. Express **10**, 1250–1257(2002) http://www.opticsexpress.org/abstract.cfm?URI=OPEX-10-22-1250
[12] K. Matsushima, H. Schimmel, and F. Wyrowski, "Fast calculation method for optical diffraction on tilted planes by use of the angular spectrum of plane waves," J. Opt. Soc. A. **20**, 1755–1762 (2003).
[13] S. De Nicola, A. Finizio, G. Pierattini, P. Ferraro, and D. Alfieri, "Angular spectrum method with correction of anamorphism for numerical reconstruction of digital holograms on tilted planes," Opt. Express **13**, 9935–9940 (2005) http://www.opticsinfobase.org/abstract.cfm?URI=oe-13-24-9935.
[14] E. Cuche, F. Bevilacqua, and C. Depeursinge, "Digital holography for quantitative phase-contrast imaging," Opt. Lett. **24**, 291–293 (1999).
[15] P. Ferraro, S. Grilli, D. Alfieri, S. De Nicola, A. Finizio, G. Pierattini, B. Javidi, G. Coppola, and V. Striano, "Extended focused image in microscopy by digital Holography," Opt. Express **13**, 6738–6749 (2005) http://www.opticsinfobase.org/abstract.cfm?URI=oe-13-18-6738
[16] S. E. Fraser, "Crystal gazing in optical microscopy," Nat. Bio. **21**, 1272–1273 (2003)
[17] L. Mertz, "Transformation in Optics," **101** (Wiley, New York, 1965).
[18] J.W. Goodman and R. W. Lawrence, "Digital image formation from electronically detected holograms," Appl. Phys. Lett. **11**, 77–79 (1967).
[19] T. H. Demetrakopoulos and R. Mitra, "Digital and optical reconstruction of images from suboptical patterns," Appl. Opt. **13**, 665–670 (1974).
[20] L. P. Yaroslavsky, N.S. Merzlyakov, "Methods of digital holography," (Consultants Bureau, New York, 1980).
[21] D. Gabor, "Microscopy by reconstructed wave-fronts," Proc. Royal Society A **197**, 454–487 (1949).
[22] G. Hausler, "A method to increase the depth of focus by two step image processing," Opt. Commun. **6**, 38 (1972).
[23] R.J. Pieper and A. Korpel, "Image processing for extended depth of field," Appl. Opt. **22**, 1449–1453 (1983).
[24] For example, description of EFI capability and process in optical microscopes can be found into the web sites of two important manufacturers: http://www.olympusamerica.com/seg˙section/msfive/ms5˙appmod.asp; http://www.zeiss.de/ C12567BE0045ACF1/InhaltFrame/DA8E39D74AA60 C49412568B90054EDD2.
[25] E. R.Dowski, Jr., W.T. Cathey, "Extended depth of field through wavefront coding," Appl. Opt. **34**, 1859–1866 (1995).

[26] D.L. Barton, et al. "Wavefront coded imaging system for MEMS analysis," Presented at international Society for testing and failure analysis meeting. (Phoeneics, AZ, USA, Nov. 2002).
[27] D. L. Marks, D.L. Stack, D.J. Brady, and J. Van Der Gracht, "Three-dimensional tomography using a cubic-phase plate extended depth-of-field system," Opt. Lett. **24**, 253–255 (1999).
[28] S. Grilli, P. Ferraro, S. DeNicola, A. Finizio, G. Pierattini, and R. Meucci, "Whole optical wavefields reconstruction by digital holography," Opt. Express. **9**, 294–302 (2001).
[29] U. Schnars and W.P.O. Juptner, "Digital recording and numerical reconstruction of holograms," Meas. Sci. Technol. **13**, R85–R101 (2002).
[30] S. Cuche, F. Bevilacqua, and C. Depeursinge, "Digital holography for quantitative phase-contrast imaging," Opt. Lett. **24**, 291–293 (1999)
[31] P. Ferraro, G. Coppola, D. Alfieri, S. DeNicola, A. Finizio, and G. Pierattini, "Controlling image size as a function of distance and wavelength in Fresnel transform reconstruction of digital holograms," Opt. Lett. **29**, 854–856 (2004).
[32] P. Ferraro, S. DeNicola, A. Finizio, G. Coppola, S. Grilli, C. Magro, and G. Pierattini, "Compensation of the inherent wave front curvature in digital holographic coherent microscopy for quantitative phase-contrast imaging," Appl. Opt. **42**, 1938–1946 (2003).
[33] G. Coppola, P. Ferraro, M. Iodice, S. De Nicola, A. Finizio, and S. Grilli, "A digital holographic microscope for complete characterization of microelectromechanical systems," Meas. Sci. Technol. **15**, 529–539 (2004).

# 15

# Extraction of Three-dimensional Information from Reconstructions of In-Line Digital Holograms

Conor P. McElhinney, Bryan M. Hennelly, Bahram Javidi, and Thomas J. Naughton

## 15.1 Introduction

Holography, the science of recording and reconstructing a complex electromagnetic wavefield, was invented by Gabor in 1948 [1]. This initial invention concerned itself with electron microscopy and predated the invention of the laser. With the onset of the laser E. Leith and J. Upatnieks [2,3] appended the holographic principle with the introduction of the offset reference wave. This enabled the separation of the object wavefield from the other components that are generated in the optical reconstruction process, namely the intensities of the object and reference wavefields, and the so-called "ghost" or conjugate image. Holography may also be employed to describe the science of optical interferometry [4], which incorporates important industrial measurement techniques. We note that holography is at the heart of countless optical and nonoptical techniques [5].

Using photosensitive recording materials to record holograms is costly and inflexible. Digital holography [6,7,8,9,10,11], refers to the science of using

C.P. McElhinney
Department of Computer Science, National University of Ireland, Maynooth, County Kildare, Ireland
e-mail: conormce@cs.nuim.ie

B. Javidi et al. (eds.), *Three-Dimensional Imaging, Visualization, and Display*,
DOI 10.1007/978-0-387-79335-1_15, 

discrete electronic devices, such as CCDs, to record the hologram. In this case reconstruction is performed numerically by simulating the propagation of the wavefield back to the plane of the object. One major advantage of DH over material holography is the ability to use discrete signal processing techniques on the recorded signals [12,13,14,15]. In recent years DH has been demonstrated to be a useful method in many areas of optics such as microscopy [16], deformation analysis [17], object contouring [18], particle sizing and position measurement [19]. "In-line" or "on-axis" DH refers to the implementation of the original Gabor architecture in which the reference wavefield travels in the same direction as the object wavefield. As in the continuous case this method suffers from poor reconstructed image quality, due to the presence of the intensity terms and the conjugate image that contaminates the reconstructed object image. While it is possible to remove the intensity terms with efficient numerical techniques [20], it remains difficult to remove the conjugate image. This may be achieved using an off-axis recording setup equivalent to that used by Leith and Upatnieks [2,3]. However, this increases the spatial resolution requirements, and limits the system significantly which is undesirable when one considers the already limited resolution of digital cameras. An alternative approach known as phase-shifting interferometry [21] has been introduced which allows an in-line setup to be used with at least two successive captures and enables separation of the object wavefield from all of the other terms.

A disadvantage of holographic reconstructions is the limited depth-of-field. When a digital hologram is reconstructed, a distance value $d$ is input as a parameter to the reconstruction algorithm. Only object points that are located at the input distance $d$ from the camera are in-focus in the reconstruction. Complex 3-D scenes, scenes containing multiple objects or containing multiple object features located at different depths, lead to reconstructions with large blurred regions. By applying focus measures to sets of reconstructions autofocus algorithms have been implemented on computer generated DHs [22] and those of microscopic objects [23,24].

In this chapter we develop an approach for the estimation of surface shape of macroscopic objects from digital holographic reconstructions using multiple independently focused images. We can estimate the focal plane of such a DH by maximizing a focus metric, such as variance, which is applied to the intensity of several 2-D reconstructions, where each reconstruction is at a different focal plane. Through the implementation of our depth-from-focus (DFF) technique we can create a depth map of the scene, and this depth information can then be used to perform tasks such as focused image creation[25], background segmentation [26] and object segmentation [27]. Using the segmentation masks output by our process we can segment different DH reconstruction planes into their individual objects. By numerically propagating a complex wavefront and superposing a second wavefront at a different plane we can create synthetic digital holograms of real-world objects. These can then be viewed on conventional three-dimensional displays [28].

The structure of this chapter is as follows. In Sect. 15.2 we discuss the recording process for PSI DHs and our experimental setup. In Sect. 15.3 we introduce focus and focus detection for DHs. The algorithms for calculating a depth map using an overlapping DFF approach are discussed in detail. Section 15.4 presents a sequential discussion of our different data extraction algorithms, namely, (i) depth map extraction, (ii) extended focus image (EFI) creation, (iii) segmentation, and (iv) synthetic digital holographic scene creation, and we conclude in Sect. 15.5.

## 15.2 Digital Holographic Recording

Standard photography uses a lens to focus light diffracted from a scene onto a photographic film, or CCD, which records a focused image. This scene is illuminated using incoherent light (i.e., sunlight). Light is diffracted from an object in the scene and this wavefront propagates towards the recording medium. A camera records only the intensity of the wavefront. In general, to record a hologram we split the laser beam into an object beam and a reference beam. The object beam illuminates the object, and the diffracted object wavefront propagates a distance $d$ to the recording medium [13,10,29,31]. The recording media are holographic film in the case of optical holography, and a CCD in the case of digital holography. The reference beam propagates uninterfered to the recording medium, where it interferes with the diffracted object beam. This interference pattern is then recorded, and is called an interferogram or a hologram. It is the interference of the reference beam (which has a constant amplitude and phase) with the diffracted object wavefront (which has an unknown amplitude and phase) that produces an intensity pattern containing amplitude and phase information. This phase information allows different perspectives of the scene to be viewed by tilting the holographic film or changing the inputs to the numerical reconstruction algorithm in digital holography. To reconstruct an optical hologram, the film is illuminated using the reference beam used to record the hologram. This creates a real image at a distance $d$ and a virtual image at a distance $-d$ from the film. A digital hologram can be reconstructed using a discretized version of the Fresnel-Kirchoff integral [8]. The Fresnel-Kirchoff integral models the propagation of light. Given a wavefront and a distance $d$, it approximates the wavefront after propagation of the distance $d$.

There are two forms of digital holographic recording: in-line and off-axis. In an in-line setup, shown in Fig. 15.1a, the reference beam and the diffracted object beam are both at the same angle with respect to the CCD. In an off-axis setup, shown in Fig. 15.1b, the reference beam is at a different angle with respect to the CCD than that of the diffracted object wavefront. Single exposure holograms are recorded using one interferogram, and in both setups shown above they suffer from a corruptive error source, the DC-term. However, this can be suppressed using a host of approaches such as low-pass filtering

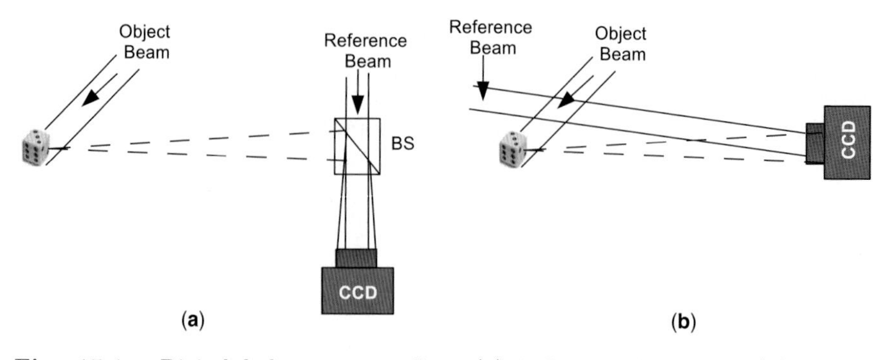


**Fig. 15.1.** Digital hologram recording: (**a**) in-line configuration, (**b**) off-axis configuration. BS: beam splitter

of the Fourier domain [20] or subtraction of the recorded reference beam [32], amongst others [9,11,33,34]. Another error source in DHs is the twin-image. In off-axis holography this can be suppressed through filtering at the hologram plane [33–34]. However, in in-line DHs this source of error cannot be suppressed easily although some approaches do exist [36]. The in-line setup has many advantages over an off-axis setup, including) a less developed speckle noise and a more efficient use of the CCD bandwidth [37]. The increased CCD bandwidth efficiency of the in-line setup is due to the shorter recording distance requirement. This is required to fulfill the sampling theorem [38], and can be calculated based on information about the CCD, the laser source, and the object to be recorded. To demonstrate this we have plotted the size of the object to be recorded as a function of the minimum recording distance for both setups in Fig. 15.2 (based on a CCD with $2,048 \times 2,048$ pixels of size 7.4 μm and a 632.8 nm laser source). This shows that the distance required to record macroscopic objects using an in-line setup is much shorter than that required for off-axis setups. In fact the in-line configuration slope is one-quater that of the off-axis configuration [37]. As the object size increases so does the gap between recording distances for in-line and off-axis setups.

PSI [21] is a digital holographic technique that calculates in-line holograms free of the twin-image and DC-term; however, it requires multiple recordings. The hologram is calculated from a set of multiple interferograms where a phase shift has been introduced to the reference beam. We use a PSI setup and a four-frame PSI algorithm to record in-line DHs. The PSI setup we use is described in Sect. 15.2.1. In Sect. 15.2.2 we discuss the theory of PSI and detail the proof.

### 15.2.1 PSI Setup

We record whole Fresnel fields with an optical system [12,39] based on a Mach-Zehnder interferometer (see Fig. 15.3). A linearly polarized

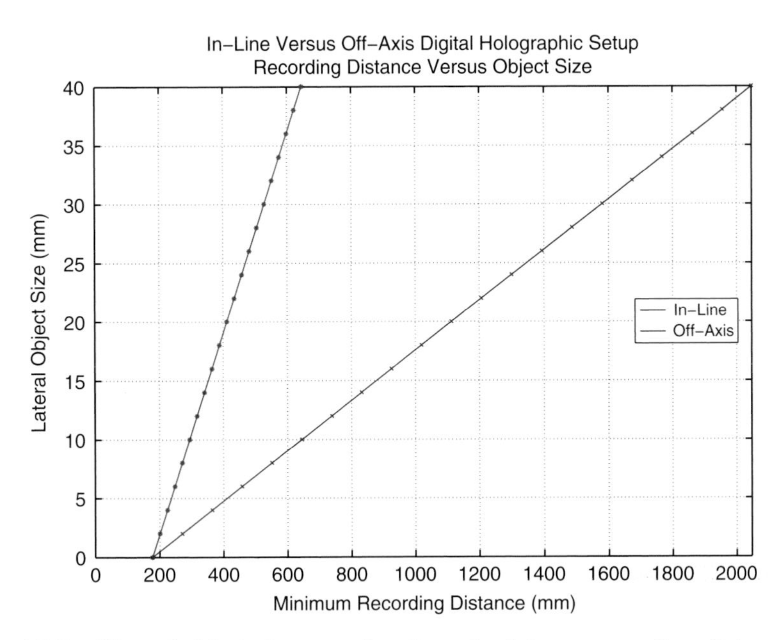


**Fig. 15.2.** Plot of object size as a function of minimum recording distance for in-line and off-axis holographic setups

Helium-Neon (632.8 nm) laser beam is expanded and collimated, and divided into object and reference beams. The object beam illuminates a reference object placed at a distance of approximately $d = 350$ mm from a CCD camera with $2,048 \times 2,032$ pixels. Let $U_0(x, y)$ be the complex amplitude distribution immediately in front of the 3-D object. The reference beam passes through $\mathrm{RP}_1$ and $\mathrm{RP}_2$, and by selectively rotating the plates we can achieve four phase shift permutations. For each one we record an interferogram. We use these four real-valued images to compute the camera-plane complex field $H_0(x, y)$ using the PSI [21,40] algorithm. We call this computed field a DH.

A DH $H_0(x, y)$ contains sufficient amplitude and phase information to reconstruct the complex field $U(x, y, z)$ in a plane at any distance $z$ from the camera [8,21,12]. This can be calculated from the Fresnel approximation [31] as

$$U(x, y, z) = \frac{-\mathrm{i}}{\lambda z} \exp\left(\mathrm{i}\frac{2\pi}{\lambda} z\right) H_0(x, y) \star \exp\left[\mathrm{i}\pi \frac{(x^2 + y^2)}{\lambda z}\right], \tag{15.1}$$

where $\lambda$ is the wavelength of the light and $\star$ denotes a convolution. At $z = d$, and ignoring errors in digital propagation due to pixelation and rounding, the discrete reconstruction $U(x, y, z)$ closely approximates the physical continuous field $U_0(x, y)$.

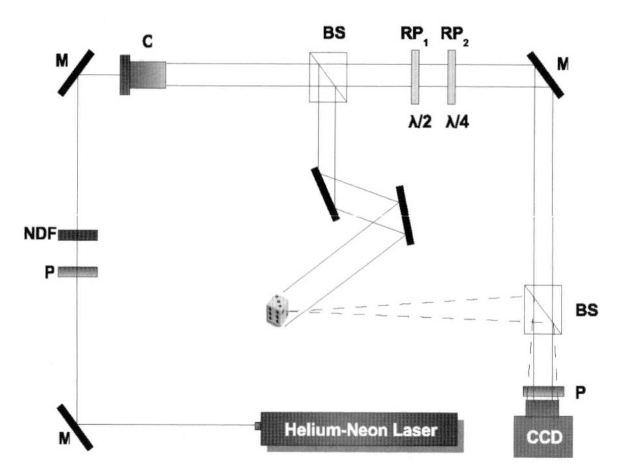


**Fig. 15.3.** Experimental setup for PSI: P, Polarizer; NDF, Neutral Density Filter; C, Collimator; BS, beam splitter; RP, retardation plate; M, mirror

### 15.2.2 PSI Theory

To calculate a hologram using a PSI setup, multiple interferograms with different phase shifts need to be recorded. Different techniques exist to create holograms using different numbers of interferograms and different phase shifts. We use a four-interferogram PSI algorithm [21]. We define the reference wavefront as,

$$R_{\Delta\phi}(x, y) = A_{R_{\Delta\phi}}(x, y) \exp\{i[\phi_{R_{\Delta\phi}}(x, y) + \Delta\phi]\}, \tag{15.2}$$

and the object wavefront is defined as,

$$O(x, y) = A_0(x, y) \exp\{i[\phi_0(x, y)]\}. \tag{15.3}$$

Where $A(x, y)$ is the wavefront's amplitude, $\phi(x, y)$ is the wavefronts phase and $\Delta\phi$ is the phase shift. When the reference wavefront and object wavefront interfere at the CCD they create an interferogram defined as,

$$\begin{aligned} I_{\Delta\phi}(x, y) &= |O(x, y, 0) + R_{\Delta\phi}(x, y, 0)|^2 \\ &= A_0^2(x, y) + A_{R_{\Delta\phi}}^2(x, y) + 2A_0(x, y)A_{R_{\Delta\phi}}(x, y) \\ &\quad \mathrm{x} \cos[\phi_0(x, y) - \phi_{R_{\Delta\phi}}(x, y) - \Delta\phi]. \end{aligned} \tag{15.4}$$

PSI assumes that the initial reference wavefront's phase $\phi_{R_{\Delta\phi}}(x, y)$ and amplitude $A_{R_{\Delta\phi}}^2(x, y)$ are constant for all $\Delta\phi$, giving $\phi_{R_{\Delta\phi}}(x, y)$ a value of zero and $A_{R_{\Delta\phi}}^2(x, y)$ a value of 1. This is because the reference beam should be a plane wave and should not affect the reference wavefront's original phase and amplitude. We can now rewrite Eqn. 15.4 which gives us

$$\begin{aligned} I_{\Delta\phi}(x,y) = & \quad A_0^2(x,y) + 1^2 + 2A_0(x,y)(1)\cos\left[\phi_0(x,y) - 0 - \Delta\phi\right] \\ = & \quad A_0^2(x,y) + 1 + 2A_0(x,y)\cos\left[\phi_0(x,y) - \Delta\phi\right]. \end{aligned} \tag{15.5}$$

For simplicity we define $P_{\Delta\phi}$ to be

$$P_{\Delta\phi}(x,y) = \cos\left[\phi_0(x,y) - \Delta\phi\right] \tag{15.6}$$

therefore,

$$I_{\Delta\phi}(x,y) = A_0^2(x,y) + 1 + 2A_0(x,y)P_{\Delta\phi}(x,y). \tag{15.7}$$

The amplitude of the object wavefront, $A_0(x,y)$, is calculated using four interferograms with phase shifts of $\Delta\phi = 0, \frac{\pi}{2}, \pi, \frac{3\pi}{2}$

$$\begin{aligned} A_0(x,y) = & \quad \frac{1}{4}\sqrt{|I_0(x,y) - I_\pi(x,y)|^2 + |I_{\frac{\pi}{2}}(x,y) - I_{\frac{3\pi}{2}}(x,y)|^2} \\ A_0(x,y) = & \quad \frac{1}{4}\sqrt{|\alpha|^2 + |\beta|^2}. \end{aligned} \tag{15.8}$$

We recover the phase of the object wavefront, $\phi_0(x,y)$, with the same interferograms

$$\begin{aligned} \phi_0(x,y) = & \quad \arctan\left\{\frac{I_{\frac{\pi}{2}}(x,y) - I_{\frac{3\pi}{2}}(x,y)}{I_0(x,y) - I_\pi(x,y)}\right\} \\ \phi_0(x,y) = & \quad \arctan\left\{\frac{\beta}{\alpha}\right\}, \end{aligned} \tag{15.9}$$

where $\alpha = I_0(x,y) - I_\pi(x,y)$ and $\beta = I_{\frac{\pi}{2}}(x,y) - I_{\frac{3\pi}{2}}(x,y)$.

### 15.2.3 PSI Proof

We prove that these equations return the correct amplitude and phase we first need to solve $\alpha$ and $\beta$

$$\begin{aligned} \alpha = & \quad I_0(x,y) - I_\pi(x,y) \\ = & \quad [A_0^2(x,y) + 1 + 2A_0(x,y)P_0] - [A_0^2(x,y) + 1 + 2A_0(x,y)P_\pi] \\ = & \quad 2A_0(x,y)[(P_0 - P_\pi) \end{aligned} \tag{15.10}$$

$$\begin{aligned}\beta = &\ I_{\frac{\pi}{2}}(x,y) - I_{\frac{3\pi}{2}}(x,y) \\ = &\ [A_0^2(x,y) + 1 + 2A_0(x,y)P_{\frac{\pi}{2}}] - [A_0^2(x,y) + 1 + 2A_0(x,y)P_{\frac{3\pi}{2}}] \\ = &\ 2A_0(x,y)(P_{\frac{\pi}{2}} - P_{\frac{3\pi}{2}}).\end{aligned} \tag{15.11}$$

Using the cosine sum formula

$$\cos[A - B] = \cos[A]\cos[B] + \sin[A]\sin[B] \tag{15.12}$$

and

| X | 0 | $\frac{\pi}{2}$ | $\pi$ | $\frac{3\pi}{2}$ |
|---|---|---|---|---|
| cos[X] | 1 | 0 | -1 | 0 |
| sin[X] | 0 | 1 | 0 | -1 |

we are able to simplify all $P_{\Delta\phi}$ to

$$\begin{aligned}P_0 = &\ \cos[\phi_0(x,y)], & P_\phi = &\ \cos[\phi_0(x,y)]\cos[\pi] \\ & & &+ \sin[\phi_0(x,y)]\sin[\pi] \\ & & = &\ -\cos[\phi_0(x,y)] \\ P_{\frac{\pi}{2}} = &\ \cos[\phi_0(x,y)]\cos[\frac{\pi}{2}] & P_{\frac{3\phi}{2}} = &\ \cos[\phi_0(x,y)]\cos[\frac{3\pi}{2}] \\ &+ \sin[\phi_0(x,y)]\sin[\frac{\pi}{2}], & &+ \sin[\phi_0(x,y)]\sin[\frac{3\pi}{2}] \\ = &\ \sin[\phi_0(x,y)], & = &\ -\sin[\phi_0(x,y)].\end{aligned} \tag{15.13}$$

We substitute the resulting $P_{\Delta\phi}$ into Eqs. 15.10 and 15.11

$$\begin{aligned}\alpha = &\ 2A_0(x,y)\cos[\phi_0(x,y)] + \cos[\phi_0(x,y)], \\ = &\ 4A_0(x,y)\cos[\phi_0(x,y)] \\ \beta = &\ 2A_0(x,y)\sin[\phi_0(x,y)] + \sin[\phi_0(x,y)], \\ = &\ 4A_0(x,y)\sin[\phi_0(x,y)].\end{aligned} \tag{15.14}$$

Using these reduced $\alpha$ and $\beta$ terms we are able to show that

$$|\alpha|^2 + |\beta|^2 = 16A_0^2(x,y)\cos^2[\phi_0(x,y)] + 16A_0^2(x,y)\sin^2[\phi_0(x,y)] \tag{15.15}$$

$$= 16A_0^2(x,y)\cos^2[\phi_0(x,y)] + \sin^2[\phi_0(x,y)], \tag{15.16}$$

$$= 16A_0^2(x,y). \tag{15.17}$$

Substituting this into Eqs. 15.8 we obtain

$$A_0(x, y) = \frac{1}{4}\sqrt{16A_0^2(x, y)} \tag{15.18}$$

$$= A_0(x, y). \tag{15.19}$$

Proving that, by assuming that the reference wavefront's amplitude and phase are constant, we are able to calculate the object wavefront's amplitude.

Using the $\alpha$ and $\beta$ from Eq. 15.14 we get

$$\frac{\beta}{\alpha} = \frac{4A_0(x, y)\sin\left[\phi_0(x, y)\right]}{4A_0(x, y)\cos\left[\phi_0(x, y)\right]} \tag{15.20}$$

$$= \frac{\sin\left[\phi_0(x, y)\right]}{\cos\left[\phi_0(x, y)\right]} \tag{15.21}$$

and substituting this into Eq. 15.9 we get

$$\phi_0(x, y) = \arctan\left\{\frac{\sin\left[\phi_0(x, y)\right]}{\cos\left[\phi_0(x, y)\right]}\right\} \tag{15.22}$$

which also verifies that (if the amplitude and phase of the reference wavefront are constant) we are able to recover the object wavefront's phase.

## 15.3 Focus Detection

### 15.3.1 Focus and Imaging

All imaging systems generally have a finite depth-of-field. The recorded image can either be in-focus or out-of-focus. The objects which lie within the depth-of-field of the imaging system are in-focus (appearing sharp) while the objects which lie outside the depth-of-field of the system are out-of-focus (appearing blurred). There are many applications which rely on imaging systems and the ability to accurately determine the level of focus of an image, or an image region, from a set of images of the same scene. In incoherent imaging systems some of these applications include robot navigation [41], visual surveillance [42], shape measurement [43], image fusion [44], camera focus control [45,46,47] and astronomical imaging [48,49].

To determine if an image is in-focus the accepted procedure is to record a set of images with either a dynamic scene or to record a set of images with a static scene where the focal plane of the camera is varied. A function is then applied to each image and the image which maximizes this function is taken as the in-focus image. Focus measures are also known as sharpness functions originating from the work of Muller, et al. [48] where they were the first to apply a function to an image to determine focus, which they called an "image sharpness function." Modern focus measures are based on the assumption that edges are more defined in focused images and are, therefore, more sharp (have higher spatial frequencies). Numerous sharpness based focus measures have been developed [49,50] and evaluated [44,46,51] for incoherent imaging which generally satisfy the following requirements:

- independent of image content
- low computation complexity
- focus measure should be unimodal
- there should be a large variation in returned value with respect to blurring
- robust to noise

While the development and application of focus measures in incoherent imaging has been well studied, there is no definitive criterion for finding the focal plane of a scene or finding the focal distance for a region within a scene. In the field of digital holography, the study of focus measures has not seen the same interest. Focus techniques have been applied to both reconstructions of digitally recorded holograms [22,24,26,54,53] and to the digitized reconstructions of optically recorded holograms [54,55,56,57,58]. These employ focus metrics such as self-entropy [22], phase changes [52], wavelet analysis [23], gray-level variance [54,55,57], and integrated amplitude modulus [24] among others. Using these metrics, applications such as the detection of the focal plane [23,24,52] in digital holographic microscopy, the measurement of 3-D objects in the digitized reconstructions of physical holograms [27,55,57], the segmentation of macroscopic objects [26] and the creation of extended focused images [25,53] have been demonstrated. The application of focus measures to the field of digital holography are reviewed in this section.

### Focus Measures in Digital Holography

Gillespie and King [22] were the first to propose a focus measure for digital holography. They proposed the use of self-entropy as their focus measure calculated on the phase of a digital hologram's numerical recontruction. They aimed to develop a function that could be used to autofocus the reconstructions of a digital hologram. However, they used computer generated holograms in their experiments which did not suffer from speckle noise. Liebling and Unser [23] developed fresnelets, a new format for storing digital holograms which use a wavelet approach to numerically reconstruct. Their holograms were recorded using digital holographic microscopy [59] and are again free of speckle noise. In their work fresnelets are evaluated as a focus metric compared to some popular image processing focus metrics – such as the Laplacian of the reconstructed intensity and intensity squared – and it was found that fresnelets out-performed both image processing metrics.

Other methods have been developed for application to the digitized images of optically reconstructed holograms [54,55,56,57,58]. Yin, et al. [54] performed a survey of the application of four focus measures to digitized reconstructions of acoustic holograms. They selected gradient magnitude, modulus difference, modified Laplacian and gray-level variance as their four focus measures. Using digitized reconstruction of a hologram of a prepared specimen they examined the output of these focus measures and determined that gray-level variance achieved the best results. With an estimation of the point spread function of their experimental setup they then applied spatial filtering to their reconstructions to reduce blurring and recover a focused image of the scene. In this work they assumed a linear relationship between the focus depth and the reconstruction depth. This can result in incorrectly focused images for objects with complex shapes. The first method for reconstructing a DH at the in-focus plane using a Depth from Focus (DFF) technique was proposed by Ma, et al. [55], who used variance as their focus measure. DFF is an image processing approach for estimating surface shape in a scene using multiple independently focused images. DFF approaches estimate the focal plane of a DH by maximizing a focus metric which is applied to the intensity of several 2-D reconstructions where each reconstruction is at a different focal plane. Depth maps can be calculated using DFF approaches through computing a focus measure on the overlapping blocks of each reconstruction. The depth of each block is estimated by finding the reconstruction depth which maximizes the focus measure. DFF has been successfully applied to the segmentation of a digital hologram into object and background [26] and to create low-resolution depth maps of digitized physical holograms [55]. Similar to the approach adopted by Ma, Thelen, et al. [57] have applied gray-level variance to the extraction of shape information from the digitized reconstructions of optically reconstructed holograms. They illuminate their scene with a speckle pattern to increase accuracy of their focus measure at the in-focus planes. They have successfully produced accurate high-resolution depth maps of human faces.

Burns, et al. [58] have developed an algorithm for computing the focal plane of plankton in digitized reconstructions of optically reconstructed holograms. They advocate the use of the Tenegrad function, which is a variation of the energy of the image gradient. Malkiel, et al. [56] processed sequences of digitized reconstructions to create a depth map. The depth of every pixel is estimated by selecting the depth which results in the highest intensity in the digitized reconstruction. In both of these approaches the objects are near 2-D objects, leaving approximately one focal plane per object. We use variance as our focus measure as it has been shown to be a sound focus measure and has been successfully applied to DH reconstructions.

### 15.3.2 Evaluation of Two Focus Measures

We now proceed to compare two focus measures, variance and a measure based on the high spatial frequency content of an image block. Focus measures are functions which attempt to determine the relative level of focus of sets of images. The accepted image property maximized by these functions is the high spatial frequency energy of the image [46], which is most accurately computed with the Fourier transform [31]. The Fourier transform of an image has been proven to be a sound focus measure [46], but is computationally expensive. The two-dimensional Fourier transform can be calculated on a row of an image using

$$F(k,l) = \sum_{m=0}^{N-1}\sum_{n=0}^{N-1} I(m,n) \exp\left[-\frac{2\pi ikm}{N}\right] \exp\left[-\frac{2\pi iln}{N}\right] \quad k,l = 0, ..., N-1 \tag{15.23}$$

where $I$ is the image, of size $N \times N$. To compute our Fourier focus measure (FFM) using only the high spatial frequency we set the center $M \times M$ pixels in $F(k,l)$ to be zero and sum

$$\text{FFM}(I) = \sum_{k=0}^{N-1}\sum_{l=0}^{N-1} \mid (F(k,l) \mid \tag{15.24}$$

$FFM(I)$ is, therefore, a focus value for the image, or image region, $I$.

Variance is a focus measure calculated on the intensity of an image using

$$V(I) = \frac{1}{N \times N} \sum_{i=1}^{N}\sum_{j=1}^{N} \left[I_k(i,j) - \overline{I_k}\right]^2 \tag{15.25}$$

where $I$ is an image, of size $N \times N$, indexed by pixel locations $i$ and $j$ and where $\overline{I_k}$ is defined as

$$\overline{I_k} = \frac{1}{N^2} \sum_{x=i-\lfloor\frac{n-1}{2}\rfloor}^{i+\lceil\frac{n-1}{2}\rceil} \sum_{y=j-\lfloor\frac{n-1}{2}\rfloor}^{j+\lceil\frac{n-1}{2}\rceil} I_z(x,y). \tag{15.26}$$

$V(I)$ is, therefore, a focus value for the image, or image region, $I$. Variance is a popular focus measure because it satisfies all the requirements identified for good focus measures particularly, compared to the FFM, computational complexity. It has been proven to be a sound focus measure [46] but, to the our knowledge, its application to digitally recorded holograms where speckle noise reduces the effectiveness of focus measures has not been evaluated. We have recently employed variance as a focus measure [26,27] for macroscopic objects recorded using digital holography. While the FFM is a more accurate measure of the spatial frequencies in an image the time taken to calculate the FFM places it at a disadvantage when compared to variance. In this section we will quantify the increase in computation time required to use the FFM as a focus measure and demonstrate that variance returns qualitatively accurate results.

To evaluate the performance of variance as a focus measure compared with the FFM we compared depth values returned by both measures and the computation time required for both on two DHs, a high and low contrast object. The first hologram contains two screws positioned approximately 15 mm apart and the second hologram contains a lego block. We selected two $81 \times 81$ object blocks, one on the two screws object ($OB_1$) and one on the lego block ($OB_2$). We qualitatively determined the in-focus planes for these blocks to be 356.2 mm and 298 mm, respectively. A numerical reconstruction of the the DHs with $OB_1$ and $OB_2$ highlighted is displayed in Figs. 15.4a and 15.5a. We numerically reconstructed each DH over a range of depths and applied a speckle reduction technique to each reconstruction [60]. We then calculated

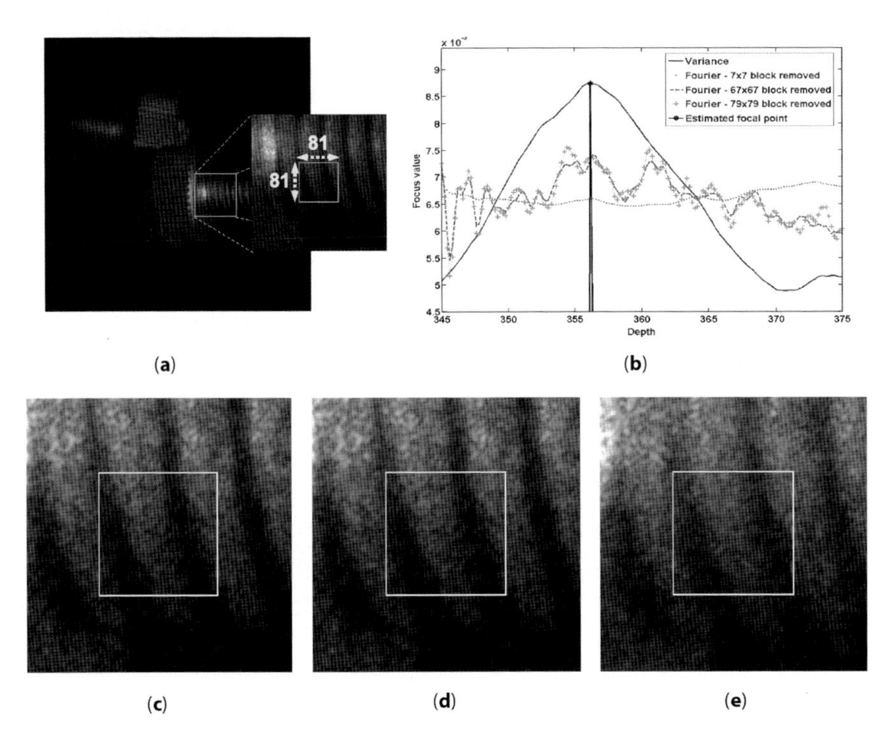


**Fig. 15.4.** Two screws object DH: (**a**) numerical reconstruction with $OB_1$ highlighted, (**b**) focus measure plots for $OB_1$, (**c**) zoomed in numerical reconstruction of $OB_1$ at depth estimated by variance, (**d**) zoomed in numerical reconstruction of $OB_1$ at depth estimated by FFM, (**e**) zoomed in numerical reconstruction of $OB_1$ at depth 5 mm away from estimated focal plane

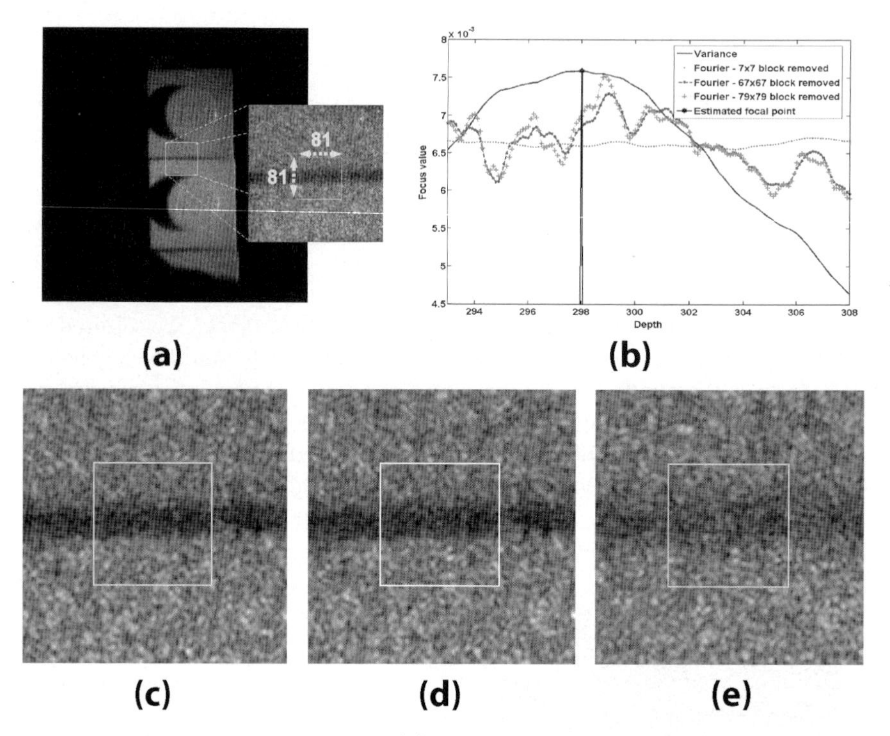


**Fig. 15.5.** Lego block object DH: (**a**) numerical reconstruction with $OB_1$ highlighted, (**b**) focus measure plots for $OB_1$, (**c**) zoomed in numerical reconstruction of $OB_1$ at depth estimated by variance, (**d**) zoomed in numerical reconstruction of $OB_1$ at depth estimated by FFM, (**e**) zoomed in numerical reconstruction of $OB_1$ at depth 5 mm away from estimated focal plane

variance and the FFM on both object blocks for each reconstruction in the range. By varying the size of the $M \times M$ block to be removed from $F(k, l)$ prior to the calculation of FFM($OB_1$) and FFM($OB_2$) we can determine which block size returns the more accurate results. We removed block sizes ranging from $7 \times 7$ up to $79 \times 79$ from the Fourier transform of the $81 \times 81$ object block. In Figs. 15.4b and 15.5b we plot the variance focus measure and selected FFMs applied to $OB_1$ and $OB_2$ as a function of depth. It is apparent that by only removing a small block, or too large a block, in the Fourier transform the FFMs, estimation of depth is negatively affected. We found that for the block sizes in the range of $31 \times 31$ up to $75 \times 75$ the estimated depth was never more than 0.2 mm away from our qualitatively estimated depth. The is equivalent to the interval between successive reconstructions in our experiment. Using multiple DHs we qualitatively selected $67 \times 67$ as the best block size to remove for use with the FFM calculated on an $81 \times 81$ object block. It is evident from the focus measure plots that variance returns unimodal curve, while the FFM has multiple local maxima. This makes variance a better focus measure based on the recommended focus measure requirements.

**Table 15.1** Focus measure computation time (seconds)

| Block Size | Variance | FFM |
|---|---|---|
| $21 \times 21$ | 0.000069 | 0.000425 |
| $41 \times 41$ | 0.000099 | 0.001205 |
| $51 \times 61$ | 0.000145 | 0.002317 |
| $81 \times 81$ | 0.000211 | 0.002425 |
| $101 \times 101$ | 0.000295 | 0.005964 |
| $121 \times 121$ | 0.000398 | 0.006998 |
| $141 \times 141$ | 0.000528 | 0.016544 |

We have also investigated the length of time required to calculate both of the focus measures. We selected seven blocks, each of a different size, from a DH reconstruction and calculated, and timed, variance on the block 10,000 times. We then carried out the same experiment using the FFM as the focus measure. The average time for both focus measures is shown in Table. 15.1, which demonstrates a 12:1 ratio in computation time for the block size of $81 \times 81$. We have shown why variance may be considered a better focus measure than the FFM in terms of a lower computational complexity and a returned unimodal focus plot.

### 15.3.3 Autofocus

Using variance as a focus measure, and image region and a set of numerical reconstructions it is possible to focus a DH using a simple algorithm. We select an image region on the desired object from a reconstruction and also select a range of depths to reconstruct. By computing variance on this image region for each reconstruction we can estimate the depth of that region. In Fig. 15.6 we demonstrate the application of this simple autofocus algorithm to our two screws object. We selected four ($281 \times 281$) object blocks, one on the threads, and head, of the front and back screw, and these blocks are highlighted in Fig. 15.6b–15.6e. The larger block size is due to a plot of variance applied to these blocks as a function of depth is displayed in Fig. 15.6 a. The reconstruction for each block at the estimated depth is shown in Fig. 15.6b–15.6e, and it is evident that the respective object blocks are sharp in each of these reconstructions. By estimating the depth of multiple blocks in reconstructions of a DH it is possible to compute a depth. One method for creating high resolution depth maps is to use a DFF approach which we detail in the next section.

### 15.3.4 Depth from Focus

An approach for the recovery of 3-D shape information from digitized physical holograms was proposed by Ma, et al. [55]. By calculating variance on

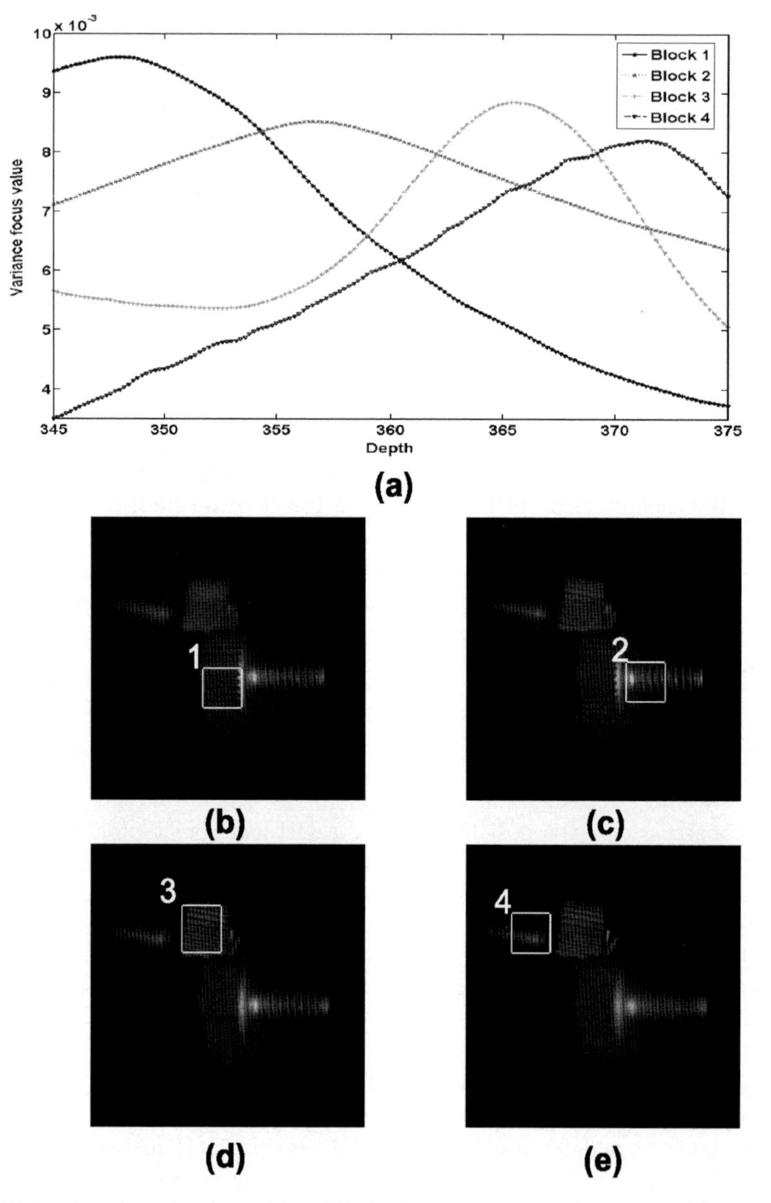


**Fig. 15.6.** Autofocusing four object blocks in a two screws object DH: (**a**) variance plot for the four blocks, (**b**) numerical reconstruction at the estimated depth for object block 1, (**c**) numerical reconstruction at the estimated depth for object block 2, numerical reconstruction at the estimated depth for object block 3, (**e**) numerical reconstruction at the estimated depth for object block 4

non-overlapping blocks from reconstructions of a DH at different depths they recovered depth information from a lower-resolution version of the sensed object. We choose to extend this variance measurement approach in order to classify each 1-D vector $(x, y)$ in the reconstruction volume (each line of pixels parallel to the optical axis) as either belonging to the object or belonging to the background. The decision is taken as follows: if vector $(x, y)$ contains an in-focus pixel from the object at any depth $z$ then $(x, y)$ is an object pixel, otherwise it is a background pixel.

Each reconstruction $\mathrm{I}_z = |U_z|^2$ is of size $M \times N$ pixels. This algorithm requires five input parameters: a DH, a block size $n \times n$, a start depth $z_{\mathrm{min}}$, an increment $z_{\mathrm{step}}$ and an end depth $z_{\mathrm{max}}$. The initial reconstruction depth $z$ is set to the starting depth, $z = z_{\mathrm{min}}$. The algorithm involves the following four steps as illustrated in Fig. 15.7:

**Step 1:** The input DH is reconstructed at depth $z$ and a speckle reduction technique is applied. The output reconstruction's intensity is stored in $\mathrm{I}_z(k, l)$.
**Step 2:** We then calculate variance for each pixel by calculating variance on $n \times n$ pixel overlapping blocks approximately centered on each pixel, and address each block with $(k, l)$ where $k \in [0, M-1], l \in [0, N-1]$. Variance of each overlapping block at each depth $z$ is calculated with function $V_z$ : $\Re^{n \times n} \rightarrow \Re^{+}$ defined by

$$V_z(k,l) = \frac{1}{n^2} \sum_{x=k-\lfloor \frac{n-1}{2} \rfloor}^{k+\lceil \frac{n-1}{2} \rceil} \sum_{y=l-\lfloor \frac{n-1}{2} \rfloor}^{l+\lceil \frac{n-1}{2} \rceil} \left[ I_z(x,y) - \overline{I_z(k,l)} \right]^2, \tag{15.27}$$

$V$ is, therefore, a volume storing a 2-D variance image for each depth $z$.

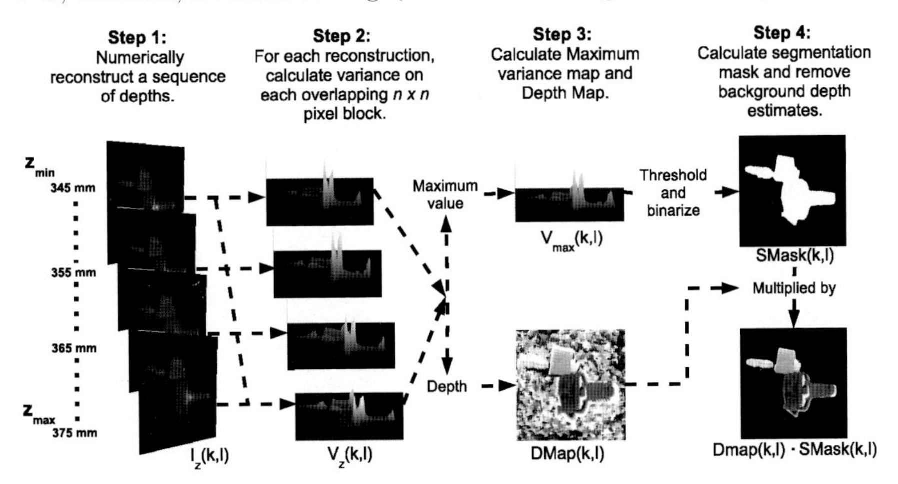


**Fig. 15.7.** DFF process for calculating depth values for object pixels in the DH reconstructions

**Step 3:** The maximum variance map is calculated from the $V_z(k,l)$ volume by finding the maximum value for each pixel $(k,l)$

$$V_{\max}(k,l) = \max_z \left[V_z(k,l)\right]. \tag{15.28}$$

For each pixel $(k,l)$, the depth where the maximum variance value occurs is stored in the depth map $\mathrm{DMap}(k,l)$.
**Step 4:** The segmentation mask is created by selecting a variance threshold $\tau$ and applying this to $V_{\max}(k,l)$

$$\mathrm{SMask}(k,l) = \begin{cases} 1, & \text{if} V_{\max}(k,l) \geq \tau \\ 0, & \text{if} V_{\max}(k,l) < \tau. \end{cases} \tag{15.29}$$

In this algorithm object blocks are labelled with a 1 and background blocks are labelled with a 0. We apply a mathematical morphology erosion operation (with neighborhood $\lceil n/2 \rceil \times \lceil n/2 \rceil$) to SMask to shrink the boundaries of the object; our use of overlapping blocks uniformly enlarges the mask. Noise in $\mathrm{DMap}(k,l)$ introduced from estimating the depth of background blocks is removed through the following operation: $\mathrm{DMap}(k,l) \cdot \mathrm{SMask}(k,l)$, where $\cdot$ means pointwise product.The output of this algorithm is a segmentation mask $\mathrm{SMask}(k,l)$ and a segmented depth map $\mathrm{DMap}(k,l) \cdot \mathrm{SMask}(k,l)$. In all the experiments in this chapter, we use a block size of $81 \times 81$ as input to the DFF algorithm. Larger block sizes have the advantage of estimating the general shape of an object with low error, but the shape of finer object features is lost. Conversely, the smaller block sizes have the advantage of estimating the shape of finer object features, but at the cost of high error in the estimate of the general shape of the object. We use the recommended block size of $81 \times 81$ for DHs containing macroscopic objects [26].

## 15.4 Extraction of Data from Digital Holographic Reconstructions

In this section we demonstrate the different types of information that can be extracted from a DH: (i) We first demonstrate how 3-D shape information can be extracted using two depth maps from two indpendent perspectives of a DH in Sect. 15.4.1. (ii) Section 15.4.2 demonstrates how, using one depth map, an EFI can be calculated where all the objects in the reconstructed image are in-focus. (iii) Our results from segmenting DH reconstructions into objects or object regions are detailed in Sect. 15.4.3. (iv) We segment an object′s full complex wavefront at the object plane by combining our autofocus, DFF and segmentation algorithms. The process for creating a superposed DH by combining two of these wavefronts at different planes is described in

Sect. 15.4.4. This digital holographic synthetic scene contains 3-D information and demonstrates how we can create synthetic scenes from objects extracted from different DHs for viewing on three-dimensional displays.

### 15.4.1 Extraction of Shape Information

As with conventional holography, a digital hologram encodes different views of a 3-D object from a small range of angles [29,31]. In order to reconstruct a particular 2-D perspective of the object, an appropriate windowed subset of pixels is extracted from the hologram and subjected to simulated Fresnel propagation [8,12,21] as given by Eq. 15.1. As the window explores the field, a different angle of view of the object can be reconstructed. The range of viewing angles is determined by the ratio of the window size to the full CCD sensor dimensions. Our CCD sensor has $2,048 \times 2,048$ pixels and approximate dimensions of $18.5 \times 18.5$ mm and so a $1,024 \times 1,024$ pixel window has a maximum lateral shift of 9 mm across the face of the sensor. With an object positioned $d = 350$ mm from the camera, viewing angles in the range $\pm 0.74°$ are permitted. Smaller windows will permit a larger range of viewing angles at the expense of image resolution at each viewpoint.

Our DFF technique extracts a $2\frac{1}{2}$D depth map from sets of reconstructions from a single perspective of the DH. Two depth maps are calculated of the Bolt object DH, see Fig.15.8a. Thirty-one reconstructions from 378 mm to 396 mm with a step size of 0.5 mm of the DH from two different perspectives are calculated. The first perspective used all $2,048 \times 2,048$ pixels to reconstruct from the optical axis perspective and the second perspective used only the $1,024 \times 1,024$ pixels centered 512 pixels above the center of the hologram plane, as shown in Fig. 15.8b. Our DFF technique is applied to both sets of reconstructions to calculate two depth maps shown in Fig.15.8c and d. The relative change of viewing angle between these two depth maps is $0.5°$. We chose the same point set of depth data along the threads of the bolt from both depth maps, L1 and L2 as displayed in Fig.15.8 and d, and plotted them in Fig. 15.8e. We proceeded to fit a 2nd order polynomial to both point sets to display the slope of point sets L1 and L2. As expected there is a more acute angle between the optical axis and the line fitted to L2 than the line fitted to L1, demonstrating the extraction of different depth information from the same object by calculating depth maps from reconstructions of different perspectives. With the advent of techniques to create super-resolution DHs[61], for window sizes of $1,024 \times 1,024$, holograms with viewing angles in the range of $\pm 1.5°$ have been demonstrated with the prediction of a much greater range in the future. By calculating depth maps on these super-resolved DHs from different perspectives and with knowledge of the camera geometry a 3-D shape model can be computed using range data registration techniques [62]. With the promise of a greater range of perspectives and the inherent capacity of digital holography to generate any perspective within that range we predict

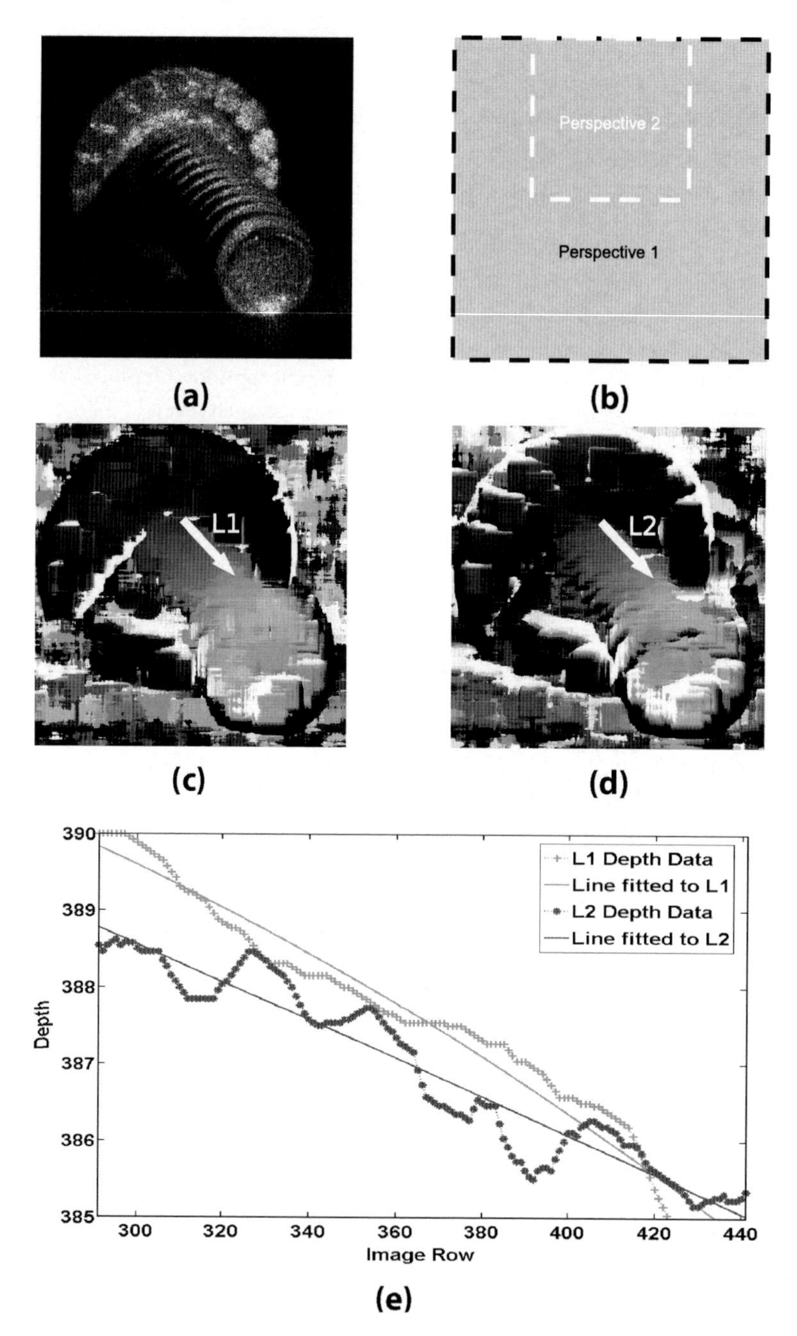


**Fig. 15.8.** Bolt object DH: (**a**) numerical reconstruction, (**b**) hologram plane showing perspective 1 and 2, (**c**) depth map computed from reconstructions from perspective 1, (**d**) depth map computed from reconstructions from perspective 2, (**e**) plot of depth data from L1, L2 with fitted lines

the application and development of digital holographic multi-view imaging algorithms [62].

### 15.4.2 Extraction of Extended Focused Image

We now present a method for creating extended focused images (EFIs) from sets of digital holographic reconstructions of macroscopic objects. To begin we create a depth map using our DFF technique. This depth map is combined with reconstructions to create an EFI. In [25] a number of approaches for creating an EFI are detailed. We focus on the pointwise approach where every pixel in the DMap$(k, l)$ corresponds to a point from $I_z(k, l)$, where $z = DMap(k, l)$. To calculate the EFI$(k, l)$ the depth for each pixel from DMap$(k, l)$ is recorded and the intensity value of the corresponding pixel from $I_z(k, l)$ for that depth in EFI$(k, l)$ is stored. We calculate EFI$(k, l)$ with the following function

$$\mathrm{EFI}(k, l) = I_{\mathrm{DMap}(k,l)}(k, l) \tag{15.30}$$

Using the depth map created from the two screws object shown in Fig. 15.7 we can compute an EFI. The segmentation mask for this hologram has been computed and the depth map only contains depth values at points identified as object points. Figure 15.9a displays a front focal plane reconstruction, Fig. 15.9b shows a back focal plane reconstruction, and in Fig. 15.9c the EFI$(k, l)$. From this figure it is clear that the EFI contains all in-focus regions from both of these objects, and that both objects are in-focus.

### 15.4.3 Extraction of Objects from Digital Holographic Reconstructions

In this section we outline a two stage technique for segmenting a DH into independent object regions at different depths. In the first stage (discussed in Sect. 15.3.4), an overlapping block based DFF algorithm is used to calculate a segmentation mask for the scene. This segmentation mask is then applied to the depth map so that it only contains depths of object regions. The second stage (discussed in Sect. Extraction of Multiple Objects) from Reconstructions applies a depth segmentation algorithm to the depth map, and shows how object regions are identified and segmented based on homogeneous regions in the depth map.

#### Extraction of Objects from Background in Digital Holographic Reconstructions

The first stage in the object segmentation technique is segmentating background from the object using the algorithm detailed in Sect. 15.3.4 and in [26]. We numerically reconstruct a single perspective of each object at a range

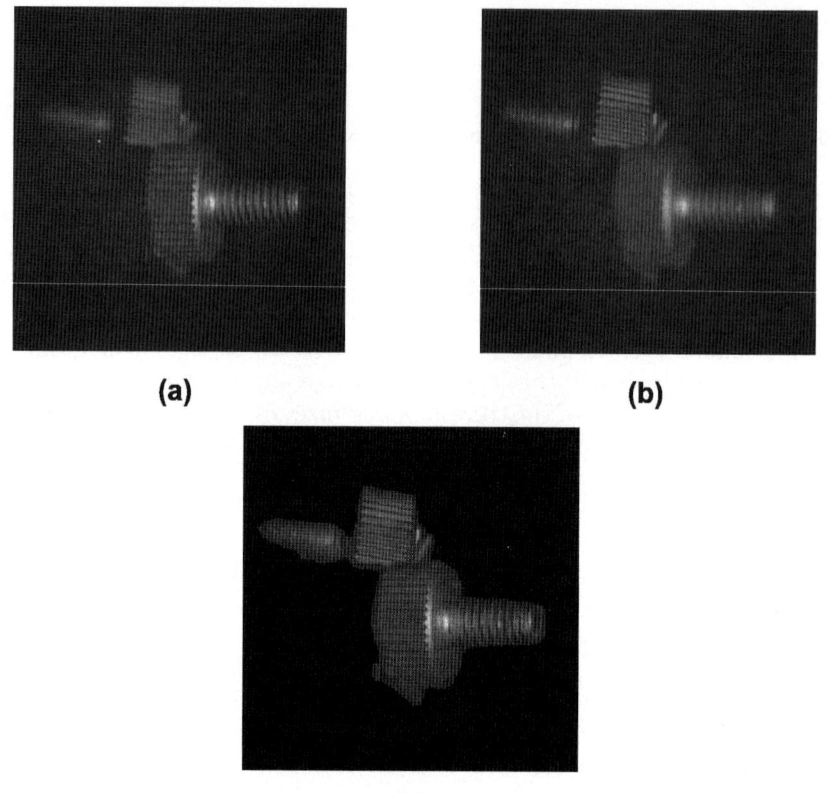


**Fig. 15.9.** Extended focused image of two screws object hologram: (**a**) front focal plane reconstruction, (**b**) back focal plane reconstruction and (**c**) EFI

of depths. At each point in the digital wavefront we calculate variance about a neighborhood. The maximum variance at each point over all depths is thresholded to classify it as an object pixel or a background pixel.

We verify our technique using a DH of a real-world object. The knight object is 2 cm×2 cm×0.6 cm and was positioned 371 mm from the camera. A sequence of reconstructions at different depths is computed from a single perspective with a uniform interval of 1 mm between successive values of $z$. The DFF technique is applied to this sequence of reconstructions to obtain SMask. We present the results of applying our DFF technique to a DH of a low contrast object in Fig. 15.10: a knight object whose reconstruction is shown in Fig. 15.10a and the resulting segmentation mask is shown in Fig. 15.10b. A segmented reconstruction at a single depth $z$ [Fig. 15.10c], obtained from $I_z(x, y) \cdot \mathrm{SMask}(x, y)$ where $\cdot$ means pointwise product, illustrates how the object can be successfully segmented from the background.

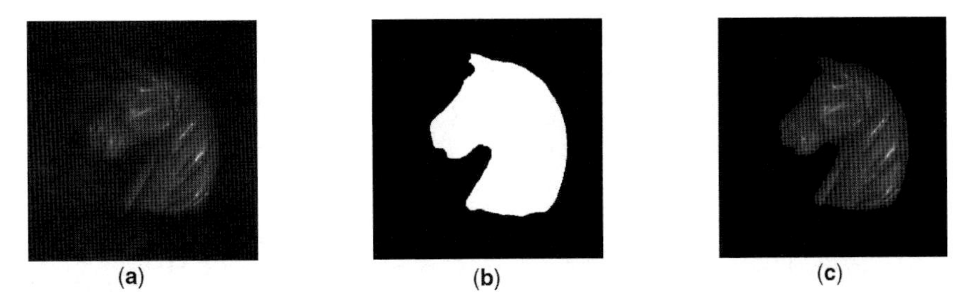


**Fig. 15.10.** Segmentation of knight object DH: (**a**) numerical reconstruction, (**b**) segmentation mask obtained, (**c**) segmented reconstruction

## Extraction of Multiple Objects from Reconstructions

The second stage of our segmentation technique is our depth segmentation algorithm which takes two inputs: a depth map DMap and the desired number of object segments $i$. A histogram of a depth map is a plot of the frequency of each depth value as a function of the depth values. We qualitatively select a histogram's modes by identifying the clustered regions within the histogram plot. These histogram modes are the basis of our depth segmentation technique. The algorithm involves the following steps:

**Step 1:** The $i$ modes with the largest area are selected from a histogram of DMap. The starting depth and ending depth of each mode is stored in MStart($i$) and MEnd($i$). The segment index $n$ is given an initial value of 1 and the segmentation image SImage is initially set to be a matrix of zeros of size $M \times N$, where a hologram's reconstruction is $M \times N$ in size.
**Step 2:** The $nth$ SImage segment is created by selecting the pixels from DMap that belong to the $ith$ mode of the histogram. We label the corresponding pixels from SImage with the segment index $n$ as follows;

$$\mathrm{SImage}(k,l) = n, \quad \text{if } \mathrm{MStart}(n) \geq \mathrm{DMap}(k,l) \geq \mathrm{MEnd}(n) \tag{15.31}$$

**Step 3:** The step counter $n$ is then incremented. Step 2 is repeated as long as the step counter $n$ is less than or equal to the desired number of segments $i$. Otherwise we progress to Step 4.
**Step 4:** The object pixels in DMap may not be are labelled in SImage during Steps 2 and 3. In this final step those unlabeled object pixels are labelled in SImage. The euclidean distance from each unlabeled object pixel to the center of mass of each of the $i$ segments is calculated, and the pixel is labelled with the index of the closest segment. The output of this algorithm is the segmented image SImage where every identified object pixel in our segmentation mask SMask has been labelled.
We apply our depth segmentation to the two screws object DH shown in Fig. 15.7 and Sec. 15.4.2. The first stage is the application of the DFF algorithm using a

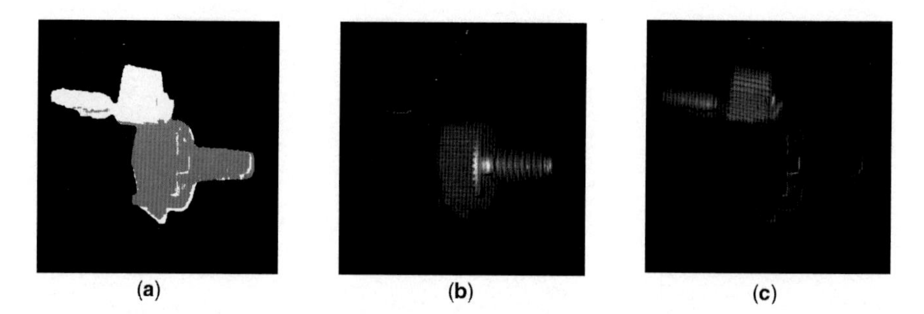


**Fig. 15.11.** Depth segmentation of two screws object hologram: (**a**) depth segmentation mask, (**b**) segmented region 1, (**c**) segmented region 2

sequence of 150 depths with an interval of 0.2 mm between successive depths. To create the scenes' SMask and DMap (Fig. 15.7), we used the block size of $81 \times 81$ [26]. Two obvious segments in the scene were observed: the front screw and the back screw. We input DMap and $i = 2$ into the second stage of our segmentation technique. A reconstruction of the first segment region at its focal plane is shown in Fig. 15.11b and a reconstruction of the second segment region at its focal plane is shown in Fig. 15.11c. SImage and the segment reconstructions illustrate that multiple objects can be automatically segmented from a digital holographic scene using depth information. This algorithm is described in more detail in Ref. [27]. Our segmentation algorithm can be used to compute a mask for the purpose of extracting an object's full complex wavefront at a chosen plane a distance $d$ away from the hologram plane.

### 15.4.4 Synthetic Digital Holographic Scene Creation

We now proceed to the creation of synthetic digital holographic scenes created by combining our data extraction algorithms. A depth map and segmentation mask are created from reconstructions of the object encoded in a DH. Our autofocus algorithm is employed to determine the distance at which we will segment the full complex wavefront. In Fig.15.12 the simulated experimental

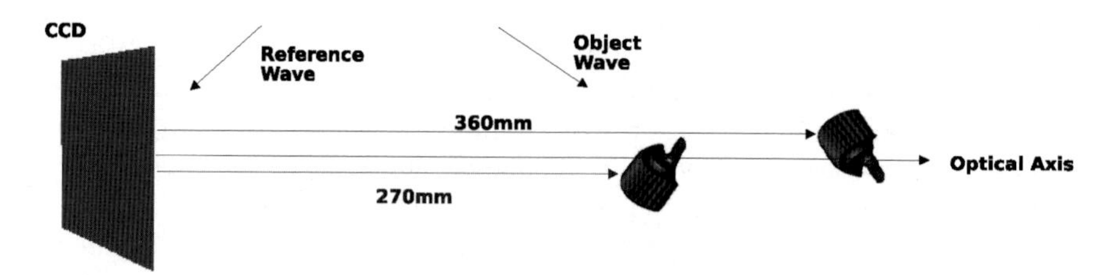


**Fig. 15.12.** Simulated experimental setup for the superposed hologram, with a second object superposed a distance of 90 mm from the original object

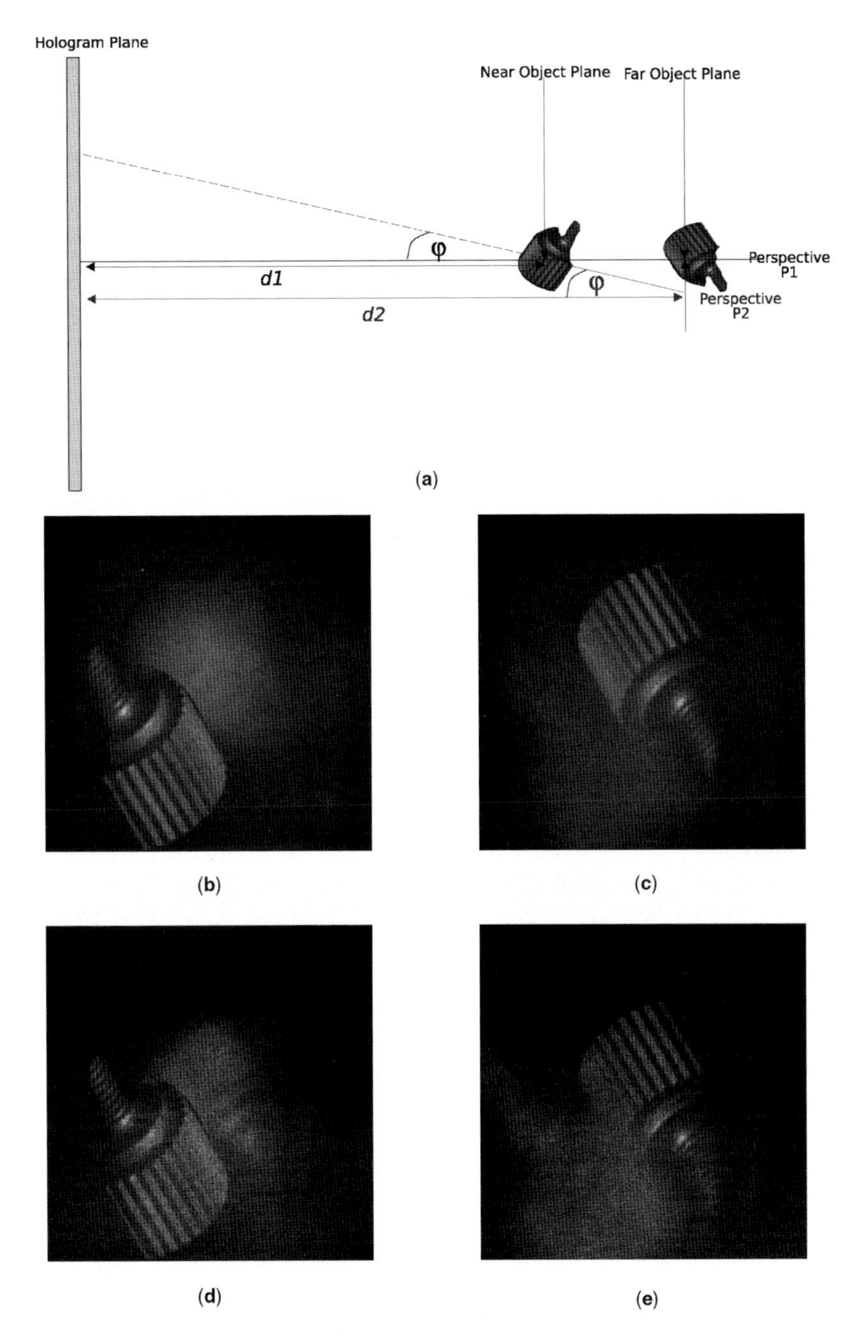


**Fig. 15.13.** Superposed DH: (**a**) diagram showing perspectives P1, along the optical axis, and P2, from above the optical axis, (**b**) near plane reconstruction at depth d1 from perspective P1, (**c**) far plane reconstruction at depth d2 from perspective P1, (**d**) near plane reconstruction at depth d1 from perspective P2, and (**e**) far plane reconstruction at depth d2 from perspective P2

setup is illustrated. We segment an object's wavefront at the in-focus plane for the object and digitally propagate the wavefront a distance of 90 mm. We proceed to add a second segmented object to the wavefront and digitally propagate a further distance of 270 mm back to the hologram plane. We now have a synthetic DH containing two objects extracted from different DHs.

To demonstrate the 3-D information in our synthetic DH we calculate four reconstructions from two perspectives at the near object plane and the far object plane. We selected the perspective from the optical axis, P1, and one from above the optical axis focused on the near object plane, P2, as shown in Fig. 15.13a. This equates to a change in the viewing angle of $0.7^{\circ}$. Reconstructions from perspective P1 are shown in Fig. 15.13b and c while reconstructions from perspective P2 are shown in Fig. 15.13d and e. As can be seen when comparing the reconstructions from perspective P1 and P2, both objects change orientation, but there is also an observable parallax between the reconstructions caused by the change of angular position and physical position of the far object. This allows us to construct synthetic holograms of real scenes, or combine real and computer generated holograms to create overlays. The synthesized result can be used in three-dimensional display systems the same way as traditional computer generated holograms, for instance using spatial light modulators as described in [63, 64].

## 15.5 Conclusions

In this chapter the field of focus detection and its recent application to digital holographic reconstructions has been discussed in this paper. We have presented a DFF technique for extracting shape information from a digital hologram and have detailed a method for creating EFIs from this depth map where all objects are in-focus. Using DHs of real-world three-dimensional macroscopic objects we have experimentally verified our technique. A novel segmentation technique has been discussed which we believe will be successful in segmenting digital holographic scenes containing macroscopic objects into object and background. However, our focus metric relies on sharpness in the reconstruction plane to determine depth. If the object encoded in the DH is planar, or if large object regions are planar with a uniform color, it may be necessary to illuminate the object with a speckle pattern, similar to Thelen, et al. [57], to correctly, estimate the depth. Our algorithm would not be successful for scenes containing pure phase objects. Also in microscopic scenes, we advise using a phase unwrapping approach for the creation of the depth maps [52]. A synthetic digital hologram containing two real-world objects encoded in two different holograms has been created. We have demonstrated it contains the 3-D information required to compute new perspectives. Our segmentation algorithm allows us to construct synthetic holograms of real scenes from sets of DH reconstructions. With our algorithms for segmenting all object pixels and for segmenting individual objects and object regions we can synthesize holograms from DHs containing single and multiple objects. These synthetic holograms are then fit for viewing on conventional three-dimensional display systems.

**Acknowledgments** This chapter has emanated from research conducted with the financial support of Science Foundation Ireland, Enterprise Ireland, the Embark Initiative of the Irish Research Council for Science, Engineering, and Technology under the National Development Plan, and the European Commission Framework Programme 6 through a Marie Curie Fellowship. The authors would also like to thank Jonathan Maycock and Lukas Ahrenberg for their help and contributions to this work.

## References

[1] D. Gabor, "A new microscope principle," *Nature* **161**, 77–79 (1948).
[2] E. N. Leith and J. Upatnieks, "New techniques in wavefront reconstruction," *Journal of the Optical Society of America A* **51**, 1469–1473 (1962).
[3] E. N. Leith and J. Upatnieks, "Wavefront reconstruction with continuous-tone objects," *Journal of the Optical Society of America* **53**, 1377–1381 (1963).
[4] O. Bryngdahl and A. Lohmann, "Interferograms are image holograms," *Journal of the Optical Society of America* **58**, 141–142 (1968).
[5] P. Hariharan, *Basics of holography*, Cambridge University Press, Cambridge (2002).
[6] J. W. Goodman and R. W. Lawerence, "Digital image formation from electronically detected holograms," *Applied Physics Letters* **11**, 777–778 (1967).
[7] T. Kreis, M. Adams, and W. Jüptne, "Methods of digital holography: a comparison," *Proc. SPIE Optical Inspection and Micromeasurements II* **3098**, 224–233 (1999).
[8] L. Onural and P. Scott, "Digital decoding of in-line holograms," *Optical Engineering.* **26**, 1124–1132 (1987).
[9] U. Schnars and W. P. O. Jüptner, "Direct recording of holograms by a ccd target and numerical reconstruction," *Applied Optics* **33**, 179–181 (1994).
[10] T. Kreis, *Handbook of holographic interferometry*, WILEY-VCH GmbH and Co. KGaA, Weinheim, first ed. (2005).
[11] U. Schnars and W. Jüptner, *Digital holography: digital hologram recording, numerical reconstruction, and related techniques*, Springer, Berlin (2004).
[12] B. Javidi and E. Tajahuerce, "Three-dimensional object recognition by use of digital holography," *Optics Letters* **25**, 610–612 (2000).
[13] T. J. Naughton, Y. Frauel, B. Javidi, and E. Tajahuerce, "Compression of digital holograms for three-dimensional object reconstruction and recognition," *Applied Optics* **41**, 4124–4132 (2002).
[14] J. Maycock, C. P. McElhinney, B. M. Hennelly, T. J. Naughton, J. B. McDonald, and B. Javidi, "Reconstruction of partially occluded objects encoded in three-dimensional scenes by using digital holograms," *Applied Optics* **45**, 2975–2985 (2006).
[15] A. Stern and B. Javidi, "Sampling in the light of wigner distribution," *Journal of the Optical Society of America A* **21**, 360–366 (2004).

[16] G. Pedrini, P. Froning, H. Tiziani, and F. Santoyo, "Shape measurement of microscopic structures using digital holograms," *Optics Communications* **164**, 257–268 (1999).
[17] S. Schedin, G. Pedrini, H. Tiziani, A. Aggarwal, and M. Gusev, "Highly sensitive pulsed digital holography for built-in defect analysis with a laser excitation," *Applied Optics* **40**, 100–117 (2001).
[18] T. Kreis, M. Adams, and W. Jüptner, "Digital in-line holography in particle measurement," *Proc. SPIE* **3744** (1999).
[19] N. Y. S. Murata, "Potential of digital holography in particle measurement," *Optics and Laser Technology* **32**, 567–574 (2000).
[20] T. Kreis and W. Jüptner, "Suppression of the dc term in digital holography," *Optical Engineering* **36**, 2357–2360 (1997).
[21] I. Yamaguchi and T. Zhang, "Phase-shifting digital holography," *Optics Letters* **22**, 1268–1270 (1997).
[22] J. Gillespie and R. King, "The use of self-entropy as a focus measure in digital holography," *Pattern Recognition Letters* **9**, 19–25 (1989).
[23] M. Liebling and M. Unser, "Autofocus for digital fresnel holograms by use of a fresnelet-sparsity criterion," *Journal of the Optical Society of America A* **21**, 2424–2430 (2004).
[24] F. Dubois, C.Schockaert, N. Callens, and C. Yourassowsky, "Focus plane detection criteria in digital holography microscopy," *Optics Express* **14**, 5895–5908 (2006).
[25] C. P. McElhinney, B. M. Hennelly, and T. J. Naughton, "Extended focused imaging for digital holograms of macroscopic three-dimensional objects," *Applied Optics* **47**, D71–D79 (2008).
[26] C. P. McElhinney, J. B. McDonald, A. Castro, Y. Frauel, B. Javidi, and T. J. Naughton, "Depth-independent segmentation of three-dimensional objects encoded in single perspectives of digital holograms," *Optics Letters* **32**, 1229–1231 (2007).
[27] C. P. McElhinney, B. M. Hennelly, J. B. McDonald, and T. J. Naughton, "Multiple object segmentation in macroscopic three-dimensional scenes from a single perspective using digital holography," *in preparation* (2007).
[28] T. M. Lehtimäki and T. J. Naughton, "Stereoscopic viewing of digital holograms of real-world objects," in *3DTV Conference 2007 – Capture, Transmission and Display of 3D Video*, IEEE Press, New York, (Kos, Greece) (2007). article no. 39.
[29] H. J. Caulfield, *Handbook of optical holography*, Academic Press, New York (1979).
[30] E. N. Leith and J. Upatnieks, "Wavefront reconstruction with diffused illumination and three-dimensional objects," *Journal of the Optical Society of America* **54**, 1295 (1964).
[31] J. Goodman, *Introduction to Fourier optics*, Roberts and Company, Englewood, Colorado, 3rd ed. (2005).
[32] D. Kim and B. Javidi, "Distortion-tolerant 3-D object recognition by using single exposure on-axis digital holography," *Optics Express* **12**, 5539–5548 (2004).
[33] E. Cuche, P. Marquet, and C. Depeursinge, "Spatial filtering for zero-order and twin-image elimination in digital off-axis holography," *Applied Optics* **39**, 4070–4075 (2000).

[34] Y. Zhang, W. Lu, and B. Ge, "Elimination of zero-order diffraction in digital off-axis holography," *Optics Communications* **240**, 261–267 (2004).
[35] T. Poon, T. Kim, G. Indebetouw, B. Schilling, M. Wu, K. Shinoda, and Y. Suzuki, "Twin-image elimination experiments for three-dimensional images in optical scanning holography," *Optics Letters* **25**, 215–217 (2000).
[36] S. Lai, B. Kemper, and G. vonBally, "Off-axis reconstruction of in-line holograms for twin-image elimination," *Optics Communications* **169**, 37–43 (1999).
[37] L. Xu, J. Miao, and A. Asundi, "Properties of digital holography based on in-line configuration," *Optical Engineering* **39**, 3214–3219 (2000).
[38] R. Bracewell, *The fourier transform and its applications*, McGraw-Hill (1986).
[39] Y. Frauel, E. Tajahuerce, M. Castro, and B. Javidi, "Distortion-tolerant three-dimensional object recognition with digital holography," *Applied Optics* **40**, 3887–3893 (2001).
[40] J. H. Bruning, D. R. Herriott, J. E. Gallagher, D. P. Rosenfeld, A. D. White, and D. J. Brangaccio, "Digital wavefront measuring interferometer for testing optical surfaces and lenses," *Applied Optics* **13**, 2693–2703 (1974).
[41] I. R. Nourbaksh, D. Andre, C. Tomasi, and M. R. Genesereth, "Mobile robot obstacle avoidance via depth from focus," *Robotsics and Autonomous Systems* **22**, 151–158 (1997).
[42] V. Murino and C. S. Regazzoni, "Visual surveillance by depth from focus," *IEEE International Conference on Industrial Electronics, Control and Instrumentation* **2**, 998–1002 (1994).
[43] F. Chen, G. Brown, and M. Song, "Overview of three-dimensional shape measurement using optical methods," *Optical Engineering* **39**, 10–22 (2000).
[44] W. Huang and X. Jing, "Evaluation of focus measures in multi-focus image fusion," *Pattern Recognition Letters* **28**, 493–500 (2007).
[45] A. Erteza, "Sharpness index and its application to focus control," *Applied Optics* **15**, 877–881 (1976).
[46] M. Subbarao and T. Choi, "Focusing techniques," *Optical Engineering* **32**, 2824–2836 (1993).
[47] K. Takahashi, A. Kubota, and T. Naemura, "A focus measure for light field rendering," *IEEE International Conference on Image Processing* **4**, 2475–2478 (2004).
[48] R. A. Muller and A. Buffington, "Real-time correction of atmospherically degraded telescope images through image sharpening," *Journal of the Optical Society of America A* **64**, 1200–1210 (1974).
[49] J. Kautsky, J. Flusser, B. Zitová, and S. Šimberová, "A new wavelet-based measure of image focus," *Pattern Recognition Letters* **27**, 1431–1439 (2006).
[50] V. H. Bove. Jr, "Entropy-based depth from focus," *Journal of the Optical Society of America A* **10**, 561–566 (1993).
[51] M. Subbarao and J. Tyan, "Selecting the optimal focus measure for autofocusing and depth-from-focus," *IEEE Transactions on Pattern Analysis and Machine Intelligence* **20**, 864–870 (1998).
[52] P. Ferraro, G. Coppola, S. Nicola, A. Finizio, and G. Peirattini, "Digital holographic microscope with automatic focus tracking by detecting sample displacement in real time," *Optics Letters* **28**, 1257–1259 (2003).
[53] P. Ferraro, S. Grilli, D. Alfieri, S. D. Nicola, A. Finizio, G. Pierattini, B. Javidi, G. Coppola, and V. Striano, "Extended focused image in microscopy by digital holography," *Optics Express* **13**, 6738–6749 (2005).

[54] R. Yin, P. Flynn, and S. Broschat, "Position-dependent defocus processing for acoustic holography images," *International Journal of Imaging Systems and Technology* **12**, 101–111 (2002).
[55] L. Ma, H. Wang, Y. Li, and H. Jin, "Numerical reconstruction of digital holograms for three-dimensional shape measurement," *Journal of Optics A: Pure Applied Optics* **6**, 396–400 (2004).
[56] J. A. E. Malkiel and J. Katz, "Automated scanning and measurement of particle distributions within a holographic reconstructed volume," *Measurement Science and Technology* **15**, 601–612 (2004).
[57] A. Thelen, J. Bongartz, D. Giel, S. Frey, and P. Hering, "Iterative focus detection in hologram tomography," *Journal of the Optical Society of America A* **22**, 1176–1180 (2005).
[58] N. Burns and J. Watson, "Data extraction from underwater holograms of marine organisms," *Proceedings of Oceans 07, Aberdeen* (2007).
[59] T. Colomb, E. Cuche, P. Dahlgen, A. Marian, F. Montfort, C. Depeursinge, P. Marquet, and P. Magistretti, "3D imaging of surfaces and cells by numerical reconstruction of wavefronts in digital holography applied to transmission and reflection microscopy," *Proceedings of IEEE – International Symposium on Biomedical Imaging* , 773–776 (2002).
[60] J. Maycock, B. M. Hennelly, J. B. McDonald, T. J. Naughton, Y. Frauel, A. Castro, and B. Javidi, "Reduction of speckle in digital holography by discrete fourier filtering," *Journal of the Optical Society of America A* **24**, 1617–1622 (2007).
[61] B. M. Hennelly, T. J. Naughton, J. B. McDonald, Y. Frauel, and B. Javidi, "A method for superresolution in digital holography," *Proceedings of SPIE Optics and Photonics, San Diego* **6311**, 63110 J (2006).
[62] D. Forsyth and J. Ponce, *Computer vision: a modern approach*, Prentice Hall (2003).
[63] M. Lucente, "Interactive three-dimensional holographic displays: seeing the future in depth," *SIGGRAPH Computer Graphics* **31**, 63–67 (1997).
[64] C. Slinger, C. Cameron, and M. Stanley, "Computer-generated holography as a generic display technology," *IEEE Computer Magazine* **38**, 46–53 (2005).

# 16

# Polarimetric Imaging of 3-D Object by Use of Wavefront-Splitting Phase-Shifting Digital Holography

Takanori Nomura and Bahram Javidi

**Abstract** A polarimetric imaging method of a 3-D object by use of wavefront-splitting phase-shifting digital holography is presented. Phase-shifting digital holography with a phase difference between orthogonal polarizations is used. The use of orthogonal polarizations can make it possible to record two phase-shifted holograms simultaneously if we use pixelated polarimetric optical elements. By combining the holograms with the distributions of a reference wave and an object wave, the complex field of the object wavefront can be obtained. The polarimetric image results from a combination of two kinds of holographic imaging using orthogonal polarized reference waves . An experimental demonstration of a 3-D polarimetric imaging is presented.

## 16.1 Introduction

Digital holography is a useful technique for recording the fully complex field of a wavefront [1,2,3,4]. In line with advances in imaging devices such as CCDs, digital holography is accessible. A phase-shifting technique [2] is mandatory for effective utilization of the number of pixels and pixel size of an imaging device. Digital holography has been used for lots of applications, including encryption [5,6] and 3-D object recognition [7,8]. On the other hand, the polarization information provides important information about the surface of the material, birefringence of an anisotropic medium, or photoelastic effect [9,10,11]. In polarimetric imaging by use of digital holography, [10] the polarization state

T. Nomura
Department of Opto-Mechatronics, Wakayama University, 930 Sakaedani, Wakayama 640-8510 Japan
e-mail: nom@sys.wakayama-u.ac.jp

B. Javidi et al. (eds.), *Three-Dimensional Imaging, Visualization, and Display*,
DOI 10.1007/978-0-387-79335-1_16, © Springer Science+Business Media, LLC 2009

(Jones vector) of a 2-D transparent object has been obtained using two off-axis reference waves with orthogonal polarizations. Because reconstruction of a 3-D object is one of the most important aspects of holography, it is useful if a polarimetric imaging system of a 3-D object is accomplished. Namely, we know both the polarimetric information and the shape of a 3-D object simultaneously. One of the authors has proposed 3-D polarimetric imaging based on integral imaging [12]. However, digital holographic 3-D imaging with polarization has not been reported. To implement polarimetric imaging of a 3-D object, we propose using two kinds of digital holograms obtained by orthogonal polarized reference waves.

## 16.2 Wavefront-Splitting Phase-Shifting Digital Holography with a Phase Difference Between Orthogonal Polarizations

### 16.2.1 Phase Analysis by Two-Step Method

Let $A_o(x, y)$ and $A_r(x, y)$ denote an object wave and a reference wave on a digital hologram plane, respectively. Using the amplitude distribution of the object wave $a_o(x, y)$, the phase distribution of the object wave $\phi_o(x, y)$, and those of the reference wave, $a_r(x, y)$ and $\phi_r(x, y)$, we can write the two waves as

$$A_o(x, y) = a_o(x, y) \exp\left[i\phi_o(x, y) - \alpha\right], \tag{16.1}$$

$$A_r(x, y) = a_r(x, y) \exp\left[i\phi_r(x, y)\right], \tag{16.2}$$

respectively, and the parameter $\alpha$ denotes the phase-shifting quantity of the object wave. Then we have a digital hologram written as

$$I(x, y, \alpha) = a_o^2(x, y) + a_r^2(x, y) + 2a_o(x, y)a_r(x, y) \cos\left[\phi_o(x, y) - \phi_r(x, y) - \alpha\right]. \tag{16.3}$$

The digital holograms that use 0 and $\pi/2$ phase-shifted object beams are

$$I(x, y, 0) = a_o^2(x, y) + a_r^2(x, y) + 2a_o(x, y)a_r(x, y) \cos\left[\phi_o(x, y) - \phi_r(x, y)\right], \tag{16.4}$$

$$I(x, y, \pi/2) = a_o^2(x, y) + a_r^2(x, y) + 2a_o(x, y)a_r(x, y) \sin\left[\phi_o(x, y) - \phi_r(x, y)\right], \tag{16.5}$$

respectively. If we know the two intensity distributions, $I_o(x, y)$ and $I_r(x, y)$, given by

$$I_o(x, y) = |a_o(x, y)|^2, \tag{16.6}$$

$$I_r(x, y) = |a_r(x, y)|^2, \tag{16.7}$$

respectively, the fully complex information from the object wave can be obtained by

$$a_o(x, y) = \sqrt{I_o(x, y)} \tag{16.8}$$

$$\phi_o(x, y) = \tan^{-1} \frac{I(x, y, \pi/2) - I_o(x, y) - I_r(x, y)}{I(x, y, 0) - I_o(x, y) - I_r(x, y)} + \phi_r(x, y). \tag{16.9}$$

Since the constant phase distribution $\phi_r(x, y)$ of the reference wave is negligible, Eq. (16.9) can be considered as the object phase distribution. From the above discussion we learn that two holograms and two intensity distributions are needed to obtain fully complex information on the object wave. This is a conventional two step phase-shifting method. In general, since the reference wave is constant and stationary, the intensity distribution $I_r(x, y)$ can be recorded beforehand. The concern is how we can obtain the intensity distribution of the object wave $I_o(x, y)$ and two phase-shifted holograms $I(x, y, 0)$ and $I(x, y, \pi/2)$ simultaneously. We may call them three intensity distributions.

### 16.2.2 Phase-Shifting Method with Orthogonal Polarizations

We propose a method in which three intensity distributions can be recorded simultaneously by using the phase difference in the orthogonal polarizations [13,14]. The optical setup is shown in Fig. 16.1.

A laser with a vertical polarization state is used as a coherent light source. The light is divided by a beam splitter (BS) into two waves. One is a reference wave and the other is an object wave. The object wave passes through a quarter-wave plate (QWP) and a polarizer (P). The fast axes of the QWP is set at an angle of 45 degrees. with respect to the horizontal axis. The transmission axis of the P is set vertically. After passing through the devices, the polarization state of the object wave is a vertical polarization. The reflected object wave from the object surface passes through the P and the QWP. After passing through the devices, the polarization state of the object wave is a right circular polarization. If we detect the object beam along with the fast and slow axes of the QWP separately, the phase difference in the two beams is equal to $\pi/2$. If we detect the object beam along with the horizontal axis,

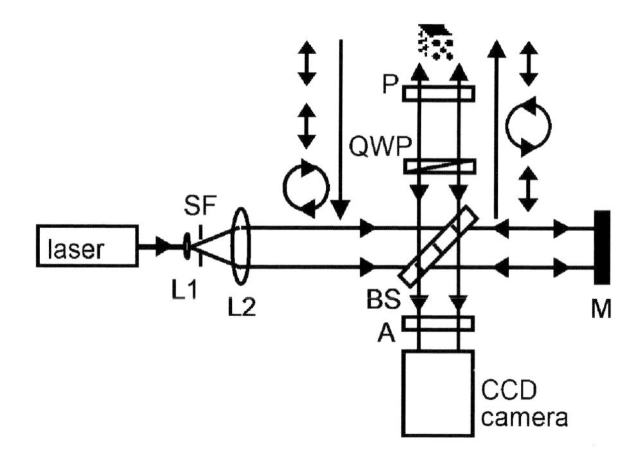


**Fig. 16.1.** Optical setup for phase-shifted digital holography when the phase difference between orthogonal polarization is used: L1, L2, lenses; SF, a spatial filter; BS, a beam splitter; M, a mirror, O, an object; P, a polarizer; A, an analyzer; QWP, a quarter-wave plate

the object intensity distribution is obtained. To detect beams, we use three kinds of analyzer (A). In this case, we can obtain the information of the object wave with a vertical polarization component. For a horizontal component, the transmission axis of the P is set horizontally.

## 16.3 Wavefront-Splitting Phase-Shifting Digital Holography

To achieve simultaneous detection of the above-mentioned three beams, we assume that the CCD camera has pixelated polarizers as shown in Fig. 16.2. Three kinds of polarizers are aligned like the Bayer & color imaging array for a color CCD. We call it a polarization CCD camera. The hologram recorded by the polarization CCD camera is called a one-shot digital hologram. For a pixel lacking data in the one-shot digital hologram, we simply interpolate the data by averaging the neighbors shown in Fig. 16.3. For the intensity distribution of the object $I_o(x, y)$, if there are no data $u$, the data are interpolated by

$$u = \frac{1}{4}(a + b + c + d). \tag{16.10}$$

For the two phase-shifting digital holograms $I(x, y, 0)$ and $I(x, y, \pi/2)$, if there are no data $u$, $v$, and $w$, the data are interpolated by

$$u = \frac{1}{4}(a + b + c + d), \tag{16.11}$$

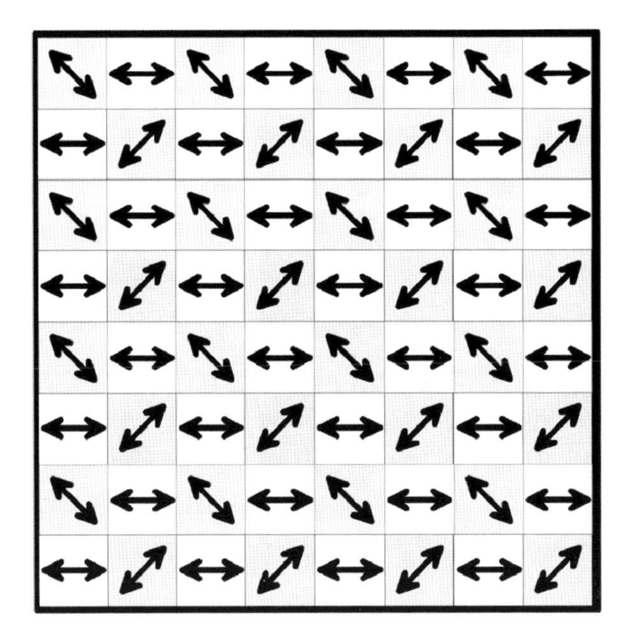

**Fig. 16.2.** Pixelated polarizers of the polarization CCD

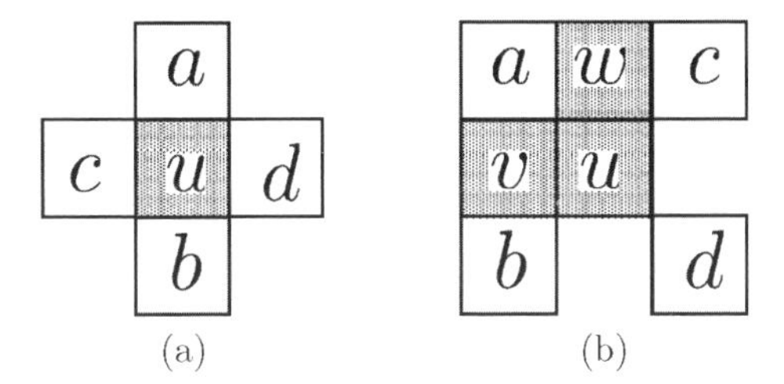


**Fig. 16.3.** Interpolation method of a one-shot digital hologram

$$v = \frac{1}{2}(a + b), \tag{16.12}$$

$$w = \frac{1}{2}(a + c), \tag{16.13}$$

respectively. The schematic diagram for obtaining three images from the one-shot digital hologram is shown in Fig. 16.4.

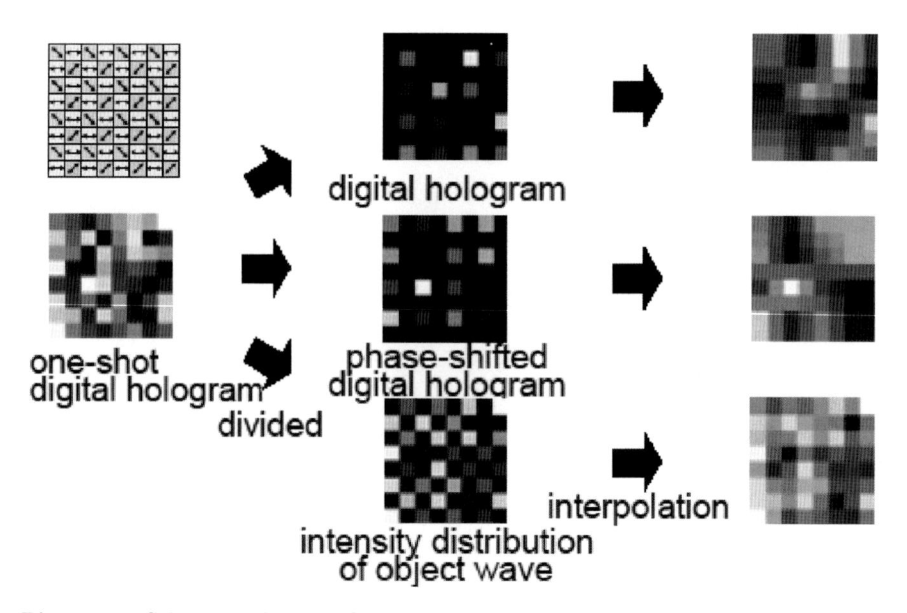


**Fig. 16.4.** Schematic diagram for obtaining three images from the one-shot digital hologram

## 16.4 Stokes Vector

If we have both horizontal and vertical components of an object wave, we can obtain the Stokes parameters of the object. Using the horizontal and vertical amplitude distributions of the reconstructed object $A_{oh}(X, Y)$ and $A_{ov}(X, Y)$, the vertical phase distributions of the reconstructed object $\Phi_{oh}(X, Y)$ and $\Phi_{ov}(X, Y)$, the Stokes vector is given by

$$\boldsymbol{S} = \begin{bmatrix} S_0(X, Y) \\ S_1(X, Y) \\ S_2(X, Y) \\ S_3(X, Y) \end{bmatrix} = \begin{bmatrix} A_{oh}^2(X, Y) + A_{ov}^2(X, Y) \\ A_{oh}^2(X, Y) - A_{ov}^2(X, Y) \\ 2A_{oh}(X, Y)A_{ov}(X, Y)\cos\varepsilon(X, Y) \\ 2A_{oh}(X, Y)A_{ov}(X, Y)\sin\varepsilon(X, Y) \end{bmatrix}, \tag{16.14}$$

where $\varepsilon(X, Y)$ denotes the phase difference distributions between a horizontal and a vertical component given by

$$\varepsilon(X, Y) = \Phi_{oh}(X, Y) - \Phi_{ov}(X, Y). \tag{16.15}$$

This vector gives the polarimetric information of the object.

## 16.5 Experimental Results

The optical setup which records the digital hologram with polarization information is the same as shown in Fig. 16.1. A diode-pumped laser (with a wavelength of 532 nm, 5 W) that is in a vertical polarization state is used as a coherent light source. The three-dimensional object, the die shown in Fig. 16.5, is used for both the randomly polarized and polarimetric case. The size of the die is 8 mm × 8 mm × 8 mm. Two orthogonal polarizers are placed in front of the polarimetric object. Arrows in Fig. 16.5b show the direction of the transmission axis of the polarizers. The left polarizer has the horizontal transmission axis and the right is vertical. The upper central portion of the two orthogonal polarizers overlap. On the other hand, there are no polarizers in the lower central portion. The CCD camera has $1024 \times 768$ pixels and 8 bits of gray-level. The pixel size of the CCD is $4.65\,\mu$m × $4.65\,\mu$m. The distance from the die to the CCD is 290 mm.

In this experiments, we do not have a polarization CCD camera. Therefore the experiments are performed as follows. First, we record the intensity distribution of a reference wave $I_r(x, y)$. Next, we record the zero phase-shifted digital hologram $I(x, y, 0)$ with an analyzer set at an angle of −45 degrees. The $\pi/2$ phase-shifted hologram $I(x, y, \pi/2)$ is recorded with a 45 degrees. analyzer. The intensity distribution of an object wave $I_o(x, y)$ is recorded with 0 degrees. analyzer. We extract the data from the three intensity distributions to obtain a one-shot digital hologram and then synthesize the data according to Fig. 16.2. This procedure is schematized in Fig. 16.6

If we rotate P to set the transmission axis along the vertical axis, we can record the vertical polarimetric information of the object.

The reconstructed objects are obtained from the holograms using computational Fresnel diffraction integral. The amplitude distributions for horizontally and vertically polarized light are shown in Figs. 16.7 and 16.8. Each figure is normalized so that black and white denote 0 and maximum values,

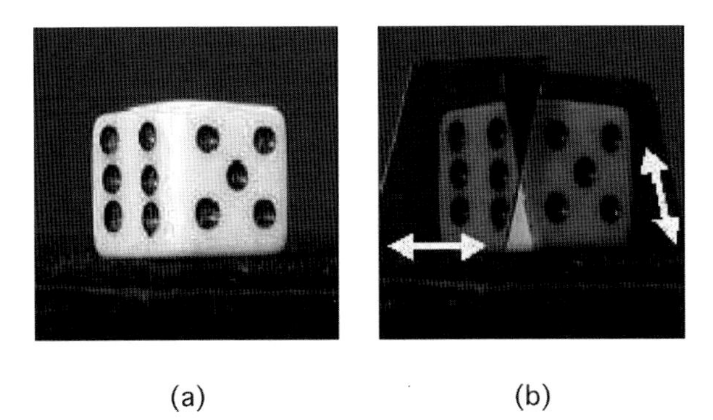

(a) (b)

**Fig. 16.5.** (a) Randomly polarized object and (b) a polarimetric objects

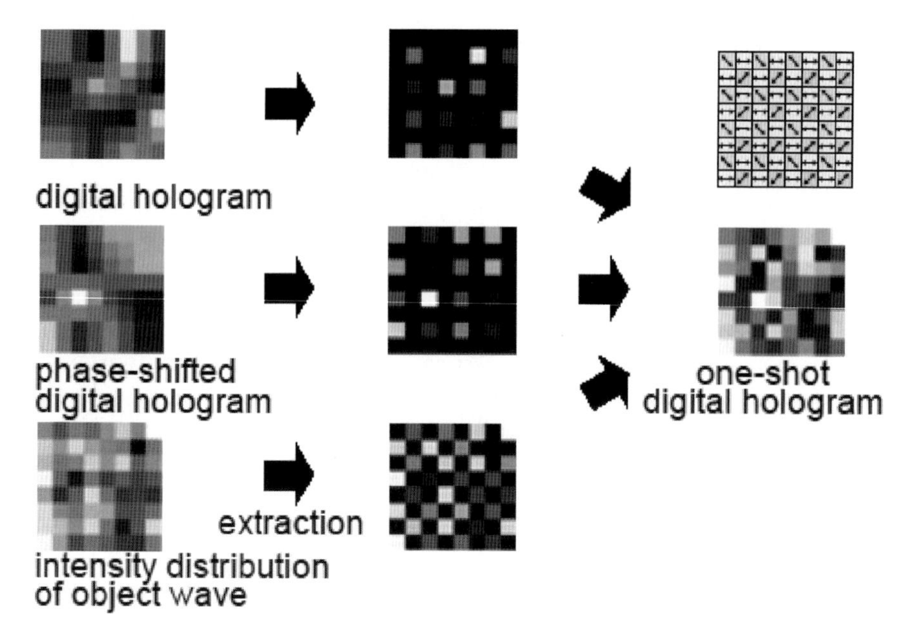


**Fig. 16.6.** Schematic diagram for obtaining the one-shot digital hologram from three images

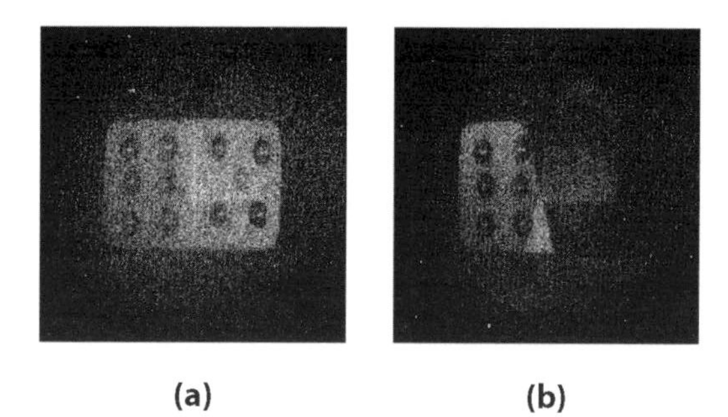


**Fig. 16.7.** The amplitude distirbutons for horizontally polarized light: (**a**) a randomly polarized object and (**b**) a polarimetric object

respectively. The phase difference distributions between horizontally polarized light and vertically polarized light are shown in Fig. 16.9. Black and white denote 0 and $2\pi$, respectively.

The grayscale images according to the Stokes parameters of the reconstructed objects are shown in from Figs. 16.10 to 16.14. In Fig. 16.10, each image is normalized so that black and white denote 0 and $S_{0\mathrm{MAX}}$, respectively, where $S_{0\mathrm{MAX}}$ denotes the maximum value of $S_0$. The $S_0(x, y)$ images are shown in Fig. 16.10. These are ordinal digital holographic reconstructed objects. In Figs. 16.11 to 16.14, each image is normalized so that black and

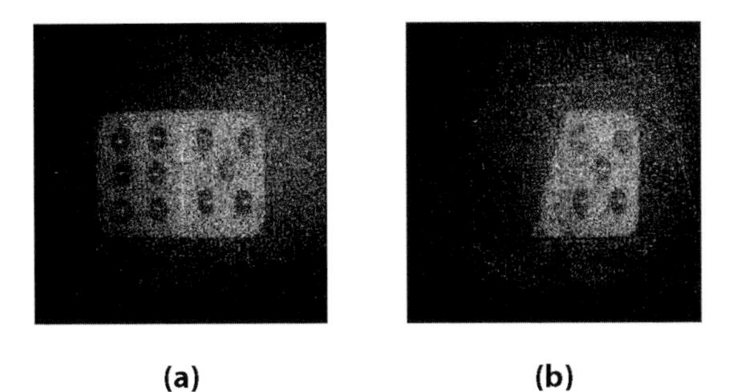

(a) (b)

**Fig. 16.8.** The amplitude distributions for vertically polarized light: (**a**) a randomly polarized object and (**b**) a polarimetric object

(a)

**Fig. 16.9.** The phase difference distributions between horizontally polarized light and vertically polarized light t: (**a**) a randomly polarized object and (**b**) a polarimetric object

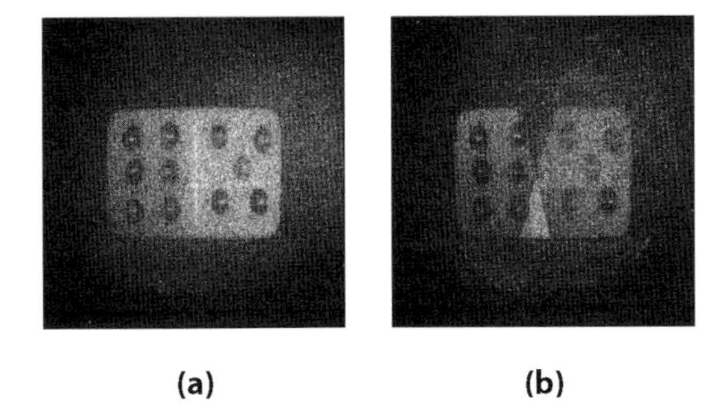

(a) (b)

**Fig. 16.10.** The grayscale images according to the Stokes parameter $S_0(X, Y)$: (**a**) a randomly polarized object and (**b**) a polarimetric object

white denote $-S_{0\mathrm{MAX}}$ and $S_{0\mathrm{MAX}}$, respectively. The $S_1(x, y)$ images are shown in Fig. 16.11. The white and black denotes the horizontal and vertical components of polarization, respectively. From Fig. 16.11, we learn that the image represents the characteristics of the polarimetric object. We also show the

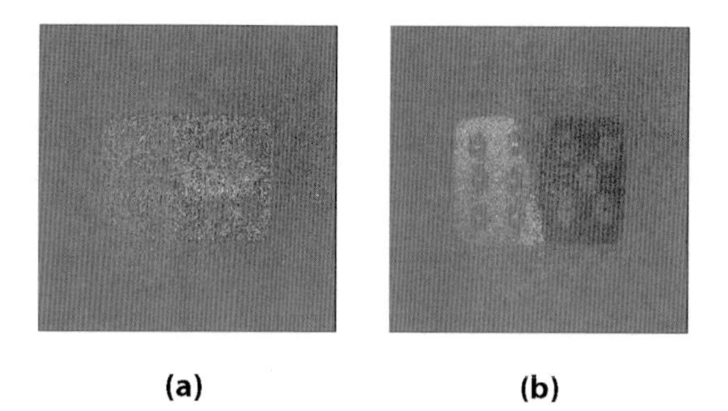



**Fig. 16.11.** The grayscale images according to the Stokes parameter $S_1(X, Y)$: (**a**) a randomly polarized object and (**b**) a polarimetric object

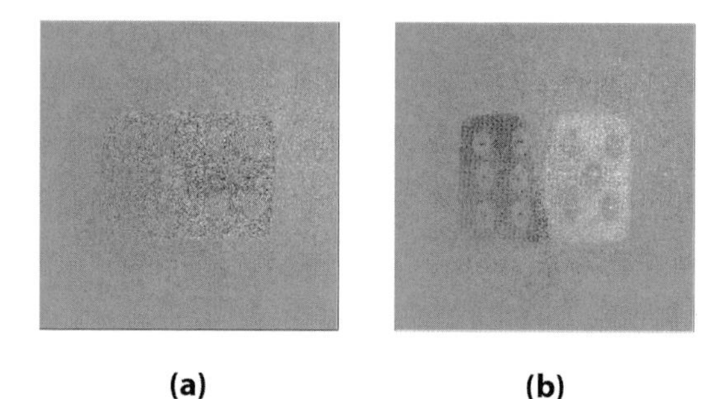



**Fig. 16.12.** The grayscale images according to the Stokes parameter $-S_1(X, Y)$: (**a**) a randomly polarized object and (**b**) a polarimetric object

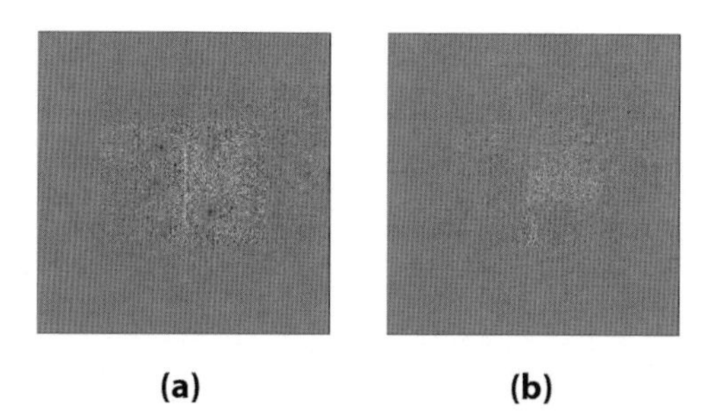



**Fig. 16.13.** The grayscale images according to the Stokes parameter $S_2(X, Y)$: (**a**) a randomly polarized object and (**b**) a polarimetric object

images in Fig. 16.12 according to the equation given by

$$-S_1(X, Y) = A_{ov}^2(X, Y) - A_{oh}^2(X, Y). \tag{16.16}$$

Figures 16.13 and 16.14 also suggest the polarimetric information of the object. These experimental results illustrate that three-dimensional polarimetric imaging is possible.

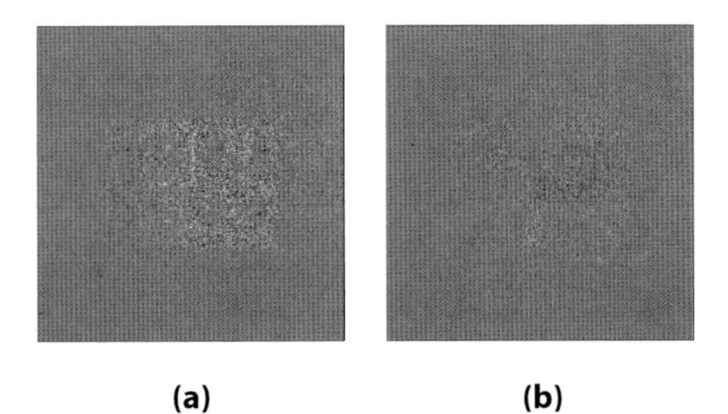

**Fig. 16.14.** The grayscale images according to the Stokes parameter $S_3(X, Y)$: (**a**) a randomly polarized object and (**b**) a polarimetric object

## 16.6 Conclusions

We have shown a method to obtain three-dimensional polarimetric imaging by use of digital holography. In the proposed methods, we used two reconstructed objects. One consists of a reference wave with a vertical polarization state and an object wave with a horizontal polarization state, the other consists of opposite polarization states. We gave some experimental results of polarimetric imaging to confirm the method.

## References


[1] Goodman, J. W., Lawrence, R. W.: Digital image formation from electronically detected holograms. Appl. Phys. Lett. **11**, 77–79 (1967).

[2] Yamaguchi, I., Zhang, T.: Phase-shifting digital holography. Opt. Lett. **22**, 1268–1270 (1997).

[3] Osten, W., Baumbach, T., Juptner, W.: Comparative digital holography. Opt. Lett. **27**, 1764–1766 (2002)

[4] Kreis, T.: Handbook of Holographic Interferometry. (Wiley VCH, Weinheim, 2005).

[5] Tajahuerce, E., Javidi, B.: Encrypting three-dimensional information with digital holography. Appl. Opt. **39**, 6595–6601 (2000).

[6] Nomura, T., Uota, K., Morimoto, Y.: Hybrid optical encryption of a 3-D object using a digital holographic technique. Opt. Eng. **43**, 2228–2232 (2004).

[7] Frauel, Y., Tajahuerce, E., Castro, M.-A., Javidi, B.: Distortion-tolerant three-dimensional object recognition with digital holography. Appl. Opt. **40**, 3877–3893 (2001).

[8] Pedrini G. ,Tiziani,H. J. : Short-coherence digital microscopy by use of a lensless holographic imaging system. Appl. Opt. **41**, 4489–4496 (2002).

[9] Sadjadi, F. A., Chun, C. L.: Automatic detection of small objects from their infrared state-of-polarization vectors. Opt. Lett. **28**, 531–533 (2003).

[10] Colomb, T., Cuche, E., Montfort, F., Marquet, P., Depeursinge, Ch.: Jones vector imaging by use of digital holography: simulation and experimentation. Opt. Commun. **231**, 137–147 (2004).

[11] Refregier, P., Roueff, A.: Coherence polarization filtering and relation with intrinsic degrees of coherence. Opt. Lett. **31**, 1175–1177 (2006).
[12] Matoba, O, Javidi, B.: Three-dimensional polarimetric integral imaging. Opt. Lett. **29**, 2375–2377 (2004).
[13] Nomura, T., Javidi, B. Murata, S., Nitanai, E., Numara, T.: One-shot digital holography by use of polarization. Proc. SPIE **6311**, 63110I-1–5 (2006).
[14] Nomura, T., Murata, S., Nitanai, E., Numata, T.: Phase-shifting digital holography with a phase difference between orthogonal polarizations. Appl. Opt. **45**, 4873–4877 (2006).

# 17

# Three-dimensional Display with Data Manipulation based on Digital Holography

Osamu Matoba

## 17.1 Introduction

In recent years, the two-dimensional display system has been enthusiatically developed to create HDTV or more. A three-dimensional (3-D) display system is still difficult for commerical products due to the large amount of data that must be handled or reconstructed 3-D information. Holography is one of the best solutions for 3-D display. For 3-D display application, digital holography is an available technique to develop holographic 3-D display systems as digital information [1, 2, 3, 4, 5, 6, 7, 8, 9, 10, 11, 12, 13, 14, 15].

In digital holograhy, an intensity distribution recorded directly by an imaging device consists of an object intensity pattern, a reference intensity pattern and an interference pattern that includes complex amplitude of a 3-D object. The recorded intensity distribution is called a digital hologram. Digital holography has been applied in many fields such as measurement, 3-D display and information processing. Digital format of recorded data can be used to analyze the 3-D object information and for data communication. This is suitable for 3-D display applications.

We have been developing 3-D display systems based on digital holography [8, 9, 11, 14, 15]. The systems consist of three sub-systems: a recording system of 3-D information, a 3-D information processing system, and

O. Matoba
Department of Computer Science and Systems Engineering, Kobe University, Rokkodai 1-1, Nada, Kobe 657-8501, Japan
e-mail: matoba@kobe-u.ac.jp

B. Javidi et al. (eds.), *Three-Dimensional Imaging, Visualization, and Display*,
DOI 10.1007/978-0-387-79335-1_17, © Springer Science+Business Media, LLC 2009

a 3-D display system. In the recording system of a real 3-D object, complex amplitude information of the 3-D object is obtained by phase-shifting digital holography or the phase retrieval method. The complex amplitude of a virtual 3-D data is obtained numerically. In the information processing system, complex amplitude information is processed to improve the quality of reconstructed data and manipulate the reconstructed information. 3-D object manipulation can be achieved by using complex amplitude information. Information processing of digital complex amplitude will open new applications in display and measurement. In the reconstruction system, the complex amplitude data or phase-only data is displayed for 3-D reconstruction.

## 17.2 Three-Dimensional Holographic Display Systems Based on Digital Holography

Figure 17.1 shows a schematic of 3-D display systems based on digital holography with information processing. The proposed system consists of three parts: a recording, an information processing, and a reconstruction system. In the recording system, a virtual or real 3-D object can be recorded as a digital hologram. When measuring real 3-D objects, a fast or instantaneous recording system is required to measure the moving object or dynamical change in the 3-D object. Usually, the reconstruction by a hologram consists of a 3-D real object as well as DC and conjugate reconstructions. These reconstructions, except for real 3-D objects, act as noise in 3-D display applications. Therefore, we present recording systems to extract a complex amplitude of 3-D objects to avoid the noise reconstruction.

In coherent readout, speckles appeared in the reconstruction. The manipulation of complex amplitude enables us to improve the quality of reconstructed a 3-D object or to eliminate, add, and transform the 3-D object. A virtual or real 3-D object can be eliminated or added in complex amplitude at the hologram plane. This information processing in complex amplitude domain can open new applications such as interactive 3-D display.

In 3-D display systems, complex amplitude or phase-only distribution is used to optically or numerically reconstruct 3-D objects. In the optical reconstruction, it was shown that the phase distribution can reconstruct a 3-D object, when a measured 3-D object scatters light enough. This is preferred to the commercially available spatial light modulator which can control only amplitude or phase distribution. The optical reconstruction and phase-only reconstruction are described in Ref. [9]. Note that it is called electro-holography when a three-dimensional object is reconstructed optically by displaying the hologram data in spatial light modulator.

In Section 17.3, we present two fast recording systems of complex amplitude of a 3-D object based on digital holography. We also present information processing to manipulate 3-D objects in Section 17.4.

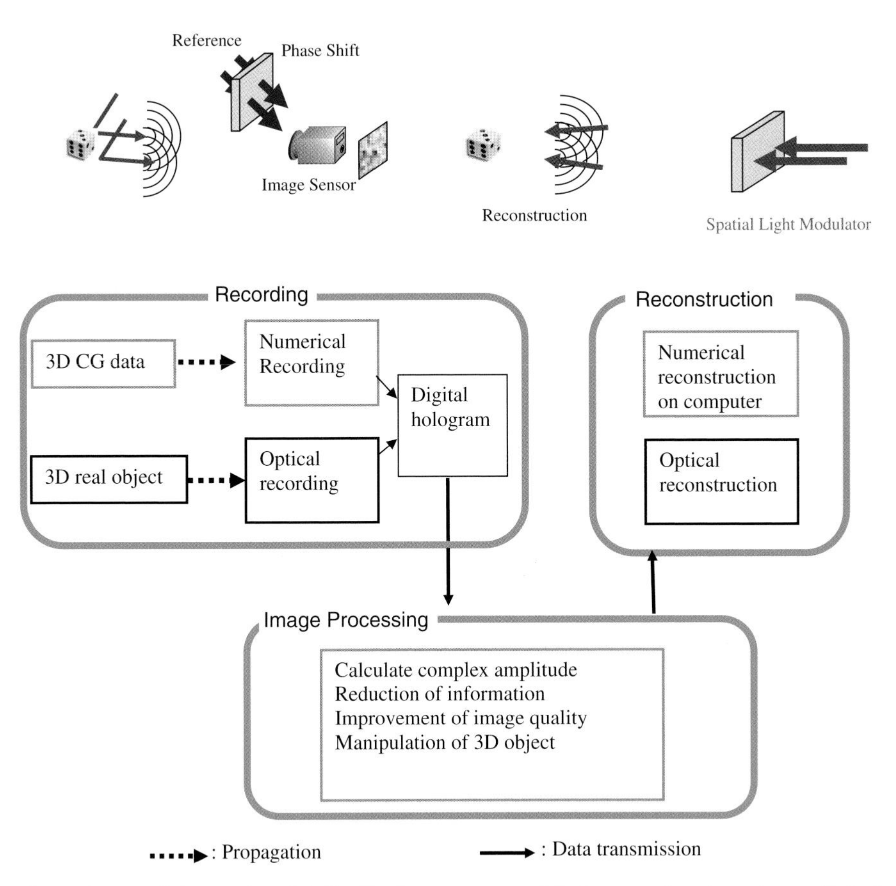


**Fig. 17.1.** Schematic of 3-D display system with interaction based on digital holography

## 17.3 Fast Recording System of Complex Amplitude of 3-D Object Based on Digital Holography

In this section, we present two fast recording systems. One recording system uses a temporal phase-shifting digital holography. In the other system, the phase distribution of the 3-D object at the hologram plane is estimated from a single hologram and object intensity information. This enables us to develop an instantaneous recording system.

### 17.3.1 Fast Recording Systems Based on Phase-shifting Digital Holography

We present a fast recording system based on phase-shifting digital holography [8]. There are papers to implement the instantaneous recording in phase-shifting digital holography by use of spatial coding [5–7]. Here we show a temporal phase shift digital holography for a 3-D object recording [4]. One of the typical setups to obtain 3-D information by phase-shifting digital holography is shown in Fig. 17.2. A laser beam is divided into two beams that are object and reference beams. In the fast recording system, an electro-optic modulator is used as a phase retarder. The electro-optic modulator can change the phase at up to 100 MHz. We also use a C-MOS image sensor with 512 $\times$ 512 pixels and 10 bit intensity resolution at each pixel. The interference pattern can be obtained by the image sensor at up to 2,000 fps. The pixel size is 16 $\times$ 16 $\mu$m. The phase modulation and the capture of the interference pattern are synchronized by a trigger signal generated by a function generator.

We briefly derive 3-D object recording in a four-step phase-shifting digital holography. In the following derivation, we assume that a 3-D object consists of many point sources with a grayscale intensity level in 3-D space. Let a point source located at the three-dimensional coordinate $(\xi,\eta,z)$ be $o_z(\xi,\eta)$. The original 3-D object is illuminated by a coherent light source and then the reflected light propagates through free space. In the paraxial region, the reflected light is given by the Hygens-Fresnel integral as follows:

$$\begin{aligned} f(x,y) &= o_z(\xi,\eta)g_z(x-\xi,y-\eta)d\xi d\eta = o_z(x,y)\otimes g_z(x,y) \\ &= f_r(x,y)\exp\{if_i(x,y)\} \end{aligned}, \tag{17.1}$$

where

$$g_z(x,y) = \exp\frac{i\pi}{\lambda z}(x^2+y^2) \tag{17.2}$$

and $\otimes$ denotes convolution. In Eqs. (17.1) and (17.2), the constant phase and amplitude are neglected. Each pixel of the 3-D object can be expressed to

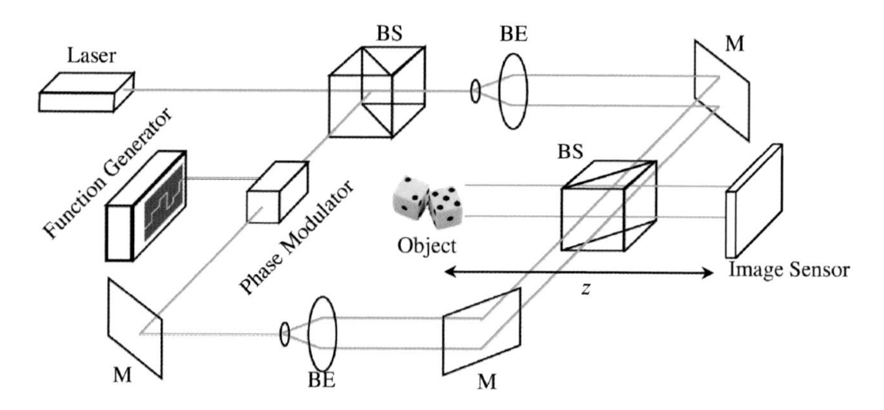


**Fig. 17.2.** Optical setup of temporal phase shift digital holography. BS; beamsplitter, BE; beam expander, M; Mirror

have different $(x, y, z)$. After the Fresnel propagation, the object pattern is recorded as a digital hologram by making an interference pattern with a plane reference wave, $r_p(x, y)$. The intensity pattern is given by

$$\begin{aligned} I_p(x, y) &= \left|f(x, y) + r_p(x, y)\right|^2 \\ &= \left|f_r(x, y)\right|^2 + \left|r_r(x, y)\right|^2 \\ &\quad + 2 f_r(x, y) r_r(x, y) \cos(f_i(x, y) - r_i(x, y) - p) \end{aligned}, \tag{17.3}$$

where

$$r_p(x, y) = r_r(x, y) \exp\{i(r_i(x, y) + p)\}. \tag{17.4}$$

In Eqs. (17.3) and (17.4), $p$ is 0, $\pi/2$, $\pi$, or $3\pi/2$, in four-step phase-shifting digital holography.

From the four interference patterns, we calculate the complex field of object wave at the hologram.

$$f_r(x, y) = \frac{1}{4 r_r(x, y)} \sqrt{(I_0(x, y) - I_\pi(x, y))^2 + (I_{\pi/2}(x, y) - I_{3\pi/2}(x, y))^2}, \tag{17.5}$$

and

$$f_i(x, y) = \tan^{-1} \frac{I_{\pi/2}(x, y) - I_{3\pi/2}(x, y)}{I_0(x, y) - I_\pi(x, y)} + r_i(x, y). \tag{17.6}$$

When the reference beam is a plane wave, the amplitude and the phase of the object wave can be calculated.

We experimentally demonstrate the fast recording. A die with 5 mm cube is used as a 3-D object. The die is located at 232 mm from the image sensor. Figure 17.3 shows the reconstructed images at a frame rate of 15, 250, and 500 Hz. We can see that the reconstruction of the 3-D object at 500 Hz was successful. The contrast of the reconstructed image worsens when the speed of the image detection is high. This is because the optical power is not sufficient in the high-speed detection. In this setup, the complex amplitude of a 3-D object can be recorded at 500 Hz.

### 17.3.2 Phase Retrieval Method for Instantaneous Recording

In the previous subsection, we show a fast temporal recording of 3-D object information by phase-shifting interferometer. There is, however, the limitation in the recording of a moving 3-D object. It is ideal if we can extact the complex amplitude information of a 3-D object from a single hologram. One method to extract only an object component from a hologram is the Fourier transform fringe analysis [16]. In this method, carrier frequency spatially

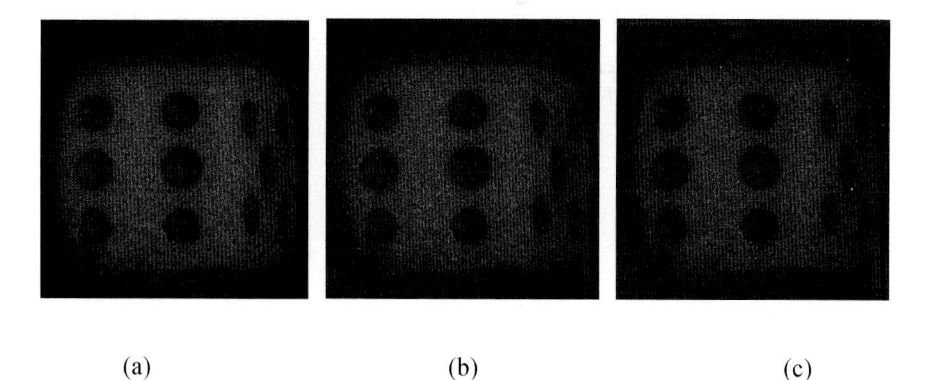

(a) (b) (c)

**Fig. 17.3.** Reconstructed images at recording rates of (**a**) 15, (**b**) 250, and (**c**) 500 Hz

separates the object component from the others. Maximum spatial frequency of the object component should be lower than the carrier frequency of the interference pattern. Another method is the phase shift method [4]. As we mentioned in the previous subsection, the spatial coding method for the phase shift method has been proposed [5, 6, 7]. This method can be applied to record a moving 3-D object, but the resolution may be distorted due to the spatial coding.

In this subsection, we present a method to determine the numerically complex amplitude of a 3-D object from a single hologram and an intensity distribution of the 3-D object at the hologram plane [15]. From the two distributions, amplitude and absolute values of phase of the 3-D object at the hologram plane are calculated. The unknown phase signs are determined by an iterative algorithm. The algorithm uses the reconstruction of a real 3-D object and defocused reconstruction of the conjugated object. The conjugated object is out-of-focus, thus it can be eliminated by a proper threshold. The iteration can provide more precise complex amplitude of the 3-D object at the hologram plane.

Figure 17.4 shows an optical setup for simultaneously recording a hologram $I_1$ and an object intensity distribution $I_2$. In Fig. 17.4, there is no additional equipment in a conventional interferometer. By using a polarization beam splitter (PBS), only an intensity distribution of the object wave can be measured by CCD2. Two image sensors are located at the same position from the 3-D object. Suppose that complex amplitudes of an object and a reference wave are $U(x, y) = A(x, y)\exp\{i\Phi(x, y)\}$ and $U_r = A_r\exp(i\Phi_r)$, respectively. The intensity patterns, $I_1$ and$I_2$, are described as

$$I_1(x, y) = A(x, y)^2 + A_r{}^2 + 2A(x, y)A_r\cos\{\Phi(x, y) - \Phi_r\} \qquad (17.7)$$

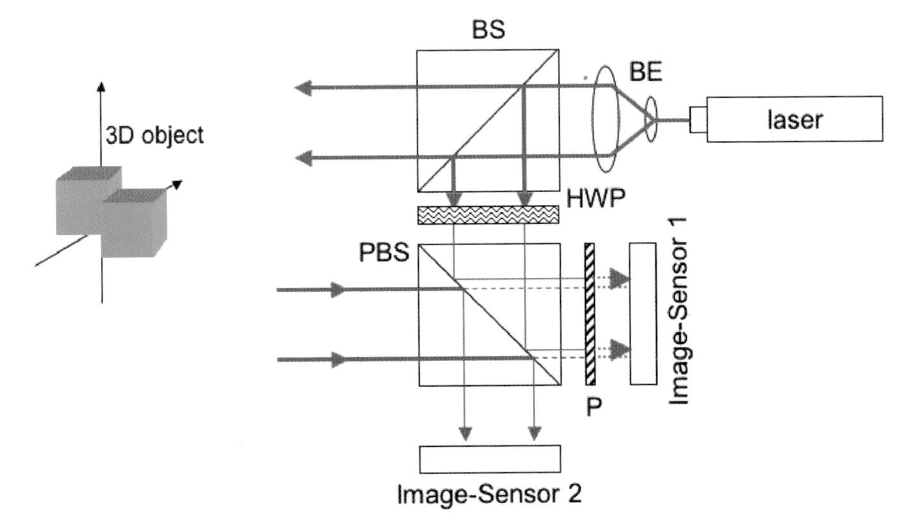


**Fig. 17.4.** Schematic of the proposed optical setup for recording simultaneously two intensity distributions

$$I_2(x, y) = A(x, y)^2. \tag{17.8}$$

The complex amplitude of object $U_0 = A_0 \exp(i\Phi_0)$ is calculated from the recorded intensity distributions:

$$A_0(x, y) = \sqrt{I_2(x, y)} \tag{17.9}$$

$$\Phi_0(x, y) = \cos^{-1} \frac{I_1(x, y) - I_2(x, y) - A_r{}^2}{2\sqrt{I_2} A_r}. \tag{17.10}$$

Equation (17.9) shows that the amplitude information is obtained by the intensity measurement of the object. In Eq. (17.10), we know the absolute value of the phase due to the arccosine function. The problem is that the phase sign cannot be determined. By using Eqs. (17.9) and (17.10) in the reconstruction, a real object and its conjugate objects are simultaneously reconstructed. If we know the phase sign, the real 3-D object is perfectly reconstructed at the appropriate distance.

To determine the phase sign in the hologram plane, we use the reconstruction of a real 3-D object to eliminate the conjugated image. Figure 17.5 shows an iterative algorithm to more precisely determine the phase signs at the hologram plane. We use two complex amplitude distributions at two planes – the hologram plane and the reconstructed plane where the real 3-D object exists. In the following description, capital letters and small letters denote the complex amplitude distributions in the hologram plane and in the 3-D object

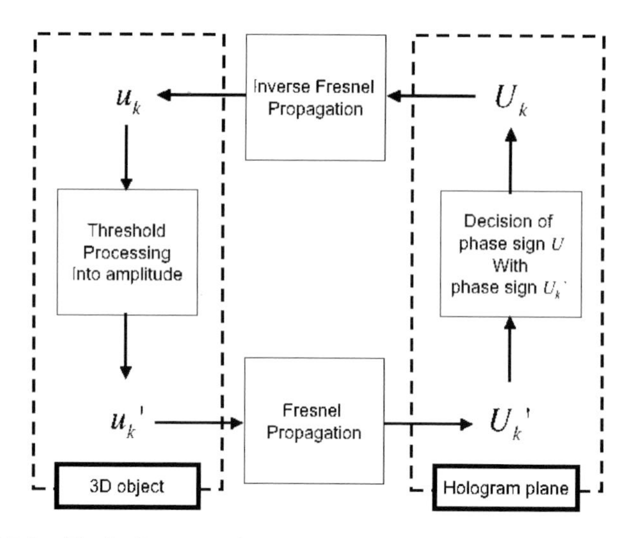


**Fig. 17.5.** Block diagram of iterative algorithm to obtain the phase sign

plane, respectively. The iterative process consists of four steps. In the first step, $k$th estimated complex amplitude in the hologram plane is described by $U_k$. $U_0$ is the initial complex distribution in the hologram plane. The complex amplitude $u_k$ in the 3-D object plane is calculated by numerical Fresnel propagation from $U_k$ [17, 18]. At the appropriate distance, a real 3-D object component of $u_k$ is in-focus and its conjugate component is out-of-focus. In the second step, by using an appropriate threshold at the intensity domain, a region where the real 3-D object exists is extracted. This process eliminates the conjugate object. In the extracted region, it is expected to obtain the complex amplitude of the real 3-D object described as $u_k'$. In the third step, the complex amplitude $U_k'$ in the hologram plane is calculated from $u_k'$ by inverse Fresnel propagation. In the fourth step, the phase sign of $U_k$ is decided by use of the phase sign of $U_k'$. We note that the absolute value of the phase is already known by Eq. (17.10). The new complex amplitude is described as $U_{k/1}$ for $(k{+}1)$th iteration. This algorithm is different from the Gerchberg-Saxton algorithm that uses two intensity patterns at the image plane and the Fourier plane. The present method uses the extracted complex amplitude of a real 3-D object by threshold to determine the sign of phase distribution in the hologram plane.

We numerically show the feasibility of the method. Figure 17.6a shows the object used in the simulation. The object is a two-dimensional picture of a bee. Figure 17.6b shows part of an enlarged image of Fig. 17.6a. We use the random phase distribution instead of a 3-D object. This random phase distribution makes the reconstruction at the 3-D located position. The wavelength of the illuminating light, the numbers of pixels and the dimensions of the pixel are $\lambda = 532$ nm, $N = 1{,}024 \times 1{,}024$, and $a = 16.4 \times 16.4$ $\mu$m, respectively.

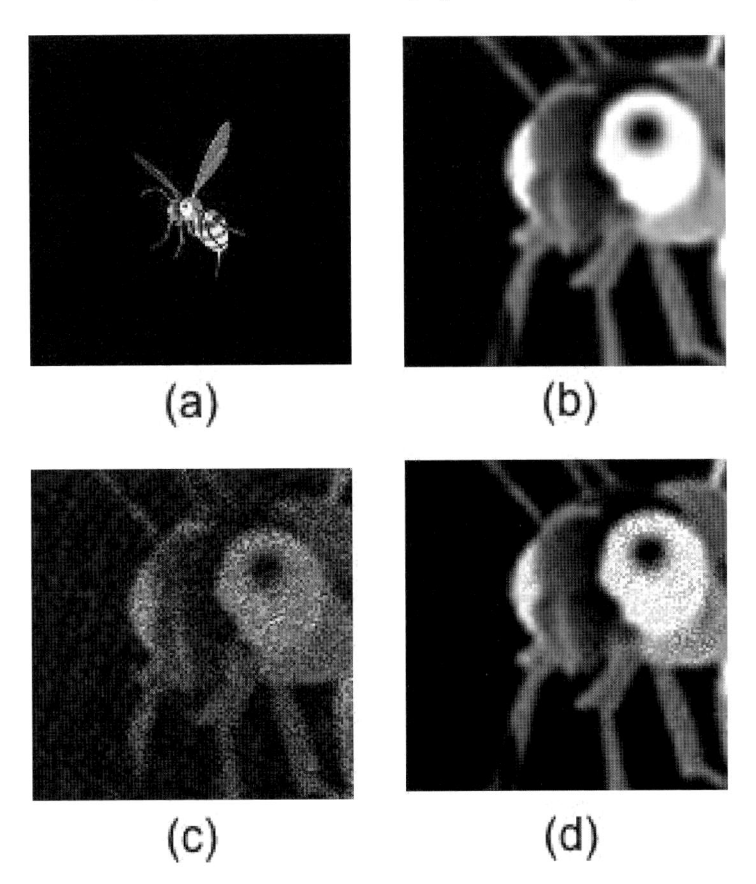


**Fig. 17.6.** Numerical results: (**a**) object used in the simulation, (**b**) an enlarged part of the object, (**c**) and (**d**) the reconstructed images before and after 39th iteration, respectively

The object is placed at a distance of 300 mm from the hologram plane. The iterative algorithm is terminated when the phase sign does not change. Figure 17.6c is the reconstructed image by using the complex amplitude given by Eqs. (17.9) and (17.10). There are many speckle noises due to the conjugated object caused by the random phase distribution. Figure 17.6d shows the estimated object after 39 iterations when the threshold value is 0.1. We can see that the image quality is extensively improved. Figure 17.7 presents the error rate in the estimation of phase sign as a function of a number of iteration. The error rate is given by

$$\text{Error} = \sum_{k=1}^{N} \frac{\left|\text{sign}(\Phi_k) - \text{sign}(\Phi)\right|}{2\,N}. \tag{17.11}$$

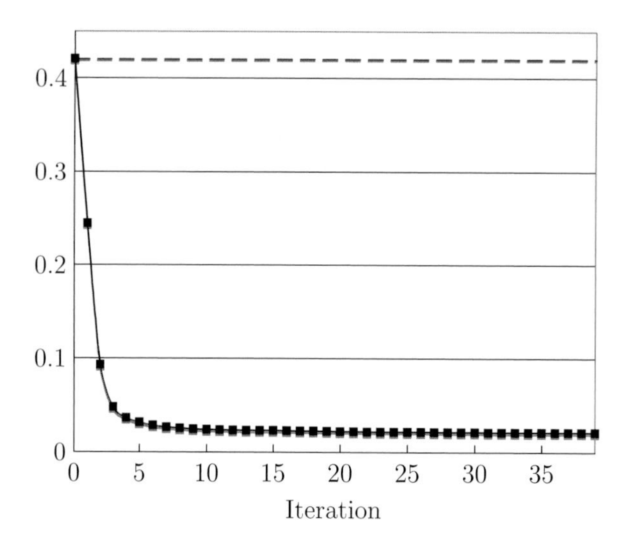


**Fig. 17.7.** Error rate as a function of iteration

The initial error rate in phase sign is 0.42. The error rate is 0.02 after 39 iterations.

## 17.4 Information Processing

For display applications, it is necessary to improve the quality of reconstruction. An example of reconstructed data is shown in Fig. 17.3. We can see a lot of speckle noise due to the rough surface of the 3-D object, coherent readout, and limited space-bandwidth product of the system. Here we present a method to improve the quality of the reconstructed image by spatially averaging the intensity distribution of the 3-D object in the reconstruction plane as shown in Fig. 17.8. In one possible approach a movable stage, such as a piezo electric actuator, can be used to implement spatial average. Figure 17.9 shows the results of spatial averaging in the numerical reconstruction. In Figs. 17.9b and c, $11 \times 11$ and $15 \times 15$ windows are used for the spatial average, respectively. Figure 17.9d shows the cross section of intensity distribution of Figs. 17.9a and c. You can see that the spatial average of the intensity distribution can improve the quality of the reconstructed 3-D object.

As different applications in 3-D display, we investigate a method to manipulate reconstructed images by processing complex amplitude in the hologram plane. Eliminating the 3-D object and adding characters in an arbitrary 3-D position are demonstrated numerically. Figure 17.10 shows the elimination of selected a 3-D object. These operations are implemented by using complex

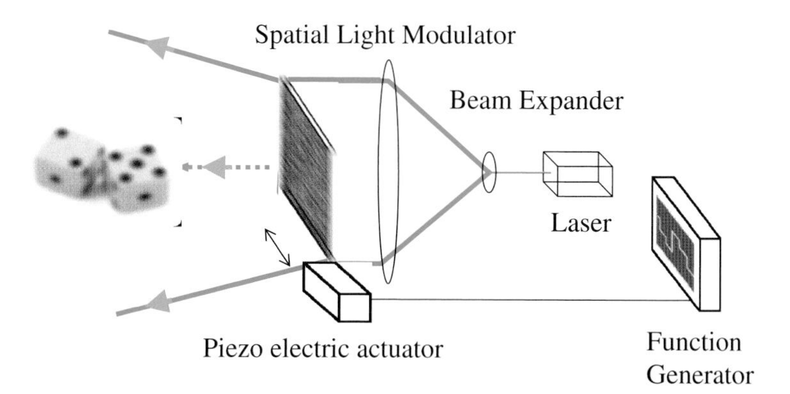


Fig. 17.8. A method to take spatial average by moving a spatial light modulator

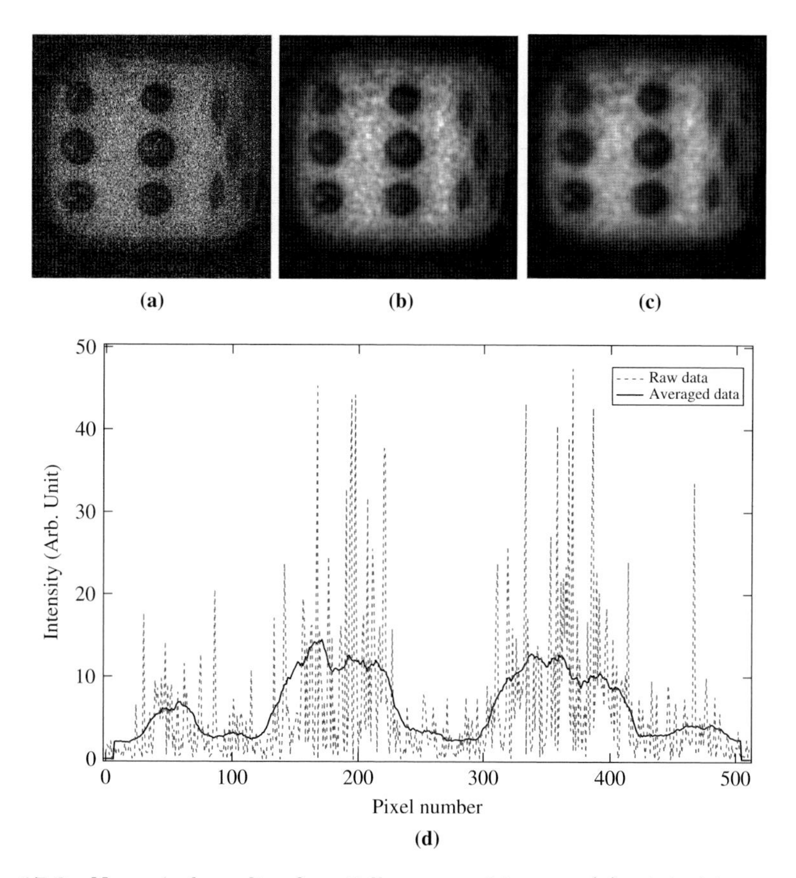


**Fig. 17.9.** Numerical results of spatially averaged images: (**a**) original image, spatially averaged images by (**b**) 11×11 window and (**c**) 15×15 window, and (**d**) cross section of intensity distribution of (**a**) and (**c**)

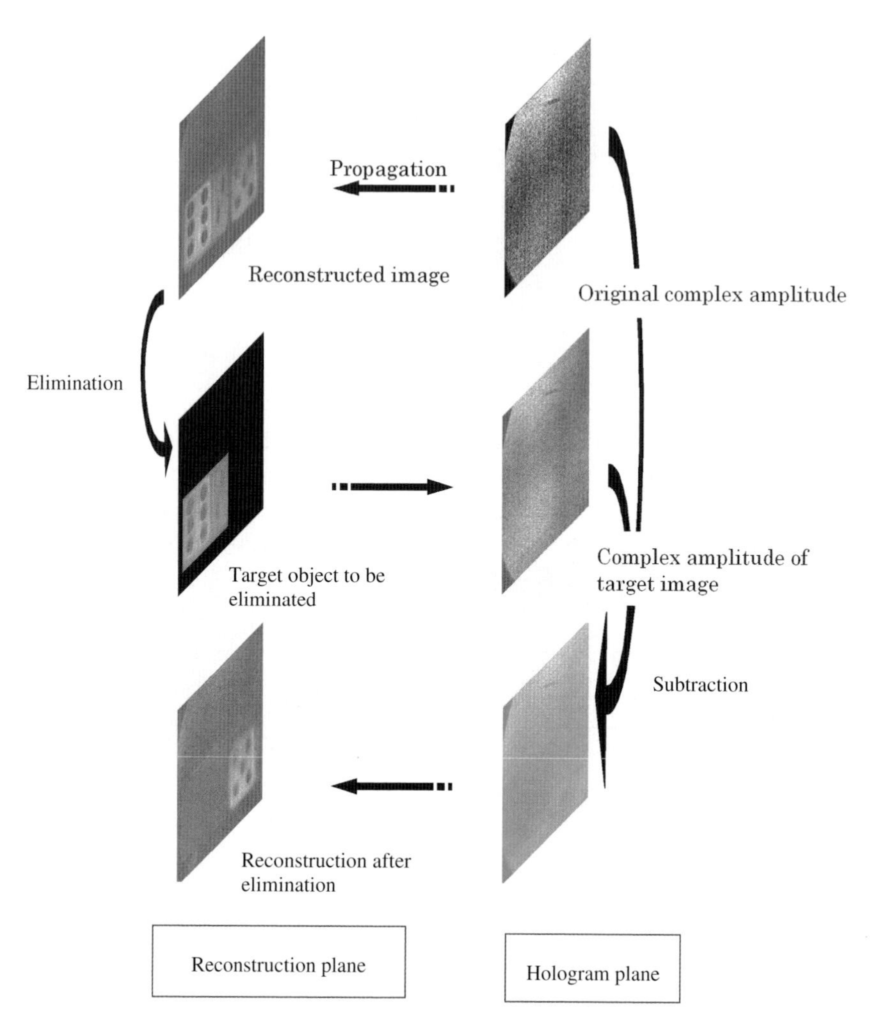


**Fig. 17.10.** Schematic of a procedure to eliminate a 3-D object

amplitude in the hologram plane. At first, the 3-D object is reconstructed by numerical backward Fresnel propagation. The propagation distance is selected enough to separate the 3-D object to be eliminated from other 3-D objects. We extract the region where the 3-D object to be eliminated exists. The extracted complex amplitude is used to calculate the complex amplitude in the hologram plane by forward Fresnel propagation. In the hologram plane, we subtract the complex amplitude of the original 3-D objects from the extracted complex amplitude. Figure 17.11 shows the result of elimination. We can see that one of the dice is clearly erased. For natural views, the background noise is added in the region of elimination before backward propagation. Figure 17.12 shows another example of the data processing by adding the

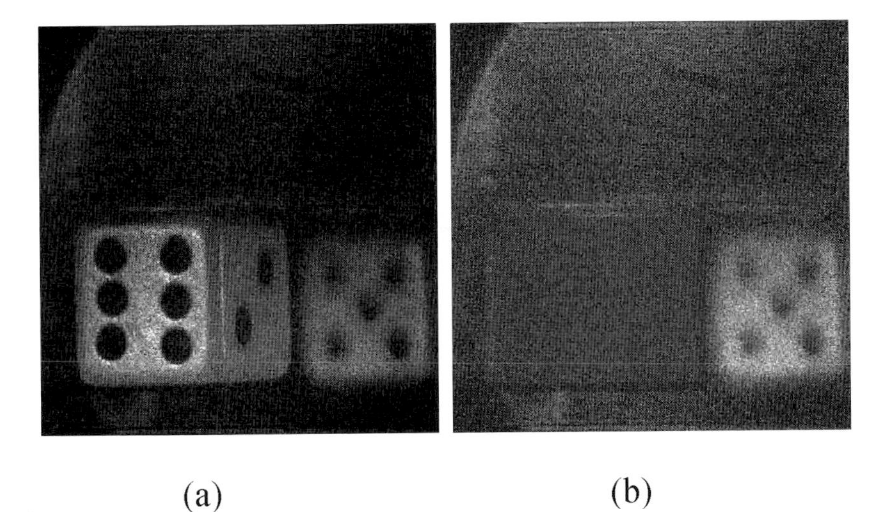


**Fig. 17.11.** Numerical results of elimination of a 3-D object. Reconstructed images (**a**) before and (**b**) after elimination

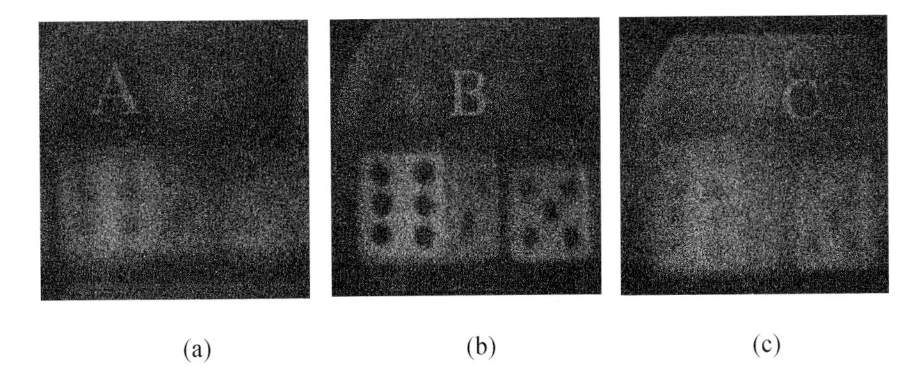


**Fig. 17.12.** Addition of characters in three different positions

character in the reconstructed 3-D object. Three characters are added in different 3-D positions. These results indicate that it is possible to manipulate a 3-D object by using complex amplitude in the hologram plane. This leads to developing new applications of digital holographic 3-D display.

## 17.5 Conclusions

In this chapter, we presented a 3-D display system based on digital holography. In the system, the complex amplitude information of a 3-D object is extracted and then the wavefront reconstruction is realized to reconstruct only a real

3-D object. We presented two types of recording systems to measure the moving 3-D object. We also presented information processing to improve the quality of the reconstructed object. The speckle patterns caused by coherent readout can be reduced by spatial averaging. We have demonstrated the manipulation of a 3-D object by modulating complex amplitude distribution in the hologram plane. These results encourage us to develop attractive 3-D display systems including data communications and interaction between the 3-D objects and humans.

**Acknowledgment** The author would like to thank Mr. K. Hosoi, Mr. T. Handa, and Mr. T. Nakamura for their calculations and experiments. The author also would like to thank Konica Imaging Science Foundation and Hyogo Science and Technology Association for supporting a part of this work.

## References

[1] U. Schnars and W. Jueptner (2005) Digital holography, Springer, New York.
[2] T. Kreis (2005) Handbook of holographic interferometry, Johnm Wiley & Sons Inc., Weinheim.
[3] U. Schnars and W. Jueptner (1994) Direct recording of holograms by CCD target and numerical reconstruction. Applied Optics 33: 179–181.
[4] I. Yamaguchi and T. Zhang (1997) Phase-shifting digital holography. Optics Letters 22: 1268–1270.
[5] Y. Awatsuji, M. Sasada, and T. Kubota (2004) Parallel quasi-phase-shifting digital holography. Applied Physics Letters 85: 1069–1071.
[6] Y. Awatsuji, A. Fujii, T. Kubota, and O. Matoba (2006) Parallel three-step phase-shifting digital holography. Applied Optics 45: 2995–3002.
[7] T. Nomura, S. Murata, E. Nitanai, and T. Numata (2005) Phase-shifting digital holography with a phase difference between orthogonal polarizations. Applied Optics 45: 4873–4877.
[8] O. Matoba, K. Hosoi, K. Nitta, and T. Yoshimura (2006) Fast acquisition system for digital holograms and image processing for three-dimensional display with data manipulation. Applied Optics 45: 8945–8950.
[9] O. Matoba, T.J. Naughton, Y. Frauel, N. Bertaux, and B. Javidi (2002) Real-time three-dimensional object reconstruction by use of a phase-encoded digital hologram. Applied Optics 41: 6187–6192.
[10] B. Javidi and T. Nomura (2000) Securing information by use of digital holography. Optics Letters 25: 28–30.
[11] O. Matoba and B. Javidi (2004) Secure three-dimensional data transmission and display. Applied Optics 43: 2285–2291.
[12] E. Tajahuerce, O. Matoba, and B. Javidi (2001) Shift-invariant three-dimensional object recognition by means of digital holography. Applied Optics 40: 3877–3886.
[13] Y. Frauel, E. Tajahuerce, O. Matoba, M.A. Castro, and B. Javidi (2004) Comparison of passive ranging integral imaging and active imaging digital holography for 3-D object recognition. Applied Optics 43: 452–462.

[14] Y. Frauel, T.J. Naughton, O. Matoba, E. Tajahuerce, and B. Javidi (2006) Three-dimensional imaging and processing using computational holographic imaging. Proceedings of the IEEE 94: 636–653.
[15] T. Nakamura, K. Nitta, and O. Matoba (2007) Iterative algorithm of phase determination in digital holography for real-time recording of real objects. Applied Optics 46: 6849–6853.
[16] M. Takeda, H. Ina, and S. Kobayashi (1982) Fourier-transform method of fringe-pattern analysis for computer-based topography and interferometry. Journal of the Optical Society of America 72: 156–160.
[17] D. Mas, J. Garcia, C. Fereira, L.B. Bernardo, and F. Marinho (1995) Fast algorithms for free-space calculation. Optics Communications 164: 233–245.
[18] J.W. Goodman (1996) Introduction to Fourier Optics 2nd Ed., McGraw-Hill, New York.

# Part IV

# Other 3-D Image Acquisition and Display Techniques, and Human Factors

18

# A 3-D Display System Using Motion Parallax

Hiroshi Ono and Sumio Yano

**Abstract** Motion parallax was described as a cue to depth over 300 years ago. Despite this long history, little research has been addressed to it in comparison to binocular parallax. In recent years experimental interest in motion parallax has increased, following the re-discovery of the idea of yoking stimulus motion to head movement. Exploiting what is now known about observer-produced parallax, we propose a novel display system in which 3-D perception results from synchronizing the movement of stimuli on a 2-D screen to observers' side-to-side head movements, or back and forth head rotations. After briefly reviewing the relevant literature, this chapter suggests how one might go about creating such a display system.

## 18.1 Introduction[1]

Motion parallax is described in most modern textbooks on perception. Most textbook descriptions of motion parallax focus on the perception of depth, but several also discuss perceived motion. Despite its widespread use, the term "motion parallax" is relatively new, but the concept itself has been applied to the perception of depth for over three centuries [2]. Although meager

[1] This chapter is not a comprehensive review of the area of motion parallax. For such a review, we recommend the chapter entitled "Depth from motion parallax" in Howard and Rogers [1]. They are currently writing a more up-to-date review that will appear soon.

H. Ono
Centre for Vision Research and Department of Psychology, York University, Toronto, Ontario, M3J 1P3 (Canada)
e-mail: hono@yorku.ca

B. Javidi et al. (eds.), *Three-Dimensional Imaging, Visualization, and Display*,
DOI 10.1007/978-0-387-79335-1_18, 

compared to the extensive literature on binocular parallax, we now have sufficient knowledge on the depth perception from motion parallax to conceive an experimental 3-D display system. In this chapter, we first discuss the background material needed to understand the basis for the system, and then discuss, in general terms, how such a system might be made.

The literature on motion parallax makes a distinction between observer-produced parallax and object-produced parallax. This chapter focuses on observer-produced parallax in making the 3-D display. See footnote 2 for a brief discussion of object-produced parallax.

## 18.2 Early Accounts of Motion Parallax and Early and Recent Experimental Studies

Early historical accounts of motion parallax are detailed in Ono and Wade [2, 3] and, therefore, will not be elaborated here. To present a sample of the early writings we quote from Wheatstone [4]. Having invented the stereoscope, he was confronted with a question regarding depth perception by those who did not have binocular vision. He wrote:

> ... the same solid object is represented to the mind by different pairs of monocular pictures, according as they are placed at different distances before the eyes, and the perception of these differences (though we seem to be unconscious of them) may assist in suggesting to the mind the distance of the object. ... The mind associates with the idea of a solid object every different projection of it which experience has hitherto afforded; a single projection may be ambiguous, from its being one of the projections of a picture, or of a different solid object; but when different projections of the same objects are successively presented, they cannot all belong to another object, and the form to which they belong is completely characterized. [4, p. 377]

Although several accounts of motion parallax were given before the time of Wheatstone, the concept did not receive the concerted experimental attention accorded to binocular parallax (for a recent discussion of the surge of experiments using a stereoscope soon after its invention, see Ono et al. [5]). It was not until the turn of the last century that the first experiment on motion parallax was conducted.

Bourdon [6, 7] demonstrated that judgments of the separation between two stimuli, equated in visual angle at different distances from an observer, were made accurately when the head moved, but not when the head remained stationary. Soon after Bourdon, Heine [8] controlled the rate of retinal image motion (or eye movement) with respect to the head movement and successfully simulated two stationary rods with two moving rods (see Fig. 18.1). He mechanically yoked his lateral body movement to the movements of the rods using a shoulder harness: the closer rod physically moved in the same lateral direction as the head/body movement and the far one physically moved

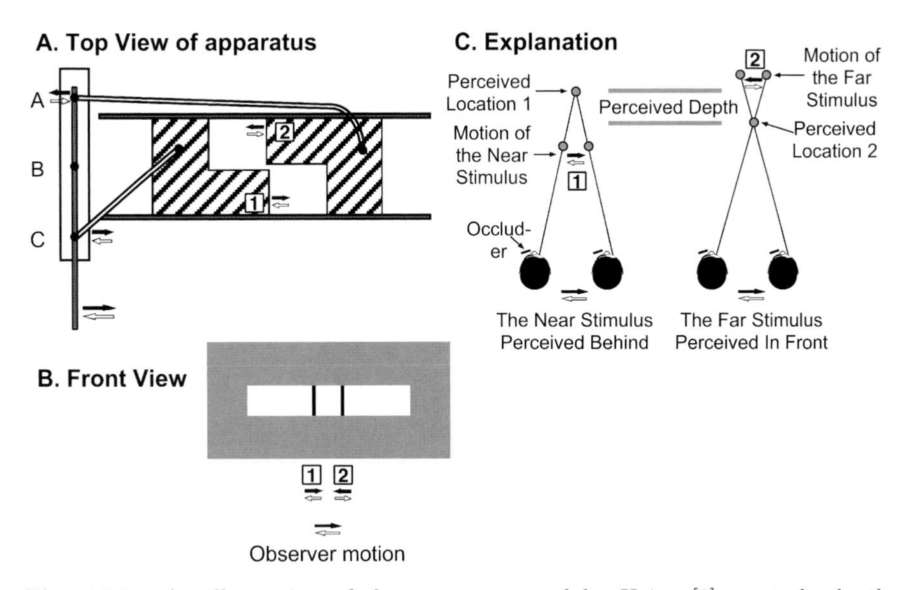


**Fig. 18.1.** An illustration of the apparatus used by Heine [8] to study depth perception from self-generated retinal motion (A and B) and the virtual reality created when an observer moves from side to side (C). The rod shown on the left side in Fig. 18.1A was attached to an observer's shoulder harness. His/her lateral movement moved the rod that pivots at B and produced the movement of Stimulus 1 and 2. Adapted from Ono and Wade [2]. See text for a discussion

in the opposite lateral direction. That is, the normal relationship between retinal image displacement and head movement was reversed. Under these conditions, the perception of depth was opposite to the actual depth; the near stimulus appeared farther than the far one, which provided convincing evidence that the retinal image motion produced by a head/body movement is a cue to depth perception. Moreover, he found that yoking or "slaving" stimulus movements to head movements leads to unambiguous depth perception. (By "unambiguous" we mean the direction of the depth perceptions is stable, unlike that of the kinetic depth effect.)

Modern counter parts of these two studies are González, et al. [9] and Rogers and Graham [10]. González, et al. used a procedure similar to that of Bourdon [6] and measured depth thresholds with and without head movement. The measured thresholds with head movement were considerably smaller than those without head movement. Rogers and Graham [10] developed a technique somewhat similar to that devised by Heine [8]. Their apparatus consisted of electronically yoking dot movements on a screen to lateral head movements. They successfully simulated different stationary surfaces (square, sine, triangular, or saw tooth) by yoking random dot movements to the observer's head

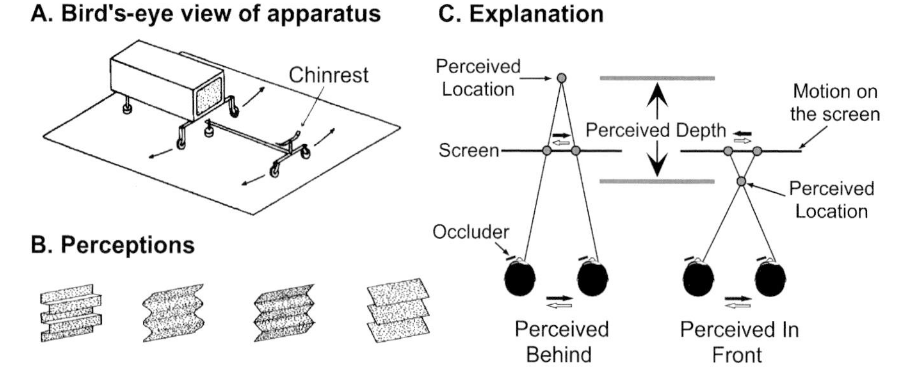


**Fig. 18.2.** An illustration of the apparatus used by Rogers and Graham [10] to study depth perception from self-generated retinal motion (18-2A) and the virtual reality created when an observer moved from side to side (18-2B and 18-2C). Adapted from Ono and Wade [16]. See text for a discussion

movement[2] (see Fig. 18.2). The perception produced was analogous to that produced by the random dot stereograms devised by Julesz [11]. As to the lower threshold of depth perception from motion parallax, Rogers and Graham's [12] data indicated that it is not as low as that of binocular parallax, but the shape of the thresholds as a function of spatial frequency of corrugation was similar. The lowest threshold was located at 0.3 cpd for both the motion and binocular parallaxes (see Fig. 18.3). Subsequently, using a similar technique, Ono and Ujike [13] showed that (a) small parallax magnitudes led to the perception of depth without motion; (b) larger magnitudes led to the perception of depth with concomitant motion—apparent motion that occurs in synchrony with a head movement, and (c) yet larger magnitudes led to no depth perception with a large concomitant motion (see Appendix for a definition of the magnitude of motion parallax and for what is called "equivalent disparity"). The last finding indicated that the range of effectiveness of motion parallax for depth perception is limited, just as the range of effectiveness of binocular parallax is limited. Also as with binocular parallax Nawrot [14] and Ono, et al. [15] found that the distance of the screen scales the magnitude of parallax; for a given parallax magnitude, the apparent depth was smaller when the distance of the screen was greater.

[2] Rogers and Graham [10] also produced these apparent surfaces by yoking the relative dot positions to the position of a moving screen (the object-produced parallax). See the arrow bars near the oscilloscope indicating the movement of the screen in Fig. 18.2. This finding suggests that a moving stimulus on a screen can be given an apparent depth by yoking parts of the stimulus with the position of it on the screen, i.e., a moving object can be made to have apparent depth in animation.

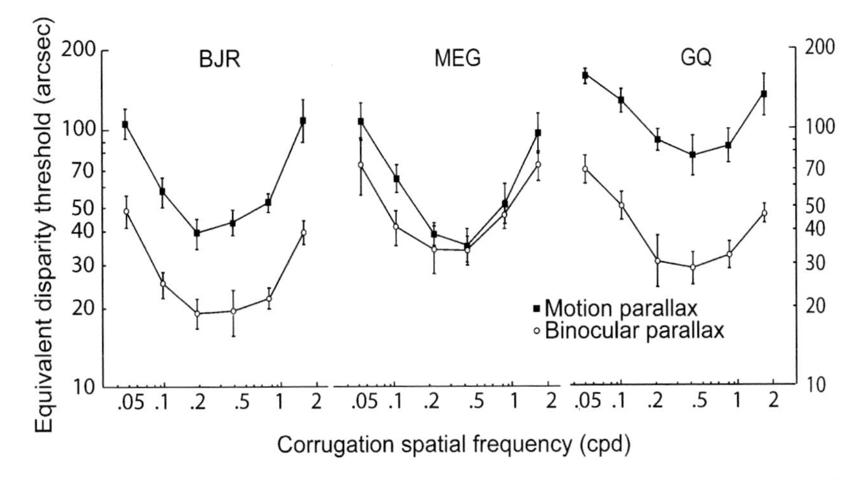


**Fig. 18.3.** The lower depth thresholds with motion parallax and binocular parallax as a function of corrugation frequency. Redrawn from Rogers and Graham [12]

Both the experimental techniques and the findings of Heine [8] and Rogers and Graham [10] are particularly relevant to the construction of a 3-D display system using motion parallax. They successfully produced "identical incoming messages," or what Ames called "equivalent configurations" from different external physical arrangements (see Ittelson [17]) and created a virtual reality of depth. Just as a stereoscope provides the "identical incoming messages" received by the two eyes from a "natural" object or scene, the apparatus of Heine and that of Rogers and Graham provided the identical incoming messages received by one eye from a natural object or scene. Note from Fig. 18.1 that whether the far (or near) simulated stationary point is fixated or the actually moving far (or near) stimulus is pursued, the extent of eye movement relative to head movement is identical. Also note that if there were two actual stationary rods located at "Perceived Location 1" and "Perceived Location 2" the incoming message would be the same as that produced by Heine's apparatus.

What has been discussed above lays the foundation upon which we base our proposed 3-D display system. Before we do that, however, readers are asked to view the demonstrations located at the Web site: *http://www.yorku.ca/hono/parallax˙demo*. The demonstrations were made for a classroom setting [16], but viewing them on a computer screen can produce the 3-D experience. Remember to use one eye only and to synchronize your head movement with the moving marker at the bottom of the demonstration.

## 18.3 Demonstration

The demonstration consists of two moving bars analogous to those used by Heine [8], or the moving dots that produced one of the four simulated surfaces analogous to those created by Rogers and Graham [10]. Moving along the

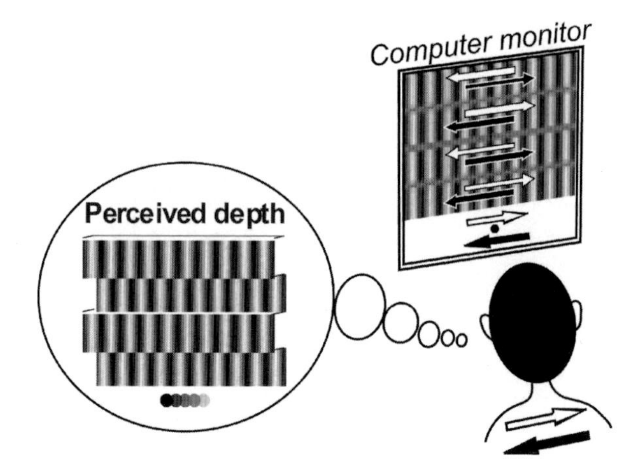


**Fig. 18.4.** An illustration of how the demonstrations are viewed with one eye

bottom of the screen is a marker used to guide the head movements (see Fig. 18.4). When you move your head from side to side in synchrony with the moving marker, the bars or the dots will move on your retina as though produced by your head movement. (The reason that these demonstrations work when viewed on the Web site is that the yoking of stimulus movements to head movements need not be as exact as in Heine's or Rogers and Graham's experiments.) The appropriate extent of head movement depends on how far you are from the screen.

For these demonstrations, the greater the magnitude of lateral head movement, the smaller the parallax magnitude, because the parallax magnitude is the ratio of retinal image displacement (or velocity) to the displacement (or velocity) of head movement. Therefore, if you are near the screen, you need to move your head by a greater extent than if you are farther from the screen to produce the same parallax magnitude. Moreover, if you perceive concomitant motion, you need to move your head a greater extent (or move farther from the screen) to see stationary bars or a stationary surface. Once you find the appropriate extent of head movement to see a stationary stimulus with depth, move your head in the opposite direction to the moving marker on the bottom of the screen; you will see a reversal of the direction of depth.

## 18.4 The Suggested 3-D Display System

To be completely faithful to Heine's [8] or Rogers and Graham's [10] idea, one can design a system where the lateral movement of the video camera is driven by the head movement. The signal from the camera can be presented on a screen in front of an observer while s/he keeps moving his/her head from side to side with one eye occluded; this arrangement eliminates the need for the moving marker on the bottom of the screen. The video signal obtained with the camera moving leftward and rightward provides the identical message as an observer viewing the scene while moving his/her head from side to side. However, this system is impractical for 3-D tele-presence or remote sensing purposes, since the video signal would be delayed relative to the head position. A delay of half a cycle of head movement would reverse the direction of depth, as the demonstration indicated. Therefore, for the first experimental 3-D display system, it is suggested that you include a moving marker to indicate the position of the camera.

The suggested 3-D display system using motion parallax is illustrated in Fig. 18.5. Figure 18.5A shows the recording system—the video camera moves

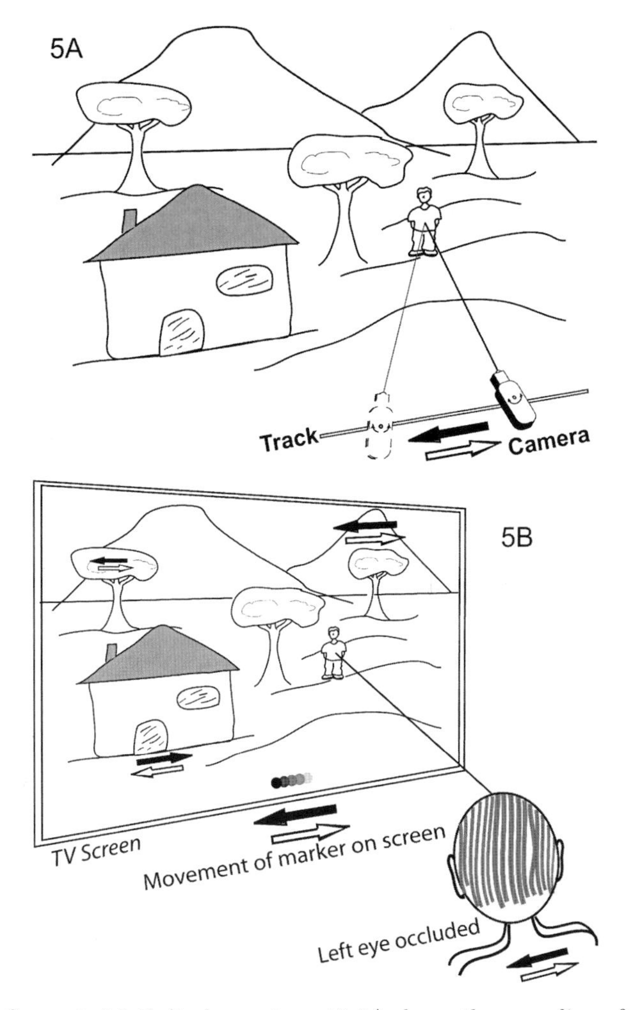


**Fig. 18.5.** Suggested 3-D display system. 18-5A shows the recording of a scene and 18-5B shows an observer viewing the scene on a television screen

from side to side on a rail while the camera is fixed on the person standing in a field. We suggest that the movement of the camera on the rail be controlled by a motorized crank system that produces reciprocating motion, but with a human operator, to select an object of interest and control the direction in which the camera is pointed. In Fig. 18.5A, the camera is pointed to the human standing in the field.

The screen for the display can be of any size; a large screen in a lecture hall as the demonstrations discussed in the previous section or a TV or computer screen. Figure 18.5B shows the display system using a relatively small TV screen. Whatever the screen size, an observer moves his head from side to side in synchrony with the moving marker on the screen. The figure also shows the movement of stimuli located at different parts of the screen; the point to which the camera was fixed remains stationary, whereas a part above (farther than) the stationary point moves in the same direction as the head and a part below (nearer than) this point moves in the opposite direction. The frequency of head movement matches that of the frequency of the reciprocating motion of the camera. The extent of head movement required by the observer depends on the extent of the excursion made by the camera, the viewing distance of the screen, and the reduction of the visual angles of the stimuli on the screen. It is recommended that these factors be combined so that the frequency and the amplitude of the required head movement will have a peak velocity greater than 15 cm/s: i.e., above $^1/_2$ Hz for 10 cm, and above 1/4 Hz above 20 cm excursion to ensure that the depth threshold is lowest with these head movements [18].

We expect that the motion parallax cue for depth produced by this system would combine with other depth cues such as the perspective cue, but how effectively it would combine remains to be determined. Also yet to be determined are the consequences of image parameters such as contrast ratio, resolution, and the color appearance of the display system. Nonetheless, this system would likely have advantages over a binocular 3-D display system in that it does not have to deal with (a) stereoblindness, (b) the visual fatigues usually associated with a binocular display system, and (c) two channels of video signals. Moreover, a 3-D experience can be created with present internet technology as seen in the demonstration. (Instead of lateral head movement, a horizontal rotation of the head can be used, as it would translate the viewing eye. See Steinbach, et al. [19]).[3]

## 18.5 Summary

Based upon what is known about motion parallax, a 3-D display system can be created. As a display for games or entertainment, having to keep moving the head from side to side may not be an appealing feature, but the suggested

[3] An up-and-down or a forward-and-backward head movement when yoked to appropriate retinal image motion is also effective in producing depth [19, 20, 21].

system may be useful as a tele-presence system. Creating the suggested experimental system would be useful for examining its cost effectiveness and exploring those stimulus situations that have an advantage over binocular display systems.

**Acknowledgements** Writing of this chapter was supported by a grant from the Natural Sciences and Engineering Research Council of Canada. We wish to thank Esther González, Linda Lillakas, Al Mapp, and Daniel Randles for their helpful comments on earlier versions of this chapter, and Linda Lillakas, for preparing the figures.

## Appendix

The geometry for specifying the magnitude of motion parallax is almost identical to that of binocular parallax, and vice versa. With this understanding, we now define the magnitude of motion parallax, which is the relative retinal image motion per head movement. If we were to specify the extent of retinal image motions in terms of visual angles (∝. and ß. in Fig. 18.6) (or velocities), the difference between the two divided by the extent (or velocity) of head movement is the parallax magnitude. If we were to compute (∝. – ß) when the head moves 6.2 cm, we would have a unit called "Equivalent Disparity" [12] that is equal to the unit of retinal disparity for the identical depth at a given distance.

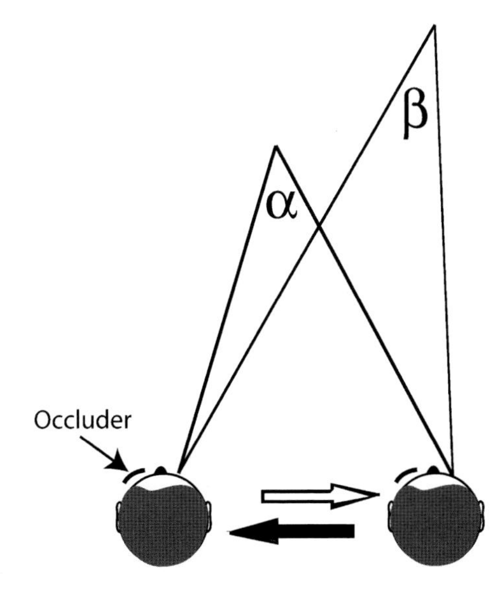


**Fig. 18.6.** An illustration for defining the magnitude of motion parallax

## References

[1] Howard IP, Rogers BJ (2002) Seeing in Depth: Vol. II. Depth perception. Toronto: Porteous Publisher

[2] Ono H, Wade NJ (2005) Depth and motion in historical descriptions of motion parallax. Percept 34:1263–1273

[3] Ono H, Wade NJ (2007) The paradoxes of parallaxes. Percept

[4] Wheatstone C (1838) Contributions to the physiology of vision – Part the first. On some remarkable, and hitherto unobserved, phenomena of binocular vision. Philosophical Transactions of the Royal Society 128:371–394

[5] Ono H, Lillakas L, Wade NJ (2007) Seeing Double and Depth with Wheatstone's Stereograms. Percept 36:1611–1623

[6] Bourdon B (1898) La perception monoculaire de la profondeur. Revue Philosophique 46:124–145

[7] Bourdon B (1902) La perception visuelle de lespace. Paris: Librairie Schleincher Freres

[8] Heine L (1905) Über Wahrnehmung und Vorstellung von Entfernungsunterschieden. Albrecht von Graefes Archiv für klinische und experimentelle Opthamologie 61:484–498

[9] González EG, Steinbach MJ, Ono H, Wolf ME (1989) Depth perception in children enucleated at an early age. Clinical Vision Sciences 4:173–177

[10] Rogers BJ, Graham ME (1979) Motion parallax as an independent cue for depth perception. Percept 8:125–134

[11] Julesz B (1971) Foundations of Cyclopean Perception. Chicago: University of Chicago Press

[12] Rogers BJ, Graham ME (1982) Similarities between motion parallax and stereopsis in human depth perception. Vision Research 22:261–270

[13] Ono H, Ujike H (2005) Motion parallax driven by head movement: conditions for visual stability, perceived depth, and perceived concomitant motion. Percept 34:477–490

[14] Nawrot M (2003) Depth from motion parallax scales with eye movement gain. Journal of Vision 3:841–851

[15] Ono ME, Rivest J, Ono H (1986). Depth perception as a function of motion parallax and absolute-distance information. Journal of Experimental Psychology: Human Perception and Performance 12:331–337

[16] Ono H, Wade NJ (2006) Depth and motion perceptions produced by motion parallax. Teaching of Psychology 33:199–202

[17] Ittelson WH (1960) Visual Space Perception. New York: Springer

[18] Ujike H, Ono H (2001) Depth thresholds of motion parallax as a function of head movement velocity. Vision Research 41:2835–2843

[19] Steinbach MJ, Ono H, Wolf M (1991) Motion parallax judgments of depth as a function of the direction and type of head movement. Canadian Journal of Psychology 45:92–98

[20] Sakurai K, Ono H (2000) Depth perception, motion perception and their tradeoff while viewing stimulus motion yoked to head movement. Japanese Psychological Research 42:230–236

[21] Yajima T, Ujike H, Uchikawa K (1998) Apparent depth with retinal image motion of expansion and contraction yoked to head movement. Percept 27: 937–949

# 19

# Dynamic Three-Dimensional Human Model

Y. Iwadate

## 19.1 Introduction

Three-dimensional (3-D) content is widely used in movies and video games, thanks to advances in computer graphics, motion capture, 3-D scanners, and so on. However, producing high quality 3-D content is expensive and involves many different technologies. The dynamic 3-D model discussed in this chapter contains shape information about a moving subject. Thus, it can express the motions of humans in detail, including movements of clothes. It is useful as a component in video synthesis, because of its flexibility in viewpoint and view direction. In addition, it can be transformed into stereoscopic images such as binocular, lenticular, integral photographic, etc.

To generate a dynamic 3-D model, 10 or more cameras are set up in a dome surrounding a subject to synchronously take video images. The model is calculated frame by frame from these video images. There are many choices from which to choose a shape restoration algorithm. Of these, the popular volume-intersection method is steady and has low calculation cost [1]. However, it cannot produce a concave surface of a subject. To make up for this fault, various methods such as stereo matching [2, 3], voxel coloring [4, 5], photo hull [6], deformable mesh [7], etc., have been proposed.

The quality of the dynamic 3-D model depends not only on the accuracy of its shape, but also the quality of texture mapping. When multiple cameras

Y. Iwadate
NHK (Japan Broadcasting Corporation) Science & Technical Research Laboratories, 1-10-11, Kinuta, Setagaya-ku, Tokyo 157-8510, Japan
e-mail: iwadate.y-ja@nhk.or.jp

B. Javidi et al. (eds.), *Three-Dimensional Imaging, Visualization, and Display*,
DOI 10.1007/978-0-387-79335-1_19, © Springer Science+Business Media, LLC 2009 

are used, the key point is deciding which camera image should be mapped on a vertex forming the surface of the model. Various techniques have been proposed for this; for instance, the best camera can be chosen according to the virtual view position [8, 9, 10].

At the NHK Science and Technical Research Laboratories, we attempted to apply dynamic 3-D modeling to archive intangible cultural assets such as traditional dramatic performances [11, 12]. This chapter explains the generation method and the display method of such dynamic 3-D models.

## 19.2 Outline of 3-D Modeling

The techniques listed below are used to generate dynamic 3-D models of performers, such as dancers, and display fine arbitrary viewpoint images of their performances.

(1) **Synchronously Capture Multiple Videos of a Performer**
A dynamic 3-D model of a performer is generated for each video frame. Thus, the video frames of the multiple videos must be completely synchronized. High-resolution CCD cameras are thus needed to generate fine arbitrary viewpoint images.

(2) **Calibration of Multiple Cameras**
The transformation between the world coordinates and the camera coordinates must be calculated. For this, we need to know the position, orientation, and focal length of each camera. This information is compiled by using a camera calibration technique.

(3) **3-D-modeling from Multiple Videos**
The shapes of the 3-D models must be accurate and stable. As mentioned later, our modeling method combines volume intersection and stereo matching of multiple videos.

(4) **Texture Mapping**
The arbitrary viewpoint images must be able to reproduce the fine surface texture of the performer's costume. To make this possible, we use a texture mapping method that not only produces a high-resolution surface whose texture depends on the viewpoint, but also enables real-time and extended continuous display.

## 19.3 Synchronous Capture of Images from Multiple Video Cameras

To capture the videos of a performer, we use a camera dome 2.5 m high and 8.0 m in diameter. Figure 19.1 shows the dome's appearance. The dome is covered with blue walls for chroma-key processing. Forty FireWire cameras are set up around the dome facing its center. Ten cameras are set on the

**Fig. 19.1.** Camera dome

ceiling, 24 at the upper level of the wall, and six at the middle level, as shown in Fig. 19.2.

FireWire color cameras with XGA resolution (1,024 x 768) are used for synchronous shooting of video at a rate of 10 frame/s. As shown in Fig. 19.3, each camera is connected to a capture PC via FireWire cable. The cameras

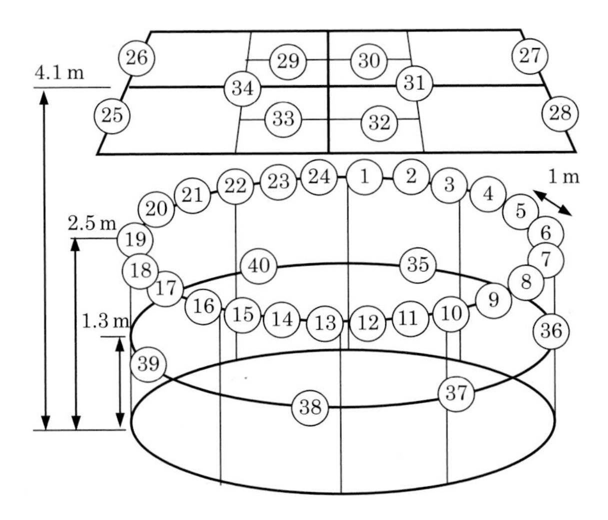


**Fig. 19.2.** Camera setup

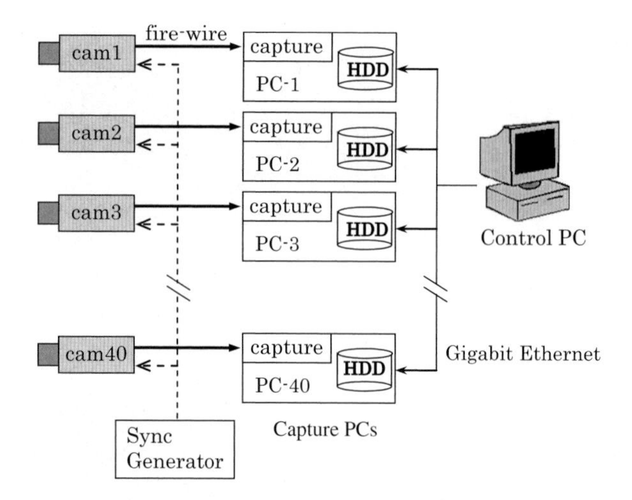


**Fig. 19.3.** Capture system

and capture PCs are controlled by a control PC via a Gigabit Ether network. The control PC issues commands to the capture PCs, when video capture starts and stops. The control PC can also display multiple videos and correct the color parameters of all cameras.

## 19.4 Camera Calibration

Generating a dynamic 3-D model from multiple videos requires the parameters of all the cameras. These parameters include the position in 3-D space, orientation, and focal length. The calibration pattern shown in Fig. 19.4 can

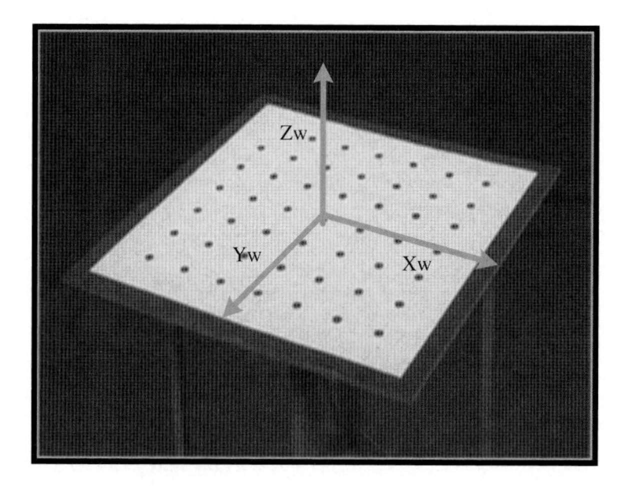


**Fig. 19.4.** Calibration pattern

be used to obtain these parameters. Specifically, each camera takes an image of this pattern on which world coordinates (3-D coordinates in real space) have been set, and the Tsai's technique [13] is applied to that image to calculate the parameters.

Given that a point $\boldsymbol{P_W} = (\boldsymbol{X_W}, \boldsymbol{Y_W}, \boldsymbol{Z_W})^T$ in 3-D space expressed in world coordinates is projected onto a coordinate point $\boldsymbol{P_g} = (\boldsymbol{x_g}, \boldsymbol{y_g})^T$ on the camera's image, the transformation between these coordinate systems is as follows.

$$\omega \cdot \begin{bmatrix} \boldsymbol{P_g} \\ 1 \end{bmatrix} = \boldsymbol{A} \cdot [\boldsymbol{R} \quad \boldsymbol{T}] \cdot \begin{bmatrix} \boldsymbol{P_W} \\ 1 \end{bmatrix} \tag{19.1}$$

$\boldsymbol{R}$: Rotation matrix (3×3)
$\boldsymbol{T}$: Translation vector (3×1)
$\boldsymbol{A}$: Intrinsic camera parameter matrix
$\boldsymbol{\omega}$ : Distance from camera to $P_W$

$$\boldsymbol{A} = \begin{bmatrix} \boldsymbol{aF} & \boldsymbol{s} & \boldsymbol{C_x} \\ 0 & \boldsymbol{F} & \boldsymbol{C_y} \\ 0 & 0 & 1 \end{bmatrix} \tag{19.2}$$

$\boldsymbol{a}$ : Aspect ratio
$\boldsymbol{s}$ : Element of lens distortion
$\boldsymbol{F}$ : Effective focal length of the pinehole camera
$(\boldsymbol{C_x}, \boldsymbol{C_y})$ : Intersection point of the camera image plane and optical axis

## 19.5 3-D Modeling

A dynamic 3-D model is created from the performer's 3-D models generated from each video frame. To generate a more accurate and stable 3-D model of a performer, our method combines the volume intersection with stereo matching. The stereo matching method is used to modify the approximate 3-D model obtained from the volume intersection method.

### 19.5.1 Volume Intersection Method

Figure 19.5 illustrates the basic principle of the volume intersection method, which uses silhouettes obtained by extracting the subject's area from images through, for example, chroma key processing. For each camera, the silhouette can be inversely projected onto world coordinates to form a visual cone in 3-D space in which the camera's optical center is the vertex and the silhouette is a cross section of that cone. Finding the visual cone for each camera in this

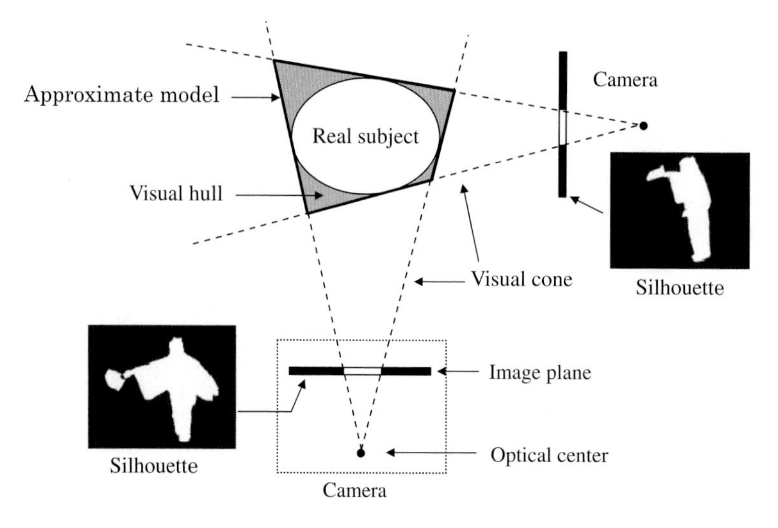


**Fig. 19.5.** Volume-intersection method

way makes it possible to determine the common area of these visual cones as a visual hull that represents the approximate shape of the subject.

To be more specific, the 3-D space where the subject is assumed to exist is divided into voxels, each of which can be projected onto the image plane of a camera using Eq. (19.1). Now, if a projected point falls within the silhouette for that camera, that voxel is considered to be included within the visual cone. The visual hull is then obtained from the set of voxels included in the visual cones of all cameras. The discrete marching cubes method [14] can now be applied to this visual hull to generate the subject's 3-D model having a surface consisting of polygons.

The volume-intersection method can reconstruct a stable 3-D shape of a subject at low computational cost. The generated model, however, is only approximate model and circumscribes the real shape of the subject. This makes it difficult to accurately reconstruct curved and other fine shapes on the subject's surface.

### 19.5.2 Modification of the 3-D Shape by Using Stereo Matching

As explained in the previous section, the volume intersection method cannot reconstruct the surface of the human body in detail. However, as the approximate model has a shape that circumscribes the real surface of the human body, it can be assumed that the real surface lies just underneath the surface of the approximate model. This property is used by the stereo matching method to modify the approximate shape.

To modify the approximate model beforehand we need the initial depth image, which consists of the distance from the camera to the surface of the approximate model. By modifying the initial depth image of each camera by stereo matching, an accurate and stable 3-D shape can be obtained. To begin with, we explain the technique to get the initial depth image.

## Initial Depth Image

An initial depth image consists of the distance from a camera to the surface of the approximate model. This distance is calculated to project the surface vertices of an approximate model onto the camera image. Note that if vertices on the reverse (occluded) side of the approximate model not captured by the camera image were to be projected, the correct initial depth could not be determined. For this reason, the visibility of surface vertices on the approximate model with respect to each camera must first be determined.

We devised a low cost technique to determine vertex visibility. Figure 19.6 shows an example of this technique. We shift a surface vertex on the approximate model slightly toward cam1 along the optical line. If, after shifting, the vertex falls inside the approximate model, it can be judged to be an occluded vertex from cam1 and, if outside, a visible vertex from cam1.

The initial depth image can be obtained by projecting only visible vertices to the camera by using Eq. (19.1). In this way, initial depth images for all cameras can be obtained as shown in Fig. 19.7.

## Stereo Matching Method

The stereo matching method is used to correct the initial depth image and obtain a more accurate depth image from the camera to the subject. The stereo matching is carried out with every pair of adjacent cameras set at the upper level of the wall, for instance a pair consisting of camera 12 and 13 in Fig. 19.2. One camera is used as a base camera from which the depth is

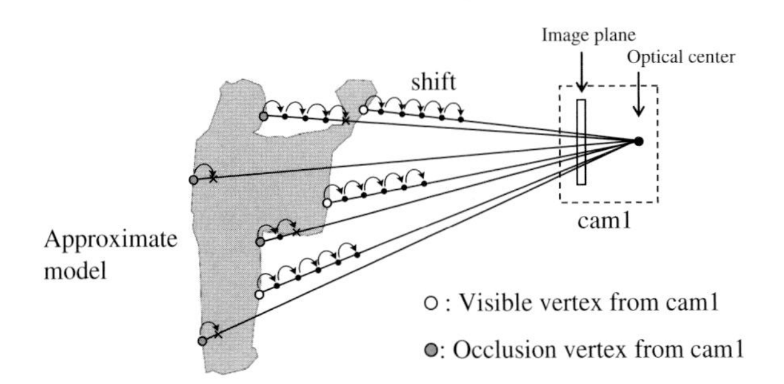


**Fig. 19.6.** Visibility of vertices

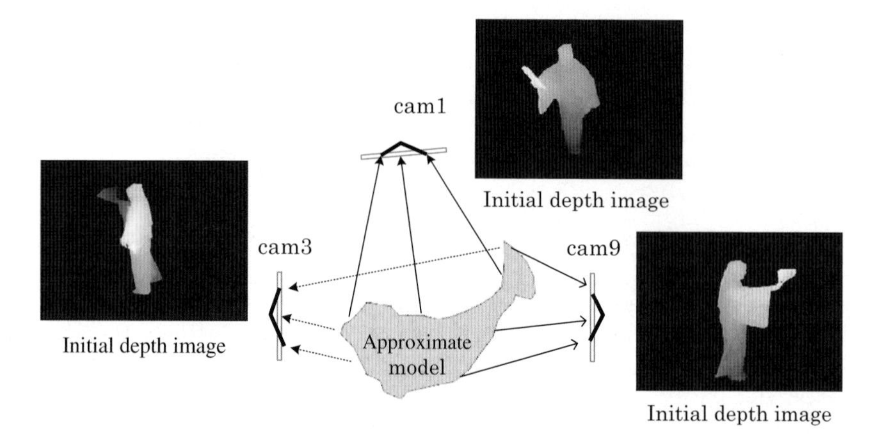


**Fig. 19.7.** Initial depth images

defined. Another is used as a reference camera. The following summarizes the procedure with reference to Fig. 19.8. The depth $\boldsymbol{\omega}$ from a target pixel on the base camera image to the subject is treated as a parameter, and the corresponding pixel on the reference camera is searched for while varying $\boldsymbol{\omega}$, starting from the initial depth and moving further inside the approximate model. Here, the initial depth is the value of the target pixel in the initial depth image of the base camera. The corresponding detection procedure is described below.

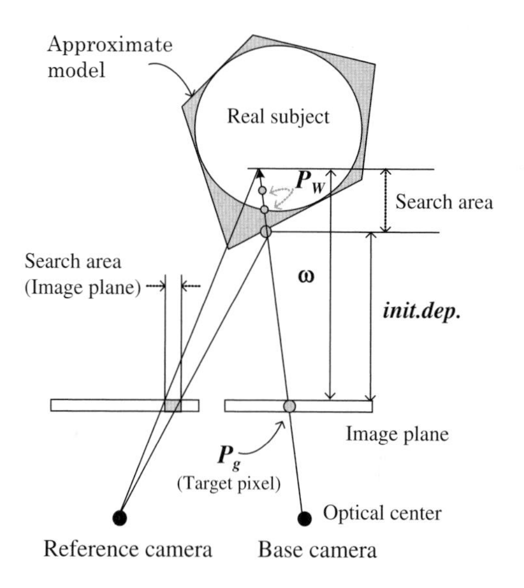


**Fig. 19.8.** Stereo matching method

(1) Define a target pixel $\boldsymbol{P}_g$ on the base camera's image.
(2) Assign a depth $\boldsymbol{\omega}$ (initial value: $\boldsymbol{\omega}$ = initial depth) to $\boldsymbol{P}_g$ and determine the reverse projection point $\boldsymbol{P}_W$ using Eq. (19.3).

$$\boldsymbol{P}_W = \omega \cdot \boldsymbol{R}^{-1} \cdot \boldsymbol{A}^{-1} \cdot \begin{bmatrix} \boldsymbol{P}_g \\ 1 \end{bmatrix} - \boldsymbol{R}^{-1}\boldsymbol{T} \tag{19.3}$$

(3) Project $\boldsymbol{P}_W$ onto the reference camera and calculate the corresponding pixel using Eq. (19.1).
(4) Determine the correlation between the target pixel and the corresponding pixel using the evaluation function of Eq. (19.4).

$$\boldsymbol{e}(\omega) = \boldsymbol{ssd}(\omega) - \boldsymbol{k}(\omega - \boldsymbol{init.dep.}) \tag{19.4}$$

$\boldsymbol{\omega}$ : Depth from the base camera to the human body
$\boldsymbol{init.dep}$ : Initial depth
$\boldsymbol{ssd}$ : Sum of the square differences of RGB
$\boldsymbol{k}$ : Experimentally obtained weighting coefficient

(5) Repeat steps (2) to (4) while incrementing $\boldsymbol{\omega}$ from the initial depth towards the inside of the approximate model. The value of $\boldsymbol{\omega}$ that minimizes Eq. (19.4) is taken to be the depth from the target pixel $\boldsymbol{P}_g$ to the subject.

The second term on the right side of Eq. (19.4) adds a weight corresponding to the distance from initial depth. This weight forces $\boldsymbol{\omega}$ to minimize Eq. (19.4) in the neighborhood of a surface of the approximate model. Therefore, the shape can be maintained even if few texture patterns exist on the subject's surface and accurate correspondence detection is difficult.

Performing this process on all pixels within the subject's area on the base camera's image gives a more accurate depth image from the camera to the subject.

### Integration of Depth Images

A 3-D model of the subject can be generated by integrating the depth images obtained by stereo matching from a set number of cameras. Here, we again divide the 3-D space in which the subject is assumed to exist into voxels and arrange the depth images for those cameras within this voxel space. We also extract the set of voxels that exist inside all depth images as a solid model. Specifically, a voxel in 3-D space that satisfies the following conditions is considered to be included in the solid model.

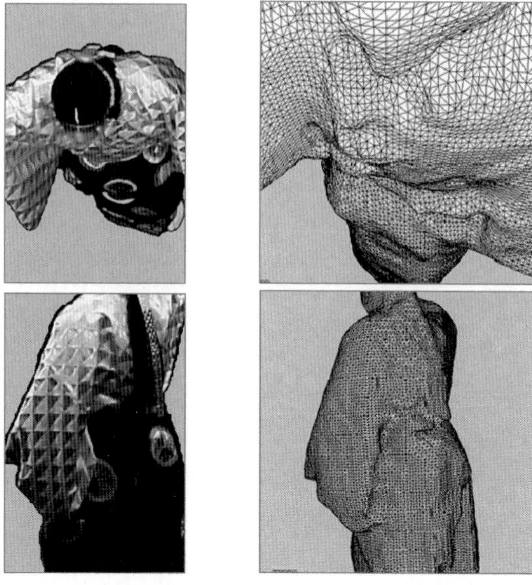

(a) Original video images. (b) Wireframe of the 3-D model.

**Fig. 19.9.** Example of 3-D model

Condition 1: Projected point on the depth image is located inside the silhouette

Condition 2: $dist_n(X_W, Y_W, Z_W) > depth_n(x_g, y_g)$
Here,

$\mathbf{dist}_n(X_W, Y_W, Z_W)$: Distance from camera (n) to a voxel
$\mathbf{depth}_n(x_g, y_g)$: Depth value of $(x_g, y_g)$ pixel on camera (n)

Finally, we apply the discrete marching cubes method [14] to the solid model and generate the subject's 3-D model with a surface consisting of polygons. This 3-D model consists of the 3-D coordinates of surface vertices and polygon indexes. Generating this 3-D model every video frame creates a dynamic 3-D model of a performer. Figure 19.9. shows a 3-D model generated from this method.

## 19.6 Texture Mapping

The surface texture of this model is very important to maintain the quality of images displayed from an arbitrary viewpoint. Below we describe a texture mapping method that can reproduce a high-resolution surface texture like

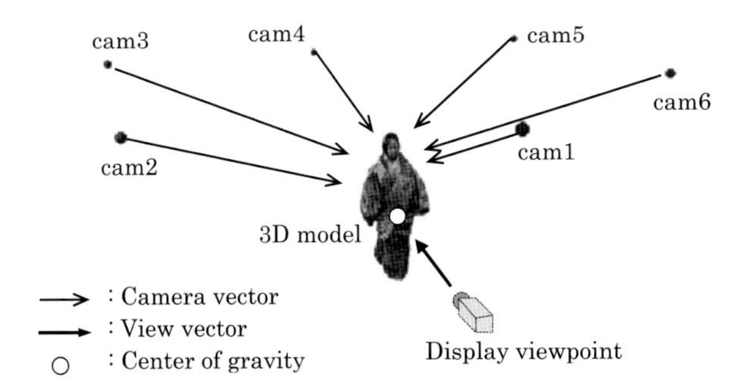


**Fig. 19.10.** Choice of texture camera

that of the original captured image. This method is suitable for real-time and continuous display.

Texture images mapped to the model's surface are chosen dependent on the viewpoint. Figure 19.10 shows the technique of choosing texture images from all camera images. To begin with, we determine the view vector that connects the center of gravity of the 3-D model with the display viewpoint. Then, from all cameras that are used to capture a performer, we choose a few texture cameras starting with the one whose camera vector has the largest inner product with the view vector. The camera vectors indicate the shooting direction of each camera and are given by the camera calibration. The images captured by these texture cameras are chosen as the texture images.

Before mapping the texture image to the surface of the 3-D model, it's possible that an erroneous mapping texture to an occluded polygon may not be captured by the texture image. We again use the technique described in Fig. 19.6 to determine the status of each polygon on the 3-D model surface beforehand with respect to each camera image, i.e., whether it is visible or occluded. In this case, the judgment as to whether the vertex after shifting falls inside or outside of the 3-D model is done by applying Condition 1 and Condition 2 in Section 19.5.2. As described in Fig. 19.11, we define that if the three vertices of a polygon are visible from the camera, the polygon is visible from the camera.

The texture mapping to the model's surface starts with the camera image having the camera vector with the largest inner product with the view vector, and continues with the image having the next smallest inner product and so on. We draw only visible polygons from the texture camera and give texture to these polygons. Figure 19.12 shows an example of texture mapping. The texture coordinates are automatically calculated using the camera parameters of the texture camera and the OpenGL function.

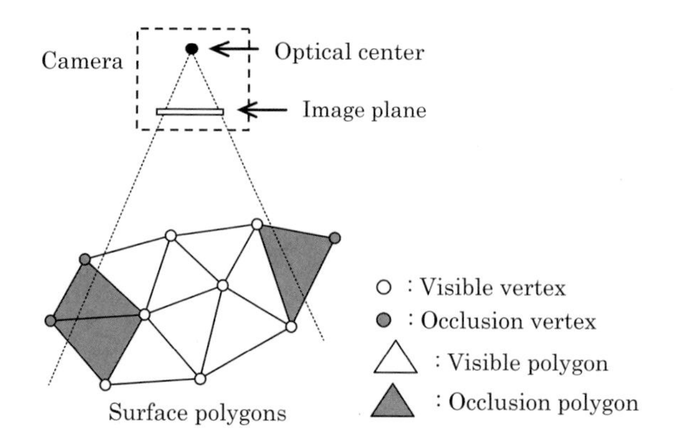


**Fig. 19.11.** Visibility of polygons

(a) Texture mapping from the camera with the largest inner product.

(b) Adding texture from the camera with the second largest inner product.

(c) Final result. Adding texture from the camera with the third largest inner product.

**Fig. 19.12.** Results of texture mapping

## 19.7 Real-Time and Continuous Display System

This texture mapping method requires large amounts of texture data and recalculation of texture cameras according to the current viewpoint. Therefore, to achieve a real-time and continuous display capability, we developed a new display system. Our system uses the following data.

### Dynamic 3-D Model

The dynamic 3-D model was created using the technique described in section 19.5.

### Texture Images

The texture images are the original captured camera images. These images must be prepared for every video frame. The original camera image has XGA resolution, so if we used the whole image, the data size of the texture images

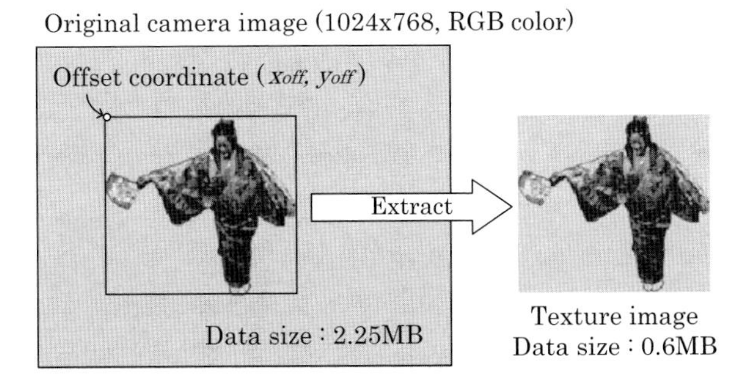


**Fig. 19.13.** Texture image

would become very large, which in turn would make real-time and continuous display difficult. However, we only need to use silhouette areas of the image for texture mapping. Thus, we can extract the smallest rectangular area, which includes the silhouette from the camera image, and keep the smallest image as the texture images (Fig. 19.13). These small texture images include offset image coordinates that are used for texture mapping. This process reduces the texture data size to about 20 or 30 percent of the camera image' s original data size.

### Information on Polygon Visibility

Whether the polygon of the 3-D model is visible or occluded can be calculated using the technique described in Section 19.6. The display system requires this information for each frame for all cameras.

### Camera Parameters

The camera parameters are used to calculate texture coordinates.

Figure 19.14 shows an outline of the display system. According to the current viewpoint indicated by the mouse pointer, the texture cameras are chosen using the technique described in Section 19.6. The texture images of the current frame are loaded from the HDD. The workload is light enough for real-time display because of the small texture images and the use of a RAID HDD. Furthermore, only the texture images of the current frame are loaded into the main memory so that our system will be capable of extended and continuous display. Next, the current frame 3-D model is drawn and the texture image is mapped to the model's surface. After this, high-resolution arbitrary viewpoint images of a performer can be displayed.

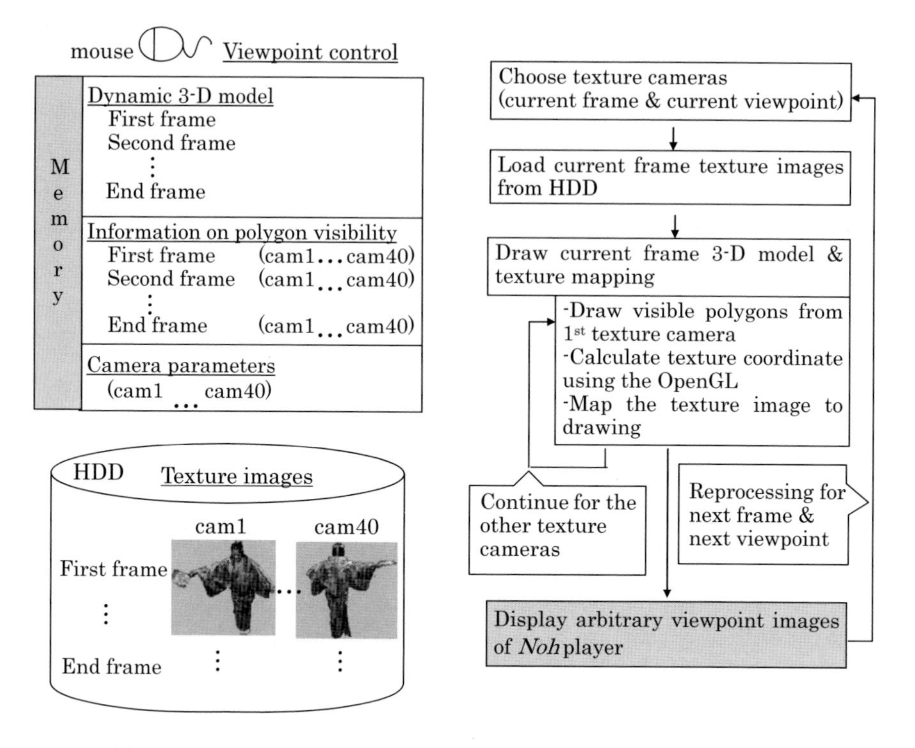


(a) Data allocation. (b) Display procedure.

**Fig. 19.14.** Outline of display system

## 19.8 3-D Video System for Archiving Japanese Traditional Performing Arts

Using the technique described above, we performed an experiment to generate a dynamic 3-D model and display arbitrary viewpoint images of a *Noh* performance [15]. *Noh* is one of Japan's traditional dramatic arts. Our subject was a *Noh* performer. We captured multiple videos of his performance using 40 FireWire cameras for three minutes. We then generated a dynamic 3-D model of the performance. The dynamic 3-D model consisted of 1,800 frames at 10 frame/sec. The 3-D model for each video frame had about 50,000 surface polygons and its data size was about 1.0 MB/frame.

The dynamic 3-D model can be synthesized with CGs. We developed an interactive and real-time synthesizing system. An outline of the system is shown in Fig. 19.15. Although the dynamic 3-D model is the main part of the virtual scene, the background, sound, and real-time user control help to give more presence to a scene. In this system, the user can control the viewpoint in the virtualized *Noh* scene. Figure 19.16 shows synthesized images.

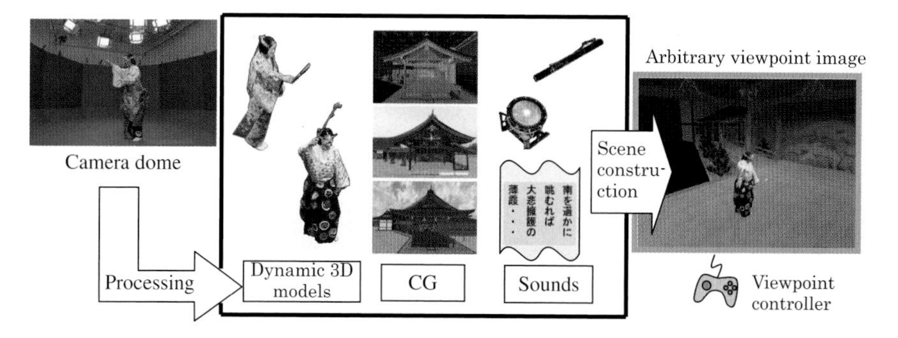


**Fig. 19.15.** Virtualized *Noh* scene generation system

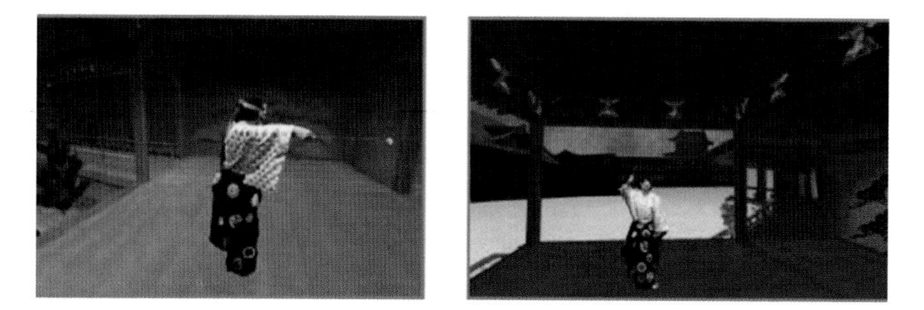

**Fig. 19.16.** Synthesizing a dynamic 3-D model with *Noh* stage CG. (Courtesy of "Kanze Kyuukoukai")

We confirmed that our methods could achieve real-time continuous display of high-resolution arbitrary viewpoint images of a *Noh* performance.

## 19.9 Conclusion

We described a method for generating a dynamic 3-D model and a texture mapping method that is dependent on the viewpoint. Using these techniques, we developed a display system that can display arbitrary viewpoint images of performances. This system is capable of real-time and continuous display of performances. An experiment that generates a dynamic 3-D model and displays arbitrary viewpoint images of *Noh*, a Japanese traditional dramatic art, confirmed that the system could display high-resolution images from arbitrary viewpoints.

**Acknowledgments** The support of the Ministry of Education, Culture and Sports, Science and Technology Japan is gratefully acknowledged.

## References

[1] A. Laurentini: "The Visual Hull Concept for Silhouette-based Image Understanding," *IEEE Trans. on PAMI*, **16**, 2, 150–162 (1994)
[2] P. Rander: "A Multi-Camera Method for 3D Digitization of Dynamic Real-World Events," *CMU-RI-TR*-9812 (1998)
[3] Y. Iwadate, M. Katayama, K. Tomiyama, and H. Imaizumi: "VRML Animation from Multi-Viewpoint Images," *Proc. of ICME2002,* **1**, 881–884 (2002)
[4] S.M. Seitz and C.R. Dyer: "Photorealistic Scene Reconstruction by Voxel Coloring," *Int. J. Computer Vis.*, **35**, 2, 1–23 (1999)
[5] W. Matusik, C. Buehler, R. Rasker, S.J. Gortler, and L. McMillan: "Image-Based Visual Hulls," *In Proc. of SIGGRAPH2000*,11–20 (2000)
[6] G. Slabaugh, R.W. Schafer, and M.C. Hans: "Image-Based Photo Hulls," *1st International Symposium on 3D Processing Visualization and transmission*, 704–708 (2002)
[7] S. Nobuhara and T. Matsuyama: "Deformable Mesh Model for Complex Multi-Object 3D Motion Estimation from Multi-Viewpoint Video," *3DPVT2006* (2006)
[8] P.E. Debvec, C.J. Taylor, and J. Malik: "Modeling and Rendering Architecture from Photographs: A Hybrid Geometry- and Image-Based Approach," *Proc. SIGGRAPH1996*, 11–20 (1996)
[9] K. Pulli, M. Cohen, T. Duchamp, H. Hoppe, L. Shapiro, and W. Stuetzle: "View-based Rendering: Visualizing Real Objects from Scanned Range and Color Data," *Proc. of 8th Eurographics Workshop on Rendering*, 23–34 (1997)
[10] P.E. Debevec, G. Borshukov, and Y. Yu: "Efficient View-Dependent Image-Based Rendering with Projective Texture-Mapping," *Proc. 9th Eurographics Workshop on Rendering*, 105–116 (1998)
[11] K. Tomiyama, M. Katayama, Y. Orihara, and Y. Iwadate: "Arbitrary View-point Images for Performances of Japanese Traditional Art," *The 2nd European Conference on Visual Production*, 68–75 (2005)
[12] M. Katayama, K. Tomiyama, Y. Orihara, and Y. Iwadate: "Archiving Performances of Japanese Traditional Dramatic Arts with a Dynamic 3D Model," *Proc. EVA 2006 Vienna*, 51–58 (2006)
[13] R. Y. Tsai: "A versatile Camera Calibration Technique for High-Accuracy 3D Machine Vision Metrology Using Off-the-Shelf TV Cameras and Lenses," *IEEE Journal of Robotics and Automation*, Vol. RA-3, No.4, 323–344 (1987)
[14] Y. Kenmochi, K. Kotani, and A. Imiya: "Marching Cubes Method with Connectivity ," *ICIP-99*, 361–365 (1999)
[15] The Japan Arts Council: "An Introduction to the World of Noh & kyogen,"http://www2.ntj.jac.go.jp/unesco/noh/en/index.html.

# 20

# Electronic Holography for Real Objects Using Integral Photography

Tomoyuki Mishina, Jun Arai, Makoto Okui, and Fumio Okano

## 20.1 Introduction

When we look at a thing ("object"), the beams that are reflected from the object ("object beams") enter our eyes, forming an image of the object on the retina through the eye lenses. This means that even if the object does not actually exist, we can see it when the same beams as the object beams that are produced by some means enter our eyes. Holography, which was invented by D. Gabor in 1948, is a technology that utilizes the diffraction and interference of light to record and reproduce the wavefronts of light [1]. By using this technology, it is possible to reproduce recorded object beams anytime. Thus, even when the object is absent, the technology permits reproducing the same object beams as when the object was present, thereby forming a virtual image of the object at the exact position where the object was situated. This virtual image is one formed in a space. Therefore, we can see the object as a 3-D image without wearing special glasses. We can also examine the object from various aspects by changing our viewing position. The virtual image is an ideal 3-D image that allows the convergence and the accommodation to work as if we are observing a natural image. The original holography employed the light of a mercury lamp. Because of the nature of the mercury lamp light, however, the images obtained were not very good. The laser that was invented in 1960 [2] dramatically improved the quality of reproduced

T. Mishina
NHK Science & Technical Research Laboratories, 1-10-11, Kinuta, Setagaya-ku, Tokyo 157-8510, Japan; National Institute of Information and Communication Technology 4-2-1, Nukui-Kitamachi, Koganei, Tokyo 184-8795, Japan, 4-2-1, Nukui-Kitamachi, Koganei, Tokyo 184-8795, Japan
e-mail: mishina@nict.go.jp, mishina.t-iy@nhk.or.jp

B. Javidi et al. (eds.), *Three-Dimensional Imaging, Visualization, and Display*,
DOI 10.1007/978-0-387-79335-1_20, 

images. Since then, holography has been in the limelight as an ideal 3-D imaging technology.

In holography, to obtain a reconstructed image which offers a wide viewing zone (i.e., an image which can be observed from various directions), it is necessary to record in a hologram the object beams that diffuse widely. The information contained in the object beam is recorded in a fringe pattern which is produced by interference between the object beam and reference beam. Since the line interval of the fringe pattern is inversely proportional to the angle formed by the object beam and reference beam, it becomes extremely narrow (1 $\mu$m or less) when the direction of propagation of the object beam forms a wide angle against the direction of propagation of the reference beam. Recording or displaying such a fringe pattern requires a recording (or display) surface which has a high resolution of 2,000 lines or more per mm. At present, light-sensitive materials (photographic plates and films) meet the above requirement [3]. Since photographic plates can be increased in size without sacrificing their resolution and are easy to handle, they are almost exclusively used to record fringe patterns. Because of the characteristics of photographic plates, most of the practical holographic methods display still images. However, since holography is an ideal 3-D image displaying method that reproduces the conditions of light under which things are perceived in the natural world, the reproduction of moving images has long been studied, too [4, 5]. To reproduce moving holographic images, it is necessary to switch the holograms one after another at a high speed. Reproducing a moving holographic image is possible by using an electronic display device (i.e., LCD panel) for displaying holograms, transferring holograms as video data to the device one after another and rewriting the holograms that sequentially appear on the display surface. Electronic holography uses such an electronic method. In 1990, a group of researchers led by Prof. Benton at the Massachusetts Institute of Technology (MIT) came up with a new method of moving holography employing an acousto-optical modulator whose refractive index can be varied by an electrical signal [6]. Since then, electronic holography has been actively studied.

In electronic holography, holograms (fringe patterns) are input as video signals to an electronic display device. It is, therefore, necessary to obtain the holograms in the form of digital data. One method of obtaining such hologram data is to capture fringe patterns as video signals by using an electronic image pickup device (i.e., CCD), in place of a photographic plate, at the hologram surface in the ordinary hologram generation process [7, 8]. The advantage of this method is that it permits recording object beams as fringe patterns on a real-time basis. The method may be suitable for high quality recording of moving objects. According to Hashimoto, et al. [7], they recorded holograms using a CCD (pixel interval 11.4 $\mu$m $\times$ 13.3 $\mu$m, number of pixels 768 (H) $\times$ 490 (V)) and reproduced the recorded hologram data on a real-time basis on an LCD panel (pixel size 30 $\mu$m (V) $\times$ 60 $\mu$m (H), number of pixels 648 (V) $\times$ 240 (H)). Reportedly, although spatial images about 1 cm in height were reproduced in their experiment, the viewing zone was almost zero. In direct recording, the high limit of spatial frequency of fringe patterns and the size (number of pixels) of holograms that can be recorded are determined, respectively, by the resolution and the number of pixels of the image pickup device used. With the performance of any of the existing image pickup devices, it is difficult to generate large holograms accurately.

Any method of directly recording holograms requires a darkroom and a laser for holography and, hence, it sets various limitations on the objects. For example, landscapes and people are unsuitable objects. As a method of generating holograms without using a laser, the using information about a given object to generate holograms of the object by calculations has been studied. An example of this is computer-generated holograms (CGHs). CGHs are holograms obtained by calculations based on the positional information about a given object. The advantage of this method is that since no optical photographic equipment is used and all holograms are obtained by calculations, it is possible to generate any holograms by freely setting the pixel interval, the number of pixels and other parameters. It is also possible to perform those types of optical processing which can hardly be done with other methods, such as limiting the scope of diffusion of the object beams. However, since it is difficult to obtain the positional information about a real object, the objects to which this method is applied at present are mostly those in computer graphics.

To obtain holograms of a real object without using a laser, there is a method which uses a collection of images of a real object photographed from various directions. In the chapter that follows, one variation of the method—calculating holograms from images (IP images) obtained by integral photography (IP) [9]—shall be described. IP, which was invented by Lippmann in 1908, is a 3-D photography employing an array of lenses. IP is characteristic in that it permits recording and reproducing 3-D images under natural light. Various attempts have been made to find practical applications of the 3-D image recording and reproducing technology of IP, as in the display of 3-D images [10, 11], the recognition of spatial images [12] and the creation of images viewed from arbitrary points [13, 14]. Reflecting upon the progress of electronic device technology in recent years, an IP system [15] that employs high-definition image pickup and display devices in combination to allow for processing at a video rate has been reported. By performing the prescribed set of calculations from moving IP images recorded by this IP system, it is possible to generate moving holograms of an actual object.

## 20.2 Calculation of Holograms from Integral Photography

The method of optically generating holograms from IP images was proposed by Pole in 1967 [16]. In this method, a laser is unnecessary since the object is photographed by IP. To transform IP images into holograms, however, a darkroom and a laser are used. Here, to generate holograms without using a laser, electrically recorded IP images are used to generate holograms by calculations [17]. Since the IP reproduced beams do not have continuous wave fronts and are a group of beams sampled in the direction of propagation, the beams reproduced from holograms generated from IP become discrete ones. This phenomenon is strikingly similar to the condition of reproducing a holographic stereogram generated from a collection of photographic images captured from many different directions [18, 19]. However, in contrast to the holographic stereogram that is obtained by photographing the object from different viewpoints one by one, the IP that photographs multipoint images at a time is suitable for capturing moving objects, though it requires high-definition photographic equipment.

### 20.2.1 Principle of Transformation from Integral Photography into Hologram

The principle of IP is shown in Fig. 20.1. For capturing, an array of convex lenses on a vertical plane is placed in front of the recording medium as shown in Fig. 20.1a. On the recording medium, small images of the object are recorded. The number of small images formed is the same as the number of lenses. Each of the small images is an image of the object observed through the center of a specific lens. Thus, the small images, called elemental images, are images of the same object viewed from different angles. Here, the collection of recorded elemental images is called an IP image. For reconstruction, the IP image and the lens array used for capturing are set in the same positions as they were in capturing as shown in Fig. 20.1b, and diffused natural light is irradiated to the IP image from behind. The diffused light whose brightness is modulated as it

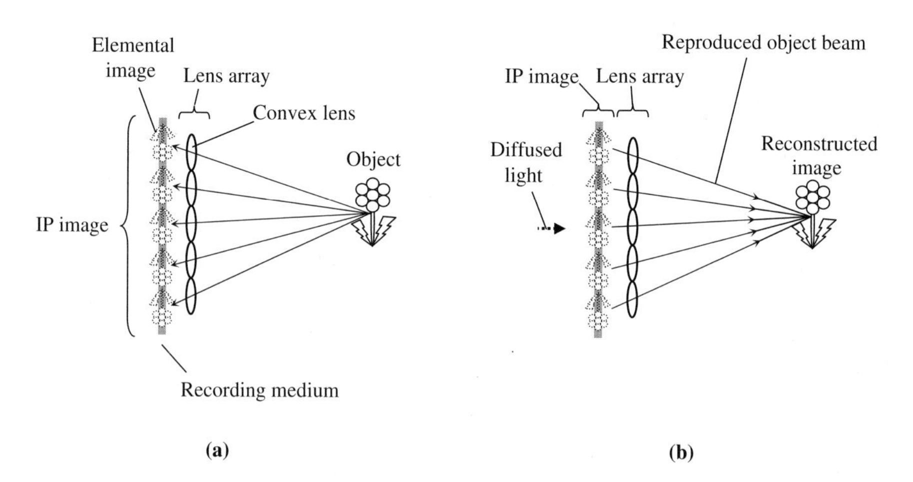


**Fig. 20.1.** Principle of IP: (**a**) capturing and (**b**) reconstruction

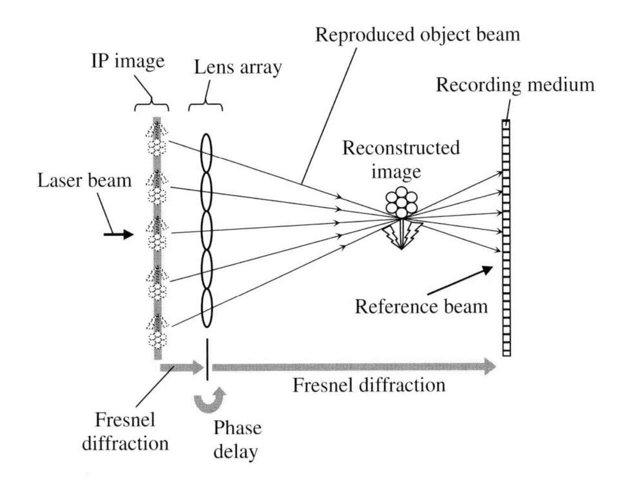


**Fig. 20.2.** Generation of holograms using an IP image

passes through the elemental images changes in the direction of propagation as it passes through the lenses. The diffused light propagates in the direction opposite to the direction of propagation of the light that reached the lens array from the object when capturing. As a result, an image equivalent to the object is formed at the position where the object was present [9].

To generate a hologram from an IP image, coherent light, such as the laser beam, is irradiated in place of natural light as shown in Fig. 20.2. A hologram can be generated by interference between the reproduced coherent object beams and the reference beam [16]. Theoretically, the hologram plane can be set at any position. Therefore, it is possible to change the positional relationship between the hologram plane and the reconstructed image.

### 20.2.2 Basic Calculation

To calculate a hologram from an IP image, it is necessary to first calculate the condition of distribution ("optical field") of object beams reproduced from the IP image at the hologram plane. The optical field can be expressed by complex amplitude. The basic concept of optical field calculation consists of simulating the propagation of beams in the IP reconstruction shown in Fig. 20.2. Here, the simulation is called the basic calculation. Specifically, the following three calculations are performed:

[I] Calculation of light propagation from the elemental image to the lens incident plane using the Fresnel diffraction
[II] Calculation of phase delay of light caused by the lens array
[III] Calculation of light propagation from the lens exit plane to the hologram plane using the Fresnel diffraction

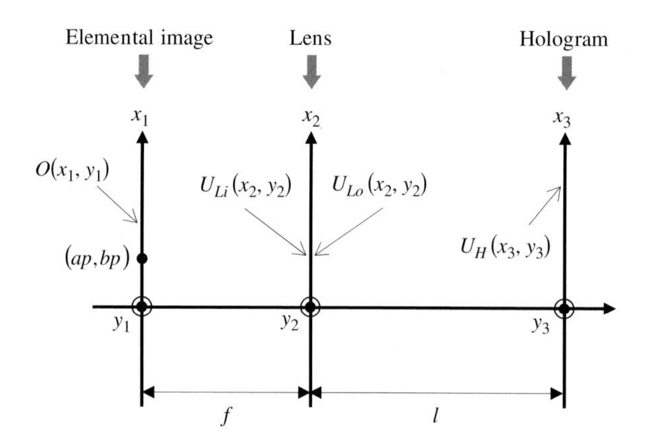


**Fig. 20.3.** Configuration used to calculate the optical field of an object beam reconstructed from an elemental image on a hologram plane

These calculations can be performed for individual elemental images. By adding up the results of calculations for the individual elemental images, it is possible to obtain the optical field of the object beams at the hologram plane. Here, the basic calculation is explained by using the configuration of a single elemental image as shown in Fig. 20.3. In calculation [I], the optical field, or the complex amplitude distribution, generated at the lens incident plane as a result of propagation of light from the elemental image is calculated. For the purpose of propagation calculation, the approximate expression (Fresnel diffraction formula) in the Fresnel region of the Fresnel-Kirchhoff diffraction formula can be used. The Fresnel region refers to the region in which the distance of propagation is comparatively short as in the case discussed here. Diffraction in this region is called Fresnel diffraction. Assume that the optical field on the elemental image is $O\,(x_1, y_1)$. Then, using the Fresnel diffraction formula, the complex amplitude distribution, $U_{Li}\,(x_2, y_2)$, at the lens incident plane can be represented by

$$U_{Li}\,(x_2, y_2) = \frac{j}{\lambda f} \exp(-jkf) \int_{-\infty}^{\infty} \int_{-\infty}^{\infty} O\,(x_1, y_1) \exp\left[-jk\left\{\frac{(x_2 - x_1)^2 + (y_2 - y_1)^2}{2f}\right\}\right] dx_1 dy_1, \tag{20.1}$$

where, $f$ is the lens focal length, $k = 2\pi/\lambda$ and $\lambda$ is the wavelength of light. Here, the distance from the elemental image to the lens is assumed as the lens focal length, $f$. The reason for this is that, in both IP capturing and reconstruction, by using the distance between the lens array and the IP image as the lens focal length, it is possible to obtain a good reconstructed image over a wide range in the depth direction [20]. Each elemental image is made up of pixels. Assume that the pixel interval is $p$, the number of pixels is $(2m+1)$ (in $x_1$ direction) $\times$ $(2n+1)$ (in $y_1$ direction), the coordinates of each pixel are $(ap, bp)$ $(-m \leq a \leq m, -n \leq b \leq n, m$ and $n$ are integers) and the pixel brightness is $C_{ab}$. Assume also that each pixel is a point light source for the purpose of simplification, although each pixel actually has a certain size. Then, optical field $O(x_1, y_1)$ of the elemental image can be represented by

$$O(x_1, y_1) = \sum_{a=-m}^{m} \sum_{b=-n}^{n} C_{ab}\delta(x_1 - ap, y_1 - bp). \tag{20.2}$$

Substituting Eq. (20.2) for Eq. (20.1), complex amplitude distribution $U_{Li}(x_2, y_2)$ at the lens incident plane becomes,

$$\begin{aligned} U_{Li}(x_2, y_2) &= \frac{j}{\lambda f}\exp(-jkf)\left(\int_{-\infty}^{\infty}\int_{-\infty}^{\infty}\sum_{a=-m}^{m}\sum_{b=-n}^{n} C_{ab}\delta(x_1 - ap, y_1 - bp)\right. \\ &\left.\exp\left[-jk\left\{\frac{(x_2 - x_1)^2 + (y_2 - y_1)^2}{2f}\right\}\right]dx_1dy_1\right) \\ &= \frac{j}{\lambda f}\exp(-jkf)\exp\left\{-jk\left(\frac{x_2^2 + y_2^2}{2f}\right)\right\} \\ &\sum_{a=-m}^{m}\sum_{b=-n}^{n}\left[C_{ab}\exp\left\{-jk\left(\frac{a^2p^2 + b^2p^2}{2f}\right)\right\}\exp\left\{jk\left(\frac{apx_2 + bpy_2}{f}\right)\right\}\right] \end{aligned} \tag{20.3}$$

In calculation [II], complex amplitude distribution $U_{Li}(x_2, y_2)$ at the lens incident plane is multiplied by the function of phase delay caused by the passing of light through the lens to obtain complex amplitude distribution $U_{Lo}(x_2, y_2)$ at the lens exit plane.

$$\begin{aligned} U_{Lo}(x_2, y_2) &= U_{Li}(x_2, y_2)\, S(x_2, y_2)\exp\left\{jk\left(\frac{x_2^2 + y_2^2}{2f}\right)\right\} \\ &= \frac{j}{\lambda f}\exp(-jkf)\, S(x_2, y_2) \\ &\times \sum_{a=-m}^{m}\sum_{b=-n}^{n}\left[C_{ab}\exp\left\{-jk\left(\frac{a^2p^2 + b^2p^2}{2f}\right)\right\}\exp\left\{jk\left(\frac{ap}{f}x_2 + \frac{bp}{f}y_2\right)\right\}\right] \end{aligned} \tag{20.4}$$

$$S(x_2, y_2) = \begin{cases} 1 & \text{(inside of convex lens)} \\ 0 & \text{(outside of convex lens)} \end{cases}, \tag{20.5}$$

where $S(x_2, y_2)$ represents an aperture similar in shape to the lens. It is a function valued 1 inside of convex lens and 0 outside of convex lens. In calculation [III], the propagation of light from the lens exit plane to the hologram plane is calculated by using the Fresnel diffraction formula to obtain complex amplitude distribution $U_H(x_3, y_3)$ at the hologram plane.

$$\begin{aligned} U_H(x_3, y_3) = &\frac{j}{\lambda l}\exp(-jkl)\int_{-\infty}^{\infty}\int_{-\infty}^{\infty} U_{Lo}(x_2, y_2) \\ &\exp\left[-jk\left\{\frac{(x_3 - x_2)^2 + (y_3 - y_2)^2}{2l}\right\}\right] dx_2 dy_2, \end{aligned} \tag{20.6}$$

where $l$ is the distance between the lens exit plane and the hologram plane. $U_H(x_3, y_3)$ is the complex amplitude distribution at the hologram plane of the light reproduced from a single elemental image. The sum of the complex amplitude distributions reproduced from all the elemental images is the complex amplitude distribution of the object beam. A hologram (fringe pattern) can be obtained by interference between this object beam and the reference beam.

### 20.2.3 Avoidance of Aliasing

When a hologram is displayed on a surface that has a pixel structure (i.e., LCD panel), the frequency component of the hologram that exceeds the frequency determined by the pixel interval on the display surface causes aliasing. Here, the conditions under which aliasing does not occur in the calculation of a hologram from an IP image are shown.

In holography, the line interval in the fringe pattern is determined by the angle formed by the direction of propagation of the object beam and that of the reference beam. Line interval $\Delta x$ in the fringe pattern is calculated by [21]

$$\Delta x = \frac{\lambda}{|\sin\theta_O - \sin\theta_R|} \cong \frac{\lambda}{|\theta_O - \theta_R|}, \tag{20.7}$$

where $\theta_O$ and $\theta_R$ are the angles formed by the direction of the object beam's propagation and that of the reference beam and the normal direction of the fringe pattern surface. ($\theta_O$ and $\theta_R$ are in the range of small values that meets these conditions: $\theta_O \cong \sin\theta_O$ and $\theta_R \cong \sin\theta_R$). From the sampling theorem, pixel interval $p$ on the hologram plane must satisfy the following condition to display a fringe pattern having line interval $\Delta x$ on the hologram plane without aliasing:

$$\Delta x \geq 2p. \tag{20.8}$$

When pixel interval $p$ on the hologram plane is given by Eqs. (20.7) and (20.8), maximum angle $\phi_M$ (maximum value of $|\theta_O - \theta_R|$) formed by the object beam and reference beam during generation of the fringe pattern is determined. From Eqs. (20.7) and (20.8), maximum angle $\phi_M$ can be represented by

$$\phi_M = \frac{\lambda}{2p}. \tag{20.9}$$

Figure 20.4 shows aliasing components in a hologram and an image reproduced from those aliasing components. When a fringe pattern obtained by interference between the object and reference beams with an angle difference exceeding $\phi_M$ is projected onto the display surface, aliasing occurs as shown in Fig. 20.4a. Since the position at which the image reconstructed from the aliasing component is formed is different from the position of the object shown in Fig. 20.4b, the image obstructs observation of the object. Therefore, when generating a hologram from an IP image, it is necessary to avoid aliasing by limiting the spread of object beams reproduced from the IP image. Assuming the distance between the IP image and the lens array as focal length $f_p$ of the lenses composing the lens array, viewing zone angle $\Omega$ of the IP display device is given by [22]

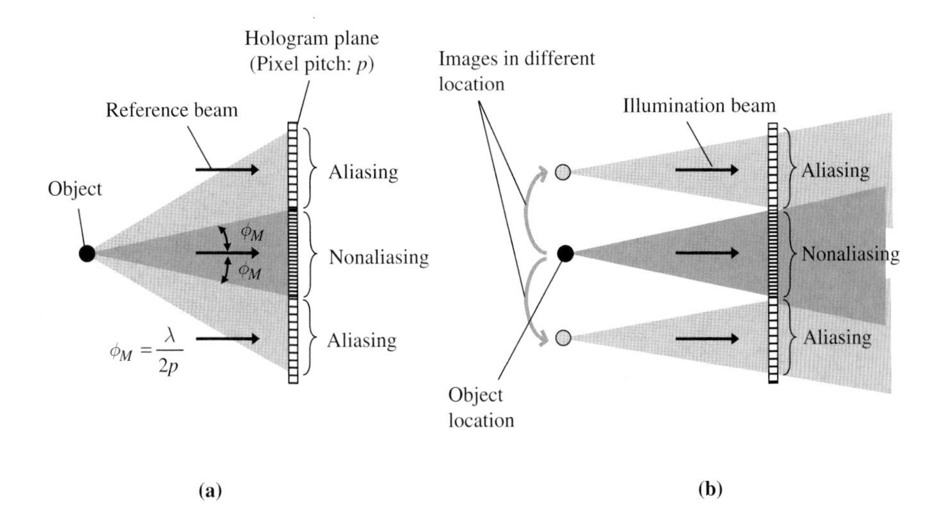


**Fig. 20.4.** Aliasing in hologram: (**a**) recording and (**b**) reconstruction

$$\Omega = 2\tan^{-1}\left(\frac{w_p}{2f_p}\right), \tag{20.10}$$

where $w_p$ is the elemental image size during reconstruction. The viewing zone angle of the IP display device given by Eq. (20.10) can be considered equivalent to the maximum angle of diffusion of the object beams reproduced from the IP image. Thus, the maximum angle of diffusion of the object beams is determined by elemental image size $w_p$ during reconstruction and focal length $f_p$ of the lens. Figure 20.5 shows the reproduction of object beams from a single elemental image and the occurrence of aliasing. Assuming that the reference beam is a plane wave which propagates perpendicularly to the hologram plane, when diffusion angle $\Omega$ of the object beams reproduced from the IP image is greater than $2\phi_M$, aliasing occurs in the fringe pattern displayed on the hologram plane as shown in Fig. 20.5a. Therefore, to avoid

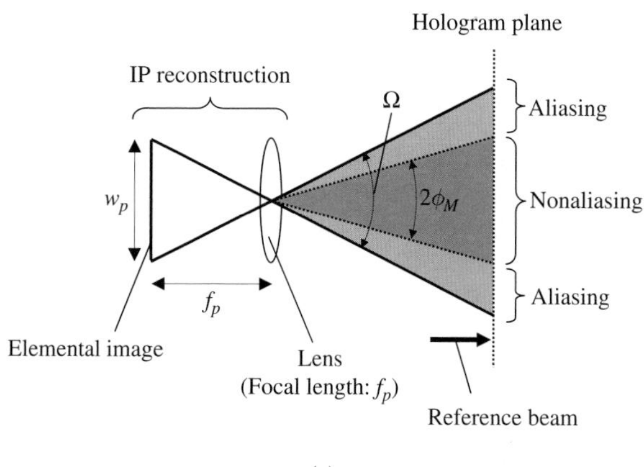


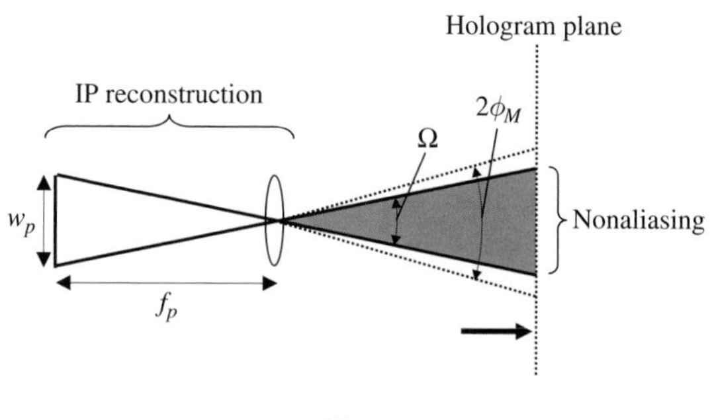


**Fig. 20.5.** Hologram produced by object beams reconstructed from a single elemental image: (**a**) when $\Omega > 2\phi_M$, aliasing occurs in a hologram; (**b**) when $\Omega \leq 2\phi_M$, aliasing does not occur in a hologram

the occurrence of aliasing as shown in Fig. 20.5b, the following condition must be satisfied:

$$\Omega \leq 2\phi_M. \tag{20.11}$$

From Eqs. (20.9) through (20.11), when calculating a hologram from an IP image, the elemental image size for IP reconstruction and the lens focal length must satisfy the following condition,

$$f_p \geq \frac{w_p}{2\tan\left(\frac{\lambda}{2p}\right)}. \tag{20.12}$$

In IP reconstruction, the elemental image size and the lens focal length can be set arbitrarily. However, if they are different from the elemental image size and lens focal length during capturing, the reconstruction space becomes distorted. For example, when the elemental image size and lens focal length during reconstruction, respectively, are m times and n times of those during capturing, the reconstructed image is greater than the object by m times laterally and n times longitudinally [22, 23]. In generating a hologram, therefore, it is necessary to set an elemental image size and a lens focal length which meet the condition shown in Eq. (20.12) to prevent geometric distortion of the reconstructed image.

### 20.2.4 Elimination of Undesired Beams

When reconstructing a hologram, not only the object beams but also conjugate beams which form a real image (conjugate image) to which depth of the object reverses, and illuminating beams (transmission beams) which pass through the hologram without being diffracted occur. When displaying a fringe pattern on a high-resolution surface (i.e., photographic plate), it is possible to give a difference in angle among the directions of propagation of the object, conjugate and transmission beams as shown in Fig. 20.6a since the angle of diffraction can be made considerably large. Therefore, it is possible to reduce the entry of conjugate and transmission beams into the observer of a direct image. In electronic holography, however, since the resolution of the electronic display surface is insufficient to display holograms, the angle of diffraction is not very wide and the difference in angle among the directions of the individual beams' propagation is small as shown in Fig. 20.6b. As a result, conjugate and transmission beams get in the eyes of the observer. In electronic holography, therefore, it is necessary to reduce the obstruction by such undesired beams.

Transmission beams are modulated by the intermodulation component (intensity distribution of object beams on the hologram plane) contained in the fringe pattern. Since this component differs according to the object shape, the transmission beams depend on the object beams in moving holography. However, when a hologram is generated by calculations, it is possible to change

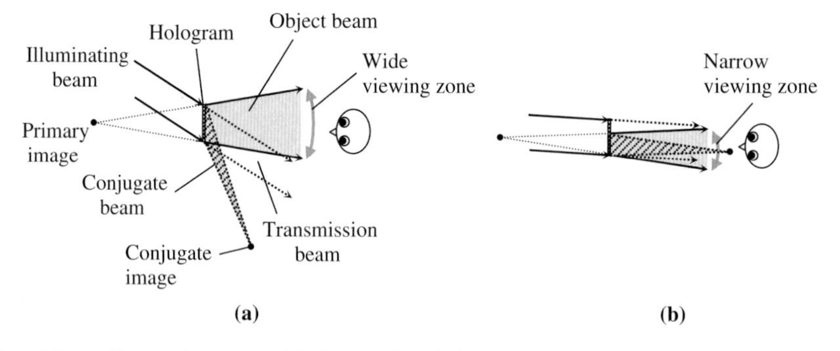


**Fig. 20.6.** Reproduction of holography: (**a**) hologram plane with high-resolution and (**b**) hologram plane with low-resolution

the transmission beams into constant beams that do not depend on the object beams by replacing the component with a suitable direct current value. In particular, when the illuminating beam (reference beam) is a plane wave, the transmission beam becomes a plane wave. In this case, the transmission beam can be eliminated by passing it through a convex lens to let it converge on a single point on the focal plane (indicated by dotted lines in Fig. 20.7b) and installing a small mask at the point of convergence.

On the other hand, the conjugate beam is phase conjugate with the object beam and, hence, they change with the object beam. In reconstruction of a moving holographic image, in particular, the conjugate beam changes so constantly that it can hardly be eliminated. The single-sideband method is a known way to eliminate conjugate beams, [24]. This method limits the band of reproduced beams to one-half in the spatial frequency range, thereby eliminating the conjugate beam component. It is effective in holography that uses a display surface that has a comparatively high resolution. When generating a fringe pattern which is to be displayed on a surface having a low resolution (i.e., LCD), it is impossible to increase the angle formed by the object beam and the reference beam. In such a case, it is effective to apply the

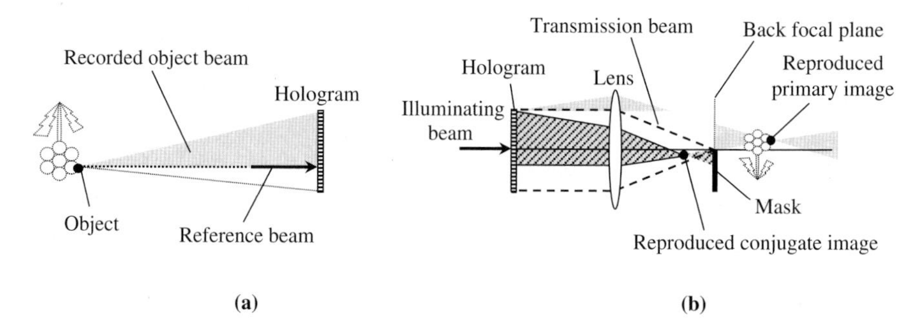


**Fig. 20.7.** Elimination of undesired beams: (**a**) recording and (**b**) reproduction

single-sideband method to a fringe pattern generated by the half zone plate processing [25, 26, 27]. The half zone plate processing, which is based on the assumption that the object is an aggregate of point light sources, limits the diffusion of light from the point light sources to one-half by a plane passing through the point light sources and containing the direction of propagation of the reference beam as shown in Fig. 20.7a. When beams reproduced from a fringe pattern created by this process are passed through a convex lens, the conjugate beams (the hatched area in Fig. 20.7b) and the object beams (the shaded area in Fig. 20.7b) are separated. At the back focal plane, in particular, the region through which the conjugate beams pass becomes invariable regardless of the positions of point light sources. By masking the region at the back focal plane, it is possible to eliminate the conjugate beams from all point light sources that make up the object.

In the basic calculation for generating a hologram from an IP image, the half zone plate processing is applied to calculate the propagation of light from the lens exit plane to the hologram plane shown in Eq. (20.6). Namely, as shown in Fig. 20.8 , the diffusion of light from the point light sources at the lens exit plane is limited to one-half by a plane which passes through the point light sources and which includes the direction of propagation of the reference beam to calculate the complex amplitude distribution at the hologram plane. By reconstructing the generated hologram by the optical system shown in Fig. 20.7b, it is possible to eliminate the undesired beams.

### 20.2.5 Image Reconstructed by Basic Calculation

Here, images optically reconstructed from holograms generated by basic calculation are demonstrated. The experimental setup of the holographic display is shown in Fig. 20.9. With holograms generated from IP images displayed on an LCD panel, a helium-neon laser beam ($\lambda$: 632.8 nm) was irradiated to observe reconstructed images. The reference beam used to generate holograms

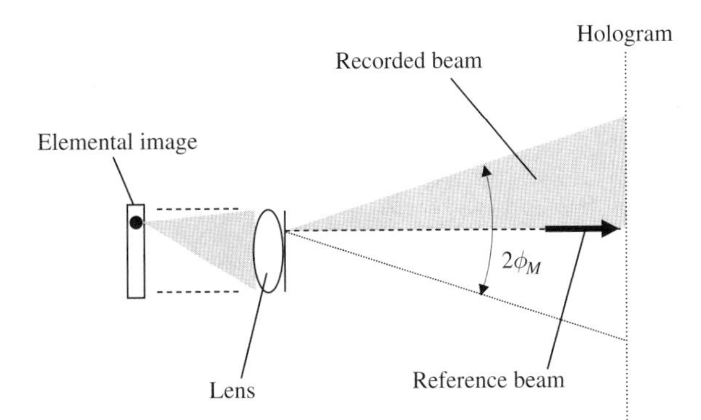


**Fig. 20.8.** Half zone plate processing for basic calculation

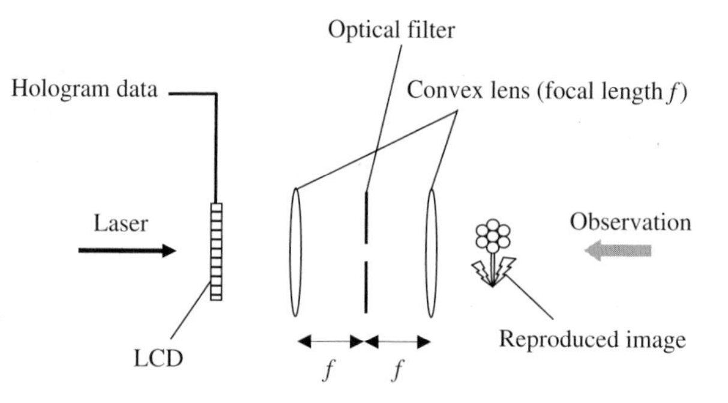


**Fig. 20.9.** Experimental setup

was a plane wave irradiated perpendicularly to the hologram plane. Therefore, the holographic display irradiated a plane wave of laser perpendicularly to the LCD panel. To observe the reconstructed images, an optical system (afocal optical system) in which two convex lenses having the same focal length ($f$: 500 mm) were arranged so that their optical axes and focal planes coincide with each other. With this optical system, only the position of image formation moves toward the observer without distorting the reconstructed image. Computer-generated flat letters "I" and "P" were used as objects. The IP images used were also computer-generated. The arrangement of the objects is shown in Fig. 20.10a. Figure 20.10b shows IP images created by calculation from the objects shown in Fig. 20.10a. The position at which the IP images were created was 50 mm in front of object "P." The parameters used to create the IP images are shown in Table 20.1. Specifications of the LCD panel used to display the holograms are shown in Table 20.2.

Since the pixel interval of the LCD panel is 10 $\mu$m, $\phi_M$ calculated by Eq. (20.9) is 1.8°. Assuming that the parameters for IP reconstruction are the same as those for IP image calculations (Table 20.1), viewing zone angle $\Omega$ is 8° from Eq. (20.10). Thus, $\phi_M$ and $\Omega$ do not satisfy the condition shown in Eq.

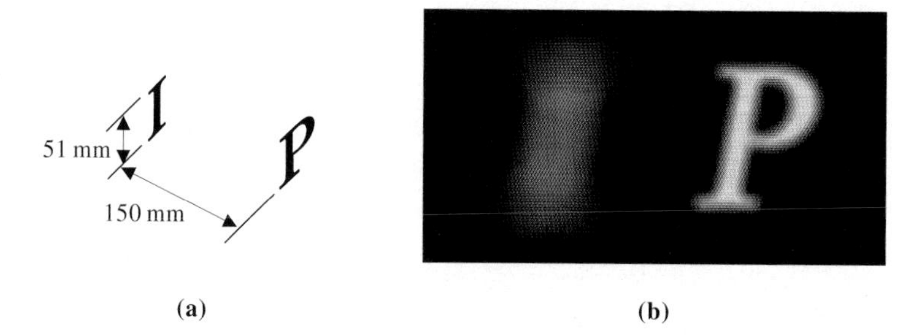


**Fig. 20.10.** Object and IP image: (**a**) locations of computer-generated objects and (**b**) computer-generated IP image

**Table 20.1.** Parameters for IP calculation of Fig. 20.9b

| | |
|---|---|
| Focal length of lens | 8.58 mm |
| Pixel pitch ($\mu$m) | 57(H), 57(V) |
| Pixel number of elemental image | 21(H) $\times$ 21(V) |
| Number of lenses | 100(H) $\times$ 60(V) |

**Table 20.2.** Specifications of LCD

| | |
|---|---|
| Screen size | 1.7 in. (diagonal) |
| Pixel pitch ($\mu$m) | 10(H), 10(V) |
| Pixel number | 3,840(H) $\times$ 2,048(V) |

(20.12). Therefore, when a hologram generated when $\Omega = 8°$ is displayed on the LCD panel (Table 20.2), an aliasing component occurs. Figure 20.11 shows an image reconstructed from a hologram generated when $\Omega = 8°$ and which contained an aliasing component, photographed with the focus on letter "P." The distance between the lens array and the hologram plane was 50 mm. In this reconstruction, only the transmission beam was eliminated. Many images were produced by the aliasing component and a conjugate image overlaps with each of those images. Since the conjugate images are formed at positions different in depth from the true image, they appear to be background noise in the photograph. To avoid the occurrence of an aliasing component, another hologram was generated using an elemental image size of 210 $\mu$m (= 10 $\mu$m $\times$ 21 pixels) and a lens focal length of 3.32 mm which satisfies the condition shown in Eq. (20.12) (Table 20.3 ). Figure 20.12 shows a photograph of images reconstructed from the hologram without aliasing. It was confirmed that both of the images produced by the aliasing component and the conjugate images had been eliminated. However, the true image and its conjugate image overlap with each other. Figure 20.13 shows photographs of images reconstructed from a hologram generated by applying the half zone plate processing. A spatial filter was installed on the focal plane of the optical system for reconstruction

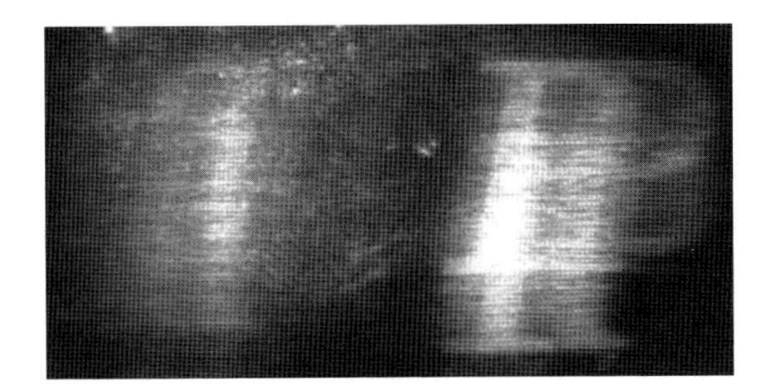

**Fig. 20.11.** Reconstructed image from a hologram with aliasing

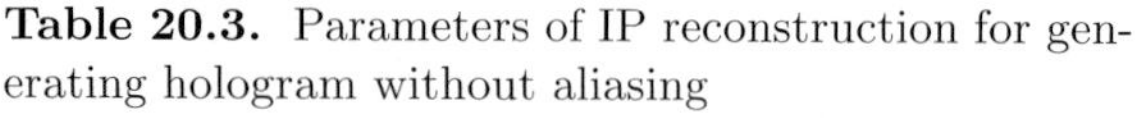

**Table 20.3.** Parameters of IP reconstruction for generating hologram without aliasing

| Focal length of lens | 3.32 mm |
|---|---|
| Pixel pitch ($\mu$m) | 10(H), 10(V) |
| Pixel number of elemental image | 21(H)$\times$ 21(V) |
| Number of lenses | 100(H) $\times$ 60(V) |

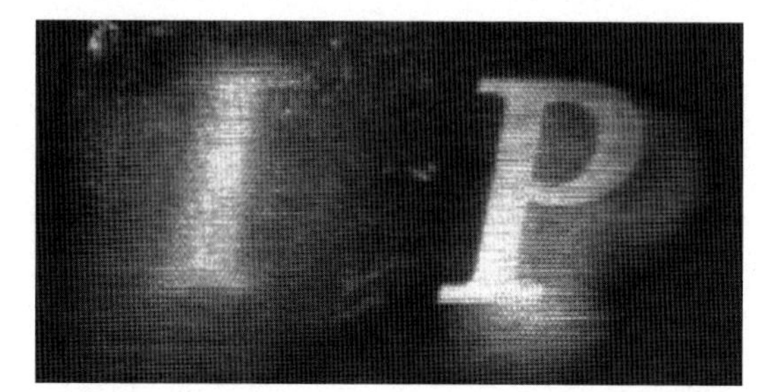

**Fig. 20.12.** Reconstructed image from a hologram without aliasing

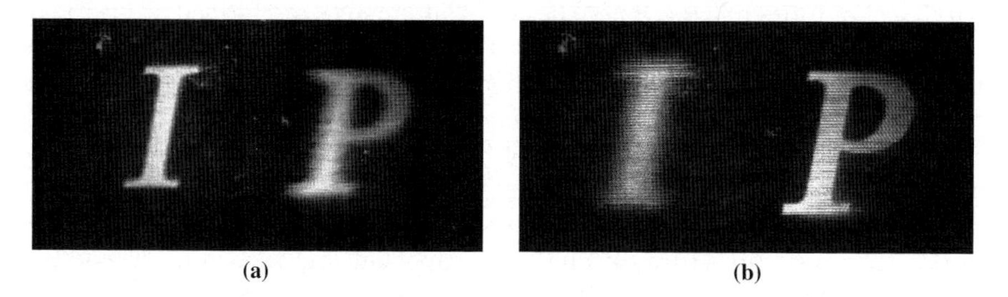

(a) (b)

**Fig. 20.13.** Reconstructed images from a hologram calculated using the basic calculation. The conjugate image is reduced in these images. (**a**) Focus is on the letter I. (**b**) Focus is on the letter P

to eliminating conjugate beams. In Fig. 20.13a, the focus was on letter "I," and in Fig. 20.13b, the focus was on letter "P." True reconstructed images from which the background noise due to conjugate images had been reduced were obtained as spatial images. Since the elemental image size and the lens focal length shown in Table 20.1 were changed to 0.18 times (= 210 $\mu$m/1.20 mm) and 0.39 times (= 3.32 mm/8.58 mm), respectively, the reconstructed images theoretically distort 0.18 times laterally and 0.39 times longitudinally, respectively. The height of the reconstructed letters was 9.1 mm (0.18 times of the original height of 51 mm) and the space between letters "I" and "P" was

60 mm (0.4 times of the original space of 150 mm). Thus, it was confirmed that the images had been reconstructed almost in accordance with theory.

## 20.3 Reducing Computing Load

Ordinarily, in Eq. (20.6) to determine the complex amplitude distribution at the hologram plane in the basic calculation, the calculation is performed on those beams from the individual pixels on the lens exit plane whose maximum diffusion angle is $2\phi_M$ ($\phi_M$ when the half zone plate processing is applied) at which aliasing does not occur in the fringe pattern displayed on the hologram plane as shown in Fig. 20.14a. The Fresnel diffraction formula used in the calculation requires a large computing load, which is in direct proportion to the square of the distance between lens and hologram. Therefore, applying the basic calculation in moving holography that requires many holograms is not easy. Because of this, positioning the hologram plane at the back focal plane of the lens to reduce the basic calculation load has been studied [28]. The complex amplitude distribution at the back focal plane is equivalent to

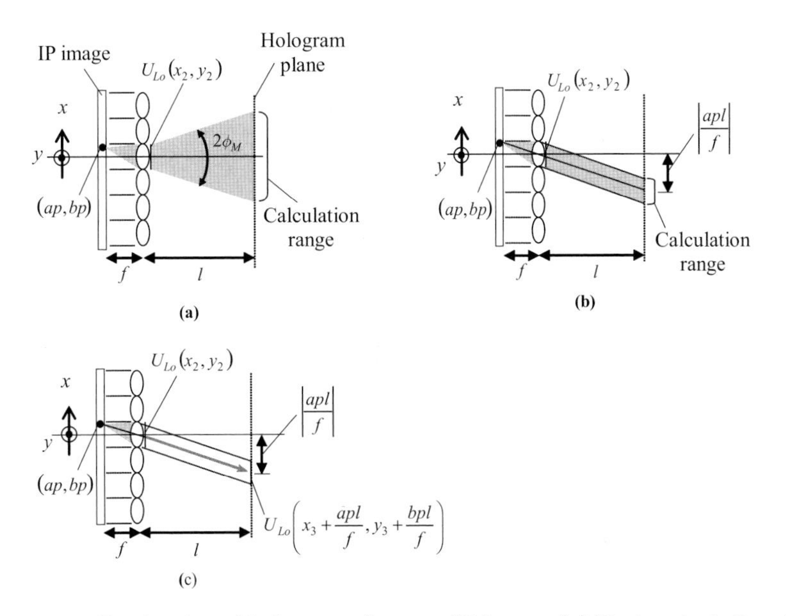


**Fig. 20.14.** Production of holograms from an IP image. (**a**) Basic calculation. The spread angle of the IP-reconstructed object beam is limited to $2\phi_M$ so as not to cause aliasing. (**b**) To reduce the calculation load, the calculation range is limited to the same size as a lens centered on $(-apl f, -bpl f)$, in which the IP-reconstructed beam is distributed. (**c**) The optical field on the exit plane of a lens is shifted to the hologram plane

the complex amplitude distribution of light at the front focal plane subjected to the Fourier transformation. Therefore, when the hologram plane is positioned at the back focal plane, that is, $l = f$, in Fig. 20.14a, it is possible to obtain the complex amplitude distribution at the hologram plane by subjecting the elemental images to Fourier transformation. By applying FFT in the Fourier transformation, the computing load can be reduced. This corresponds to the condition for generating a holographic stereogram to prepare elemental holograms from elemental images by Fourier transformation [18, 19]. This method is characteristic in that it reduces the computing load without causing holograms to deteriorate. With this method, however, the hologram plane must be positioned at the back focal plane. In the sections that follow, reducing the computing load by utilizing the features of the optical system for IP reconstruction, regardless of the hologram plane's position, shall be discussed.

### 20.3.1 Reduction of Computing Load by Limiting Range of Computation

Let us consider the complex amplitude distribution of the light produced by the elemental images at the hologram plane. From Eqs. (20.3) and (20.4), Eq. (20.6) can be rewritten as

$$\begin{aligned} U_H(x_3, y_3) = {} & \frac{-1}{\lambda^2 l f} \exp\{-jk(l+f)\} \\ & \times \sum_{a=-m}^{m} \sum_{b=-n}^{n} \left( C_{ab} \exp\left\{-jk\left(\frac{a^2p^2+b^2p^2}{2f}\right)\left(1-\frac{l}{f}\right)\right\} \exp\left\{jk\left(\frac{ap}{f}x_3+\frac{bp}{f}y_3\right)\right\} \right. \\ & \left. \times \int_{-\infty}^{\infty}\int_{-\infty}^{\infty} S(x_2, y_2) \exp\left[-jk\left(\frac{\left\{x_2-\left(x_3+\frac{apl}{f}\right)\right\}^2+\left\{y_2-\left(y_3+\frac{bpl}{f}\right)\right\}^2}{2l}\right)\right] dx_2 dy_2 \right). \end{aligned} \tag{20.13}$$

In Eq. (20.13), the second exp is a phase component which depends on the coordinates $(ap, bp)$ of the pixels of the elemental images. It can be regarded as a constant. The third exp represents the complex amplitude distribution at the hologram plane of the plane wave in the direction of propagation $(-ap\,/\,f, -bp\,/\,f)$. The double integral is the Fresnel diffraction formula, which represents the complex amplitude distribution at the hologram plane of the plane wave that passes through aperture $S(x_2, y_2)$ in the form of a lens and that propagates perpendicularly to the aperture, shifted by $(-apl\,/\,f, -bpl\,/\,f)$ within the $x_3 - y_3$ plane.

In the Fresnel region, most of the light (plane wave) that has passed through the aperture continues propagating while concentrating within the shape of the aperture. From Eq. (20.13), complex amplitude distribution $R_H(x_3, y_3)$ at the hologram plane of the light produced by a single pixel of an elemental image can be represented as

$$R_H(x_3, y_3) = \exp\left[jk\left(\frac{ap}{f}x_3 + \frac{bp}{f}y_3\right)\right] \times \int_{-\infty}^{\infty}\int_{-\infty}^{\infty} S(x_2, y_2)$$
$$\exp\left(-jk\left\{\frac{\left[x_2 - \left(x_3 + \frac{apl}{f}\right)\right]^2 + \left[y_2 - \left(y_3 + \frac{bpl}{f}\right)\right]^2}{2l}\right\}\right) dx_2 dy_2, \quad (20.14)$$

where the coordinates of the pixel are $(ap, bp)$. Note that the constant terms have been omitted. $S(x_2, y_2)$, which represents the lens shape, is assumed as a square whose size is $w \times w$. Assuming the ratio of the energy distributed within a certain area (expressed as $\alpha$ times of aperture width $w$) around coordinates $(-apl / f, -bpl / f)$ to the energy of the optical field distributed over the entire hologram plane as $E(\alpha)$, $E(\alpha)$ can be represented as

$$E(\alpha) = \frac{\int_{-\frac{bpl}{f}-\frac{\alpha w}{2}}^{-\frac{bpl}{f}+\frac{\alpha w}{2}} \int_{-\frac{apl}{f}-\frac{\alpha w}{2}}^{-\frac{apl}{f}+\frac{\alpha w}{2}} \left|R_H(x_3, y_3)\right|^2 dx_3 dy_3}{\int_{-\infty}^{\infty}\int_{-\infty}^{\infty} \left|R_H(x_3, y_3)\right|^2 dx_3 dy_3}. \quad (20.15)$$

Figure 20.15 shows the values of $E(\alpha)$ obtained when $a{=}b{=}0$, $w{=}0.2$ mm and $I$ (distance between the lens and hologram plane) = 10 mm, 50 mm and 100 mm, respectively. The figure indicates that when the value of $\alpha$ remains the same, the smaller the value of $l$, the larger the value of $E(\alpha)$, that is, the higher the degree of beam concentration. For example, when $l$ is 50 mm, nearly 80 percent of the light concentrates in the area around coordinates $(-apl / f, -bpl / f)$ that is the same in size as the aperture ($\alpha = 1$). From

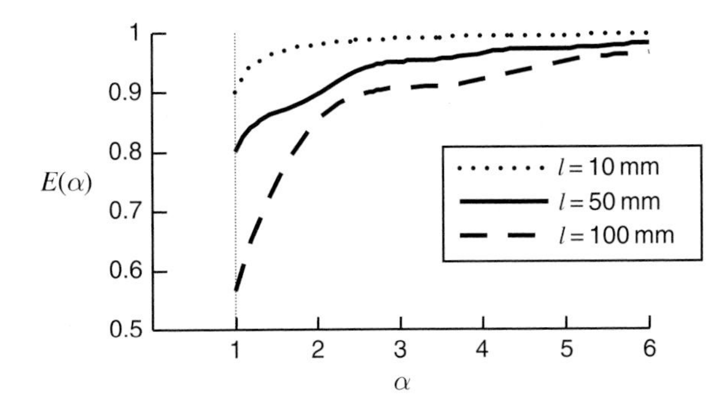


**Fig. 20.15.** Ratio of the energy of a beam from a pixel of an elemental image distributed inside a fixed range (expressed as $\alpha$ times aperture width $w$) to the energy distributed on the whole hologram plane. Distance l between the lens and hologram plane is set at 10, 50 and 100 mm

these facts, when calculating the complex amplitude distribution at the hologram plane, it is possible to reduce the computing load by limiting the range of calculation of $(x_3, y_3)$ within a certain region in which the light concentrates as shown in Fig. 20.14b, rather than using the maximum region in which aliasing does not occur in the hologram. For example, the maximum region in which aliasing does not occur is $(2\phi_M l)^2 \quad [= (\lambda l/p)^2]]$. By limiting the range of calculation to $(\alpha w)^2$, it is possible to reduce the computing load shown in Eq. (20.13) to $(\alpha w)^2 / (\lambda l/p)^2$. The implication is that the larger the value of $l$, the greater the effect to reduce the computing load. However, since limiting the range of calculation causes the higher spatial frequency components of the object beam to be eliminated, deterioration (blurring, etc.) can occur with the reproduced images.

### 20.3.2 Reduction of Computing Load by Shifting Optical Field

Geometrical optics assumes that light from point light sources on the focal plane of a lens becomes a plane wave after it passes through the lens, and that the plane wave propagates along an imaginary line connecting the point light sources with the principal point of the lens. Applying this concept, the optical field at the hologram plane of the light produced from a pixel at coordinate point $(ap, bp)$ of the elemental image can be approximated by shifting complex amplitude distribution $U_{Lo}(x_2, y_2)$ at the lens exit plane to a position around $(-apl/f, -bpl/f)$ within the hologram plane ($x_3 - y_3$ plane) as shown in Fig. 20.14c. In this case, complex amplitude distribution $U_H(x_3, y_3)$ at the hologram plane can be calculated by

$$U_H(x_3, y_3) = U_{Lo}\left(x_3 + \frac{apl}{f}, y_3 + \frac{bpl}{f}\right)$$
$$\cong \sum_{a=-m}^{m} \sum_{b=-n}^{n} \left\{ C_{ab} \exp\left[ jk \left( \frac{ap}{f} x_3 + \frac{bp}{f} y_3 \right) S\left( x_3 + \frac{apl}{f}, y_3 + \frac{bpl}{f} \right) \right] \right\}. \tag{20.16}$$

Note that the constant terms have been omitted from the above equation. When the complex amplitude distribution at the hologram plane is approximated by that at the lens exit plane as described above, it is unnecessary to calculate the Fresnel diffraction between the lens exit plane and the hologram plane and, hence, the computing load can be reduced significantly. Thus, this method is considered effective when large volumes of hologram calculations are required as in moving holograms. However, the images reconstructed from a hologram generated by this method are may deteriorate as the diffraction effect produced by the propagation of light from the lens exit plane to the hologram plane is eliminated. Therefore, in the future, the influence of the method on the quality of reconstructed images needs further study.

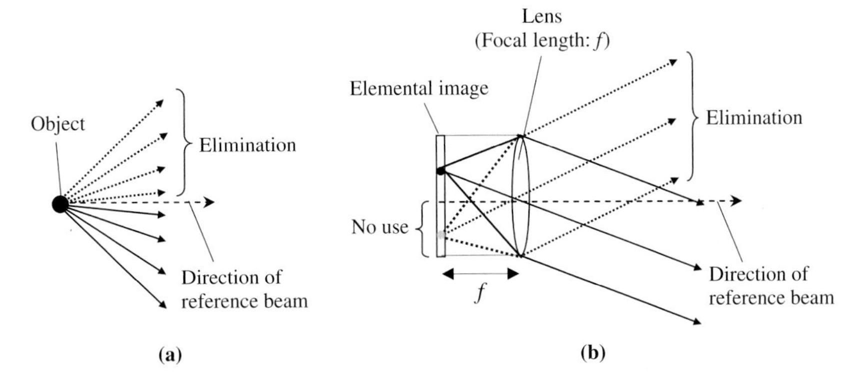


**Fig. 20.16.** Half zone plate processing: (**a**) principle of half zone plate processing and (**b**) application of half zone plate processing to the optical field shift method

Next, the application of the half zone plate processing to calculate shifting the optical field to eliminate conjugate beams is described below. In the actual half zone plate processing, as shown in Fig. 20.16a, when the reference beam is assumed to be a plane wave, only those beams from point light sources composing the object on either side (the lower side in the figure) of the plane, including the point light sources and the direction of propagation of the reference beam, are used to generate holograms. This process can be directly applied in the basic calculation to obtain the complex amplitude distribution at the hologram plane of the light from the point light sources at the lens exit plane. However, when the complex amplitude distribution at the hologram plane is approximated by that at the lens exit plane, the half-zone plate processing cannot be directly applied since the individual point light sources are not processed. Figure 20.16b shows a process equivalent to the half zone plate processing in the approximate calculation of complex amplitude distribution. The figure concerns only one elemental image. To extract only those beams which propagate downward from the direction of propagation of the reference beam, it is sufficient to generate beams only from the images on the upper side of the position where the straight line that passes through the principal point of the lens and propagates the direction of the reference beam hits against the elemental image. When only the beams from images on the upper side are used for all elemental images, it follows that all object beams are subjected to a process equivalent to half zone plate processing. Applying the above equivalent process to Eq. (20.16), the following equation can be derived:

$$U_H(x_3, y_3) \cong \sum_{a=0}^{m} \sum_{b=-n}^{n} \left[ C_{ab} \exp\left\{ jk \left( \frac{ap}{f} x_3 + \frac{bp}{f} y_3 \right) S\left( x_3 + \frac{apl}{f}, y_3 + \frac{bpl}{f} \right) \right\} \right]. \tag{20.17}$$

The only difference from Eq. (20.16) is that $a$, or the range of summation on the $x$-direction, has been halved ($a = 0, 1, \ldots, m$). Here, the range in $x$-direction has been halved as an example. Actually, the direction in which the range is halved is arbitrary. It should be noted, however, that the region through which the conjugate beams pass during reconstruction changes according to the direction in which the range is halved. The equivalent process reduces the computing load to one-half, speeding up the necessary calculations, since only one-half of the elemental images is used for hologram calculations. On the other hand, it has the drawback of reducing the diffusion of object beams, or the viewing zone, to one-half. In the holography display system described in Section 20.4, the viewing zone in the vertical direction is sacrificed by reducing the diffusion of object beams to one-half in the vertical direction; however, a sufficient horizontal viewing zone needed to observe 3-D images is secured.

### 20.3.3 Images Reconstructed by Using Method of Reducing Computing Load

Images reconstructed by using a method that reduces the computing load are shown. The IP images used to generate holograms, the specifications of the holographic display, etc. are the same as those used in the reconstruction experiment for basic calculation described in Section 20.2.5. Figure 20.17 shows images reconstructed from a hologram generated by limiting the range of calculation in Eq. (20.13) to two times ($a = 2$) the size, in both horizontal and vertical directions, of the lenses composing the lens array. Figure 20.18 shows images reconstructed from a hologram generated by shifting the complex amplitude distribution at the lens exit plane to the hologram plane. In both cases, the aliasing component and the obstruction due to conjugate and transmission beams have been eliminated. The focusing positions in Figs. 20.17a and b, and Figs. 20.18a and b are the same as in Figs. 20.13a and b. The images reconstructed from a hologram generated with the range

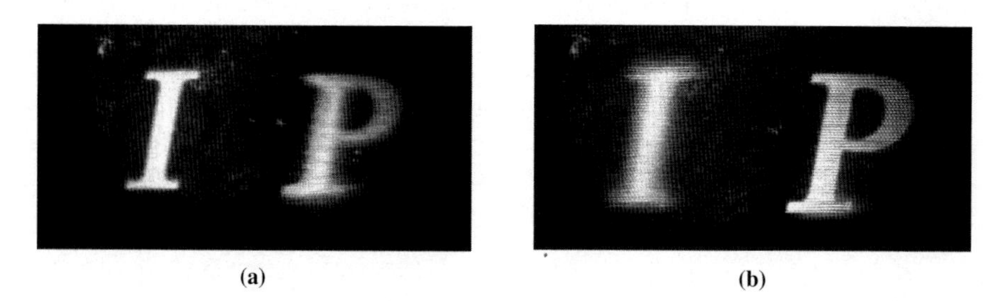


**Fig. 20.17.** Reconstructed images from a hologram calculated by limiting the calculation range to two times the size of a lens vertically and horizontally. (**a**) Focus is the same position as in Fig. 20.13a. (**b**) Focus is the same position as in Fig. 20.13b

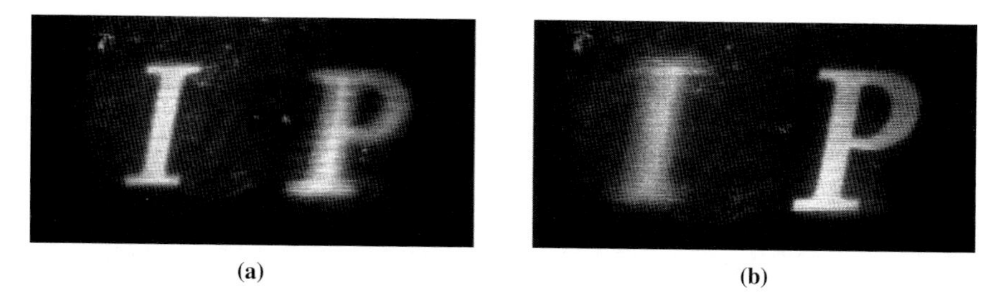

(a) (b)

**Fig. 20.18.** Reconstructed images from a hologram calculated using the field shift method. (**a**) Focus is the same position as in Fig. 20.13a. (**b**) Focus is the same position as in Fig. 20.13b

of calculation limited (Fig. 20.17) were formed at the same positions as the reconstructed images obtained by the basic calculations (Fig. 20.13). The calculation time was reduced to about 1/60 [theoretically, $1/57(= (2 \times 210\,\mu\text{m} \times 10\,\mu\text{m}/632.8\,\text{nm}/50\,\text{mm})^2)$], demonstrating the effect of reducing the computing load. However, the reconstructed images look somewhat blurred. The reason for this may be that the higher spatial frequency components of the object beams were eliminated as the range of calculation was narrowed down. Figure 20.15 shows that as the value of $\alpha$ is decreased (or the value of $l$ is increased), the degree of beam concentration decreases. Therefore, when the range of calculation is narrowed down or the distance to the hologram plane is increased, the image quality deteriorates markedly. The images reconstructed from a hologram generated by using the complex amplitude distribution at the lens exit plane (Fig. 20.18) were also formed at the same positions as those obtained by the basic calculations. However, a decline in the resolution is observed. In the basic calculations, the range of calculation widens and the computing time increases as the distance between the hologram plane and the lens array increases. By contrast, when the complex amplitude distribution at the lens exit plane is used, the computing load remains the same. When the distance between the lens array and the hologram plane was 50 mm and the complex amplitude distribution at the lens exit plane was used, the measured computing time was about 1/150, compared with the basic calculation. Thus, the computing time could be reduced more than when the range of calculation was narrowed down.

## 20.4 Electronic Holography Using Real IP Images

This section describes the results of optical experiments in which holograms were generated from IP images of real objects using calculations for IP hologram transformation. The generated holograms were displayed on electronic holography display equipment.

**Table 20.4.** Specifications of holographic display system

| IP camera | Focal length of lens | 2.65 mm |
|---|---|---|
| | Number of lenses | 54 (H) × 63 (V) |
| | Pixel pitch ($\mu$m) | 52 (H), 52 (V) |
| | Pixel number of elemental image | 21 (H) × 21 (V) |
| LCD | Screen size | 1.3 in. (diagonal) |
| | Pixel pitch ($\mu$m) | 26 (H), 26 (V) |
| | Pixel number | 1024 (H) × 768 (V) |

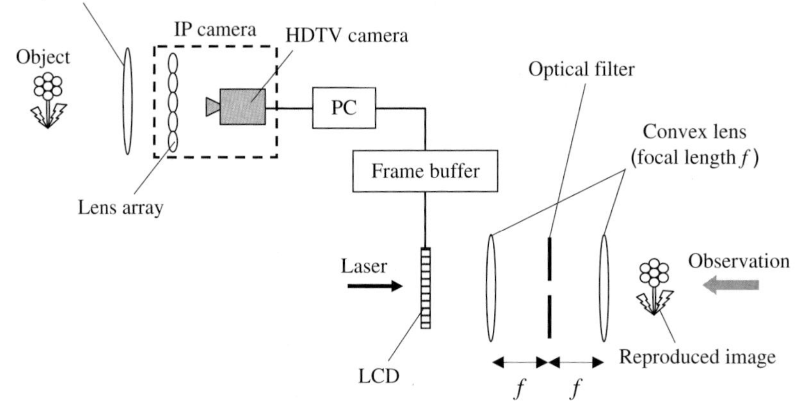


**Fig. 20.19.** Experimental setup for real IP images

Experiments were carried out on the reconstruction of holograms of real objects existing under natural light by using holograms generated from IP images. Specifications of the experimental apparatus are shown in Table 20.4, and the apparatus configuration is shown in Fig. 20.19. IP images of the objects were captured by an IP camera [15] and holograms were generated from the IP images by using a personal computer. To capture IP images, a convex lens (called a depth control lens) was placed between the object and the lens array, and an image of the object was formed at the front and back of the lens array. The reason for this is that images on the lens array can be captured and reconstructed with the highest possible resolution [20]. Generating a hologram from an IP image captured by an HDTV camera takes about 20 seconds per frame. Therefore, the calculations for hologram generation were performed off-line. The holograms were displayed on an LCD panel via a frame buffer and the reconstructed images were observed while irradiating a laser beam to them. The reference beam used in the generation of holograms was a plane wave ($\lambda$: 632.8 nm) propagating perpendicularly to the hologram plane. Therefore, in the reconstruction of holograms, a plane wave produced by a helium-neon laser beam ($\lambda$: 632.8 nm) was irradiated

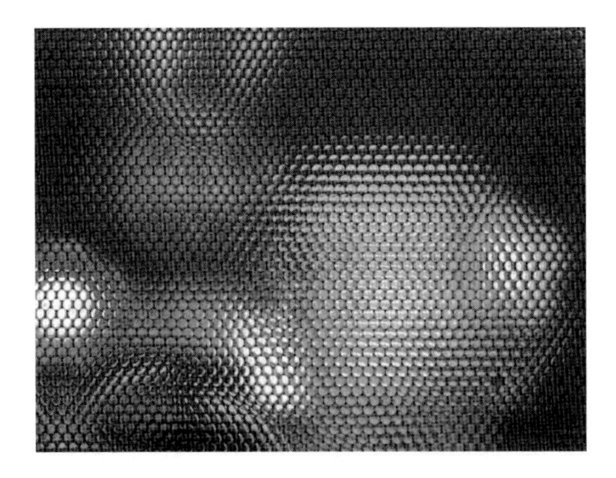

**Fig. 20.20** Real IP image captured by IP camera

perpendicularly to the LCD panel. As the optical system for reconstruction, an afocal optical system comprised of two convex lenses with the same focal length (150 mm) was used. A spatial filter for eliminating the transmission and conjugate beams was installed on the focal plane common with the two lenses.

The objects were two dolls placed one behind the other. IP images of the objects captured by an IP camera are shown in Fig. 20.20. From the pixel interval of the LCD panel used to display the holograms, the elemental image size of 546 $\mu$m and the lens focal length of 22.5 mm that satisfy the condition shown in Eq. (20.12) were selected and holograms were generated by calculations using the optical field on the lens exit plane. The distance between the hologram plane and the lens array for reconstruction was 50 mm, and the obstruction by aliasing, conjugate beam and transmission beam was eliminated. Reconstructed images are shown in Fig. 20.21. Each of the reconstructed images measures 26 mm (H) $\times$ 20 mm (V). The resolution of

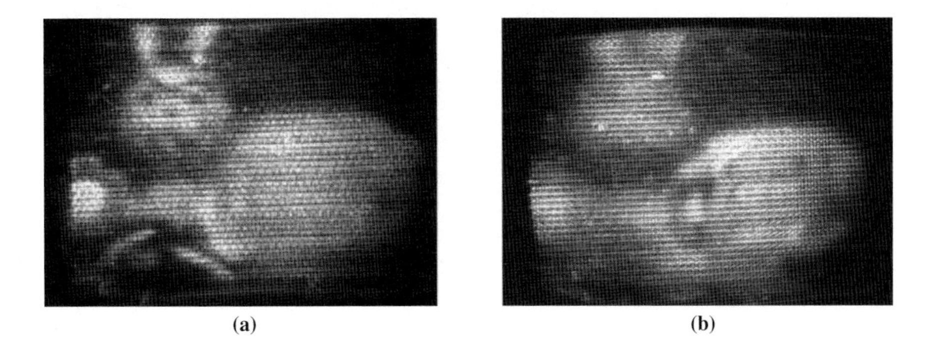


**Fig. 20.21.** Reproduced images: (**a**) Focus is on the rear doll. (**b**) Focus is on the front doll

each reconstructed image is 54 (H) $\times$ 63 (V). This corresponds to the number of lenses composing the lens array for IP capturing (see Table 20.4). In Fig. 20.21a, the focus is on the doll at the back and, in Fig. 20.21b, the focus is on the doll at the front. When the focusing position is changed, the doll that is in focus changes too. This confirmed that the spatial images of the objects reconstructed from holograms generated from the real IP image are formed at their respective depth positions.

Figure 20.22 shows reconstructed images obtained with a higher resolution by increasing the number of lenses composing the lens array of the IP camera to 178 (H) $\times$ 115 (V). The object is a moving person (woman throwing a ball). Actually, the moving images were reconstructed. Figures 20.22a and b show two photographs focusing on the person in the background and on the ball in the foreground. Figures 20.22c and d show two photographs, one taken from the left side of the viewing zone and the other taken from the right side of the

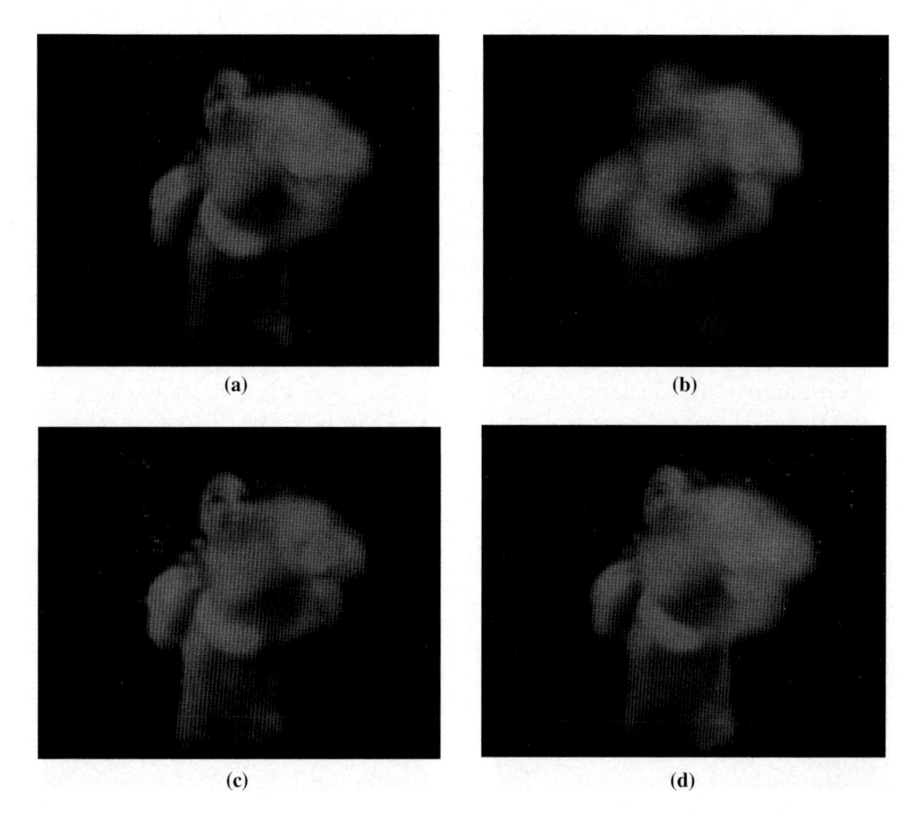


**Fig. 20.22.** Reproduced images from holograms calculated from real IP images. (**a**) Focus is on the rear human. (**b**) Focus is on the front ball. Images are taken from (**c**) left and (**d**) right of viewing zone

viewing zone (the focus is on the person in the background). The quality of the spatial images is better than that shown in Fig. 20.21.

## 20.5 Conclusion

With the aim of obtaining hologram data for an electronic holography system, this chapter described the generation of holograms from IP images by calculations. In addition to the basic method of calculations, methods for reducing the computing load were demonstrated. Furthermore, methods for eliminating aliasing components and conjugate beams suitable for the calculations were shown. Calculating holograms from IP images obtained under natural light is an effective method of generating holograms without using a laser. By using an ordinary TV camera as the IP camera, it is possible to obtain moving IP images in red, green and blue. Therefore, holograms in desired colors can be generated easily. It may be said that the method is useful also as a means of inputting real images, color images and moving holography. However, the quality of images reconstructed from holograms generated by using this method depends on the conditions under which IP images are captured. In particular, since the resolution of reconstructed images is determined by the number of elemental images in the IP images (the number of lenses composing the lens array), it will become necessary to develop an IP camera which employs a lens array made of many lenses and a high-resolution image pickup device in combination to obtain high quality 3-D images.

## References


[1] D. Gabor, "A new microscopic principle," Nature, 161, pp. 777–778 (1948)
[2] H. Maiman, "Optical and microwave-optical experiments in ruby," Phys. Rev. Lett., 4, 11, pp. 564–566 (1960)
[3] P. Hariharan, Optical Holography – Principles , techniques, and applications – Second edition (Cambridge University Press, (1996)
[4] N. Leith, J. Upatnieks, A. Kazama, and N. Massey, "Hologram visual display," J. Soc. Motion Picture Televis. Eng., 75, pp. 323–326 (1966)
[5] D. J. De Bitetto, "A holographic motion picture film with constant velocity transport," Appl. Phys. Lett., 12, pp. 295–297 (1968)
[6] P. St. Hilaire, S. A. Benton, M. Lucente, M. L. Jepsen, J. Kollin, H. Yoshikawa, and J. Underkoffler "Electronic display system for computational holography," Proc. SPIE 1212, pp. 174–182 (1990)
[7] N. Hashimoto , S. Morokawa, and K. Kitamura, "Real-time holography using the high-resolution LCTV-SL," Proc. SPIE 1461, pp. 291–302 (1991)
[8] K. Sato, K. Higuchi, and H. Katsumma, "Holographic television by liquid crystal device," Proc. SPIE 1667, pp. 19–31 (1992)

[9] M. G. Lippmann, "La photographie integral," Comp. Rend., 146, pp. 446–451 (1908)
[10] S.-H. Shin and B. javidi, "Speckle-reduces three-dimensional volume holographic display by use of integral imaging," Appl. opt., 41, 14, pp. 2644–2649 (2002)
[11] K. Choi, H. Choi, H. Kim, J. Hahn, Y. Lim, J. Kim, and B. Lee, "Viewing-angle enhanced computer-generated holographic display system combined with integral imaging," Proc. SPIE 6016, pp. 601–612 (2005)
[12] O. Matoba, E. Tajahuerce, and B. Javidi, "Real-time three-dimensional object recognition with multiple perspective imaging," Appl. Opt., 40, 20, pp. 3318–3325 (2001)
[13] T. Naemura, T. Yoshida, and H. Harashima, "3-D computer graphics based on integral photography," Opt. Express, 8, 2, pp. 255–262 (2001)
[14] H. Arimoto and B. Javidi, "Integral three-dimensional imaging with digital reconstruction," Opt. Lett., 26, 3, pp. 157–159 (2001)
[15] F. Okano, J. Arai, H. Hoshino, and I. Yuyama, "Three-dimensional video system based on integral photography," Opt. Eng., 38, 6, pp. 1072–1077 (1999)
[16] R. V. Pole, "3-D Imagery and holograms of objects illuminated in white light," Appl. Phys. Lett., 10, 1, pp. 20–22 (1967)
[17] T. Mishina, M. Okui, and F. Okano, "Calculation of holograms from elemental images captured by integral photography," Appl. Opt. 45, 17, pp. 4026–4036 (2006)
[18] T. Yatagai, "Stereoscopic approach to 3-D display using computer-generated holograms," Appl. Opt., 15, 11, pp. 2722–2729 (1976)
[19] M. Yamaguchi, H. Hoshino, T. Honda, and N. Ohyama, "Phase added stereogram: calculation of hologram using computer graphics technique," Proc. SPIE 1914, pp. 25–31 (1993)
[20] J. Arai, H. Hoshino, M. Okui, and F. Okano, "Effect of focusing on the resolution characteristics of integral photography," J. Opt. Soc. Am. -A, 20, 6, pp. 996–1004 (2003)
[21] T. Okoshi, Three-dimensional imaging techniques (Academic, New York, 1976)
[22] J. Arai, M. Okui, M. Kobayashi, and F. Okano, "Geometrical effects of positional errors in integral photography," J. Opt. Soc. Am. -A, 21, 6, pp. 951–958 (2004)
[23] J.-H. Park, H. Choi, Y. Kim, J. Kim, and B. Lee, "Scaling of three-dimensional integral imaging," Jpn. J. Appl. Phys., 44, 1A, pp. 216–224 (2005)
[24] E. N. Leith and J. Upatnieks, "Reconstructed wavefronts and communication theory," J. Opt. Soc. Am., 52, 10, pp. 1123–1130 (1962)
[25] O. Bryngdahl and A. Lohmann, "Single-sideband holography," J. Opt. Soc. Am., 58, 5, pp. 620–624 (1968)
[26] T. Takemori, "3-dimensional display using liquid crystal devices – fast computation of hologram," ITE Tech. Rep. 21, 46 , pp. 13–19 (1997) (in Japanese)
[27] T. Mishina, F. Okano, and I. Yuyama, "Time-alternating method based on single-sideband holography with half-zone-plane processing for the enlargement of viewing zones," Appl. Opt. 38, 17, pp. 3703–3713 (1999)
[28] R. Oi, T. Mishina, M. Okui, Y. Nojiri, and F. Okano, "A fast hologram calculation method for real objects," J. ITE, 61, 2, pp. 72–77 (2007) (in Japanese)

# 21

# Working Towards Developing Human Harmonic Stereoscopic Systems

Masaki Emoto and Hirokazu Yamanoue

We human beings want to see and hear what happens not only before our very eyes, but also at distant places in the world or outer space, including past events. Although we can now see images and hear sounds using electric equipment (i.e., a television set), displayed images are generally flat and have no depth. We want exciting simulated experiences that we cannot experience in daily life, where we can feel as though distant or recorded events happen before our very eyes. Stereoscopic three-dimensional (3-D) images enable us to experience the sensation of depth and to feel as though we are in the space shown by the stereoscopic video system. This feeling is called the sensation of "presence" or "being there." Generating the sensation of presence is one of the major purposes of various image systems. It is very effective to increase the sensation by enlarging the field of view of the video display, and by presenting depth information. Stereoscopic 3-D television is considered a promising broadcasting system for generating a sense of presence and is expected to add a new dimension to TV broadcasting services.

However, several issues need to be resolved, including addressing video hardware questions (i.e., the differences in characteristics between right and left cameras and displays) and human factors (i.e., visual fatigue on viewing stereoscopic images), before there will be wide public acceptance of stereoscopic TV. These issues are attributed to a mechanism specific to binocular vision (which rarely needs to be considered in two-dimensional TV), because viewers make full use of binocular visual functions to view stereoscopic images. The implementation of human harmonic stereoscopic systems

M. Emoto and H. Yamanoue
Science & Technical Research Laboratories, Japan Broadcasting Corporation (NHK)
1-10-11, Kinuta, Setagaya-ku, Tokyo 157-8510, Japan
e-mail: emoto.m-hy@nhk.or.jp

B. Javidi et al. (eds.), *Three-Dimensional Imaging, Visualization, and Display*,
DOI 10.1007/978-0-387-79335-1_21, © Springer Science+Business Media, LLC 2009

must first resolve two primary difficulties – the geometrical distortions of the represented space by a stereoscopic video system and visual fatigue on viewing stereoscopic images.

Geometrical distortion brings strange spatial perceptions to viewers. When viewing the real world, our visual system reconstructs a three-dimensional world from right and left two-dimensional retinal images [1]. Although the reconstruction of the 3-D world from right and left images is performed in the same manner when viewing a stereoscopic TV, the differences between human eye optics and stereo camera optics cause geometrical distortions of the represented space. This may be due to the close interaction that exists between the perception of distance and size in the visual system; the distance is determined by binocular parallax when viewing stereoscopic images, and perceptional inconsistency between the interacted distance and parallax-determined distance could occur.

The second issue in visual fatigue is most important for wide public acceptance of stereoscopic TV; viewing of stereoscopic TV may affect human health. Amusement park theatergoers enjoy stereoscopic programming; although the presentation length is usually short (typically 30 minutes or less), it has been found to be visually bearable. It is, however, quite common for viewers to complain about visual fatigue at some point after watching the program. Once a stereoscopic broadcasting TV system has been established, program viewing time will be extended to several hours. It is worth noting that young children still developing cranial and visual function may have the opportunity to view stereoscopic images. It is, therefore, imperative that before beginning the stereoscopic TV public broadcasting service we define the extent of any harmful effects on human health. If the health effect is not serious enough to require subsequent medical management, but produces a considerable degree of visual fatigue, stereoscopic broadcasting TV will not be widely accepted. A comprehensive study of potential visual fatigue associated with stereoscopic TV is justified in order to develop an eyestrain-free system.

First, we describe the geometrical distortions of the represented space caused by binocular parallax and stereopsis-based stereoscopic video system. The conversion rule from real space into stereoscopic images will be geometrically analyzed in Section 21.1.

Secondly, visual fatigue factors on viewing stereoscopic images will be studied. We propose a hypothesis to explain the binocular visual fatigue.

**Hypothesis**: The major cause of visual fatigue accompanied with stereopsis is the fusion efforts of the viewer in reconciling a stereopair of images with fusion difficulty.

This hypothesis is tied, by analogy, to the patients with visual function anomalies which experience visual fatigue when viewing the real world binocularly. Causes of difficulty in dealing with a fusion-difficult stereopair of images can be classified into two categories. After discussing the relationship between visual fatigue and these two categories in detail, the desirable ranges for each category of factors from a visual fatigue standpoint are covered in Section 21.2.

Lastly, based on our results of geometrical analysis and factors of visual fatigue study, we show how to avoid any undesired effects while maintaining enough depth information in 3-D TV programs (Section 21.3).

## 21.1 The Geometry of Reproduced 3-D Space and Space Perception

### 21.1.1 Setting Optical Axes in Stereoscopic Shooting

#### Shooting with Parallel Optical Axes

In this shooting method, which keeps the optical axes parallel to each other (hereafter referred to as the "parallel camera configuration"), the concept is to create a horizontal linear shift between the right and left images during image acquisition or viewing to show the object, which is at an infinite point, at a point over an infinite distance. If, as shown in Fig. 21.1, the distance between the two cameras ($d_c$) and the horizontal shift between the right and left images ($H_c$) are the same as the distance between the observer's right-left irises ($d_e$), and if the angle of view of the lenses and that of the display screen ($\theta$) are the same, distortion-free conditions are produced and the shooting space can be wholly reflected in stereoscopic images. These conditions set standards for converting real space to stereoscopic image space. It is not easy to satisfy these conditions at all times in broadcasting, where shooting and viewing conditions often vary. A more practical approach is required to deal with situations where these conditions are not easily met.

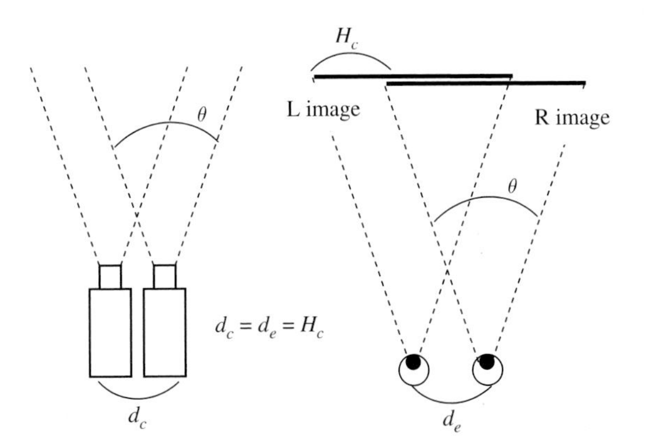


**Fig. 21.1.** An example of parallel camera configuration

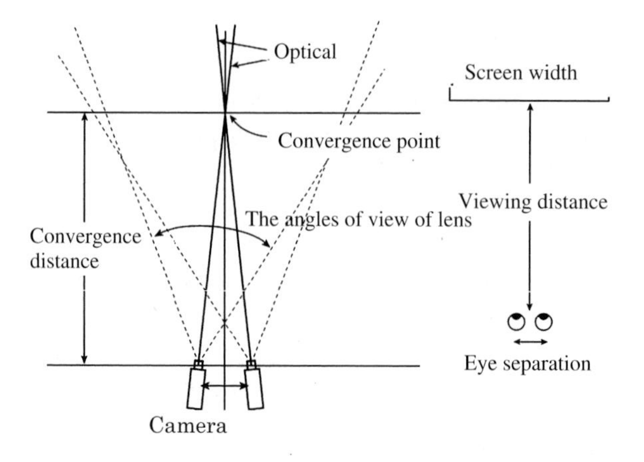


**Fig. 21.2.** An example of toed-in camera configuration

### Shooting with Crossed Optical Axes

As shown in Fig. 21.2, this shooting method with crossed optical axes (hereafter referred to as the "toed-in camera configuration") has an object at the point where the optical axes converge reproduced on the screen as a stereoscopic image. The degree of stereoscopic effects, influenced by the camera separation, varies around the point of optical axis convergence. This means that it is possible to relocate where the image forms, toward or away from the viewer, by moving the convergence point back and forth. At the same time, the camera separation can be changed to adjust the amount of stereoscopic effect (i.e., the sensation of watching the image coming at you or moving away from you). Because these stereoscopic effects can be easily manipulated, the toed-in camera configuration is frequently used in television program production. However, visual distortions such as the "puppet-theater effect" and "cardboard effect" often accompany 3-D pictures. The next section explains the conditions which give rise to these visual distortions.

## 21.1.2 Converting from Shooting Space to Stereoscopic Image Space

Figures 21.3 and 21.4 present models of the shooting and display systems. For simplicity, we use the center of an image free of keystone distortion. The parameters are shown in Table 21.1. The following equation can be established from Fig. 21.3:

$$\begin{aligned}&(L_b - L_c)\sin\theta_c : x' \\ &= \left\{(L_b - L_c)\cos\theta_c + \sqrt{\left(\tfrac{d_c}{2}\right)^2 + L_c^2}\right\} : L_s\end{aligned} \tag{21.1}$$

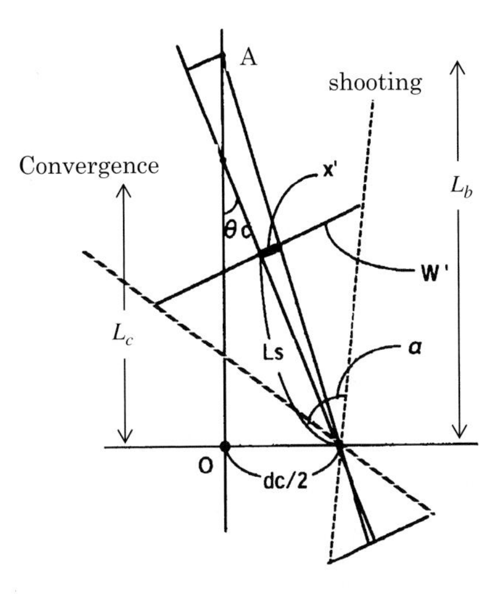


**Fig. 21.3.** A model of a shooting system

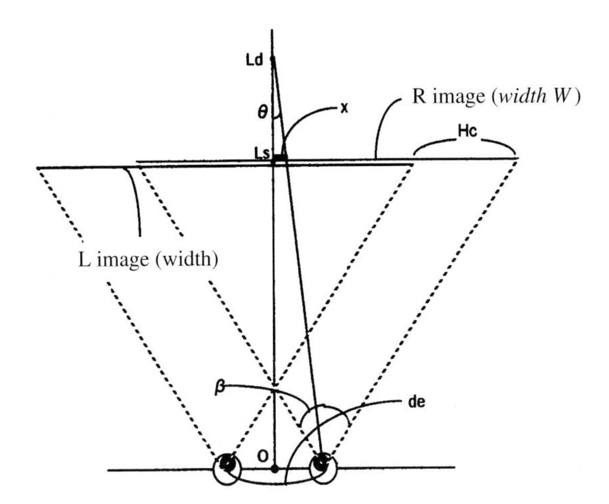


**Fig. 21.4.** A model of a display system

In case of ordinary 3-D program production, $d_c$ is several centimeters, and $L_c$ ranges from several meters to several dozen meters. Assuming that $\theta_c$ is sufficiently small, and that $\sin\theta_c \cong \frac{d_c}{2L_c}$, $\cos\theta_c \cong 1$, $\sqrt{\left(\frac{d_c}{2}\right)^2 + L_c^2} \cong L_c$,
we obtain:

$$x' = \frac{(L_b - L_c) \cdot d_c \cdot L_s}{2L_b \cdot L_c} \tag{21.2}$$

From Figs. 21.3 and 21.4, we obtain:

**Table 21.1.** Parameters in shooting and display models

| | |
|---|---|
| $d_c$ | Camera separation |
| $d_e$ | Eye separation |
| $L_b$ | Shooting distance |
| $L_c$ | Convergence distance |
| $L_s$ | Viewing distance |
| $L_d$ | Position of a stereoscopic object |
| $\alpha$ | Angles of view of lens |
| $\beta$ | Viewing angle |
| $\theta_c$ | Camera convergence angle |
| $\theta$ | Convergence angle of eye |
| $H_c$ | A horizontal gap between L and R images |
| W | The width of screen |
| W′ | The width of virtual screen at a viewing distance in the shooting model |
| x′ | The distance from the center of the virtual screen at a viewing distance in the shooting model (see Fig. 21.4) |

$$x = \frac{W}{W'}x' + \frac{H_c}{2} \tag{21.3}$$

With the angle of the eyes′ convergence being $2\theta$ when seeing a stereoscopic image, we derive:

$$\tan\theta = \frac{d_e}{2L_d} = \frac{1}{L_s} \cdot \left(\frac{d_e}{2} - x\right) \tag{21.4}$$

where we define "camera separation ratio" as

$$\frac{d_c}{d_e} = a_1 \tag{21.5}$$

and the magnification of an image on the retina as

$$\frac{W}{W'} = \frac{\tan\frac{\beta}{2}}{\tan\frac{\alpha}{2}} = a_2 \tag{21.6}$$

By substituting (21.2), (21.3), (21.5), and (21.6) into (21.4), we now have the final location, $L_d$, where the stereoscopic image is formed:

$$L_d = \frac{1}{\frac{1}{L_s} - \frac{a_1 \cdot a_2}{L_c} + \frac{a_1 \cdot a_2}{L_b} - \frac{H_c}{L_s \cdot d_e}} \tag{21.7}$$

### Parallel Camera Configuration

Here, we assume in equation (21.7) that the distance to the convergence point is infinite. As for $H_c$, the horizontal shift of right-left images, we also assume that $H_c = d_e$ as the infinite point during the shooting will also be located at an infinite point during the viewing. With these assumptions, equation (21.7) can be simplified to:

$$L_d = \frac{L_b}{a_1 \cdot a_2} \tag{21.8}$$

As this equation shows, if shooting is performed in the parallel camera configuration and the amount of right–left horizontal shift equals the distance between the observer's irises $H_c = d_e$, linearity can be retained between the shooting distance to the object and the distance at which a stereoscopic image is formed. An example is shown in Fig. 21.5. This shows the relation-

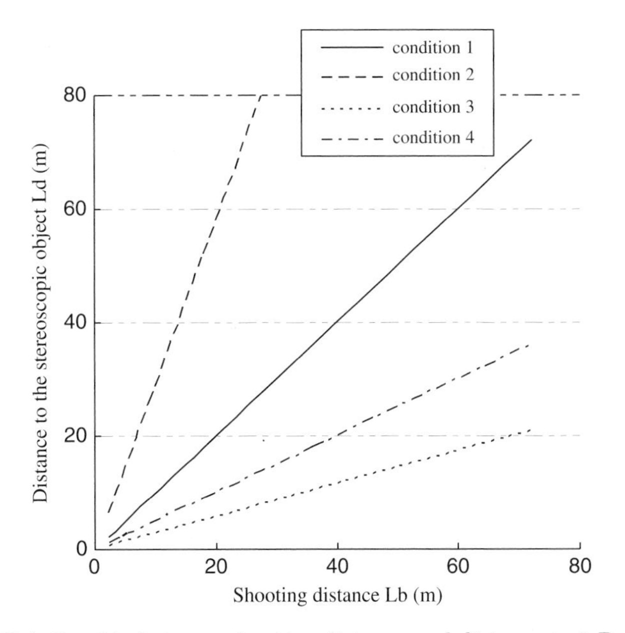


**Fig. 21.5.** Relationship between shooting distance and distance to 3-D objects shot by the parallel camera configuration

**Table 21.2.** Examples of shooting and viewing conditions

| Condition | 1 | 2 | 3 | 4 | 5 | 6 | 7 | 8 |
|---|---|---|---|---|---|---|---|---|
| Camera configuration | Parallel | | | | Toed-in | | | |
| Camera separation ($d_c$, in mm) | 65 | 30 | 90 | 130 | 65 | 65 | 32.5 | 130 |
| Angles of view of lens ($\alpha$ , in degrees) | 33.4 | 43.6 | 13.7 | 33.4 | 33.4 | 43.6 | 43.6 | 43.6 |
| Convergence distance ($L_c$, in m) | $\infty$ | | | | 4.5 | 4 | 1.8 | 6.1 |
| Viewing angle ($\beta$, in degrees) | 33.4 | | | | | | | |
| Viewing distance ($L_s$, in m) | 4.5 | | | | | | | |
| Horizontal shift ($H_c$, in mm) | 65 | | | | 0 | | | |
| $a_1$ | 1.00 | 0.46 | 1.38 | 2.00 | 1.00 | 1.00 | 0.50 | 2.00 |
| $a_2$ | 1.00 | 0.75 | 2.50 | 1.00 | 1.00 | 0.75 | 0.75 | 0.75 |
| $L_c/a_1 * a_2 * L_s$ | | | | | 1.00 | 1.19 | 1.07 | 0.90 |

ship between the shooting distance ($L_b$) and the distance to the 3-D images ($L_d$) shot by the parallel camera configuration under the conditions shown in Table 21.2. When an object is shot by this camera configuration, the projected 3-D image moves toward the viewer as the distance between the cameras widens, but it moves away from the viewer when the lens view angles expand.

### Toed-in Camera Configuration

Unlike the parallel camera configuration, this shooting method moves the convergence point back and forth without creating a horizontal shift between right and left images. Assuming that $H_c = 0$ in equation (21.7), we obtain:

$$L_d = \frac{1}{\frac{1}{L_s} - \frac{a_1 \cdot a_2}{L_c} + \frac{a_1 \cdot a_2}{L_b}} \tag{21.9}$$

Figure 21.6 shows the relationship between shooting distance ($L_b$) and the distance to the 3-D images ($L_d$) shot by the toed-in camera configuration under the conditions shown in Table 21.2. Under the toed-in camera configuration, as the graph shows, $L_d$ begins to undergo a major change when $L_c = a_1 \cdot a_2 \cdot L_s$. When $L_c < a_1 \cdot a_2 \cdot L_s$, $L_d$ cannot be calculated by equation (21.9) if the shooting distance ($L_b$) is large. In this situation the eyes do not converge, but instead diverge, as the distance between the right–left images becomes greater than the space between the right–left irises. Apparently, this situation does not immediately mean a lack of image fusion. If $L_c = a_1 \cdot a_2 \cdot L_s$

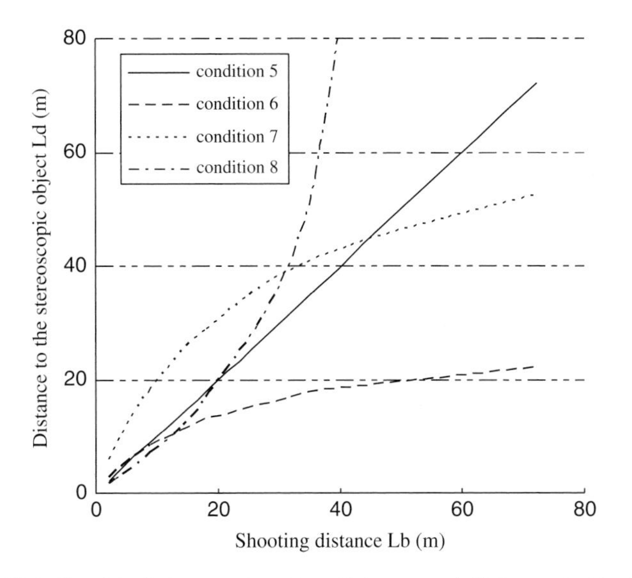


**Fig. 21.6.** Relationship between shooting distance and distance to the 3-D objects shot by the toed-in camera configuration

(Condition 5 in Table 21.2), equation (21.9) becomes the same as equation (21.8), meaning that there is no distortion at the center of the screen, but keystone distortion occurs in the periphery.

### 21.1.3 Puppet-Theater Effect

The "puppet-theater effect" is a size distortion characteristic of 3-D images that has not been clearly defined. Generally, a 3-D image of a shooting target looks unnaturally small. This effect is a problem which sometimes spoils the sensation of reality. In particular, if shooting targets are human beings, people in the 3-D image look unnaturally tiny. How observers perceive the apparent size of the object in the image is largely influenced by such factors as how familiar they are with the object, as no one can evaluate the size of an object they have never seen before. The puppet-theater effect is, therefore, not perceived as a physically measurable amount; rather, it is a concept that can be subjectively evaluated. However, in this chapter we try to evaluate the effect quantitatively, under the assumption that the observers know the shooting target well.

MacAdam [2] discusses the size distortion that occurs between an object in the foreground and one in the background of an image, focusing on the relationship between the depth information from 2-D images (perspective) and 3-D images (binocular parallax) (MacAdam 1954). He showed that, depending on the shooting distance to the object, the reproduction magnification of each image's apparent size may differ, referring to a possible link with the puppet-theater effect. This idea also applies in this chapter, where we regard the puppet-theater effect as the distortion of size caused by the fact that the reproduction magnification of the image's apparent size differs depending on the shooting distance to the object. We can calculate the relative size differences that occur between shooting objects inside the image space, arising from the fact that the reproduction magnification of an object differs between the foreground and background. The results of this calculation will be used as the reference amount when analyzing the puppet-theater effect. Absolute size may also contribute to this visual distortion, but this point is not addressed here.

Relative size can be perceived in two ways: an object looks larger, or other objects look smaller. Which of these is perceived largely depends on whether the object is in the background or foreground, the content of the object, its patterns, and other factors. In other words, the object's picture patterns and content largely determine whether the object in the background looks smaller (or larger) when the foreground serves as a reference point, or whether the object in the foreground looks smaller (or larger) when the background serves as a reference point. This chapter does not deal with these matters. Rather, it handles the puppet-theater effect as a distortion of size, as dimensional mismatching between shooting targets. However, in general 3-D programs, the puppet-theater effect often makes objects in the foreground look unnaturally small when the background serves as a reference point.

With these assumptions, this chapter determines that the puppet-theater effect is likely to occur if the reproduction magnification of an image's apparent size is dependent on the shooting distance to the object. If the shooting distance to objects in the background or foreground is known, the reproduction magnification of their images can be calculated, and the ratio of these two reproduction magnifications can then be used as a reference and predictor of the puppet-theater effect.

If $W_b$ is the real size in the shooting space (real space) of the object and $W_r$ is its apparent size in the stereoscopic image space, we obtain

$$W_r = \frac{L_d}{L_b} \cdot \frac{\tan\frac{\beta}{2}}{\tan\frac{\alpha}{2}} \cdot W_b = \frac{L_d}{L_b} \cdot a_2 \cdot W_b \tag{21.10}$$

where the reproduction magnification of image ${}^{W_r}/_{W_b}$ corresponds to the apparent value commonly known as the lateral magnification of a lens optical system.

### Parallel Camera Configuration

From equations (21.8) and (21.10), the reproduction magnification of the image (lateral magnification) $M_s \left(\cong \frac{W_r}{W_b}\right)$ can be expressed as:

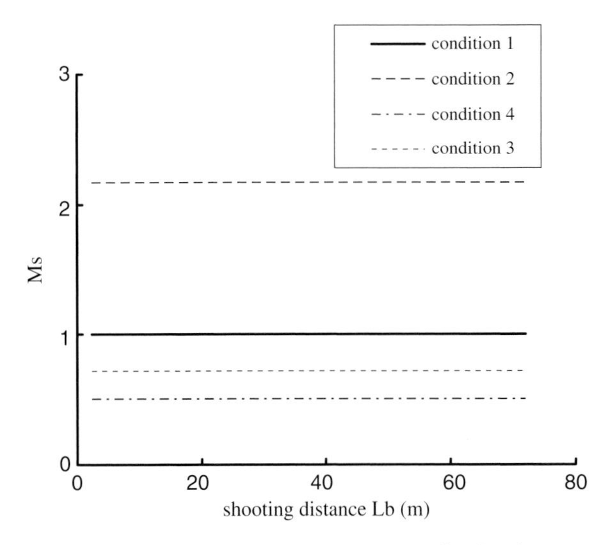


**Fig. 21.7.** The ratio of real size to apparent size of 3-D objects as shot by the parallel camera configuration

$$M_s = \frac{W_r}{W_b} = \frac{1}{a_1} \tag{21.11}$$

This means that, with a parallel camera configuration, reproduction magnification of the image is determined only by the camera interval ratio $a_1$. Under the definition of the puppet-theater effect mentioned earlier, however, this visual distortion does not occur, as the shooting distance $L_b$ is not contained. Figure 21.7 shows the relationship between $M_s$ and $L_b$ when 3-D images are shot by the parallel camera configuration under the conditions shown in Table 21.2.

## Toed-in Camera Configuration

From equations (21.9) and (21.10) we obtain:

$$M_s = \frac{W_r}{W_b} = \frac{1}{\dfrac{1}{L_s} - \dfrac{a_1 \cdot a_2}{L_c} + \dfrac{a_1 \cdot a_2}{L_b}} \cdot \frac{a_2}{L_b} \tag{21.12}$$

Figure 21.8 shows the relationship between $M_s$ and $L_b$ when 3-D images are shot by the toed-in camera configuration under the conditions shown in Table 21.2. If $L_c = a_1 \cdot a_2 \cdot L_s$, equations (21.12) and (21.11) are the same; if not, the reproduction magnification (lateral magnification) $M_s$ is dependent on the shooting distance $L_b$. The following cases should be considered: (1) $L_c > a_1 \cdot a_2 \cdot L_s$ – under Conditions 6 and 7 in Fig. 21.8, the image of the shooting target that forms the background looks smaller than that in the foreground;

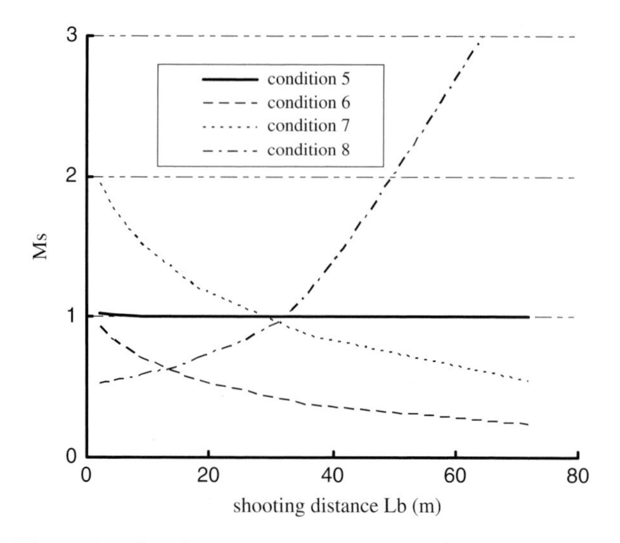


**Fig. 21.8.** The ratio of real size to apparent size of 3-D objects as shot by the toed-in camera configuration

(2) $L_c < a_1 \cdot a_2 \cdot L_s$ – under Condition 8, conversely, the image of the shooting target that forms the foreground looks smaller than that in the background. When $L_c << a_1 \cdot a_2 \cdot L_s$ in particular, the reproduction magnification of the object′s apparent size becomes greatly dependent on the shooting distance, making the puppet-theater effect more likely. Whether that effect can be recognized is largely dependent on the patterns and content of the image. For quantitative referencing, we need information on the shooting distance to the object which is either in the background or the foreground.

### 21.1.4 Cardboard Effect

The "cardboard effect" is a phenomenon in which a shooting target looks layered (i.e., consisting of flat object and flat background), though an observer can grasp the situation in front of and behind the target. A cardboard effect occurs when viewers perceive 3-D objects as being flat, although at a different depth than other objects or background. This effect has also not been clearly defined, and is a problem which spoils the sensation of reality in 3-D images. In general 3-D program production, the effect is likely to occur when the shooting target is far away and the convergence point is set at the shooting target. In this situation, narrow lens viewing angles are often used in order to locate the target in a large retina size and set the 3-D position of the target around the display or the screen. A comprehensive approach to this phenomenon as a shape perception distortion [3] is necessary.

If depth $\Delta L_b$ at the shooting distance $L_b$ in real space is reproduced as depth $\Delta L_d$ in 3-D image space, the depthwise reproduction magnification $M_d$ can then be expressed as follows:

$$M_d = \frac{\Delta L_d}{\Delta L_b} \tag{21.13}$$

To precisely reproduce the shape of an object during a conversion from real space to 3-D image space, the reproduction magnification $M_s$ (lateral magnification) of the apparent size must be the same as this depthwise reproduction magnification $M_d$. Here, the degree of thickness $E_c$, which is the reference amount of cardboard effect, is defined as follows:

$$E_c = \frac{M_d}{M_s} = \frac{\frac{\Delta L_d}{\Delta L_b}}{M_s} \tag{21.14}$$

If the reproduction magnification in the direction of depth is smaller than the size in equation (21.14), meaning that $E_c$ is small, the cardboard effect is likely to occur. Other likely factors contributing to this phenomenon are shadings, contours, textures, and motion parallax. This chapter, as mentioned earlier, focuses on geometrical analysis based on binocular parallax.

### Parallel Camera Configuration

From equations (21.8) and (21.11), we obtain:

$$E_c = \frac{\frac{dL_d}{dL_b}}{M_s} = \frac{\frac{1}{a_1 \cdot a_2}}{\frac{1}{a_1}} = \frac{1}{a_2} \tag{21.15}$$

Equation (21.15) shows that, under a parallel camera configuration, the cardboard effect is dependent on the ratio of the lens′s field of angle and the angle of viewing.

### Toed-in Camera Configuration

From equations (21.9) and (21.12), we obtain:

$$E_c = \frac{\frac{dL_d}{dL_b}}{M_s} = \frac{a_1}{L_b} \cdot \frac{1}{\frac{1}{L_s} - \frac{a_1 \cdot a_2}{L_c} + \frac{a_1 \cdot a_2}{L_b}} \tag{21.16}$$

If the image of the shooting target forms near the screen, $L_b$ equals $L_c$ and equation (21.16) can be simplified as follows:

$$E_c = a_1 \cdot \frac{L_s}{L_c} \tag{21.17}$$

Therefore, to reduce the cardboard effect with the toed-in camera configuration, it is effective to increase the distance between the cameras, shorten the distance to the point of convergence, or increase the viewing distance. But, in actual 3-D program production, it is impossible to increase the distance between the cameras to a very large distance, because general 3-D images include foreground objects or background objects around a shooting target that, due to their large binocular disparity, cannot be seen as stereoscopic objects.

Subjective evaluations of puppet-theater and cardboard effects were conducted and we found a significant correlation between subjective evaluation data and $1/a_2$, the reference/prediction amount of the puppet-theater effect obtained by analysis based on binocular parallax, and a significant correlation between subjective evaluation data and $E_c$, the reference amount of the cardboard effect obtained by analysis based on binocular parallax. In actual 3-D program production, this geometrical analysis plays one of the most important roles in the cardboard effect [4].

### 21.1.5 Summary

Converting real space into stereoscopic images was analyzed under several shooting and viewing conditions. The study focused on how optical axes are set, which has received little attention in the past, and clarified the relationship between the shooting distance to the object and the display position in 3-D image space. The results show that this relationship remains linear when objects are shot by the parallel camera configuration (optical axes are kept parallel to each other). However, depending on parameters, the relationship is radically nonlinear if objects are shot using the toed-in camera configuration (optical axes converge at some point).

We then studied the conditions that give rise to the puppet-theater and cardboard effects when using these two camera configurations. The puppet-theater effect is defined as a size-related distortion caused by the relative size disparity of shooting objects in the image space. The puppet-theater effect does not occur under parallel camera configuration conditions, as the reproduction magnification of the image is constant and not dependent on the shooting distance. Using the toed-in camera configuration, we showed that this visual distortion may occur because such magnification is sometimes dependent on shooting distance.

The cardboard effect is defined as the ratio between the reproduction magnification of the image's apparent size and depthwise reproduction magnification. We also showed that under parallel camera configuration conditions, this effect can be referred to by the ratio $1/a_2$ of the shooting angle and the viewing angle. Furthermore, with toed-in camera configurations, the inter-camera distance, distance to the convergence point, and viewing distance are closely related to the cardboard effect.

### 21.1.6 Geometry Mapping Simulation System

Based on the conversion rules from real space into stereoscopic images described above, a geometry mapping simulation system was developed [5]. This system requires the shooting, display, and viewing conditions to calculate the mapping from real space into stereoscopic images as inputs. It can be used to predict the extent of the perceived puppet-theater and cardboard effects. The magnitude of spatial distortion and the extent of the puppet-theater and cardboard effects are displayed using a space grid whose size can be estimated based on the objects′ depths, calculated from the binocular parallax of the acquired stereoscopic images. This system can also be used to predict excessive binocular parallax and excessive parallax distribution, which affect levels of comfort and visual fatigue in watching stereoscopic images, as described below.

## 21.2 Binocular Fusion, Stereopsis, and Visual Comfort

This section describes the visual fatigue factors related to viewing stereoscopic images. The scope of this study is limited to visual fatigue factors specific to stereopsis when watching stereoscopic images, although other factors such as fast movement on a wide field of view display, frequent cutaways, and low-resolution, can affect visual fatigue when viewing two-dimensional (2-D) images. When viewing the real world, our visual system reconstructs a 3-D world from our right and left 2-D retinal images [1]. For the eyes to fixate on an object, the visual axes of the right and left eye converge at the object point. When the convergence is complete, the difference between any two retinal images is zero. Viewers can then perceive the object as a single image. Other objects in a limited area near the fixated object also can be seen as single objects, in spite of having a slightly different position on the retinae; this area is called Panum's fusional area [6]. Right and left retinal images of objects outside Parnum's fusional area need eye movement (i.e., convergence or divergence) to be fused binocularly. Before eye movement, an object image is dropped to a point on the left retina, while the same object image is dropped to a slightly different point on the right retina. The difference between the retinas is called "binocular disparity;" convergent or divergent eye movement is evoked by binocular disparity [7].

To complete stereopsis adequate eye movement is needed as well as the coordination of some visual functions. The function essential to normal stereopsis can be treated as having hierarchical grades, as Worth suggested [8]:

1st grade: **Simultaneous macular perception** – defined as superimposition of dissimilar images. This is tested with a dichoptic display, such as a haploscope presenting a lion image for the left eye and a cage for the right eye.

2nd grade: **Fusion** – defined as simultaneous foveal perception after evoked convergent or divergent eye movement. This is tested with a prism bar or Risley prism in front of the eyes.

3rd grade: **Stereopsis** – defined as the blending of slightly dissimilar images from the retinae with the perception of depth.

These hierarchical layers function not separately, but together; this hierarchical understanding is very useful to address and manage visual fatigue. When we do feel visual fatigue in daily life viewing in the real world? Some people with normal visual function feel visual fatigue in daily life, as do many individuals with abnormal visual function. The causes of a patient's visual fatigue are visual malfunctions which make it difficult to attempt fusion; therefore, they lead patients to considerable fusional efforts. Patients must overcome the difficulties to fuse two images and, over a sustained period of time, the effort to do this tires the visual system (i.e., vergence system).

We have discussed several visual functions related to visual fatigue when viewing real world objects binocularly, before referring to visual fatigue in when viewing with a stereoscopic image system.

**1. Horizontal Ocular Position:** The visual function most closely linked to the degree of difficulty of fusion and the degree of effort needed to fuse right and left images, is "vergence function." When horizontal binocular parallax is presented to the visual system, which is the direct distance cue, vergence function is evoked to fuse binocularly. To fuse binocularly, the two visual axes of the left and right eyes are directed parallel each other when viewing an object at infinity, and the shorter the distance to the object, the nearer the cross point of the two visual axes. If a viewer's horizontal ocular position shifted outside of the head (i.e., exophoria or exotropia), he must make an effort at convergence to fuse near objects. If a viewer's horizontal ocular position shifts inside the head (i.e., esophoria or esotropia), he must diverge to fuse distant objects. This shift of the horizontal ocular position can cause visual fatigue due to the fusion effort needed to overcome the shift.

**2. Vertical Ocular Position:** The viewer's vertical ocular position, as well as the horizontal, can affect the degree of fusion effort. If a viewer's one ocular position shifted upwards or downwards, compared to another ocular position (i.e., hyperphoria, hypophoria, hypertropia, or hypotropia), he had difficulty fusing and had to make an effort to fuse viewing objects. Furthermore, the interaction between horizontal and vertical ocular position must be considered, because the eye movement is controlled by three pairs of extraocular muscles, which cannot control the horizontal and vertical eye movement independently. Therefore, the vertical ocular position could affect the horizontal fusion characteristic.

**3. Torsional Ocular Position:** Torsion is the rotatory ocular position and its axis is the visual axis. If one eye rotated compared to another ocular position (i.e., cyclophoria or cyclotropia), the viewer must make an effort to rotate his one eye to the opposite orientation to fuse viewing objects.

**4. Aniseikonia:** Aniseikonia is a condition in which the two retinal images have an unequal size. We need asymmetry vergence eye movement when viewing objects far from the midline, where the distances from the two eyes are different. The size of the image on the retina nearer the object to be viewed is slightly larger than the other, and this is the physiological aniseikonia, which is not a problem. When a patient, whose two eyes have unequal refractive power, corrects his vision by different powers of lenses, there is a difference in size between right and left retinal images. Viewers have a great difficulty in binocularly fusing such a pair of images with size difference, and need a considerable degree of fusion effort which results in viewer visual fatigue.

**5. Binocular Rivalry:** Binocular rivalry is a phenomenon in which visual perception alternates between right and left images and their mixture as a mosaic. This is common to color rivalry and perception is unstable. Viewers are disturbed by fusing images, and they are forced to fuse and, consequently, experience visual fatigue.

**6. Combination of a Number of Factors:** When the factors mentioned above combine, viewers need a great degree of fusion effort and tire easily.

By drawing an analogy to the factors involved in viewing real objects binocularly, we can proceed to visual fatigue factors associated with images. The major cause of visual fatigue associated with stereopsis is due to viewers′ efforts to fuse image stereopairs with fusion difficulty. The explanation for this difficulty can be put into two categories: in-principle factors and non-principle factors. In-principle factors are related to horizontal binocular parallax and to horizontal ocular position. The factors in the non-principle category are related to differences in the characteristics of right and left image equipment and to various visual functions (except horizontal ocular position). We will enumerate the factors involved in fusion difficulty when viewing stereoscopic images with numbers corresponding to those above for viewing real objects. Table 21.3 was the correspondence between the factors of the visual fatigue and the visual functions.

**1. Horizontal Ocular Position (In-principle Factor):** In viewing stereoscopic images, an object at the camera convergence point with zero horizontal binocular parallax is presented and perceived on a display screen. When the object comes close to the camera, the horizontal binocular parallax represented on the screen is increased and the viewers′ vergence point also comes close to themselves. When the horizontal binocular parallax is small enough for the vergence point to be included within the depth-of-focus, the viewer′s visual system cannot detect any blurring of images on the retinae. When the horizontal binocular parallax is large enough for the vergence point to not be included within the depth-of-focus, the viewer′s visual system can detect blur

**Table 21.3.** Correspondence between factors of visual fatigue and visual function

| No. | Factors of visual fatigue | Visual functions |
|---|---|---|
| | **In-principle factor** | |
| 1 | Horizontal binocular parallax | Horizontal ocular position |
| | **Non-principle factors** | |
| | Differences in (between right and left images) | |
| 2 | Vertical position | Vertical ocular position |
| 3 | Torsional position | Torsional ocular position |
| 4 | Size | Aniseikonia |
| 5 | Luminance<br>Contrast<br>Color(Hue) | Binocular rivalry |

of the images on the retinae and this causes the viewer to compensate for this. This relationship between vergence and accommodation functions in viewing stereoscopic images differs from viewing in the real world. The dissociation of vergence and accommodation exists, and this dissociation is inevitable to stereoscopic image systems. In order to fuse a stereoscopically displayed object with large horizontal binocular parallax, viewers must be able to fuse images under a wide range of horizontal binocular parallax, while the range is limited to a certain area. The existence of fusional limitations means that a large horizontal binocular parallax requires considerable efforts to fuse, that will last for extended periods of time.

Furthermore, in viewing stereoscopic images, an object varies not only in the lateral, but also anteroposterior position. This demands that viewers suppress their accommodation when dissociation of vergence and accommodation exists, and the demands continue over time. Cutaway in the video sequence might make horizontal binocular parallax change discontinuous, and the viewers need to renew the relationship between vergence and accommodation. Periodic renewal of the relationship between vergence and accommodation occurs in real world viewing because the viewer willingly moves his line of sight from one object to another, while they are passive in viewing cutaway of stereoscopic images. Horizontal binocular parallax drives viewer′s vergence, because the viewer cannot suppress vergence accompanied by the movement of the fixated object. Viewers′ vergence system could be overused by the excessive horizontal binocular parallax and its discontinuous changes over time. This visual image overload might cause visual fatigue in viewing stereoscopic images for extended periods of time and, after an effort, some visual functions might not be able to maintain function at the pre-load level.

**2. Vertical Ocular Position (Non-principle Factor):** Right and left images with a vertical offset presented on the stereoscopic display trigger fusion difficulty. This offset is caused by insufficient alignment of the left and right cameras when shooting, and of right and left displays when replaying, and it causes the viewers′ vertical ocular positions to shift, which is common to hyperphoric, hypophoric, hypertropic, or hypotropic patients. When the camera's stereopair is placed on a slant, horizontal binocular parallax transfers to the vertical parallax. The vertical parallax needs the vertical fusion effort to align two ocular heights. Thus, the vertical ocular position could affect the horizontal fusion characteristic.

**3. Torsional Ocular Position (Non-principle Factor):** When the roll angle of each camera in a stereopair is different, the rotatory ocular position is needed to fuse. This ocular position is common to cyclophoria or cyclotropia patients.

**4. Aniseikonia (Non-principle Factor):** When the zoom lenses of a stereopair each have different zoom ratios, a difference in size between the right and left images exists. These differently sized images are difficult to fuse and cause fusion effort, which is common to aniseikonic patients.

**5. Binocular Rivalry (Non-principle Factor):** When a stereopair of images with large differences in luminance, color, and geometric deformation cannot fuse, the viewers perceive binocular rivalry, and feel visual fatigue.

**6. Combination of a Number of Factors:** When the factors mentioned above combine, viewers need a great degree of fusion effort and tire out similar to when viewing a real object.

The most important and overriding issue of stereoscopic image systems is that they have an ability to present fusion-difficult image stereopairs to the viewers as with a real object; furthermore, they have an ability to present more fusion-difficult stereopairs of images than the real object. This ability might cause a great deal of fusion effort to fuse two retinal images which can never exist in the real world. Visual fatigue factors must be considered when in viewing stereoscopic images just as we consider factors in viewing the real world binocularly. The changes of overworked visual functions specific to the binocular stereopsis could be used as a quantitative index of the visual fatigue, though visual fatigue is inherently subjective. This is based on the idea that the decline of some visual function after viewing stereoscopic images might reflect the visual fatigue. It is not clear yet which visual functions are adequate to quantitatively evaluate the visual fatigue in viewing stereoscopic images. We must explore which visual functions are adequate to quantify the visual fatigue in viewing stereoscopic images, or which visual functions decline after viewing stereoscopic images.

First, we explore the decreased visual function after viewing stereoscopic image systems with in-principle and non-principle fusion disturbance factors to establish indices of visual fatigue. Hereafter, we call this system a "non-ideal stereoscopic image system." In Section 21.2.1, we will discuss an experiment conducted to evaluate visual fatigue in viewing a non-ideal stereoscopic image system compared to the visual fatigue produced when viewing a two-dimensional image system. The results suggest that this may be an adequate index of visual function to quantify visual fatigue.

Second, we conducted an experiment to quantify visual fatigue by a quantitative index in viewing stereoscopic image systems with only in-principle fusion disturbance factors. Hereafter, we call this system an "ideal stereoscopic image system." In Section 21.2.2, we focused attention on the presented distance, which is added to the two-dimensional image system, to study how spatial distribution of objects effects visual fatigue or visual comfort.

Third, in Section 21.2.3, we will study the effects that temporal distribution of the objects has on visual fatigue.

Fourth, we then leave in-principle fusion disturbance factors, and in Section 21.2.4 we refer to the desirable ranges of non-principle fusion disturbance factors.

Finally, Section 21.3 presents some strategies to avoid undesired effects.

### 21.2.1 Visual Functions as Indices of Visual Fatigue When Watching Stereoscopic Images

In the first study on visual fatigue specific to viewing stereoscopic images, we measured some visual functions pre- and post-exposure to stereoscopic images to establish objective and quantitative fatigue indices. A non-ideal stereoscopic image system is used to present a stereopair of images which included both in-principle and non-principle fusion disturbance factors. We compare the degree of fatigue and the changes in visual function after viewing two-dimensional and stereoscopic images. The results revealed the visual function parameters useful for quantifying the visual fatigue, because decreased visual functions post-exposure to stereoscopic images could be objective and quantitative indices of the fatigue.

Video display terminals (VDTs) [9, 10, 11], head-mounted displays (HMDs) [12], and stereoscopic TV, among other devices, have been recent targets for assessing post-viewing visual fatigue. Several of the studies measured visual discomfort or fatigue after viewing VDTs and HMDs, but rarely after viewing stereoscopic TV. In these studies, a battery of visual and laboratory parameters (as well as subjective reports), were employed, including visual acuity, stereoacuity, fixation stability, critical temporal fusion frequency, accommodative response, heterophoria, convergent eye movement, spatial contrast sensitivity, near and light pupillary reactions, blinking, tear film breaking time, electroencephalogram (EEG), pulse rate, and respiration rate. Experimental parameters for optimal visual fatigue assessment should fulfill a few basic requirements – the measurements should be simple and non-interventional, but still quantitative with high sensitivity and specificity.

A study using time sharing high-definition TV was developed. Test participants were asked to view a stereoscopic display having relatively small horizontal binocular parallax requiring 1.6 arc degrees (2.8 prism diopters) of convergence and 0.7 arc degrees (1.2 prism diopters) of divergence through a liquid crystal shutter for 24 minutes. These conditions did not present significant changes in subjective symptoms or visual function after viewing [13]. Time sharing types of display systems inevitably have leakage between images ("cross talk") which will be described in Section 21.2.4. Image cross talk affects human visual functions, causing decreased image stereopair fusional amplitudes during long presentation times, long enough, for the eyeballs to move [14]. Viewing images with a larger horizontal binocular parallax for an extended period of time is likely to affect visual function. In the present study, we designed an experiment in which test participants viewed stereoscopic images through polarizing glasses with reduced cross talk and evaluated visual fatigue by subjective assessment along with a battery of objective tests after viewing images for approximately 60 minutes.

Stereopsis, in stereoscopic display viewing, can be achieved by ocular vergence following changes in horizontal binocular parallax imaging. The action may produce strain on the vergence eye movement system and affect fusional amplitudes for binocular single vision [15]. When viewing stereoscopic displays distinct from that of real objects, the dissociation of vergence and the accommodation could affect the AC/A ratio [16, 17].

Thus, we measured fusional vergence limits and AC/A ratios before and after one hour of viewing stereoscopic images presented by a non-ideal image system and compared them with the values of two-dimensional images. Our major interests are, firstly, whether the degree of visual fatigue from viewing stereoscopic images is larger than the one caused by viewing conventional two-dimensional images which is now accepted worldwide; secondly, to determine the appropriate index of visual fatigue to quantitative analysis, and thirdly, whether viewers can recover from visual fatigue or not. Our results will reveal the appropriate index of visual fatigue in viewing stereoscopic images.

## Participants and Methods

Twelve healthy adults participated in the study. Prior to the start of the experiments, visual function screening tests were performed to ensure study participants had healthy eyes. Stereoscopic ability was examined using two charts selected from the standard stereoscopic test materials [18], Number 4 (circle test) for testing static stereopsis and stereoacuity and Numbers 9 and 10 (Maddox test) for testing heterophoria. Six participants were able to free fuse in perceiving binocular single vision, and the remaining six were to unable to free fuse by using random dot stereogram. Image displays on the two 32-inch high-definition TV monitors were polarized. Participants adjusted to viewing stereoscopic images by wearing polarizing glasses, each right-angled with polarization filters worn by each eye, and to view monoscopic images by wearing polarization glasses with same-angled polarization filters.

### Viewing Condition and Procedures

Participants viewed images under standard viewing conditions for high-definition TV [19, 20]. The images were presented at a distance of 1.2 m and had a horizontal visual angle of about 33 arc degrees.

The experimental procedure consisted of a pre-viewing session, a post-viewing session followed by response to a questionnaire, and two testing sessions, each, after taking a rest (each completed with rest periods in between). Measurements included fusional vergence limits, AC/A ratio, and the participant's self-evaluation via a questionnaire. The stimulus AC/A ratio was determined by the gradient method using a prism bar and a concave (minus) lens. Fusional vergence limits were determined using test materials Numbers 5 and 6 (bar test) [18]. The fusional vergence measurements were repeated twice for each participant, with measured values averaged. Differentials in the fusional vergence break point between the convergence and divergence sides were defined as the"fusional amplitude."

Participants were required to view a stereoscopic high-definition TV (HDTV) program displayed on the monitor screens for one hour. After exposure, participants categorized the level of their fatigue into five categories: Severe fatigue, Fatigue, Moderate fatigue, Slight fatigue, and Not at all. They also were asked to select their physical status from the symptoms shown in Table 21.4, and write down any additional comments.

### Results

Figure 21.9 shows each participant's self-evaluation of visual fatigue after viewing monoscopic and stereoscopic images. Of 12 participants, five reported a more marked visual fatigue with stereoscopic viewing, six felt a similar degree of visual fatigue after viewing stereoscopic and monoscopic images, and one claimed more visual fatigue after monoscopic viewing. Figure 21.10 describes the symptoms of participants. Table 21.5 shows viewers comments. Visual complaints are more prominent after viewing stereoscopic TV.

Two participants did not provide AC/As ratios because of insufficient accommodative ability. Figure 21.11 shows AC/A ratio data (mean and standard deviation) obtained from five participants who were able to free fuse, and five who were unable-to-free fuse. In the analysis of variance (ANOVA), there was no significant difference in the AC/A ratio between pre- and post-viewing of stereoscopic images, between participants who were able-to-free fuse and those who were not, nor between the interactions of the two main effects. Likewise, as it pertains to viewing of monoscopic images, the AC/A ratio was not significantly affected with image viewing, free fuse ability, or

**Table 21.4.** Symptom Questionnaire

| | |
|---|---|
| 1 | Severe eye fatigue |
| 2 | Moderate eye fatigue |
| 3 | Llittle eye fatigue |
| 4 | Feel heavy in eyes |
| 5 | Interior eye ache |
| 6 | Feel heavy in eyes with strain |
| 7 | Heliophobia |
| 8 | Gritty eyes |
| 9 | Tear eyes |
| 10 | Smarting eyes |
| 11 | Dry eyes |
| 12 | Bloodshot eyes |
| 13 | Blur in gazing |
| 14 | Difficulty in focusing on distant objects |
| 15 | Difficulty in focusing on near characters |
| 16 | Headache |
| 17 | Brow ache |
| 18 | Headache, if shaken |
| 19 | Temple ache |
| 20 | Shoulder stiff |
| 21 | Shoulder pain |
| 22 | Arm or finger pain |
| 23 | Back pain |
| 24 | Poor digestion condition |
| 25 | Nausea |
| 26 | Frequently taking digestives |
| 27 | Feel heavy in legs |
| 28 | Feel cold in legs |
| 29 | Sometimes feel giddy |

**Table 21.5.** Viewer Comments

| | | |
|---|---|---|
| Monoscopic viewing | Eye Symptoms | Bleary eyes<br><br>Irritating eyes<br>Feel heavy in moving eyes<br>Slight ophthalmalgia |
| Stereoscopic viewing | Eye Symptoms | Feel both eyes equally heavy<br><br>Irritating eyes |
| | Other Symptoms | Feel heavy in the head |
| | Symptoms of accommodation | Feel blur in cutchange and eye-strain<br>Feel burden in cutchange<br>Feel slight blur in cutchange<br>Feel blur in cutchange<br>Feel blur |
| | Symptoms of vergence | Diplopia in near images<br>Diplopia<br>Feel fatigue in convergence |
| | Comparison of monoscopic and stereoscopic viewing | Feel more fatigue than monoscopic viewing<br>(3 subjects)<br>Feel the images small |
| | Comments on TV programs | Feel little sensation of depth in "Christmas concert"<br>Feel easy viewing in "Waffen"<br>Feel hard in "dream media" |

cross-interaction. These results indicate that the AC/A ratio did not change after viewing TV for about one hour.

Figure 21.12 shows the fusional amplitude (mean and standard deviation) in the six able-to-free fuse participants and six unable-to-free fuse participants, measured in pre- and post-viewing of high-definition TV programs, and after taking the second rest. These graphs indicate that able-to-free fuse participants have wider fusional amplitudes than unable-to-free fuse participants do, and that the standard deviation of able-to-free fuse participants is large. To test whether the fusional amplitude varied with free fusing ability, multiple analyses of variance were performed. There was a significant difference in the fusional amplitude between the groups before both monoscopic and stereoscopic viewing, and able-to-free fuse participants had a wider fusional amplitude.

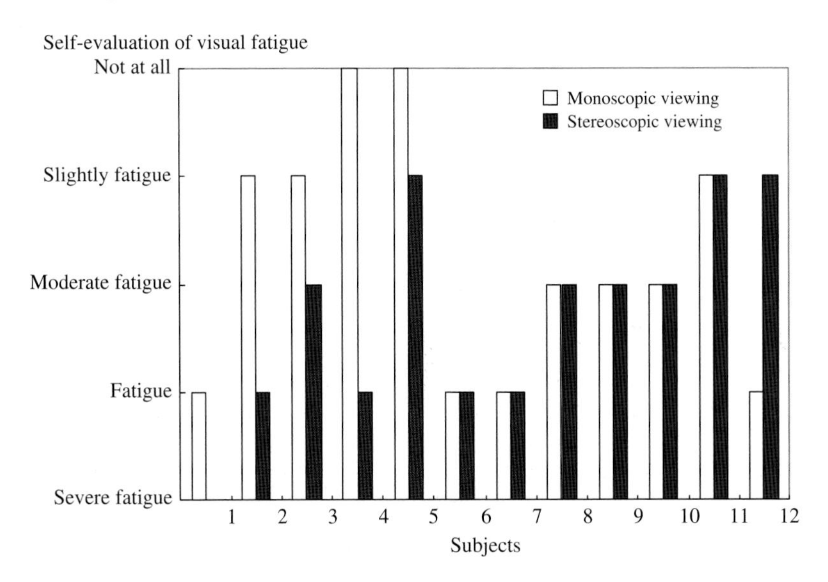


**Fig. 21.9.** Subjective evaluation of visual fatigue after viewing monoscopic and stereoscopic TV programs

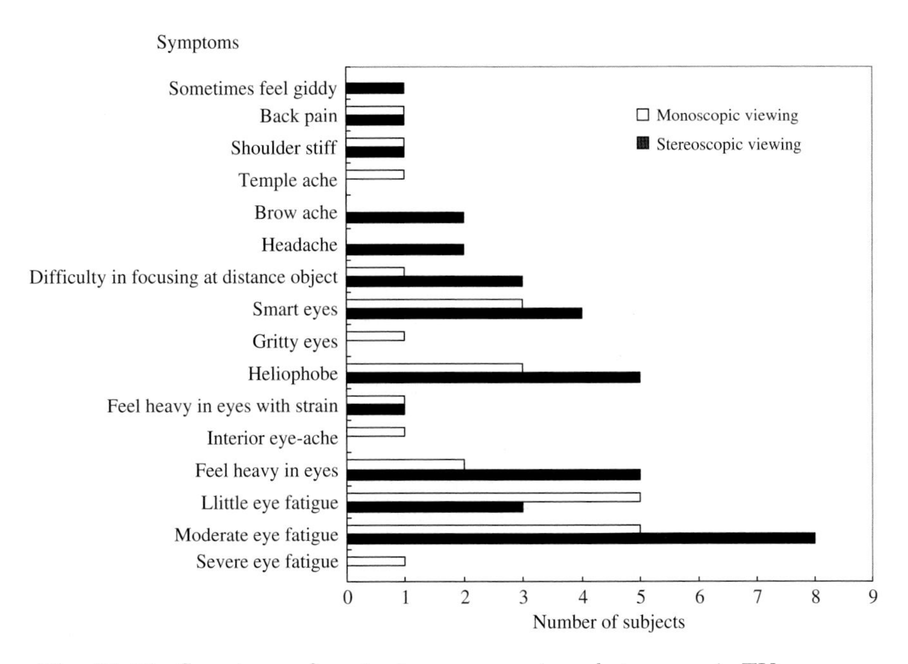


**Fig. 21.10.** Symptoms after viewing monoscopic and stereoscopic TV programs

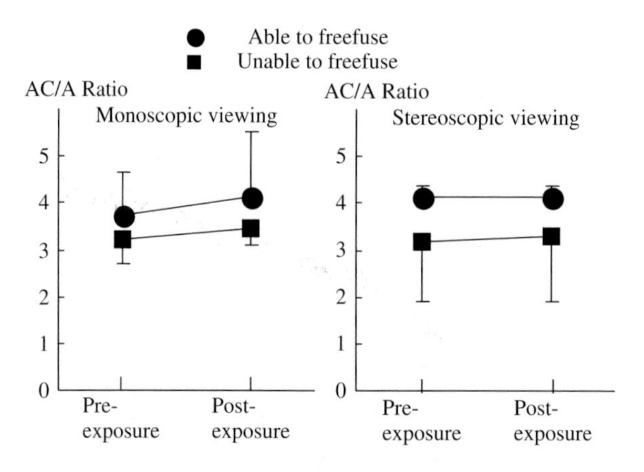


**Fig. 21.11.** Accommodative convergence per accommodative response (AC/A) ratio

To determine whether fusional amplitude changes after viewing high-definition TV programs, data were normalized to those of pre-viewing values, as illustrated in Fig. 21.13. With respect to stereoscopic viewing, a repeated measures ANOVA using normalized data revealed a significant decrease in fusional amplitude after viewing stereoscopic images and recovery after taking a rest. Although not significant, a tendency was found in terms of the ability to free fuse and the interaction of both variables. On the other hand, monoscopic viewing did not present any significant change in the fusional amplitude.

The interaction tendency in viewing stereoscopic images suggested that the pattern from the able-to-free fuse participants′ fusional amplitude change is different from that of unable-to-free fuse participants. To verify this finding, a Dunnett's multiple comparison was performed with data normalized to the pre-viewing fusional amplitude for both groups. In free fuse-able subjects, the fusional amplitude in comparison with pre-viewing did not significantly change immediately after stereoscopic viewing, after taking a first or second rest. On the other hand, subjects without the ability to free fuse showed a significant decrease of fusional amplitude after stereoscopic viewing, and there was also a decreasing trend of fusional amplitude after the first rest (although not significant). Fusional amplitude was not determined to be significantly changed after the second rest.

The overall results indicated that fusional amplitude decreases after stereoscopic viewing depending on the individual′s free fuse ability. Individuals without free fuse ability suffer from a narrowing of fusional amplitude immediately after stereoscopic image viewing, though it may recover after being relaxed. These findings may be relevant to the visual fatigue associated with viewing stereoscopic high-definition TV. These results also suggest that about one hour of stereoscopic image viewing is likely to cause a larger degree of visual fatigue than viewing monoscopic images for the same duration.

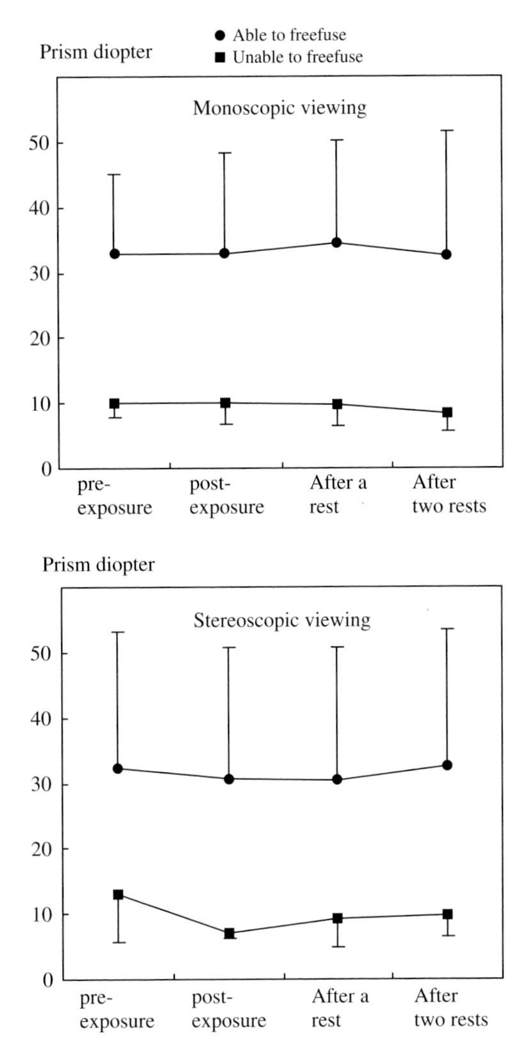


**Fig. 21.12.** Changes in fusional amplitude. *Top*: monoscopic viewing; *Bottom*: stereoscopic viewing. *Circles* = six participants with free fuse ability; *squares* = six participants without free fuse ability. Data points represent mean and bars represent standard deviation

## Conclusions from this Experiment

Stereoscopic viewing causes more serious visual fatigue than monoscopic viewing. It was also determined that fusional amplitude can be utilized as an index of visual fatigue. Decreased fusional amplitude, after viewing stereoscopic images, recovered to pre-viewing levels after taking a short relaxation pause. Stereoscopic TV images sometimes include a larger amount of horizontal

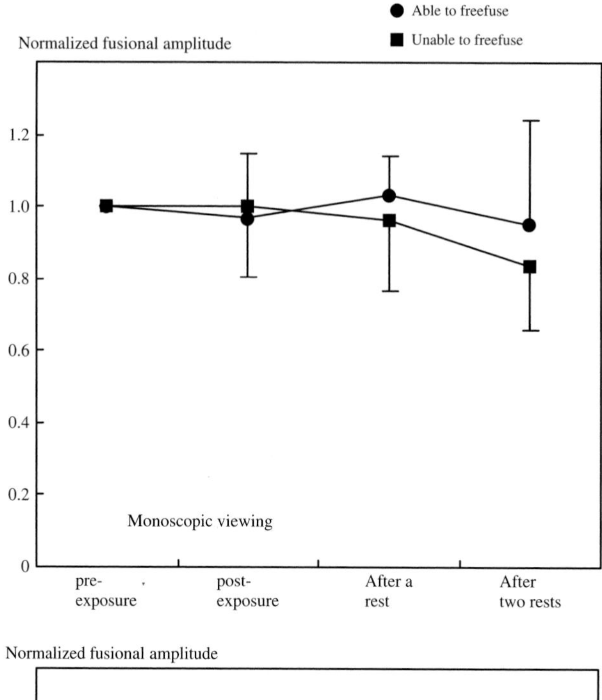


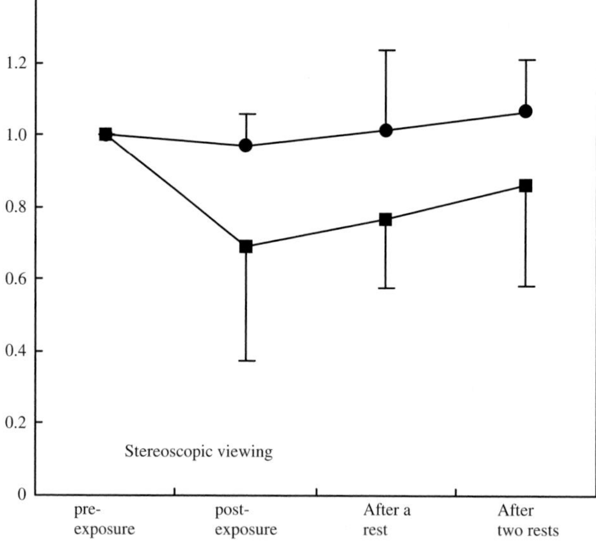


**Fig. 21.13.** Changes in normalized fusional amplitude. *Top*: monoscopic viewing; *Bottom*: stereoscopic viewing. Data (mean and standard deviation) for six participants with free fuse ability (*circles*) and six participants without free fuse ability (*squares*) are normalized to their mean pre-viewing measurements

binocular parallax to stimulate and provide entertainment to viewers. But the fusion effort required to appreciate stereoscopic images with excessive horizontal binocular parallax is thought to cause severe visual fatigue.

Why does viewing a stereoscopic image cause visual fatigue? In the next section, we will try to answer this question by studying the relationship between the horizontal binocular parallax of image stereopairs and visual fatigue. In the stereoscopic image system, this relationship is equivalent to the one between the distance information and visual fatigue. We will conduct an experiment to measure visual fatigue by using a quantitative index when viewing stereoscopic image systems with only in-principle fusion disturbance factors. The in-principle factors should be investigated first, because non-principle factors will fade as hardware for stereoscopic image systems is developed in the future.

### 21.2.2 Spatial Distribution in Depth of Objects

By studying the relationship between spatial distribution of horizontal binocular parallax of image stereopairs and visual functions, we will try to clarify why viewing stereoscopic images causes visual fatigue. To quantify the degree of visual fatigue when viewing an ideal stereoscopic system with a certain horizontal binocular parallax, but without its distribution (i.e., the distribution of the horizontal binocular parallax is zero), the degree of changes of some visual functions after about one hour of viewing are measured. This viewing condition is equivalent to the condition in which the viewers see one object in a still stereoscopic image without any change in its distance. As mentioned above, we hypothesized that the major cause of the visual fatigue accompanied with stereopsis is the viewers′ efforts to fuse the image stereopair with fusion difficulty. We also mentioned above that the factors of a fusion-difficult stereopair of images can be categorized as eithre in-principle factors, in which there is horizontal binocular parallax, and non-principle factors, where there are differences between right and left images except for the horizontal binocular parallax.

To quantify the relationship between the degree of visual fatigue and horizontal binocular parallax we must establish adequate quantitative indices to evaluate visual fatigue, and control the amount of horizontal binocular parallax presented by the stereoscopic image system.

As to Problem 1, a helpful index has been presented in the previous experiment, where fusional amplitude could be utilized as a quantitative index of visual fatigue. We will describe Problem 2 in this section – how to control the amount of horizontal binocular parallax which burdens the viewer′s vergence system. The maximum amount of horizontal binocular parallax is described in several studies [12, 13]. It is difficult to know the amount of horizontal binocular parallax loaded to viewers under experimental conditions, because it is necessary to control the image viewing position or measure where the viewers see and then calculate the amount of horizontal binocular parallax by stereo matching [21]. In spite of this difficulty, it is essential to control the amount of horizontal binocular parallax loaded to viewers.

Another important aspect of horizontal binocular parallax is that it decides the amount of vergence and accommodation dissociation in viewing stereoscopic images. A plausible hypothesis says that one of the major factors of visual fatigue in viewing stereoscopic images may be a dissociation of vergence and accommodation. The difference in visual function between viewing real objects and viewing stereoscopic images has been noted. Figure 21.14a shows vergence and accommodation when viewing a real object; the vergence point is included within the depth-of-focus. Figure 21.14b shows what happens when viewing stereoscopic images; the vergence point is sometimes outside the depth of focus when the horizontal binocular parallax is large. As illustrated in Fig. 21.14c, we simulated the visual functions typical to viewing stereoscopic images using fixed and variable prisms in front of viewers′ eyes to control vergence load, which caused fixed and variable amounts of dissociation of vergence and accommodation. As seen by comparing these figures (Figs. 21.14a–c), we can simulate typical stereoscopic image viewing vergence and accommodation because the physical effects are the same as shown in Fig. 21.14b. Controlling the power of prisms enables us to control the amount of horizontal binocular parallax loaded to viewers, and this solves the difficulty of horizontal binocular parallax control in experiments. The direct purpose of the next experiment is to clarify the relationship between visual fatigue and the amount of vergence load (i.e., horizontal binocular parallax, or the amounts of dissociation of vergence and accommodation) which dissociates from accommodation.

In principle, it is clear that attempts to view stereoscopic images would cause dissociation; however, the human visual system can tolerate some level of dissociation without difficulty. The tolerance range for varying binocular vergence with almost no change in accommodation is called the "range of relative vergence," which was referred to as “fusional amplitude” in the previous section. The "area of comfort" is the range of relative vergence one can see for extended viewing without visual fatigue. Percival defined the zone between the limits of 0 diopters (infinity) and 3 diopters (3.3 meters away), or the middle third of the range of relative vergence, as the area of comfort [22]. Figure 21.15 shows an example of the range of relative vergence and area of comfort at 0.83 diopters (1.2 m) as bilateral horizontal arrows. The validity of Percival's area of comfort has been verified empirically, but no experimental evidence exists. For example, visual fatigue when using a head-mounted display (HMD) was assessed, and the result shows that severe visual fatigue was not associated with horizontal binocular parallax within the area of comfort [12]. This result only shows that a small degree of horizontal binocular parallax does not cause severe visual fatigue, rather than the validity of the area of comfort.

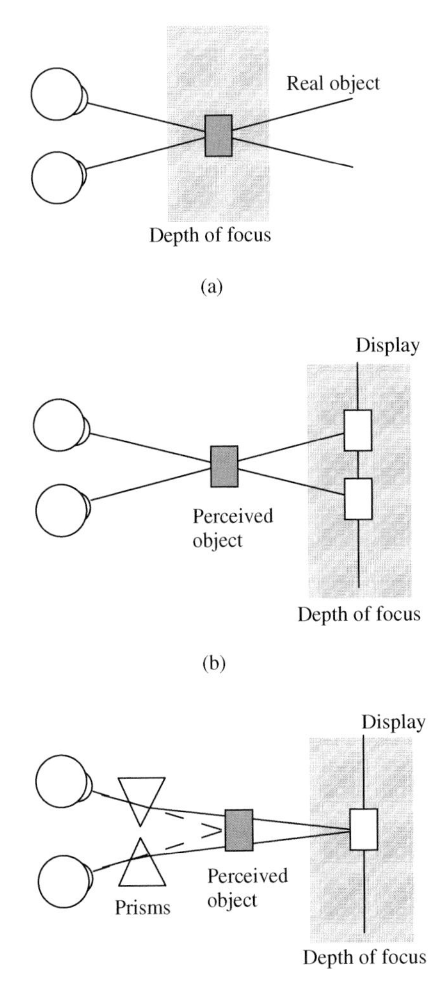


**Fig 21.14.** (a) Vergence and accommodation in viewing a real object (b) Vergence and accommodation in viewing stereoscopic images (c) Vergence and accommodation in viewing image through prisms

It is necessary to show that a degree of parallax beyond the area of comfort would lead to visual fatigue in addition to the results from within the area of comfort. It is also a quantitative evaluation of the relationship between visual fatigue and the amount of vergence load, which dissociates from accommodation. In this study, we verify the validity of Percival's area of comfort in viewing stereoscopic images by an experiment using a visual function simulator. Participants viewed a high-definition television through the simulator with fixed prisms for approximately one hour. The powers of the fixed

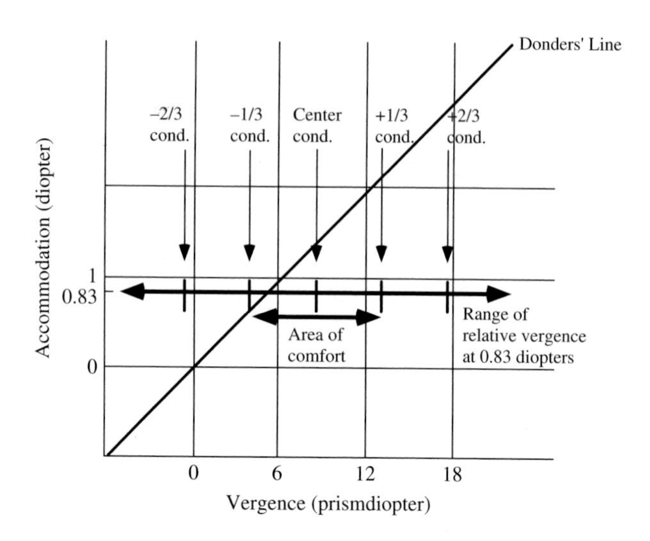


**Fig. 21.15.** Range of relative vergence and area of comfort

prisms were set within and beyond Percival's area of comfort for each participant as the vergence load and visual fatigue were evaluated. Hereafter, we call these experimental conditions "fixed prism conditions." The evaluation indices of visual fatigue are subjective symptoms: the range of relative vergence, accommodation step response, and visual evoked cortical potentials (VECP). They are the fatigue index for the extraocular muscle, intraocular muscle, and central nerve (or brain), and interact with each other. Previous studies have found changes in the range of relative vergence [13, 23], accommodation step response [24], and VECP [25] after subjects viewed stereoscopic images.

## Methods

A simulator was developed to imitate the visual functions typical to viewing stereoscopic images, using rotary prisms in front of the viewers' eyes [26]. Vergence load was set according to each participant's area of comfort after measuring relative vergence break points. General viewing conditions of television (including the one in this study) satisfy the accommodation condition of the area of comfort, which is within 3 diopters. However, the degree of horizontal binocular parallax of stereoscopic images could be beyond the area of comfort. This is why we control only the power of the prisms of the visual function simulator. Before all experiments, the range of relative vergence for each participant was measured to determine their area of comfort. Each participant's base-in (BI) and base-out (BO) break points of relative vergence through rotary prisms in viewing a '+' target with approximately 2.0 arc degree of visual angle on a display surface at 1.2 m were measured by a vergence test procedure described by Borish [27]. The range of relative vergence is referred to as the "fusion field." Five powers of fixed prisms were burdened based on the range of relative vergence of each participant.

Under the "center" condition, the prism powers were at the center of the ranges of relative vergence (i.e., center of the area of comfort). In the "–1/3" condition, the power of the BI prisms was within the middle third of the range of relative vergence (i.e., divergence side of the area of comfort). In the "–2/3" condition, the power of the BI prisms was the average of the –1/3 condition and the BI limit (i.e., divergence side of out of the area of comfort). Under the "+1/3" and "+2/3" conditions, the corresponding powers of the BO prisms were used (convergence side). Figure 21.15 shows these conditions. When the powers of the prisms were calculated from –2.0 prism diopters to 2.0 prism diopters, which is in the range of physiologic horizontal phoria, a 0 prism diopter load was burdened; too little prism power is not a load for the visual system, and it is important to compare with 0 prism diopter as the traditional two-dimensional display system. The results of the corrected conditions were also used as the two-dimensional condition.

### Procedure

Break points and recovery points were measured by the vergence test procedure (mentioned above) at four time points – before exposure, immediately after exposure, after taking a rest, and after taking a second rest. Flash VECPs were measured two times, at pre- and post-exposure, and averaged 50 waves without artifact. Accommodation step responses were measured five times at pre- and post-exposure using an infrared optometer. Each participant sat on a chair with his/her chin fixed on a chin support and watched the inner target of the optometer. The inner target is a black starburst with 3 arc degrees in visual angle against a uniform green background with 8 arc degrees in visual angle. The optical position of the target alternated between 0 and 5 diopters every five seconds. Viewing conditions of the TV were set according to published standards [28]. The viewing distance was 1.2 m, (three times the image height), and the horizontal field of view was approximately 33 arc degrees. The program viewed contained almost one hour of the stereoscopic version of the second curtain section of Johann Strauss′ opera 'Die Fledermaus.' Three pairs of high-definition television stereo cameras with a 12-cm inter-camera distance shot the stage from the back, right, and left of the viewers′ perspective seat. The convergence point for the cameras was a big vase on the stage. Camera-mounted zoom lenses were supplied, but the zoom ratio was identical in an image cut. The speaking and singing were in German, with Japanese translations superimposed on the bottom of the display. Participants viewed the

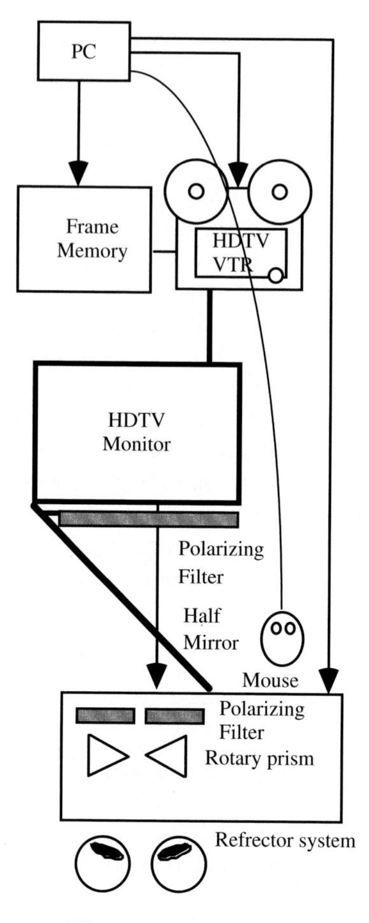


**Fig. 21.16.** Apparatus

program through prisms presented on a monitor. After viewing the program, subjective reports of symptoms were collected by five categories of ratings (i.e., 'I feel so' ∼ , 'Not at all') for Questionnaire No. 1, and Numbers 4–29 in Table 21.4.

Figure 21.16 illustrates the apparatus used in the experiment, equipped with high-definition digital VTR for movable image display, frame memory for still image display, 32-inch high-definition TV monitor, and a computer for system control. Half-mirror and polarizing filters were used only for the compatibility to 3-D conditions to be described in the next section. The computer was employed to control the high-definition VTR, to alter frame memory-produced still images, to control the power of prisms in the refractor system, and to record the participants' responses.

## Results

### *Rating of the Subjective Reports of Symptoms*

Five categories of ratings were assigned, from 1 to 5 (i.e., 'I feel so': 1, ~ , 'Not at all': 5). Figure 21.17 shows the results for the "severe eye fatigue" question. Participants tended to feel more severe eye fatigue when the vergence load was heavy or it changed over time, though the ANOVA showed no significant difference between experimental conditions.

The range of relative vergence, which is defined as the range between BI and BO break points, was calculated at four times. The individual ranges were normalized by each subject's pre-exposure value. Figure 21.18 shows the ratios of the range of relative vergence under fixed prism conditions at three time points compared to the pre-exposure time point; measurements were made pre-exposure, post-exposure, after one rest period, and after two rest periods. The ranges appeared to decrease temporarily post-exposure and then recover to pre-exposure levels under all experimental conditions. We found that the degree of decrement was more severe under conditions outside the area of comfort. Test results of repeated measures ANOVA revealed a significant change in range of relative vergence over time in the −2/3, −1/3, +2/3 conditions. To verify this finding, Dunnett's multiple comparison was performed. The ratios of the range of relative vergence compared with pre-exposure changed significantly post-exposure under the −2/3, −1/3, and +2/3 conditions. The result under the conditions, in which vergence load was heavy, suggests that a large degree of the load beyond the area of comfort led to the visual fatigue. It suggests a certain validity of the Percival's area of comfort.

Figure 21.19 shows five traces of accommodation responses to the step change of accommodation demand from alternating 0 to 5 diopters. The straight line beginning with 0 diopters of the ordinate indicates the movement

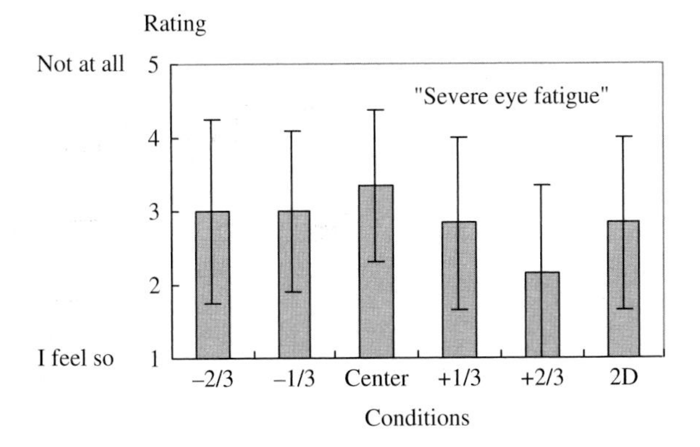


**Fig. 21.17.** Rating of "Severe Eye Fatigue." Data presented as means and standard deviations

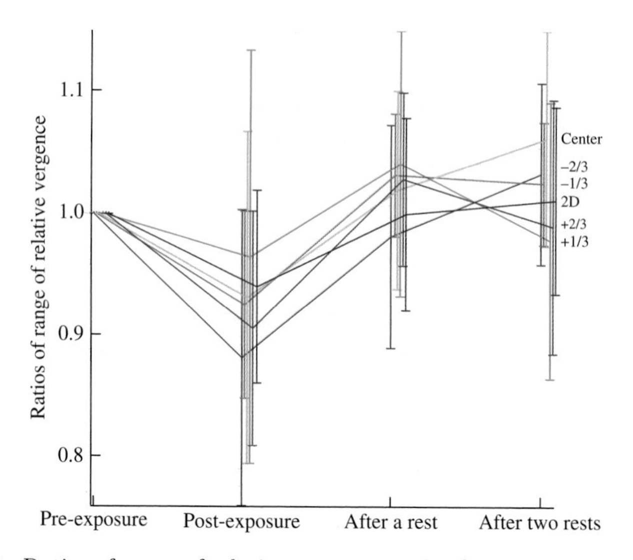


**Fig. 21.18.** Ratios of range of relative vergence under fixed prism conditions. Data presented as means and standard deviations

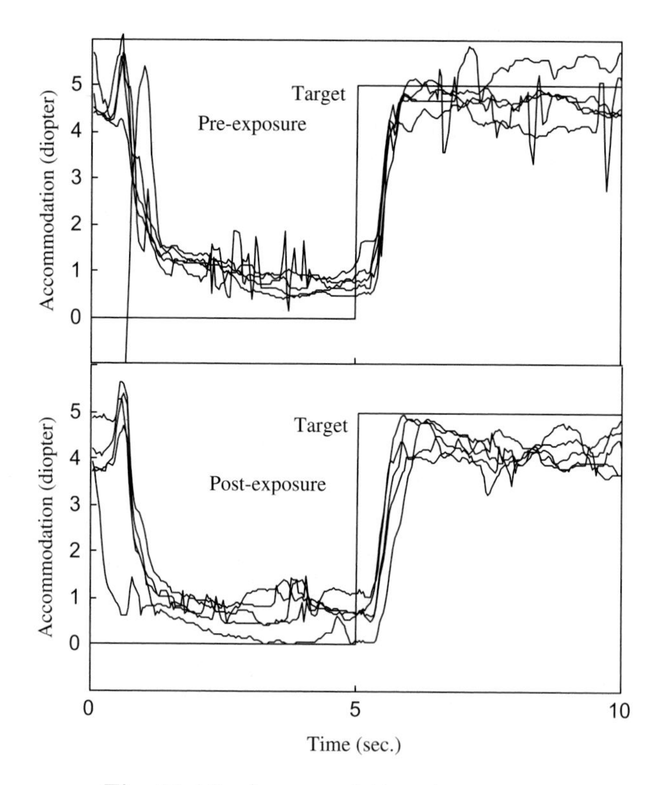


**Fig. 21.19.** Accommodation step responses

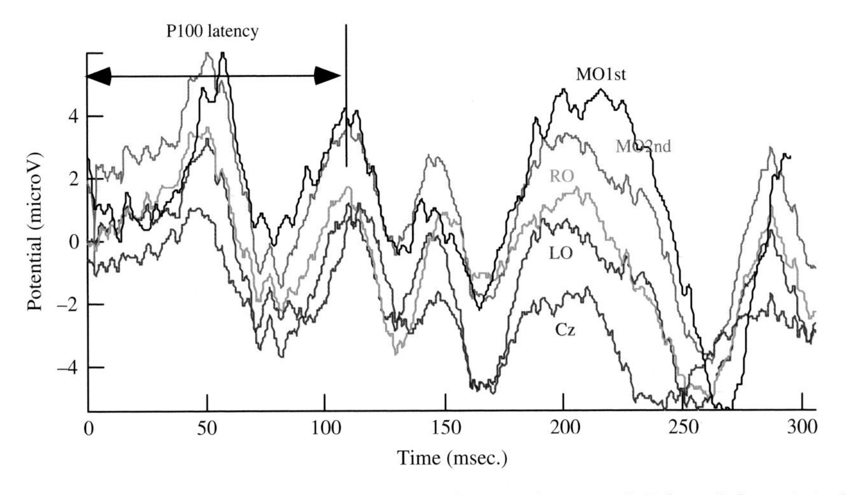


**Fig. 21.20.** VECP waves. Measured at MO = midoccipital, LO = left occipital, RO = right occipital, and Cz = vertex

of the accommodation target, and the five lines beginning at around 5 diopters of the ordinate, indicate the accommodation responses. There was no systematic change common to all participants.

Figure 21.20 shows examples of the waves of the VECP. Four waves were acquired from the four points (MO = mid-occipital, LO = left occipital, RO = right occipital, Cz = vertex) in the first measurement, and one wave labeled MO 2nd acquired in the second measurement was added to show the repeatability of the measurements. We took the positive component (P100) with about 100 msec peak latency, which corresponds to Cigánek's wave IV [29], as a fatigue index according to the previous study [25]. Paired t-tests revealed no significant delay of P100 latency under all fixed prism conditions.

## Conclusions of this Experiment

Our findings lead us to conclude that the major factor of visual fatigue in viewing stereoscopic images is principally the viewer effort required to fuse right and left images with excessive horizontal binocular parallax, which causes the decline of the range of relative vergence. A plausible hypothesis, which supposed that the major factor of fatigue may be a dissociation of vergence and accommodation, was verified. In this experiment, we controlled horizontal binocular parallax without its distribution. Does distribution of horizontal binocular parallax, which is excluded in this experiment, affect visual fatigue?

Generally, with still image stereopairs, some objects have different distances from the viewer and, therefore, a range of distribution of horizontal binocular parallax exists. Viewers must move their visual axis from one object to another with their vergence changing to view each object in the scene. Visual comfort when viewing images presented by stereoscopic high-definition TV (HDTV) with distribution of horizontal binocular parallax has been studied. Nojiri, et al. determined the amount of the distribution of horizontal binocular parallax in a stereopair of still images by calculating the correlation of the fast Fourier transform (FFT) phase information [21]. They conducted an experiment to assess visual comfort when viewing images presented by stereoscopic HDTV with various distributions. They concluded that viewers reported their visual comfort when viewing stereopairs of still images with less than 60 arc minutes of distribution of horizontal binocular parallax. Figure 21.21 shows their results, in which each symbol and error bar indicates mean position and range of horizontal binocular parallax of a still image stereopair. The stereo matching results of 48 stereopairs of still images are plotted in the figure and it is clear that viewers feel visual comfort when the distribution of horizontal binocular parallax was narrow. They described their comfortable range of the distribution of horizontal binocular parallax as equivalent to the range of the depth–of-focus when a viewer's pupil diameter was 5 mm. This suggested a mechanism where it is not necessary to change accommodation when the vergence change is small enough to fit into the depth–of-focus, although accommodation and vergence systems have a close interaction.

Another aspect of accommodation is that accommodation response depends on target blur. Okada, et al. demonstrated the vergence-accommodation interaction, which includes the fact that viewers' accommodation points shift toward the vergence point when visual targets were blurred under the circumstance of the dissociation of vergence and accommodation [30].

When visual stimuli are not blurred, which is common in stereoscopic image systems, we must take care of the case when distribution of horizontal binocular parallax exceeds 60 arc minutes from the standpoint of the vergence load. Viewers have to not only move the convergence point of their visual axes from one object to another object in the scene, but also keep their accommodation on the display surface. Such changing vergence demand might be a

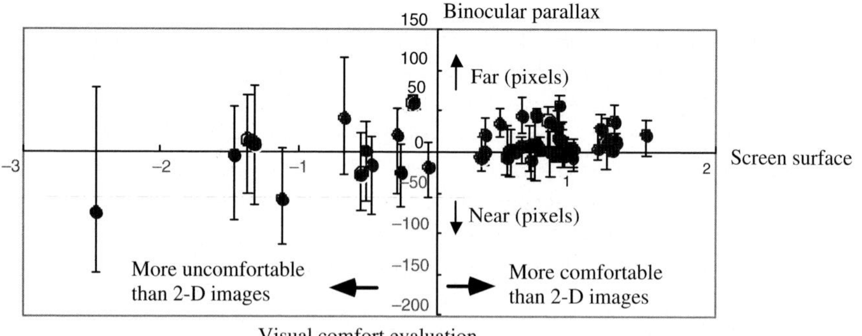

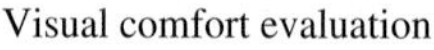


**Fig. 21.21.** Visual comfort and distribution of horizontal binocular parallax

burden to the viewer's vergence system and it might affect visual fatigue. The changes of vergence demand of a stereopair of image sequences also occur when anteroposterior movement of the objects and cutaways are included in sequences. We should study visual fatigue in such changing vergence demand cases. In Section 21.2.3, therefore, we will describe the effects of time-varying changes of the anteroposterior position of objects on the visual fatigue.

### 21.2.3 Temporal Distribution in Depth of Objects

Here we will describe the effects of time-varying change of the anteroposterior position (i.e., distance) of the objects on the visual fatigue. In the previous experiment, we quantified the degree of visual fatigue after about one hour's viewing using an ideal stereoscopic system with certain horizontal binocular parallax, but without distribution by the degree of changes of some visual functions after about one hour of viewing. Now we can proceed to the time aspect of the time-varying vergence load caused by a change of anteroposterior positions of the objects and cutaways included in the stereopair of sequences, and of the visual fatigue which might be caused by the eye following the time-varying vergence load. We conducted an experiment with the time-varying vergence load presented on the same apparatus as the previous experiment for about one hour, using a visual function simulator. To compare the degree of visual fatigue to the previous results when viewing an ideal stereoscopic system with time-varying horizontal binocular parallax, we measured the degree of change of some visual functions after about one hour of viewing.

It is well-known that continuous viewing through fixed prisms, as in the previous experimental conditions (i.e., fixed prism conditions), causes prism or phoria adaptation. In viewing stereoscopic programs, the viewers would not have enough time for adaptation, because the degree of horizontal binocular parallax changes over time according to the cutaways or depth change of the object and, therefore, the changing vergence load demands viewers to gain a new adaptation status. This may be a new factor of visual fatigue in addition to the plausible hypothesis that one of the major factors of fatigue may be vergence and accommodation dissociation when stereoscopic images are viewed.

#### Methods

We simulated the time change of horizontal binocular parallax using the visual function simulator with variable prisms. Participants viewed an HDTV through the simulator with variable prisms for approximately one hour. The power of the variable prisms changed within and beyond Percival's area of comfort for each participant (as vergence load), and visual fatigue was evaluated. Hereafter, we define these experimental conditions as "variable prism conditions." The evaluation indices of visual fatigue are subjective

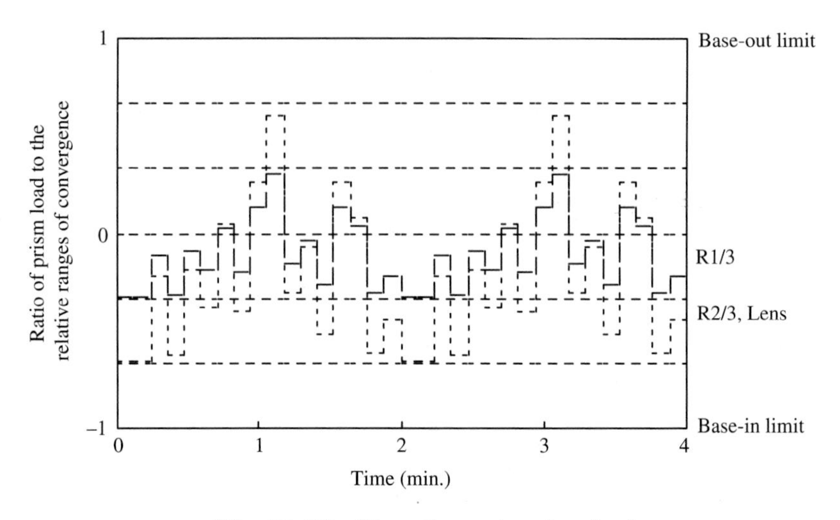


**Fig. 21.22.** Time change in prism load

symptoms: the range of relative vergence, accommodation step response, and visual evoked cortical potentials (VECP), as seen in the previous experiment.

Under "Random1/3" (R1/3) conditions, the prism powers changed randomly over time within the area of comfort for both BI and BO. Under "Random 2/3" (R2/3) conditions, the prism powers changed at random over time outside the area of comfort for both BI and BO. Under "Lens" conditions, the prism powers changed similar to what was seen under R2/3 conditions; compensation lenses were inserted to enable accommodation demand to match the vergence point. Under these conditions, the duration of any power of prisms was also random; this resulted in 16 changes in prism power every two minutes, with continued durations from three to 12 seconds. This two minute cycle was repeated 29 times over approximately one hour, and all 29 cycles were identical. Figure 21.22 shows the time change in prism load for two cycles. Additionally, a stereoscopic program was presented using dichoptic displays without prisms in front of the viewers′ eyes under "3-D" conditions. Figure 21.23 shows the apparatus used in these experiments.

## Results

Figure 21.24 shows the result of viewer ratings of "severe eye fatigue." Participants tended to feel eye fatigue was more severe when the vergence load was changed over time, though the ANOVA showed no significant difference between experimental conditions. The range of relative vergence, which is defined as a range between the BI and BO break points, was calculated at four time points and the ranges were normalized by the pre-exposure value.

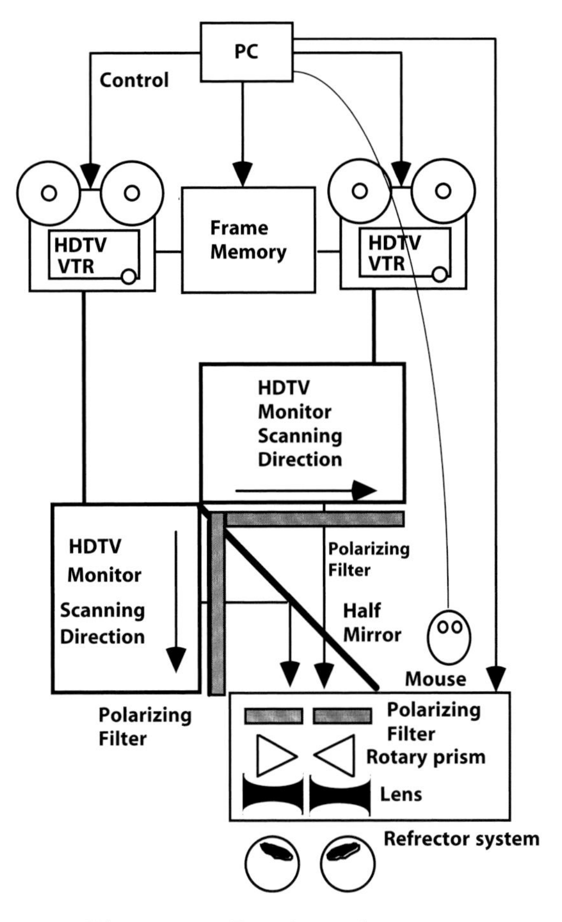


**Fig. 21.23.** Experimental apparatus

Figure 21.25 shows the ratios of the range of relative vergence under variable prism conditions and 3-D conditions at three time points compared to the pre-exposure time point; measurements were made pre-exposure, post-exposure, after one rest period, and after two rest periods. The ranges seemed to decrease temporarily post-exposure and then recover to pre-exposure levels under all experimental conditions. We found that the degree of decrement was more severe under conditions outside the area of comfort. The repeated measures ANOVA revealed a significant change in the range of relative vergence over time in 3-D, R1/3, and R2/3 conditions. To verify this finding, Dunnett's multiple comparison test was performed. The ratios of the range of relative vergence compared with pre-exposure changed significantly post-exposure under the 3-D and R2/3 conditions, and after the second rest period under R1/3 conditions. The result under the R1/3 condition suggests that repeated adaptation to a vergence load within the area of comfort has a training effect on the

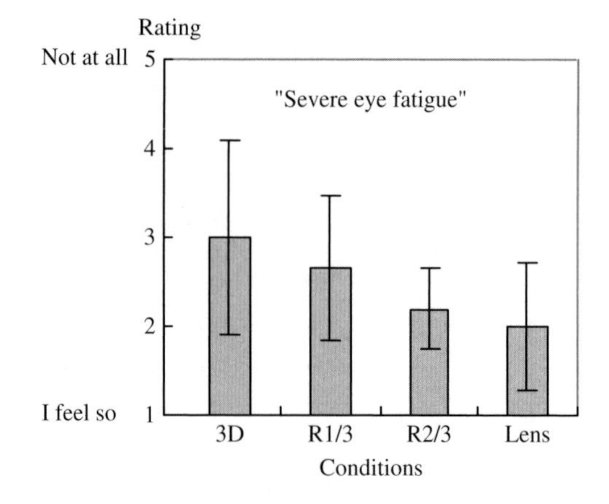


**Fig. 21.24.** Rating of "Severe Eye Fatigue." Data presented as means and standard deviations

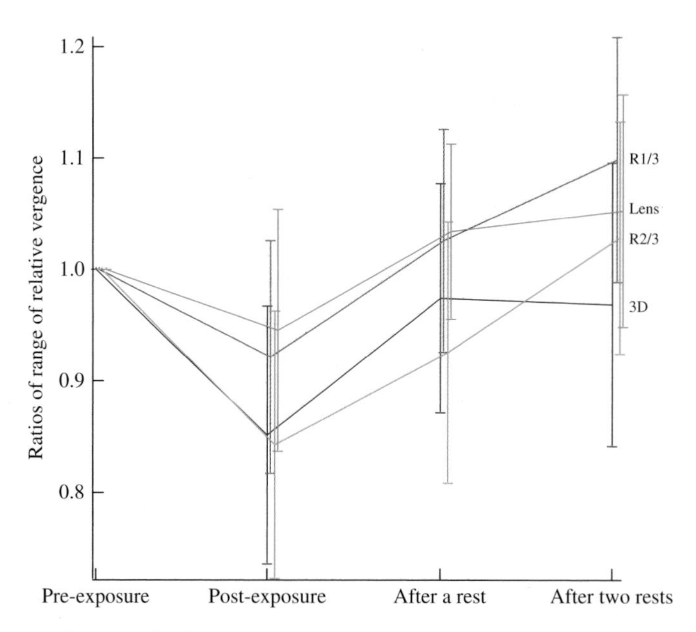


**Fig. 21.25.** Ratios of relative vergence under variable prism and 3-D conditions. Data presented as mean and standard deviation

range of relative vergence. The result under these conditions, where vergence load changed over time, suggests that the temporal change of the load beyond the area of comfort led to visual fatigue, and it suggests a certain validity of Percival's area of comfort.

Some participants showed an accommodation decline under several experimental conditions, while no systematic change common to all participants was

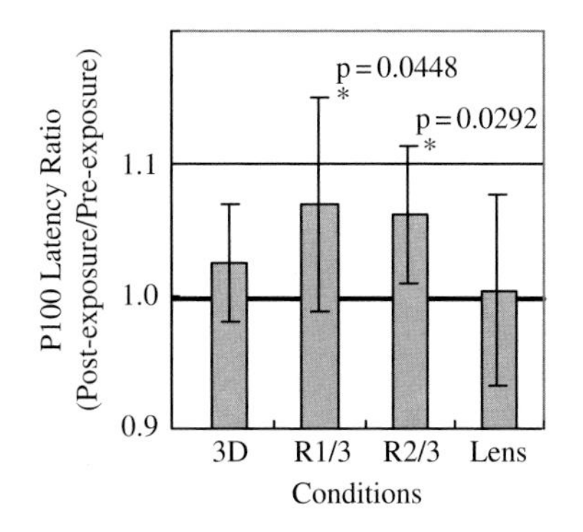


**Fig. 21.26.** Ratios of post-exposure P100 latency to pre-exposure value. Error bar indicates standard deviation. "*" indicates significant difference ($p<0.05$) using the paired t-test

found. The possibility of lessening the accommodation demand on a participant, who had strong accommodative power, was guessed from the unchanged result of the participant′s response.

Figure 21.26 shows the ratio of post-exposure P100 latency to pre-exposure values. Paired t-tests revealed the significant delay of P100 latency under R1/3 and R2/3 conditions. This result suggests that the temporal change of horizontal binocular parallax would result in the delay of transmission of visual information in neurons.

### Conclusions of this Experiment

Our findings lead us to conclude that the major factor of visual fatigue in viewing stereoscopic images is the viewers′ efforts to fuse right and left images with time-varying horizontal binocular parallax, or the viewers′ continuous adaptation to new vergence demand inconsistent with accommodation demand. They cause a decline of the range of relative vergence, accommodation response, and conductivity of the neuron. A plausible hypothesis, which supposes that the major factor of fatigue may be a dissociation of vergence and accommodation, was verified and a new temporal factor is suggested. We found that not only dissociation, but also temporal discontinuous changes in dissociation also led to visual fatigue.

Eye movements following discontinuous change of horizontal binocular parallax by the cutaway were measured [31]. It takes approximately 0.6 sec and 0.35 sec until the completion of the vergence response to the 120 arc minutes, and 40 arc minutes of the change of the vergence load. The results suggested that vergence eye movement to follow significant discontinuous change of the horizontal binocular parallax affect the visual comfort. The allowable range of discontinuous changes of horizontal binocular parallax is less than 60 arc minutes, which is the same as the allowable range of distribution of horizontal binocular parallax [16]. This shows not only the comfortable range for viewing an image stereopair, but also may have something to do with visual fatigue associated with lengthy viewing of stereoscopic images.

Under more practical TV viewing conditions, Yano, et al. studied a change of visual function in reading sentences presented with horizontal binocular parallax on a high-definition stereoscopic TV, based on the same idea as the study described in this section [32]. The results showed that the accommodation step response for some participants declined after viewing the discontinuous change of the horizontal binocular parallax.

These findings lead us to conclude that it is possible to establish a stereoscopic TV system without producing severe visual fatigue by having adequate control of the temporal distribution of horizontal binocular parallax in producing stereoscopic programs.

### Non-Principle Factors and Their Desirable (Allowable) Ranges

In this section, we refer to the desirable ranges of non-principle fusion disturbance factors mentioned in Section 21.1.2, Table 21.3. Non-principle fusion disturbance factors, caused by differences in the characteristics of stereopair image equipment or hardware and their desirable ranges, are listed. Although non-principle factors will resolve with development of new hardware for future stereoscopic image systems, we should control these factors adequately until the ideal hardware is available. The factors of concern include differences in size, horizontal position, roll position, luminance, and cross talk.

An experiment assessing the degree of distortion of the represented space and the degree of annoyance in binocular fusion presented by the stereoscopic HDTV system was conducted using a five-point impairment scale [33, 34]. The scale consists of "imperceptible," (5), "perceptible, but not annoying" (4), "slightly annoying" (3), "annoying" (2), and "very annoying" (1).

Results showed that acceptable limits of size, vertical position, and roll position, which were defined as subjective evaluation values corresponding to 3.5 after the converted interval scale (placed at the point between "perceptible, but not annoying" = 4 and "slightly annoying" = 3) were determined.

The size of the left image was reduced to (100–a) percent, while the right image was expanded into (100+b) percent ; the acceptable limit of the size difference (a+b) was 2.89 percent .

The vertical position of the left image was shifted upward by (c) percent of the image height, while the right image was shifted downward by (d) percent of the image height; the acceptable limit of the vertical position (c+d) was 1.45 percent .

The roll position of the left image was rolled counterclockwise by (e) degrees, while the right image was rolled clockwise by (f) degrees; the acceptable limit of the roll position (e+f) was 1.14 degrees.

From the viewpoint of binocular fusion or stereopsis, the tolerance limit of the difference in size or aniseikonia is about 5–13 percent [35]. A difference in size up to 3.0 percent shows binocular summation in the visual evoked response of the brain to a reversal pattern; binocular summation started to decrease at 3 percent aniseikonia, and there was no significant binocular summation by 5 percentaniseikonia. At higher aniseikonia percentages, binocular inhibition replaced binocular summation [36].

The size difference is usually caused by the difference in zoom ratio between right and left zoom lenses. It is common for camera stereopairs with pairs of zoom lenses to be driven by two separate motors, but this difference could be eliminated by driving two zoom lenses with one motor [37].

From the viewpoint of binocular fusion, the limit of the vertical position or vertical fusional range of vergence cannot be defined by one numerical value; it depends on the size of the presented stimuli [38], viewing distance [39], and retinal position [40].

Though the limit of the fusional range of roll positions or cyclovergence also cannot be defined by one numerical value, several degrees of cyclovergence might be a limit [41, 42].

The difference in luminance, especially a black-and-white clip level of a video signal (which is specific to TV systems) was studied [43], because the difference of clip area in the images might disturb viewers' binocular fusion. They concluded that the black clip level should be carefully controlled to maintain the viewer′s visual comfort, while the white clip level affects little in the use of the stereoscopic system.

Cross talk is the leakage from the left to right image and vice-versa, and it degrades image quality. The causes of leakage are insufficient separation of right and left rays. For example, it is common that a stereopair of images is intended to be presented with their directions of linear polarization being at right angles to each other, but the birefringence of the screen surface disturbs the polarization of the rays keeping the images at right angles to each other, and it converts polarization from linear into elliptic polarization [44]; this causes the cross talk. Another example is that stereoscopic systems have caused cross talk, including insufficient shutter speed of the viewer′s LCD glasses or extended CRT phosphor persistence [45].

Hanazato, et al. studied the imperceptible limit for cross talk, which was defined as the subjective evaluation value corresponding to 4.5 on the 5-point impairment scale described above [46]. The imperceptible limit for cross talk was 55 dB (0.18%) in the worst case. They also determined the acceptable limit for cross talk, which was defined as the subjective evaluation value corresponding to 3.5 on the 5-point impairment scale described above. Three stereopair still images were evaluated, and they found that the acceptable limit is 28–20 dB (4–10%), depending on the contents of the images. They concluded that viewers could perceive cross talk when contrast is high or horizontal binocular parallax is large. This result is basically similar to the one by Pastool, et al. [47].

Other studies have dealt with the desirable ranges of non-principle fusion disturbance factors [48–53]. The most systematic study might be the one by Kooi, et al. [54]. They studied many aspects of the differences between the image stereopair by manipulating binocular images, and proposed threshold values, which were defined by a median score of 2 (viewing comfort is "a bit reduced" for 50% of the observers) or an upper quartile score of 3 (viewing comfort is "reduced" for 25% of the observers). Their threshold values (2.5% in size, 1 prism diopter or 34.4 arc minutes in vertical position, 1 arc degree in roll position, and 5% in cross talk) are almost comparable with the acceptable limits defined by Yamanoue (2.89% in size, 1.45% of the images height or 16.3 arc minutes in vertical position, and 1.14 arc degrees in roll position) [26], and the acceptable limits presented by Hanazato, et al. (4–10% in cross talk) [46].

It is difficult to compare these results because the experimental conditions differ between the studies. But the proposed desirable (allowable) ranges are valid references when designing a new stereoscopic image system without severe visual discomfort or visual fatigue. Based on the results of these studies, we will show a way to avoid undesired visual effects in the next section.

## 21.3 How to Avoid Undesired Effects

### 21.3.1 How to Avoid Spatial Distortion of Represented Space by Stereoscopic Image Systems

Based on previous discussion here, the conversion rules from real space into stereoscopic represented space are now clear. We should avoid distortions, which cause unnaturalness of the represented space, by utilizing numeric predictions based on our conversion rules. It is easy to predict undesired distortions such as the "puppet-theater effect" and "cardboard effect" by using a geometry mapping simulation system when shooting stereoscopic images. It is very important to know viewing conditions, such as display size and viewing distance when shooting, because these conditions determine adequate shooting conditions.

With a geometry mapping simulation system, we can manage the distortion of represented space in a positive way. We can produce the scene by managing the distance or depth of the object of interest. For example, we can emphasis the depth of a stereo aerial photograph so that mountain height is detectable. Even if a geometry mapping simulation system is not available, we should take care to consider the following points:

- Shoot image stereopairs using a parallel camera configuration, because this configuration helps to minimize shooting distortion.
- Narrow the inter-camera distance. It is desirable to set the inter-camera distance to the mean interpupillary distance of potential viewers.
- Set the zoom ratios of camera zoom lens stereopairs to be the same. It is also effective to control the zoom ratio by using one motor.

- Set the camera stereopair's alignment according to the following sequence: horizontal levels, vertical position, and then zoom ratio.

### 21.3.2 How to Avoid Visual Fatigue in Viewing Stereoscopic Images

Now, our principle hypothesis has been verified; the major cause of visual fatigue accompanied by stereopsis is the viewer's efforts to fuse an image stereopair with accompanying fusion difficulty. Although the cause of visual fatigue is common to individuals with visual function anomalies and viewers of stereoscopic image systems, the important stereoscopic image system feature is its ability to present more fusion-difficult image stereopairs than the real object. This ability might cause a great deal of viewer effort to fuse two retinal images, which would not exist when viewing real objects.

In viewing stereoscopic images, causes of fusion efforts can be classified into two categories. Factors in the non-principle category are related to differences in the characteristics of right and left image equipment; these differences should be minimized enough to fall into the range proposed to avoid visual discomfort or fatigue. This is a hardware issue and, therefore, non-principle factors will resolve as new stereoscopic image systems are developed in the future. The factors in the in-principle category are related to horizontal binocular parallax; to avoid visual discomfort or fatigue, excessive horizontal binocular parallax and its discontinuous change should not be presented. This is a software issue and, therefore, in-principle factors must be avoided in all phases of producing stereoscopic contents (including shooting, editing, transmitting, and displaying stereoscopic contents). With the geometry mapping simulation system mentioned above, we can manage excessive horizontal binocular parallax and its discontinuous change. Even without a geometry mapping simulation system, we should address the following points after minimizing the in-principle factors:

- Excessive horizontal binocular parallax should be avoided, with a balance between depth expression and visual fatigue.
- Discontinuous changes of horizontal binocular parallax should be avoided. Arrange the degree of horizontal binocular parallax of the object of interest to be comparable between pre-cutaway and post-cutaway, or set the convergence point of the camera stereopair at the same distance.

### Future Work

A significant issue still remains after eliminating non-principle factors, the worst case being from the viewpoint of vergence load. The geometry mapping simulation system mentioned above suggests that interpupillary distance plays an important role in geometry mapping, especially in distance conversion.

This means horizontal binocular parallax – from a moderate degree in wide interpupillary distance, to a considerable degree of load in narrow interpupillary distance. When stereoscopic sequences made by adults were viewed by children with short interpupillary distance [55], the sequences burdened children with heavy vergence load. To make matters worse, the long-term effects on the children, whose visual functions and brain are still developing, is not yet clear. An extreme case has been reported of a four-year-old child who became manifestly esotropic after viewing a stereoscopic movie [56]. The long-term effect on developing children must be studied.

Recently, some undesirable biomedical conditions associated with the effects of moving images on viewers (known as "image safety") came to our attention. In December 2004, the International Organization for Standardization (ISO) held a workshop on image safety in Japan. The focus of the workshop was to determine means to protect the public from potential physiological and psychological image hazards, such as photosensitive seizures, visually induced motion sickness, and visual fatigue by viewing stereoscopic images. Issues related to image safety are becoming increasingly recognized worldwide. After finalizing the future ISO standard, it is very important to enlighten image creators.

## Note

- Section 21.1 is based on 'Geometrical Analysis of Puppet-Theater and Cardboard Effects in Stereoscopic HDTV Images', by Hirokazu Yamanoue, Makoto Okui, and Fumio Okano which appeared in IEEE Transactions on Circuits and Systems for Video Technology, Vol. 16, No. 6, pp. 744–452. [2006] IEEE.
- Section 21.2.1 is based on the article published in 'Displays', Vol. 25, No. 2–3, Masaki Emoto, Yuji Nojiri, and Fumio Okano, 'Changes in Fusional Vergence Limit and its Hysteresis after Viewing Stereoscopic TV', pp. 67–76, Copyright Elsevier (2004).
- Sections 21.2.2 and 21.2.3 are based on 'Repeated Vergence Adaptation Causes the Decline of Visual Functions in Watching Stereoscopic Television', by Masaki Emoto, Takahiro Niida, and Fumio Okano which appeared in Journal of Display Technology, Vol. 1, No. 2, pp. 328–340. [2005] IEEE.

## References

[1] D. Marr: (Freeman, San Francisco 1982)
[2] D.L. MacADAM: SMPTE J. 62, pp. 271–289 (1954)
[3] M.S. Landy, J.A. Movshon: MIT press, pp. 306–330 (1991)
[4] H. Yamanoue, M. Okui, I. Yuyama: IEEE Trans Circuits and Systems for Video Technol, 10(3), (2000)

[5] K. Masaoka, A. Hanazato, M. Emoto, H. Yamanoue, Y. Nojiri, F. Okano: J Electronic Imaging 15(1), 013002-1-12 (2006)
[6] P.L. Panum: (Schwerssche Buchhandling, Kiel 1858)
[7] E. Maddox: J Anat Physiol, 20(Pt 3), pp. 475–508 (1886)
[8] C. Worth: Squint (Blakistoon, Philadelphia 1908)
[9] K. Tsubota, et al.: Arch. Ophthalmol, 113, pp. 155–158 (1995)
[10] K. Tsubota, K. Nakamor: N Eng J Med, 328, p. 584 (1993)
[11] B. Piccoli, et al.: Ergonomics, 33, 12, pp. 1433–1441 (1990)
[12] E. Peli: Vis Res, 38, pp. 2053–2066 (1998)
[13] A. Oohira, M. Ochiai: Ergonomics, 39, pp. 1310–1314 (1996)
[14] Y.Y. Yeh, L.D. Silverstein: Human factors, 32, pp. 45–60 (1990)
[15] R. Jones, G.L. Stevens: Horizontal fusional amplitudes. Invest Ophthalmol Vis Sci, 30, pp. 1638–1642 (1989)
[16] M. Alpern: The Eye, 3(2), p. 141 (Academic Press, London and New York 1969)
[17] C.M. Schor, T.K. Tsuetaki: Invest Ophthalmol Vis Sci, 28, pp. 1250–1259 (1987)
[18] ITU-R Rec. BT.1438 (2000)
[19] ITU-R Rec. BT.500-10 (2000)
[20] ITU-R Rec. BT.710-4 (1998)
[21] Y. Nojiri, H. Yamanoue, A. Hanazato, F. Okano: Proc SPIE, 5006, pp. 195–205 (2003)
[22] A.S. Percival: Ophthal Rev 11, pp. 313–328 (1892)
[23] M. Emoto, Y. Nojiri, F. Okano: Displays 25, pp. 67–76 (2004)
[24] T. Inoue, H. Ohzu: Appl Opt, 36(19), pp. 4509–4515 (1997)
[25] T. Yamazaki, K. Kamijo, S. Fukuzumi: Proc SID, 31(3), pp. 245–247 (1990)
[26] S.D. Risley: Trans Am Ophthalmol Soc 5, pp. 412–413 (1889)
[27] I.M. Borish (Ed.): Clinical Refraction 3rd Edition, pp. 226–229 (Professional Press Inc., Chicago, IL 1975)
[28] ITU-R Rec. BT.1438 (2000)
[29] L. Cigánek: Electroenceph Clin Neurophysiol 13, pp. 165–172 (1961)
[30] Y. Okada, K. Ukai, J.S. Wolffsohn, B. Gilmartin, A. Iijima, T. Bando: Vision Res 46, pp. 475–484 (2006)
[31] A. Hanazato, Y, Yamanoue, Y. Nojiri, F. Okano: ITE Technical Report 27, 23, pp. 37–40 HIR2003-94 (2003)
[32] S. Yano, M. Emoto, T. Mitsuhashi: Displays, 25(4), pp. 141–150 (2004)
[33] Y. Yamanoue et al.: J80-D-2, 9, pp. 2522–2531 (1997)
[34] Y. Yamanoue: Tolerance for Geometrical Distortions Between L/R Images in 3-D-HDTV, Systems and Computers in Japan, 29(5) (1998)
[35] S. Duke-Elder: Ophthalmic Optics and Refraction, pp. 513–534 (Henry Kimpton, London 1970)
[36] O. Katsumi, T. Tanino, T. Hirose: Invest Ophthalmol Vis Sci, 27, pp. 601–604 (1986)
[37] Y. Yamanoue, et al.: Proc IEICE General conference, Information and Systems, D-11-152, p. 152 (1997)
[38] A.E. Kertesz: J Opt Soc Am, 71(3), pp. 289–293 (1981)
[39] N. Hara, H. Steffen, D.C. Roberts, D.S. Zee: Invest Ophthalmol Vis Sci, 39(12), pp. 2268–2276 (1998)
[40] I.P. Howard, X. Fang, R.S. Allison, J.E. Zacher: Exp Brain Res, 130, pp. 124–132 (2000)
[41] I.P. Howard, L. Sun, X. Shen: Exp Brain Res, 100, pp. 509–514 (1994)

[42] I.P. Howard, M. Ohmi, L. Sun: Exp Brain Res, 97, pp. 349-355 (1993)
[43] A. Hanazato, M. Okui, F. Okano: Proc ITE Annual Convention, 4–5, pp. 49–50 (2001)
[44] M. Kanazawa, R. Sasage, S. Watanabe: ITE Technical Report 21, 63, pp. 31–36 (1997)
[45] M. Ihara, K. Ohno, T. Kusunoki, K. Takayanagi: Proc. of the 5th Sony Research Forum, pp. 562–565 (1995)
[46] A. Hanazato, Y. Yamanoue, M. Okui, I. Yuyama: 3-D Conference, 10–13 (1999)
[47] A. Pastool: HHI Report (1996)
[48] B. Choquet: TAO 1st Int Symp (1993)
[49] CCIR: CCIR XI/22-E, (Moscow 1958)
[50] Beldie, Kost (HHI): SPIE, 1457, p. 242 (1991)
[51] J. Fournier (CCETT): SPIE, 2177, p.45
[52] J. Konrad: IEEE Trans Image Processing (1998)
[53] K.C. Huang, J.C. Yuan, C.H. Tsai, W.J. Hsueh: 3-D Image Conf, P-9, p. 149 (2001)
[54] F.L. Kooi, A. Toet: Displays, 25, pp. 99–108 (2004)
[55] N.A. Dodgson: Proc SPIE, 5291, pp. 36–46 (2004)
[56] S. Tsukuda, Y. Murai: Japn Orthoptic J, 16, pp. 69–72 (1988)

22

# Development of Time-Multiplexed Autostereoscopic Display Based on LCD Panel

Dae-Sik Kim, Sergey Shestak, Kyung-Hoon Cha, Jae-Phil Koo, and Seon-Deok Hwang

## 22.1 Introduction

The main advantage of time-multiplexed stereoscopic display is that high-resolution stereoscopic images can be displayed using single monitor. In a time-multiplexed stereoscopic system left and right images of a stereopair consequently appear on the screen of 2-D display. The viewer is able to see the left and right images separately due to special shutter glasses or passive polarized eyeglasses combined with a polarization switch, sequentially occluding the left and the right eyes, thus preventing the viewer's eyes from seeing wrong images. Since conventional 2-D images and components of the stereopair may be displayed with the same full resolution of the image panel, 2-D / 3-D switching of the display is potentially unnecessary. Both 3-D and 2-D content can be displayed on the same screen simultaneously.

Time-multiplexed stereoscopic systems, utilizing polarization shutter glasses and polarization switchable screens are well-known and have been the objective of much research [1,2,3,4,5]. The shutter glasses system has two wearable optical switches, capable of switching their transparency thus blocking the left or right eye. The passive polarized eyeglasses system has one large (screen size) optical polarization switch switchable between two

D.-S. Kim
Samsung Electronics Co., Ltd. Suwon, Republic of Korea
e-mail: daesikkim@samsung.com

B. Javidi et al. (eds.), *Three-Dimensional Imaging, Visualization, and Display*,
DOI 10.1007/978-0-387-79335-1_22, 

orthogonal polarizations of output light, thus enabling the left or the right eye channel.

In the no-glasses type of the display (usually called autostereoscopic) such an optical switch comprises a projection system that provides switchable zones of selective vision (or viewing zones) so that the viewer can see the screen image only by placing his (her) eye in a certain area in front of the display screen. At the same time the viewer's other eye does not see any image unless the optical switch is switched in a status, enabling the image for another eye.

Time-multiplexed stereoscopic systems developed so far are mostly intended to be used with CRT displays. LCD displays have certain advantages over CRT displays, but were considered unsuitable due to their slow response rate. Recently LCD display technology made great strides in improving dynamic performance. Various techniques to reduce motion blur in LCD panels, developed so far, make it possible to apply LCD panels in time-multiplexed stereoscopy. Now the LCD displays with response times ranging from 2–3 ms are commercially available. However, before introducing a new type of display into time-multiplexed stereoscopy certain problems, inherent in LCD displays, should be solved.

This chapter is primarily devoted to the application of TN LCD panel with response time compensation in time-multiplexed stereoscopic and autostereoscopic displays.

The chapter's structure is as follows. Section 22.2, following the introduction, presents a concept, structure and principle of operation of the developed autostereoscopic display based on the LCD panel. Section 22.3 describes dynamic properties of the parts, methods to overcome their drawbacks and the related design features. Practical implementation and the experimental results are presented in Sect. 22.4. In Sect. 22.5 we show that the developed design principle can also be applied in an eyeglasses type stereoscopic system. Section 22.6 is a summary.

## 22.2 Autostereoscopic Time-Multiplexed Display

### 22.2.1 Concept of the Display

One of the critical requirements applied to modern autostereoscopic display is 2-D / 3-D compatibility. This is due to the fact that space multiplexed 3-D displays are unable to display 2-D images with full resolution of the image panel. During the last decade several viable design concepts of 2-D / 3-D switchable displays have been found. There are two different approaches to provide both 2-D and 3-D operation in autostereoscopic display. One is to design the parallax barrier or lenticula for space multiplexed display switchable on and off, thus sacrificing the resolution in 3-D mode and providing full resolution only in 2-D mode [6,7]. Another way, which we will follow, is to make a time-multiplexed display capable of displaying a full resolution 3-D image [8,9]. Since time-multiplexed displays offer left and right images in a sequence, the resolution of the 3-D image can be the same as the resolution of the 2-D image, making 2-D / 3-D switching virtually unnecessary. Such a display is capable of displaying 2-D and 3-D applications simultaneously without loss of resolution. A 2-D / 3-D switchable display can also be provided with such an option; however, the dedicated area for the 3-D image should be chosen by the switching of specially arranged electrodes [10]. In contrast, time-multiplexing display does not require any switching so that the 2-D or 3-D mode of visualization is defined by the displayed content, the 3-D window has no restrictions on size or position on screen and can be moved by mouse dragging as well as any other desktop 3-D object like for instance icon or cursor.

Another requirement to consider is commercial availability and the low cost of the LCD panel in the display. Since the TN LCD panels, having the shortest response time among the mass production LCDs, are applied mostly in desktop computer monitors, the applied design principle and methods described here is currently limited to this type of display.

### 22.2.2 Optical Layout

Layout of the autostereoscopic system [11] is shown in Fig. 22.1. A key element of the system is a switchable directional backlight. The directional backlight is composed of a conventional diffuse backlight, a polarization switch and a combination of a patterned stripe retarder and a lenticular lens array. Directional backlight, switchable between two states, in one state illuminates the LCD panel so that the screen image can be seen only from a certain viewer's position within the so-called "viewing zone" and only with one (e.g., left) eye. For the right eye of the viewer the LCD panel appears not illuminated and no image is seen. In another state the directional backlight provides the panel illumination separately for the right eye. The best part of the viewing zone is located at 70 cm from the LCD panel and is shaped as 65 mm wide vertical stripes alternating with the same width gap between them as shown in Fig. 22.1. By the applying a driving signal to the polarization switch the position of the viewing zones can be changed so that a new viewing zone appears instead of the gap and vise-versa. Since the individual viewing zone width of 65 mm is equal to the average interocular distance, the viewer can find a position in which his left eye sees the image on the panel only when the signal is applied, while the right eye sees the image when the signal is removed. Time-multiplexed autostereoscopic display operates by periodically applying and removing the driving signal in synchronism with the changing right and left images displayed on the panel.

The elements of the microretarder array are shaped as thin vertical stripes having alternating half-wave and zero retardance. The retarder film on the glass substrate has been manufactured using the technology, providing the

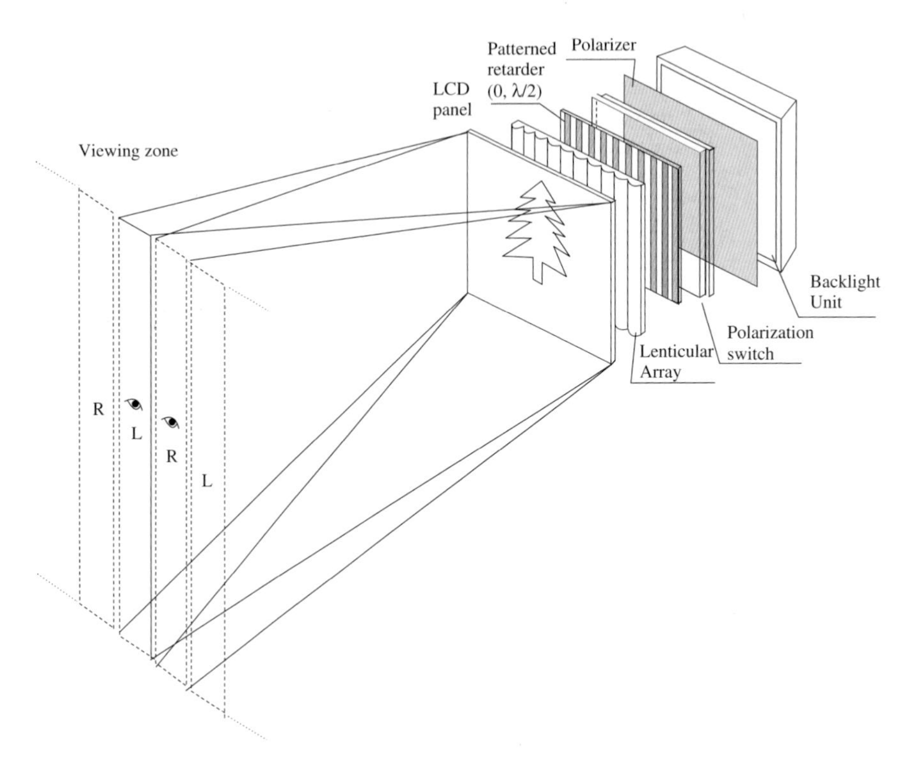


**Fig. 22.1.** Layout of the display

absence of any gap between the retarder stripes in order to provide seamless formation of the viewing zone.

Figure 22.2 shows the optical principle of viewing zone formation. Each lens of the lenticular array acts as an objective of the image projector in the horizontal plane, thus producing an image of vertical retarder stripes at the viewing distance of 700 mm. These images serve as exit pupils of the directional backlight. The output surface of the lenticular array is uniformly illuminated, thus representing a good backlight for the LCD panel. Before the light rays, emitted by the backlight and polarized by the polarizer, reach the input polarizer of the LCD they pass through the polarization switch, and a stripe of microretarder. Depending on the status of the polarization switch the input polarizer of the LCD panel blocks the light passing through the odd or even stripe of microretarder in accordance with the stripe retardance. The work of microretarder array and polarization switch, placed between crossed polarizers, is illustrated by the microphotograph, shown in Fig. 22.3. The upper part of the photo shows the microretarder itself while in the lower part the additional half-wave film is inserted between crossed polarizers. A PI-cell, used as a polarization switch, plays the role of switchable retarder. In fact the

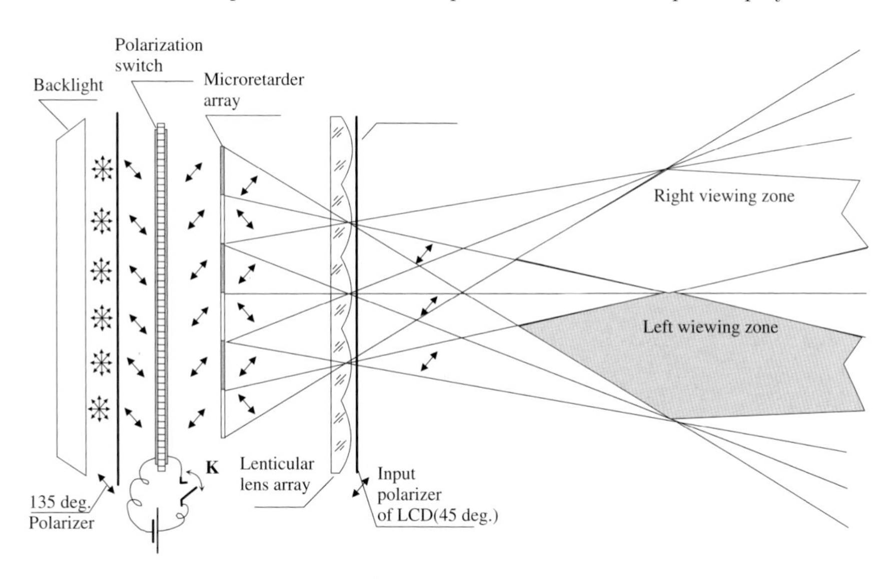


**Fig. 22.2.** Viewing zone formation by directional backlight

**Fig. 22.3.** Appearance of microretarder array in crossed polarizers

microretarder, together with the polarization switch, works as a switchable slit mask with opaque regions and transparent slits.

Let for instance the input polarizer transmits light polarized at 135 degrees, as shown in Fig. 22.2, odd and even stripes of microretarder provide zero and half-wave phase shifts, respectively, and the polarization switch provides a minus half-wave phase shift with a low level of the applied drive voltage. Then the light incident to the lens array and input polarizer experiences $-\lambda/2$ ($-\lambda/2+0=-\lambda/2$) phase shift and thus appears polarized at 45 degrees i.e., parallel to the transmission axis of the polarizer. The light, passed through the even stripes, in contrast, experiences zero phase shift ($-\lambda/2+\lambda/2=0$) and thus appears polarized at 135 degrees. As a result the light rays, passing through the odd stripes, serve as a backlight for the LCD panel, while the rays, passed through the even stripes, are blocked by the input polarizer of the LCD panel. A lenticular array collects the rays passed through the odd stripes in the left eye viewing zone (marked grey) so that the left eye of the viewer can see the displayed image. Since the right eye viewing zone, where the rays passed through the even stripes are collected, does not receive light rays, the displayed image appears invisible to the right eye. When the right eye image appears on the panel, the controller applies high-level voltage to the polarization switch that switches the inserted retardance to zero. Now the light, passed through the odd stripes of the microretarder, experiences zero phase shift ($0+0=0$) and is blocked by the input polarizer of the panel. The light, passed through the even stripes experiences $\lambda/2$ phase shift ($0+\lambda/2=\lambda/2$) changes in its polarization into orthogonal and passes through the input polarizer. As a result the displayed image becomes visible to the right eye, positioned inside the right viewing zone and invisible to the left eye. Thus, by switching the polarization switch synchronously with the displayed images, autostereoscopic time-multiplexed display is provided.

## 22.3 Dynamic Properties of the Display Parts

### 22.3.1 Scan-and-Hold Properties of LCD Panel

Both CRT and LCD monitors employ similar line-by-line methods of image update. The radical difference between CRT and active matrix LCD monitors lies in the persistence of the just updated image. Each pixel of an image on a CRT display starts to decay immediately after its excitation by the electron beam. Unlike CRT, the LCD display's pixel after updating maintains practically the same value of luminance until the next update. This difference is reflected in the definition of the CRT display as an *impulse type display,* and the LCD display as a *scan-and-hold* or simply *hold type display.* The difference between CRT and LCD displays is illustrated by instant photographs, presented in Fig. 22.4. The photographs are taken while the same graphic video signal, having alternating green and red frames, is applied to LCD and CRT displays. When applying this to time-multiplexed stereoscopy red and green frames represent left and right components of a stereopair. It could be clearly seen that while the left (red) field can be seen on a CRT monitor without green (left), both left and right images on a LCD monitor present at almost any moment of time. This why attempts to use stereoscopic shutter-

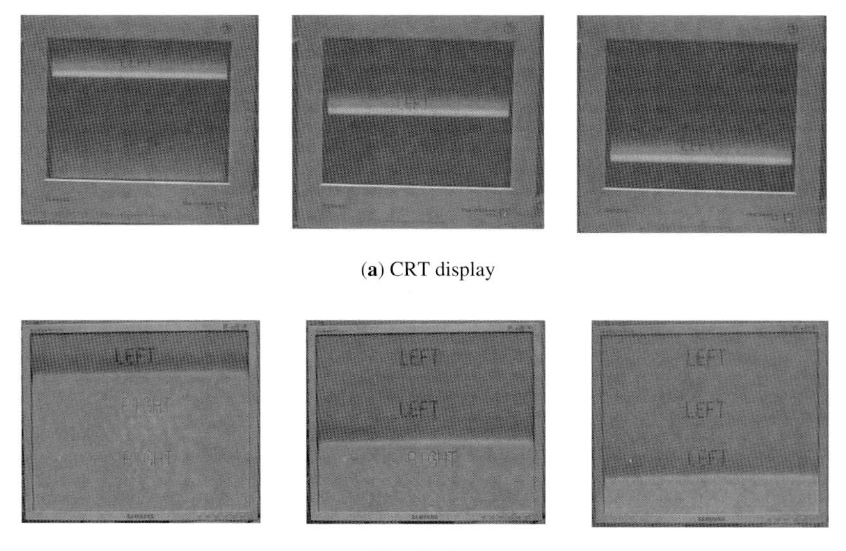


(**a**) CRT display

(**b**) LCD display

**Fig. 22.4.** Comparison of CRT and LCD displays

glasses with LCD displays generate high levels of cross talk, even if the display is characterized with a very low response time.

It is convenient to study the problem using the time-height diagram shown in Fig. 22.5a. The diagram represents the events on screen as points in time-height coordinates. For instance the update of an arbitrary line of an image is represented as one point in the diagram. The process of updating a line of image takes about 10 us; however, we can display the process as a point

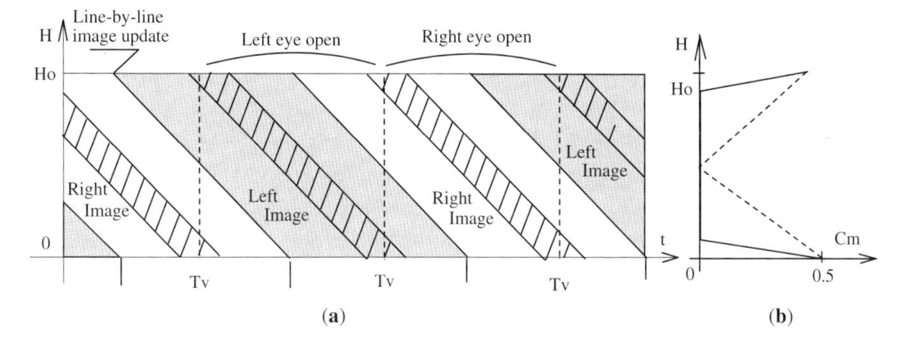


**Fig. 22.5.** (**a**) Time-height and cross talk-height diagrams of CRT (hatched area) and LCD displays (*grey* and *white* areas) operation with optical switching(*vertical dashed lines*); (**b**) cross talk-height diagram for CRT (*solid line*) and LCD (*dashed line*) displays

in the diagram because it lasts at least three orders of magnitude less than the frame period. Slanted lines in the diagram, separated with a frame period $T_V$ from each other, represent line-by-line updating of a screen image. Each frame displayed by the LCD display is confined within a time-space interval between these slanted lines. The operation of a CRT display can be presented in the diagram as narrow slanted stripes (hatched in the diagram) separated with a frame period. The frames on the CRT display are confined within these hatched stripes and separated with large time-space intervals with no image on the screen.

### 22.3.2 Time-Mismatch Cross Talk

Switching of the polarization switch to address the screen image to the right or left eye of viewer can be represented in the time-height diagram as vertical lines (see dashed lines in Fig. 22.5). For instance, the switching of the polarization switch may be delayed at half the vertical frame period in respect to the upper line update. Only in the central part of the LCD display can left and right images be separated perfectly. In the upper and lower parts of the screen wrong images present on screen almost within one-half of a frame period. As a result the viewer sees mixed left and right images in these parts of the screen, experiencing stereoscopic ghosting or cross talk. When the level of crosstalk between left and right images is too high this reduces the viewer's comfort and weakens stereoscopic sensation.

The crosstalk appeared because of time mismatch between the updating of a given LCD line and switching of the polarization switch, we refer to as "time-mismatch cross talk."

The level of the time-mismatch cross talk Cm can be defined as the ratio between the brightness of the wrong image and the full image brightness. Since the human eye averages the image brightness within the frame period, the cross talk is proportional to the portion of the period within which the wrong image is displayed. In the central section of the screen the optical switch is switching in the same moment of time as the image update that corresponds to zero time-mismatch cross talk. In the upper and the lower part the wrong image is displayed within almost half of the frame period that corresponds to 50 percent cross talk. The corresponding cross talk-height diagram is shown by a dashed line in Fig. 22.5b. One can readily see that the assumed half period delay of switching the polarization switch provides the lowest peak level of cross talk.

For comparison, a CRT based stereoscopic system with passive polarizing glasses has no cross talk in most of the screen, except the very upper and the lower parts as shown in the diagram (see Fig. 22.5b) with solid lines. The high cross talk level in the most of the screen makes the conventional approach to a LCD based time-multiplexed stereoscopic system impracticable.

Time–mismatch cross talk can be avoided if the polarization switch has a large number of segments, individually switchable in the same line-by-line order as the LCD image rendering. A similar approach is employed in high-end CRT based stereoscopic systems with passive eyeglasses [3], which are equipped with a polarization switch that has five sequentially switchable segments. It is likely that five segments in the polarization switch is enough to reject the residual phosphor afterglow on the CRT screen, but then the question emerges: how many segments are enough to work with the LCD panel?

A switching diagram of seven segmented polarization switches and an LCD panel is shown in Fig. 22.6. Slanted lines in the diagram represent line-by-line image update, with polygonal lines that represent the switching of the segmented shutter.

Considering instant updating of each line of the image, one can see that the local cross talk $C_\mathrm{n}$ within the line of the image is proportional to the temporal mismatch between the moment of updating of the line $T(n)$ and the moment of switching the polarization switch $Ts$:

$$C_n = \frac{T_S - T(n)}{T_0}, \tag{22.1}$$

where n is a number of the line.

To minimize the peak cross talk each segment of the polarization switch should be switched at the moment of updating the line, corresponding to the middle of the given segment. Hence, in the center of each segment the time-mismatch cross talk is zero while near the upper and lower boundaries the cross talk reaches its maximum:

$$C_{\max} = \frac{1}{2N}. \tag{22.2}$$

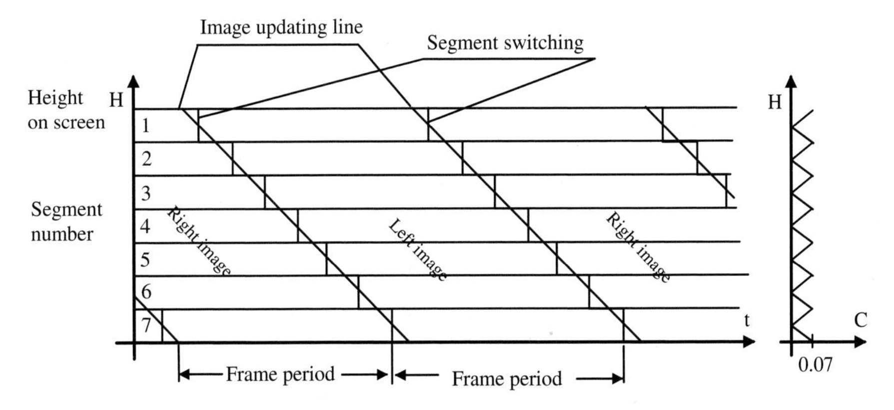


**Fig. 22.6.** Time-height diagram of switching segmented polarization switch and corresponding cross talk-height diagram

The cross talk, averaged within the full screen area, is lower than the maximum value $C_{max}$ by factor 2. The cross talk diagram corresponding to the expression (22.1) is shown in the right part of Fig. 22.6.

We can now find the number of segments, which satisfies our cross talk requirement. For instance, to reduce time-mismatch cross talk to a level less than 2.5 percent one should apply a polarization switch with 20 or more segments. Though the number of segments in the polarization switch is much lower than the number of lines in the LCD panel it appears that time-mismatch cross talk can be greatly reduced with just several tens of segments.

The photograph in Fig. 22.7 shows an appearance of the black channel image in a black-white cross talk test. The cross talk maximums are clearly seen as gray areas near the boundaries of the segments.

The Pi-cell based polarization switch that has 20 individually switchable segments has been applied in 19" autostereoscopic and stereoscopic time-multiplexed displays, reported in this chapter.

Applying a segmented polarization switch in a CRT based stereoscopic system might just improve its performance from "not so bad" to "very good." The usage of the segmented polarization switch in the system with an LCD monitor, in contrast, could greatly improve the stereoscopic image quality.

### 22.3.3 Driving of Pi-Cell

Pi-cell is a nematic liquid crystal optical modulator capable of electrically controllable birefringence [1]. A capability of fast switching has defined an application of Pi-cells in time-multiplexing displays including stereoscopic displays. One of the most widely known applications of Pi-cells is a polarization switch in which optical birefringence of the cell is electrically switchable between zero and half-wave (or $\pi$) optical retardance. Stereoscopic systems, based on CRT

**Fig. 22.7.** Appearance of the black field in a Black-White test

displays, utilizing polarization shutter glasses and polarization switches are well-known and have been the subject of much research [2–5]. A conventional method of driving a Pi-cell in a stereoscopic system may include various shapes of applied electric signal [3], but the common characteristic is that the activation period (when the voltage is applied) is equal to the relaxation period (when the voltage is removed). When the Pi-cell works together with an LCD display there are certain peculiarities, related to its dynamic properties, which have not been previously discussed.

Dynamic properties of a polarization switch based on the Pi-cell can be described by a time-transmission diagram, shown in Fig. 22.8a for the right channel, and in Fig. 22.8b for the left channel, respectively. The diagram shows how a normalized transmission in the left $T_L$ and right $T_R$ channels of a stereoscopic system changes in response to a symmetrical driving signal. Figure 22.8c shows the corresponding stereoscopic synchronization signal, generated in synchronism with vertical sync.

We do not specify a specific shape of the driving signal here, but it is assumed that the activation period (when high level is applied) is equal to the relaxation period (when low level is applied) that corresponds to 50 percent of the duty cycle of simple square wave. We also assume instant switching of the stereoscopic content between the left and right components of stereopair

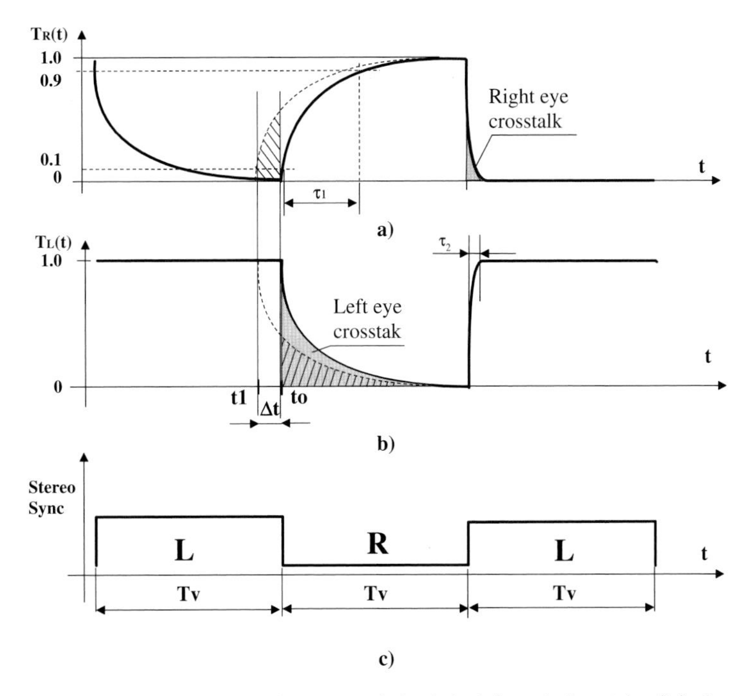


**Fig. 22.8.** Time-transmission diagram of the left (**a**) and the right (**b**) channels of stereoscopic system

and synchronous switching of the Pi-cell. Finite response time of the Pi-cell when driving voltage is applied may cause certain leakage of the right image to the left eye and vise-versa. A well-known characteristic of the Pi-cell is an activation response time ($\tau_2$ in Fig. 22.8) that is much shorter than the relaxation response time ($\tau_1$ in Fig. 22.8). If the right channel of the stereoscopic system is open during the relaxation period (see Fig. 22.8a), then the left channel is open during the activation period (see Fig. 22.8b). Leakage of the left image to the right channel and vise-versa is represented by the areas marked in grey on the diagram. As can be readily seen, great asymmetry in response time causes great asymmetry in cross talk. There are several good reasons to evenly distribute the cross talk portions between the left and right channels:

(a) It can be expected that if the cross talk is evenly distributed between the left and right channels there will be less degradation in the quality of the stereoscopic image than if the double cross talk is concentrated in one channel.
(b) Due to nonlinearity of the curves the level of symmetrized cross talk is more than twice lower than the level of cross talk concentrated in one channel.
(c) Cross talk compensation is simpler if it is symmetrical.

Unlike an LCD display, the above asymmetry would not be a problem in a CRT based stereoscopic system because of the impulse type of monitor. Also stereoscopic system with shutter glasses can be designed so that both left and right shutters are normally open or normally closed thus providing symmetrical responses for left and right eye.

To symmetrize the cross talk by equally distributing the cross talk between left and right channels we have proposed to simultaneously reduce the activation period and extend the relaxation period, thus maintaining their sum [13]. If the activation is switched off at an earlier moment $t_1$ instead of $t_0$ as conventionally applied, the cross talk in the left channel is reduced at the expense of the right channel where cross talk is increased. The time difference between $t_1$ and $t_0$, $\Delta t$ can be chosen to equalize the cross talk in the right and left channels. The corresponding areas, representing the cross talk, are shown hatched in the diagram. To estimate the value of $\Delta t$ the mathematical model of switching, developed for vertically aligned (VA) cells [12], can be applied. Though the relaxation process in the Pi-cell may differ from the same in the VA cell, the observed curve of the Pi-cell optical response does not differ much from the predicted model. Hence, the referred model switching between activation and relaxation periods can be expressed as follows:

$$T_L(t) = Sin^2\left\{(\pi/2)\cdot\exp\left[\frac{1.356\cdot(t-t_1)}{\tau_1}\right]\right\} \tag{22.3}$$

$$T_R(t) = 1 - Sin^2\left\{(\pi/2)\cdot\exp\left[\frac{1.356\cdot(t-t_1)}{\tau_1}\right]\right\} \tag{22.4}$$

Neglecting the crosstalk, which appeared due to the finite value of $t_2$ which is much smaller than $t_1$ the time shift can be found from the following equation:

$$\int_{t_0}^{t_0+T_V} T_L(t)dt = \int_{t_1}^{t_0} T_R(t)dt \tag{22.5}$$

For the given relaxation response time of 3 ms that is characteristic of the applied Pi-cell, the numerical solution of the equation (22.5) gives $\Delta t$ =1.77 ms. The symmetrized cross talk is 4.8 percent while the asymmetric cross talk corresponding to $\Delta t$ =0 is 17.6 percent.

In terms of the duty cycle to symmetrize cross talk the duty of activation periods should be reduced to 41 percent at a frame rate 100 Hz, and 39 percent at a frame rate of 120 Hz.

### 22.3.4 LCD Response Time, Response Time Acceleration Technique and Dynamic Cross Talk

There is a certain difference between the operation of the LCD panel, displaying time sequential stereoscopically and displaying conventional 2-D images. In the most important parts of the image, representing stereoscopic disparity, the liquid crystal cells are switching periodically between two different levels of luminance. So it could be interesting to study the optical response of liquid crystal cells on square wave drive signal.

Typical optical response of the modern TN LC panel on a square wave drive signal is shown in Fig. 22.9. The ideal response of the LC cell shown with a dashed line is a square wave switching between two levels of luminance $I_1$ and $I_2$ corresponding to the left and right images. Real response of the LC cell is distorted from ideal, as shown by a thin line in the diagram.

Since the left and right images are updating 60 times per second the human eye averages the pixel luminance. The luminance, perceived by the left eye and the right eye, is proportional to the area under the curve within Frame 1 and under the curve within Frame 2, respectively. The hatched areas S1 and S2 in Fig. 22.9a represent the difference between the ideal and perceived luminance in the right and left eyes of a viewer, respectively. In the shown example the image, represented to the left eye is darker than it should be in the ideal case, whereas the image represented to the right eye is brighter. Thus, the luminance of a given pixel of a stereoscopic image is represented by the LCD with a certain amount of error due to slow switching. This error in representing the luminance in the current frame is strongly dependent on

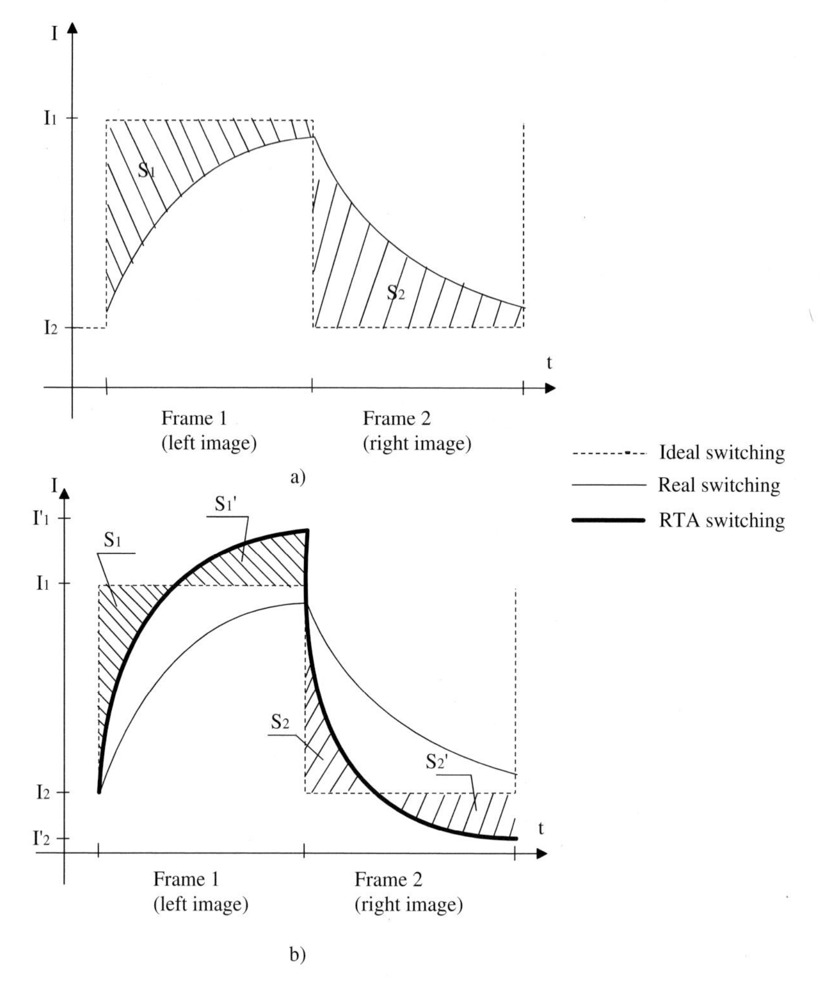


**Fig. 22.9.** LCD response time and RTA

the luminance of the previous frame. In a time-multiplexed stereoscopy this is perceived as a leakage of the left image to the right channel or vise-versa. This kind of cross talk between the left and right channels in a stereoscopic system is a time-variable parameter; thus it can be called "dynamic cross talk." The shorter the response time the lower the dynamic cross talk.

The LCD display's response time depends on design parameters, like the cell gap, as well as on the type and characteristics of the applied liquid crystal mixture, e.g., a rotational viscosity. Response time of a particular display also varies with the initial and destination levels of gray-to-gray transition. Ferroelectric liquid crystal displays have a very short response time, which can satisfy time-multiplexed application; however, their production has been discontinued due to high manufacturing costs. Custom LCD displays with comparatively short response times (3.5 ms and 0.5 ms for on and off switching, respectively) were applied for time-multiplexed stereoscopy more than one decade ago [14]. However, just recently fast TN LCD displays have become widely available.

The fast response of the latest TN LCD panels is achieved both due to the application of new liquid crystal mixtures and the overdrive method of the LC cell control [15].

The principle of the overdrive method implies that to dynamically switch the cell of the LCD panel from the first level of luminance to the second level of luminance one should apply a driving voltage higher or lower than is normally applied in a steady-state. If the target level of the luminance is higher than the current level, the applied voltage should be chosen higher than the voltage needed to produce the same target level in a steady-state (overshoot). If the target level of luminance is lower, the applied voltage should be lower than the steady-state voltage (undershoot). Since the response time of an LCD display strictly depends on both the starting and target levels the decision regarding the overdrive level can be made when both values are taken into account. To provide the above principle the LCD display controller is equipped with a previous frame memory, storing the data of previous frame levels, and a look up table (LUT) which stores the modified values of driving signals that correspond to a number of gray-to-gray transitions. The controller compares the previous gray level command to each pixel with the current gray level command and chooses a predetermined with alternative target level from a LUT.

An example of such modified levels is presented in Table 22.1 representing a look up table (LUT) with the variety of starting levels $G_1$ in the first column and the variety of target levels $G_2$ in the first row. The modified level, which should be applied to an LCD cell to perform a $G_1$-$G_2$ transition, can be found in the intersection of the corresponding row and column. Further development of the method has created a variety of particular techniques. One of them is Response Time Acceleration (RTA) [16]. RTA is originally directed to speed up switching of the LCD cells. In this section we are going to show that the overdrive method can be applied to complete compensation of both the above error in average luminance and the dynamic cross talk.

The RTA method has been designed to accelerate switching of LC cells to reach at least the target level of luminance within one frame period. In modern TN LCD displays, with the response time shorter than the frame period, the RTA is able to boost the LC cells to reach the end of the period at a level even higher than the target level of luminance. This is illustrated in Fig. 22.9b by a thick line. The RTA levels $I'_1$ and $I'_2$ in the diagram are set higher than the target level $I_1$ and lower than the target level $I_2$, respectively. The optical response boosted by RTA in the first half of the left image period has a lack of luminance, while in the second half the luminance exceeds the target level $I_1$. The areas $S_1$ and $S'_1$ in the diagram represent the lack and excess of light, accumulated within a period.

The method of complete cancellation of cross talk equalizes the average lack and excess of the luminance ($S_1 = S'_1$) by adjusting the RTA level $I'_1$,

**Table 22.1.** Example of modified overdrive levels vs. gray-to-gray transitions

| | 0 | 16 | 32 | 48 | 64 | 80 | 96 | 112 | 128 | 144 | 160 | 176 | 192 | 208 | 224 | 240 | 256 |
|---|---|---|---|---|---|---|---|---|---|---|---|---|---|---|---|---|---|
| 0 | 0 | 20 | 40 | 62 | 87 | 108 | 125 | 151 | 175 | 198 | 219 | 236 | 241 | 251 | 255 | 255 | 255 |
| 16 | 0 | 16 | 36 | 58 | 79 | 102 | 121 | 140 | 162 | 188 | 208 | 230 | 238 | 247 | 255 | 255 | 255 |
| 32 | 0 | 10 | 32 | 53 | 71 | 96 | 116 | 137 | 157 | 178 | 199 | 223 | 233 | 243 | 255 | 255 | 255 |
| 48 | 0 | 8 | 29 | 48 | 67 | 90 | 111 | 130 | 152 | 175 | 195 | 212 | 230 | 240 | 253 | 255 | 255 |
| 64 | 0 | 4 | 25 | 44 | 64 | 85 | 106 | 124 | 147 | 170 | 192 | 208 | 228 | 237 | 251 | 255 | 255 |
| 80 | 0 | 0 | 23 | 43 | 59 | 80 | 101 | 120 | 141 | 165 | 186 | 206 | 225 | 235 | 249 | 255 | 255 |
| 96 | 0 | 0 | 22 | 41 | 54 | 76 | 96 | 116 | 135 | 160 | 180 | 202 | 222 | 232 | 247 | 255 | 255 |
| 112 | 0 | 0 | 21 | 39 | 54 | 73 | 92 | 112 | 132 | 155 | 174 | 195 | 217 | 230 | 247 | 255 | 255 |
| 128 | 0 | 0 | 20 | 37 | 53 | 69 | 88 | 108 | 128 | 150 | 169 | 189 | 212 | 227 | 245 | 255 | 255 |
| 144 | 0 | 0 | 19 | 35 | 51 | 65 | 84 | 104 | 123 | 144 | 164 | 184 | 207 | 226 | 242 | 255 | 255 |
| 160 | 0 | 0 | 18 | 33 | 48 | 63 | 80 | 100 | 118 | 140 | 160 | 180 | 202 | 222 | 240 | 255 | 255 |
| 176 | 0 | 0 | 16 | 30 | 46 | 60 | 82 | 96 | 113 | 135 | 156 | 176 | 197 | 217 | 239 | 255 | 255 |
| 192 | 0 | 0 | 14 | 29 | 44 | 58 | 80 | 92 | 110 | 130 | 148 | 169 | 192 | 212 | 234 | 255 | 255 |
| 208 | 0 | 0 | 13 | 27 | 42 | 55 | 75 | 87 | 108 | 123 | 143 | 162 | 184 | 208 | 229 | 250 | 255 |
| 224 | 0 | 0 | 11 | 25 | 40 | 51 | 68 | 82 | 100 | 118 | 138 | 158 | 176 | 198 | 224 | 245 | 255 |
| 240 | 0 | 0 | 8 | 21 | 34 | 46 | 61 | 75 | 91 | 109 | 129 | 147 | 189 | 189 | 216 | 240 | 255 |
| 256 | 0 | 0 | 5 | 17 | 28 | 41 | 53 | 67 | 81 | 97 | 114 | 133 | 151 | 172 | 201 | 232 | 255 |

thus eliminating the error in perceived luminance. In the same manner one can equalize the areas $S_2 = S'_2$ by adjusting of the RTA level $I'_2$, thus eliminating the error in average luminance perceived by the right eye. Once the target average luminance is displayed perfectly without errors there is no reason for a dynamic cross talk; it is cancelled. The compensation of cross talk is achieved without drawing attention to the rise time with RTA, which is usually applied as a characteristic of the display speed in the specks. What is more important, the LCD response time without overdrive should be short enough. For instance the response time may vary within 8 ... 4 ms. Another parameter critical to successful cancellation of the dynamic cross talk is the vertical refresh period. The longer the response time and the shorter the vertical period, the greater the cross talk level to be cancelled.

RTA and other overdrive techniques are unable to compensate a cross talk in transition to the levels close to the maximum (gray-white) or minimum (gray-black) limits of brightness. When applying "cross talk cancellation" we can only consider medium contrast images or images with reduced dynamic range. The higher the level of cross talk we are going to compensate, the higher the difference between the original signal and the RTA signal, and the narrower the dynamic range of the luminance, which can be presented in the image without cross talk.

### 22.3.5 Other Methods for Correction of the LCD Shortcomings

To fight the above-discussed problems inherent in LCD displays, at least two rather effective methods have been introduced so far. One of the methods is scanning (or blinking) backlight [17, 18], in which the backlight is divided on sections, capable of individual switching on and off. Another is the method of inserting a black frame between image frames [19]. The duration of the black frame can be shorter than the usual frame, thus reducing signal bandwidth demand. The above methods applied in conventional LCD displays reduce motion blur and do not affect still images. On the contrary, when they are applied in time-multiplexed stereoscopic displays they are able to improve the quality of even still stereoscopic images. Both methods, applied separately or simultaneously, as illustrated by the time-height diagram in Fig. 22.10, can be used to reduce both dynamic cross talk and time-mismatch cross talk.

To reduce time-mismatch cross talk the segmented backlight should provide sequential blinking of each segment (instead of conventional CW operation) in synchronism with an image update. Thus operating the LCD display becomes more or less similar to the operation of a CRT display depending on the duty cycle of the backlight blinking. Operating a six-segmented scanning backlight with about 25 percent duty is shown in the time-height diagram (Fig. 22.10) as a zigzag area. The image on the screen of the display with scanning backlight is seen only within the periods, covered with that zigzag area, and the screen appears black in other time intervals. For instance no one image can be seen on the screen within the time-space intervals, shown in the diagram with gray and black colors. These inactive intervals can be used to hide stepwise switching of the segments of polarization switch, thus eliminating time-mismatch cross talk. If the duty of the backlight is short enough the simple (not segmented) polarization switch becomes capable of separating

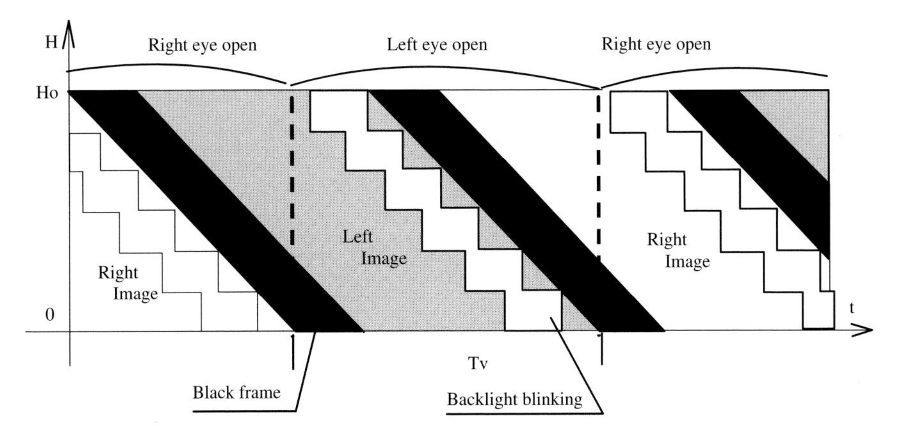


**Fig 22.10.** Application of black frame insertion and scanning techniques for crosstalk reduction

left and right images. Switching of the polarization switch is represented by vertical dashed lines in Fig. 22.10. The display, characterized with low duty of the backlight, can also be used for stereoscopic imaging with shutter glasses.

Scanning backlight can also be applied to reduce dynamic cross talk caused by slow switching of LCD cells, just as it is applied for motion blur reduction [20]. Namely, each segment of the backlight is off during certain parts of the period until the optical response of the cell is stabilized.

Better cross talk compensation can be achieved with simultaneous application of a scanning backlight and RTA. As noted in the previous section the cross talk cannot be cancelled if the target luminance is close to the luminance limits. This situation is illustrated in Fig. 22.11. The target levels of luminance $I_1$ in the left and $I_2$ in the right image are so close to the limits I=255 and I=0, respectively, that the dynamic cross talk cannot be cancelled by RTA. More specifically, the lack of average luminance in the first period of the left image (S1) cannot be compensated by the maximum excess ($S_1$") of luminance, which can be provided by RTA, setting the RTA level at 255. A similar problem has a right image: $S_2$ remaining larger than $S_2$'even if the RTA level is set to zero. Blinking of the backlight can be synchronized so that during the first part of each period the backlight is off and the worst period of slow switching is hidden from the viewer. The blinking backlight is shown in the diagram by the square wave, drawn with a thin solid line. It is apparent that, with the blinking backlight, the lack of displayed luminance in the left image is defined by the area $B_1$ which is much smaller than $S_1$ and can be now compensated by the excess of luminance $S_1$'. In a similar manner to compensate the right image for cross talk the area $S_2$' should be equal to $B_2<S_2$.

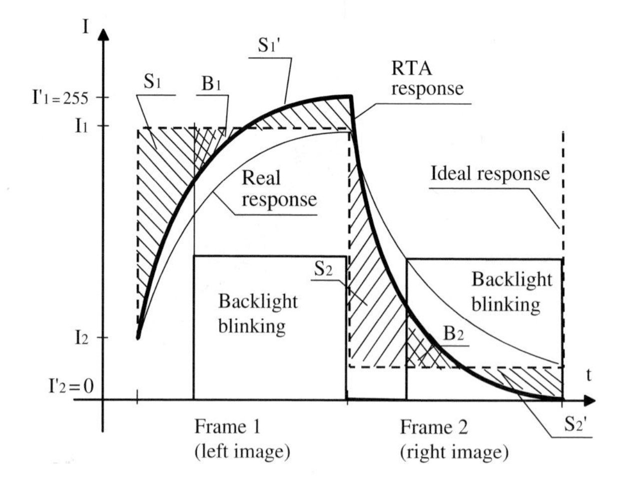


**Fig. 22.11.** RTA cross talk canceling, assisted by the blinking backlight

The role of the blinking backlight in the cross talk cancellation increases with the vertical refresh rate.

The black frame insertion can be applied for dynamic cross talk reduction [14]. This method does not compensate slow switching of the LCD cells. However, since the previous frame is always uniformly black, the dependence on a new image frame from the previous one disappears as well as the dynamic cross talk.

One more problem that can be solved with black frame insertion and scanning backlight is the asymmetry of Pi-cell optical response, discussed in Sect. 22.3.3. Both of these methods provide inactive time-space intervals in which no images are seen on the screen. These intervals can be used to hide the Pi-cell asymmetry just as we could hide slow switching of LCD or stepwise switching of the polarization switch.

### 22.3.6 Frame Rate of LCD Panel

In a impulse type CRT display the screen brightness pulses during a period equal to a vertical frame rate. When the vertical frequency is too low an undesirable flicker occurs. Since the image flicker may cause eyestrain and headache many CRT displays support high refresh rates up to 100–120 Hz, and some of them up to 160 Hz. The LCD display is a hold type display whose screen brightness is hardly pulses and does not produce a flicker even with a low frame rate. Most of the commercially available LCD computer monitors can run at just 60 Hz, 75 Hz, and sometimes 85 Hz, which is considered high enough to watch moving images. But in time-multiplexed stereoscopy the optical shutter periodically blocks the left or right eye of a viewer with one-half a frame frequency. If the optical shutter's frequency is lower than 50–60 Hz, viewers will feel a flicker. Therefore, the optical switch should operate at a frequency higher than 50–60 Hz. To provide synchronous left and right images the vertical frame rate should be not lower than 100–120 Hz. A motion blur reduction strategy, applied by most manufacturers of LCD TVs, includes 100–120 Hz operation, and some of the recent LCD TVs support this high vertical frequency. However, most TN monitors do not support a frequency higher than 75 Hz and we modified the monitors, applied in experiments, to provide operation up to 120 Hz.

## 22.4 Practical Implementation and Experimental Results

We have built 17" and 19" time-multiplexed autostereoscopic displays, following the above strategy.

A 17" TN LCD panel with a resolution of 1,280 × 1,024 has been equipped with a controller, modified for operation at a frame rate up to 120 Hz. It was found that 100 Hz is the highest operational frequency for the applied RTA circuit. The RTA is assisted by five-segmented backlight driving with 30 to 50 percent duty. Dynamic switching of left-right viewing zones is performed by a five-segmented polarization switch. The cross talk level in the 17" display does not exceed several percents even with factory settings of RTA due to the assistance of the scanning backlight. However, if the RTA is switched off, the cross talk increases significantly.

The 19" autostereoscopic display has a 20-segmented polarization switch and does not have a scanning backlight. On-off switching of the RTA mode has shown that the cross talk is greatly reduced by the RTA. However, the RTA's factory setting did not appear optimized for complete cross talk cancellation and the LUT of the RTA circuit has been modified to cancel visual cross talk when watching test images with various G-G transitions.

Table 22.2 represents the visual appearance of the crosstalk for various G-G transitions between left and right images, estimated by viewer, watching stereoscopic test images. The crosstalk appearance is marked with the numbers 0,1,2, placed in the intersections of row and column corresponding to various starting/target levels. If there is no visible crosstalk the mark is 0, if the crosstalk is slightly visible-1 and 2 if the crosstalk is noticeable. As can be seen the most gray-to-gray transitions can be compensated for crosstalk by this method. The exceptions are the transitions from gray level to white

**Table 22.2.** Residual crosstalk after compensation vs. g-g transitions

| | 0 | 16 | 32 | 48 | 64 | 80 | 96 | 112 | 128 | 144 | 160 | 176 | 192 | 208 | 224 | 240 | 256 |
|---|---|---|---|---|---|---|---|---|---|---|---|---|---|---|---|---|---|
| 0 | 0 | 0 | 0 | 0 | 0 | 0 | 0 | 0 | 0 | 0 | 0 | 0 | 0 | 1 | 2 | 2 | 2 |
| 16 | 0 | 0 | 0 | 0 | 0 | 0 | 0 | 0 | 0 | 0 | 0 | 0 | 0 | 0 | 1 | 2 | 2 |
| 32 | 0 | 0 | 0 | 0 | 0 | 0 | 0 | 0 | 0 | 0 | 0 | 0 | 0 | 0 | 0 | 2 | 2 |
| 48 | 0 | 0 | 0 | 0 | 0 | 0 | 0 | 0 | 0 | 0 | 0 | 0 | 0 | 0 | 0 | 2 | 2 |
| 64 | 0 | 0 | 0 | 0 | 0 | 0 | 0 | 0 | 0 | 0 | 0 | 0 | 0 | 0 | 0 | 2 | 2 |
| 80 | 0 | 0 | 0 | 0 | 0 | 0 | 0 | 0 | 0 | 0 | 0 | 0 | 0 | 0 | 0 | 2 | 2 |
| 96 | 0 | 0 | 0 | 0 | 0 | 0 | 0 | 0 | 0 | 0 | 0 | 0 | 0 | 0 | 0 | 2 | 2 |
| 112 | 0 | 0 | 0 | 0 | 0 | 0 | 0 | 0 | 0 | 0 | 0 | 0 | 0 | 0 | 0 | 2 | 2 |
| 128 | 0 | 0 | 0 | 0 | 0 | 0 | 0 | 0 | 0 | 0 | 0 | 0 | 0 | 0 | 0 | 1 | 2 |
| 144 | 0 | 0 | 0 | 0 | 0 | 0 | 0 | 0 | 0 | 0 | 0 | 0 | 0 | 0 | 0 | 0 | 2 |
| 160 | 0 | 0 | 0 | 0 | 0 | 0 | 0 | 0 | 0 | 0 | 0 | 0 | 0 | 0 | 0 | 0 | 2 |
| 176 | 0 | 0 | 0 | 0 | 0 | 0 | 0 | 0 | 0 | 0 | 0 | 0 | 0 | 0 | 0 | 0 | 2 |
| 192 | 1 | 0 | 0 | 0 | 0 | 0 | 0 | 0 | 0 | 0 | 0 | 0 | 0 | 0 | 0 | 0 | 2 |
| 208 | 1 | 0 | 0 | 0 | 0 | 0 | 0 | 0 | 0 | 0 | 0 | 0 | 0 | 0 | 0 | 0 | 2 |
| 224 | 1 | 1 | 0 | 0 | 0 | 0 | 0 | 0 | 0 | 0 | 0 | 0 | 0 | 0 | 0 | 0 | 2 |
| 240 | 1 | 1 | 0 | 0 | 0 | 0 | 0 | 0 | 0 | 0 | 0 | 0 | 0 | 0 | 0 | 0 | 1 |
| 256 | 1 | 1 | 0 | 1 | 1 | 1 | 1 | 1 | 1 | 0 | 0 | 0 | 0 | 0 | 0 | 0 | 0 |

which crosstalk cannot be compensated by the RTA technique. As for the transitions from gray to black, that also cannot be compensated, the residual crosstalk is fairly low because in TN panels the corresponding response time is much shorter than the response time of gray-to-white transitions.

Both 17" and 19" displays are capable of displaying 2-D and 3-D images with the same resolution. 2-D images in autostereoscopic displays can be seen without restricting the viewer's position, but to see full screen orthoscopic 3-D images the viewer should keep an optimum distance from the screen (700 mm and 850 mm, respectively) and maintain a proper head position. The light distribution in the right channel at the 700 mm viewing distance is indicated by the solid line in Fig. 22.12. The dashed line in the diagram represents the simultaneous left-right channel distribution of light in 2-D mode. There are no visible artifacts on the screen when a 2-D image is viewed from an arbitrary distance.

A stereoscopic image can be displayed in either a full size window or in a smaller, draggable window (see Fig. 22.13). The smaller the window, the greater the viewer's freedom in the z dimension. The display is characterized by low cross talk between left and right channels (see Fig. 22.14). Low cross talk (below 1 %) is achieved by segmenting the polarization switch, modifying the LUT of the Rise Time Accelerator and a scanning backlight (only in a 17" display).

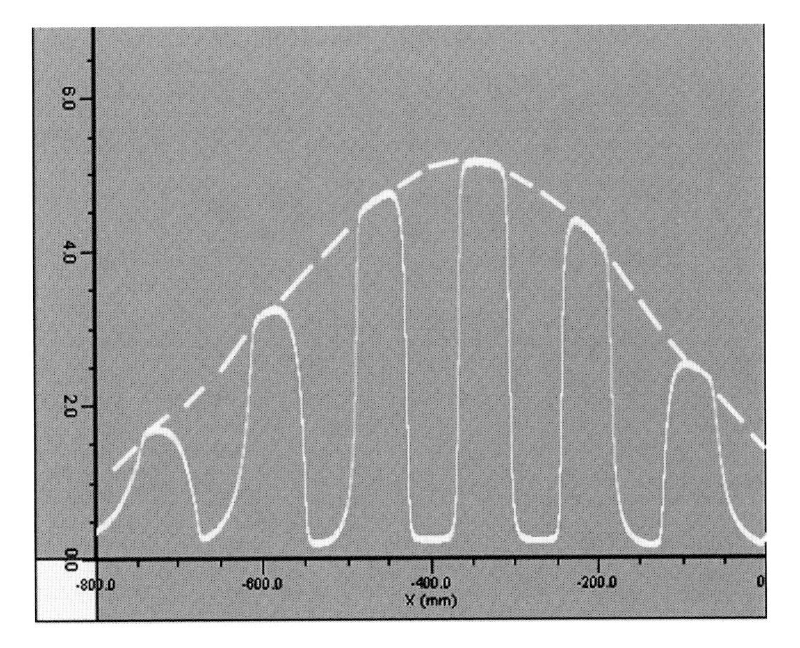


**Fig. 22.12.** Light distribution at a viewing distance

**Fig. 22.13.** Appearance of the autostereoscopic display screen with two different stereoscopic applications on 2-D Windows desktop

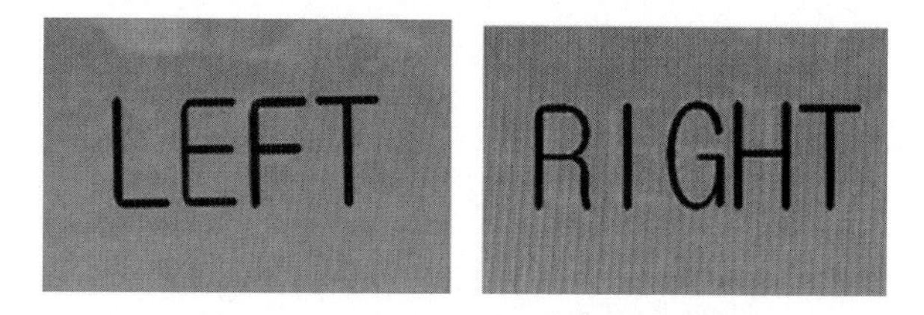


**Fig. 22.14.** Low cross talk screen images

## 22.5 Extension of the Developed Technique to Passive Eyeglasses Type Stereoscopic System

The simplified layout of the eyeglasses system is shown in Fig. 22.15. The system is composed of an LCD panel with an output polarizer, a polarization switch, based on a large area segmented Pi-cell, and two mutually crossed output polarizers mounted in eyeglasses. The polarizers could be either linear or circular. For circular polarizers a quarter-wave retarder film should be

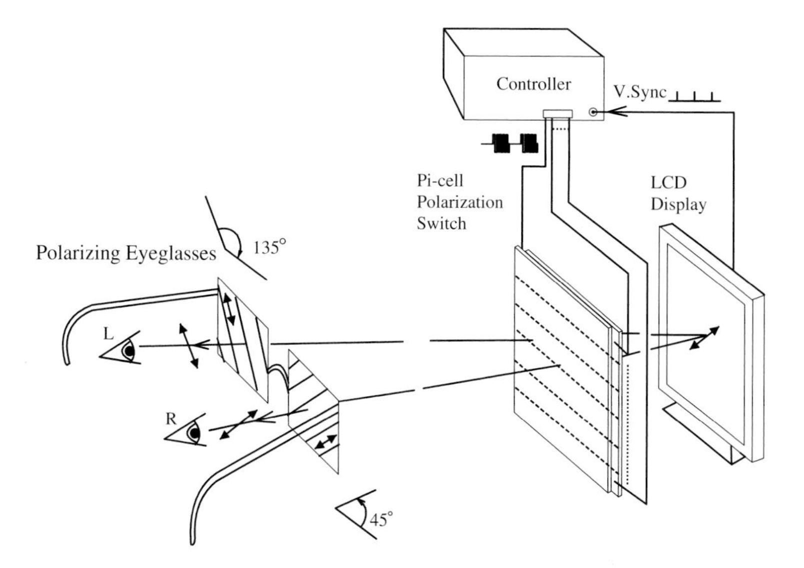


**Fig. 22.15.** Stereoscopic system with passive polarized eyeglasses

**Fig. 22.16.** Appearance of 2-D / 3-D LCD stereoscopic monitor

placed in some section between the eyeglasses and the display. The segmented polarization switch can be used either as an attachment to an arbitrary TN display or be incorporated into the monitor.

Unlike the stereoscopic system with shutter glasses, which can be driven with a low duty cycle [20] to reduce time-mismatch cross talk, the system with passive polarizing glasses cannot be switched with the duty, significantly different than 50 percent, because the optical switch transmits the image in both states. That is why other methods of crosstalk reduction, including the above described, gain their importance.

Since most TN displays do not have black insertion and a scanning backlight, the only method to reduce cross talk is the optimization of RTA LUT, that has been modified as described in the previous section. The driving signal of the polarization switch has been modified as recommended in Sect. 22.3.3.

We have built the system using a standard 19" TN LCD display which has an operational frame rate extended up to 120 Hz and a 20-segmented polarization switch incorporated in the LCD monitor along with the controller. The displays's appearance is shown in Fig. 22.16. The monitor can be used either as a time-multiplexed stereoscopic display with passive polarized eyeglasses, or as a conventional LCD monitor.

## 22.6 Summary

Though the method of image rendering, applied in LCD panels, is not suitable for time-multiplexed imaging and the response time is not short enough, modern TN LCD panels can be applied in time–multiplexed stereoscopic and autostereoscopic displays. We have developed low cross talk high-resolution time-multiplexed autostereoscopic and stereoscopic displays based on mass produced TN LCD panels.

To provide a low level of stereoscopic cross talk a set of measures has been applied:

(a) a polarization switch has been divided on 20 individually switchable segments in the vertical direction to reduce time-mismatch cross talk
(b) RTA LUT has been modified to cancel the dynamic cross talk
(c) a scanning backlight technique in combination with RTA technique has been applied to cancel cross talk at high operation frequency
(d) a driving signal polarization switch has been modified to symmetrize its optical response

## References


[1] P. J. Bos, K. Roehler/Beran "The pi-cell: a fast liquid-crystal optical-switching device," Molecular Crystals and Liquid Crystals 113, 329–339, 1984
[2] P. Bos "Stereoscopic imaging system with passive viewing apparatus," US Patent 4.719.507, 1988

[3] L. Lipton, J. Halnon, B. Dorworth, J. Wuopio "An improved Byatt modulator," Stereoscopic Displays and Virtual Reality Systems V, SPIE Proc. Vol. 3295, 121–126, 1998
[4] L. Lipton, J. Halnon, J. Wuopio, B. Dorworth "Eliminating pi-cells artifacts," Sterescopic Displays and Virtual Reality Systems VII, Vol. 3295, 264–270, 2000
[5] L. Lipton, A. Berman, L. Meyer "Achromatic liquid crystal shutter for stereoscopic and other applications," US Patent 4.884.876, 1989
[6] T. Dekker et al. "2-D/3-D switchable displays," Liquid Crystal Devices and Applications, Proceedings of SPIE 6135, pp.142–152, 2006
[7] G. J. Woodgate "High efficiency reconfigurable 2-D/3-D autostereoscopic display," SID'03 Digest LP1
[8] A. Yuuki et al "A new field sequential stereoscopic LCDs by use of dual-directional-backlight," Journal of Informational Display, 5(2), 6–9, 2004
[9] K.-W. Chien, H.-P. D. Shieh, "Time-multiplexed three-dimensional displays based on directional backlights with fast-switching liquid-crystal displays," Applied Optics 45, 3106–3110, 2006
[10] M.G.H. Hiddink et al. "Locally switchable 2-D/3-D displays," SID 06 Digest, pp. 1142–1145, 2006
[11] S. Shestak, D. Kim, K. Cha, J. Koo "Time-multiplexed autostereoscopic display with content-defined 2-D-3-D mode selection" "Stereoscopic displays and applications," Proceedings of SPIE Vol. 6490, 1B-1-1B-7, 2007
[12] H. Wang, T.X. Wu, X. Zhu, S.-T. Wu "Correlation between liquid crystal director reorientation and optical response time of a homeotropic cell," Journal of Applied Physics, 95 (10), 5502–5508, 2004
[13] S. Shestak, D. Kim "Application of Pi-cells in time multiplexed stereoscopic and autostereoscopic displays, based on LCD panels," Stereoscopic Displays and Applications, Proceedings of SPIE, 6490A-25, 2007
[14] J. Eichelaub, D. Hollands, J. Hutchins "A prototype flat panel hologram-like display that produced multiple perspective views at full resolution," Proceedings of SPIE, 2409, Stereoscopic displays and virtual reality systems II, pp. 102–112, 1995
[15] B.-W. Lee, C. Park, S. Kim, M. Jeon, J. Heo, D. Sagong, J. Kim, J. Souk, "Reducing gray-level response to one frame: dynamic capacitance compensation," SID 01 Digest, 1260–1263, 2001.
[16] T.-K. Kim, S.-J. Park, B.-Y. Chung "Design of a response time accelerator for an LCD panel," Journal of the Korean Physical Society, 43 (5), 858–862, 2003
[17] A.A.S. Sluyterman, E.P. Boonekamp "Architectural choices in a scanning backlight for large LCD TVs," SID05 Digest, pp. 996–999, 2005
[18] T. Nauta, N. Fisecivich, et al. "Improved motion picture quality of AM LCDs using scanning backlight," Proceedings of Asia Display/IDW 01, pp. 1637–1640, 2001
[19] K. Nishiyama, M. Okita, S. Kawaguchi "32" WXGA LCD TV using OCB mode, low temperature p-Si TFT and blinking backlight technology," SID 05 Digest, 36 (1), 132–135, 2005
[20] A.J. Woods, K.L. Yuen "Compatibility of LCD monitors with frame-sequential stereoscopic 3-D visualization," IMID/IDMC '06 Digest, 98–102, 2006

# 23

# 3-D Nano Object Recognition by Use of Phase Sensitive Scatterometry

Daesuk Kim, Byung Joon Baek, and Bahram Javidi

## 23.1 Introduction

Optical correlation techniques have proven to be very useful in various two- and three-dimensional pattern recognition applications. Recently, there has been increased interest in various imaging based 2-D and 3-D optical information sensing and recognition [1,2,3,4,5]. In particular, 3-D information processing based on digital holography has been proposed to extend optical correlation techniques to 3-D object recognition [4,5]. However, all such approaches were mainly based on an imaging technique which has an inherent optical diffraction limit and, for that reason, recognizing nano-size patterns smaller than around 1 $\mu$m could not be handled through such an imaging based optical approach. So far, one possible option is the SEM (Scanning Electron Microscope) based 2-D imaging approach. However, because SEM technology can only provide 2-D imaging information and it requires a vacuum environment, it has not been the preferred method in industrial fields. Recently, to complement these drawbacks of the SEM based nano object measurement approach, a more practical optical method called scatterometry emerged in the semiconductor industry. While traditional optical imaging techniques cannot resolve features smaller than the wavelength of the illumination beam, scatterometry enables physical parameters of sub-wavelength periodic structures to be extracted from the spectroscopic signature [6,7,8].

D. Kim
Division of Mechanical & Aero System Engineering, Chonbuk National University, Jeonju 561-756, Republic of Korea
e-mail: dashi.kim@chonbuk.ac.kr

B. Javidi et al. (eds.), *Three-Dimensional Imaging, Visualization, and Display*,
DOI 10.1007/978-0-387-79335-1_23, 

In this study, we propose a novel recognition method that can be applied for nano-size 3-D objects. For obtaining both amplitude and phase information through the reflected spectrum, we employ spectroscopic ellipsometry. Here, we deal with only grating-like periodic nano-patterns since scatterometry can only be applied to periodic structures. Although we prove the recognition capability of this study only for a one-dimensional periodic case, we expect two-dimensional periodic patterns could also be handled with this concept. In Section 23.2, we first describe the basic principle of scatterometry. Then, the rigorous coupled-wave analysis (RCWA) theory is introduced to explain how the diffraction of electromagnetic waves from the periodic surface of grating structures can be handled accurately with numerical analysis [9]. In Section 23.3, we deal with the proposed 3-D nano object recognition method which is based on phase sensitive scatterometry. Finally, we show that the proposed method can provide a moderate 3-D nano object recognition capability for various nano object shapes.

## 23.2 RCWA Based Scatterometry Theory

Spectroscopic ellipsometry depicted in Fig. 23.1 has been recognized as a highly suitable tool for determining the 3-D shape of nano-patterns in semiconductors and photonic crystals. The incident light diffracts into positive and negative orders.

Only the 0th order diffracted beam is collected by a spectroscopic system. The collected light is a linear combination of two linearly polarized components with a phase difference between p and s polarization. When the electric field is in the direction parallel to the grating lines the polarization

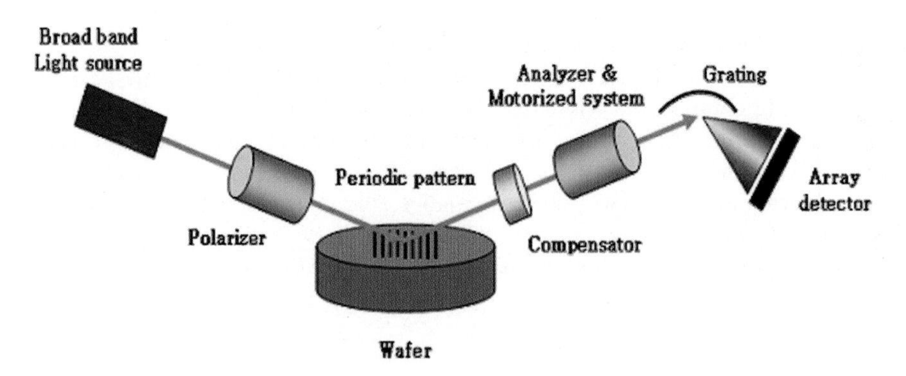


**Fig. 23.1.** Schematic of spectroscopic ellipsometer which consists of broadband light source, polarizer, compensator, analyzer and spectrometer

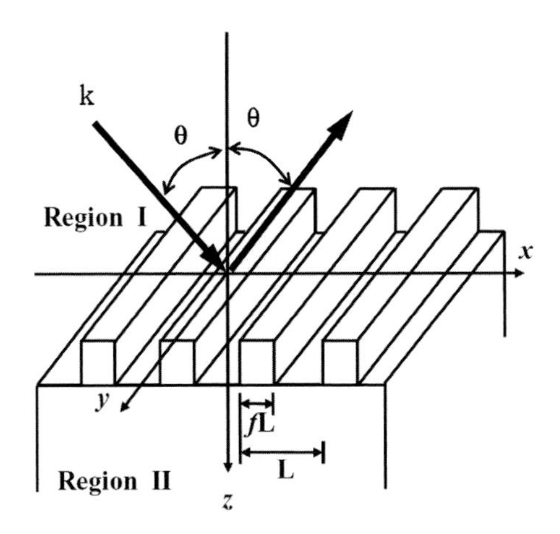


**Fig. 23.2.** 3-D geometry of nano-patterns used for scatterometry

mode is called the TE mode, and the polarization mode when the electric field is in the direction perpendicular to the grating lines is the TM mode.

RCWA is a rigorous analysis algorithm utilizing Maxwell's equations with some boundary conditions. Figure 23.2 illustrates the geometry of the diffraction configuration of a 1-D periodic pattern for the RCWA. The whole structure can be divided into an incident region (Region I), a grating (or patterned) region, and an exit region (Region II). The electric fields can be obtained from Maxwell's equations by using boundary conditions of the grating region [9]. In this grating region ($0 < z < d$), the periodic dielectric function is expandable with Fourier series with period $L$ as

$$\varepsilon(x) = \sum_h \varepsilon_h \exp\left(j\frac{2\pi h}{L}x\right) \tag{23.1}$$

where $\varepsilon_h$ is the $h$-th Fourier component of the dielectric function in the grating region.

For TE mode, the electric field in region I and II can be represented as follows:

$$\begin{aligned} E_{I,y} &= E_{inc,y} + \sum_i R_i \exp\left[-j(k_{xi}x - k_{I,zi}z)\right], \\ E_{II,y} &= \sum_i T_i \exp\{-j[k_{xi}x + k_{II,zi}(z-d)]\}, \end{aligned} \tag{23.2}$$

where $k_{xi}$ is determined from the Floquet condition and is given by

$$k_{xi} = k_0[n_I \sin\theta - i(\lambda_0/L)] \tag{23.3}$$

where,

$$k_{l,zi} = \begin{cases} k_0[n_l^2 - (k_{xi}/k_0)^2]^{1/2} & k_0 n_l > k_{xi}, \\ -jk_0[(k_{xi}/k_0) - n_l^2] & k_{xi} > k_0 n_l \quad l = I, II. \end{cases} \tag{23.4}$$

$R_i$ in Eq. (23.2) is the normalized electric field amplitude of the $i$-th backward diffracted (reflected) wave in Region I while $T_i$ is the normalized electric field amplitude of the forward diffracted (transmitted) wave in Region II. By applying the Maxwell's equations in the grating region and matching the boundary conditions at the interfaces of the three regions, one can determine the unknown amplitudes $R_i$ and $T_i$ of the diffracted waves. Those parameters are related to the ellipsometric parameters $\Psi$ and $\Delta$ as

$$\rho = \frac{r_p}{r_s} = \left|\frac{r_p}{r_s}\right| e^{i(\delta_p - \delta_s)} = \tan\Psi e^{i\Delta} = \frac{TM}{TE} \tag{23.5}$$

where, $r_p$ and $r_s$ are reflection coefficients of TE and TM polarization, respectively. Also, $\delta_p$ and $\delta_s$ represent phase shifts of the TE and TM polarization modes, respectively. The phase difference between p and s polarization, $\Delta$ means $\delta_p$–$\delta_s$.

## 23.3 Three-Dimensional Nano Object Recognition

Figure 23.3a shows the reference pattern we use for testing the object recognition capability. Figure 23.3b and c are two object patterns (upper width: 470 nm and 0 nm) among eight different object patterns that have various upper widths, i.e., 490 nm, 480 nm, 470 nm, 400 nm, 300 nm, 200 nm, 100 nm, and 0 nm. However, for simplicity, the period of all patterns was set to be 1,000 nm with a fill factor of 0.5, and all the heights were set to be 55 nm. First, for each pattern, we calculate Spectroscopic Ellipsometery (SE) parameters $\Psi(\lambda)$ and $\Delta(\lambda)$ by using the RCWA for using them in testing recognition capability. The angle of incidence defined by the angle between the normal axis of the pattern sample and the incident light is set to be 10 degrees just for simulating at a specific condition. However, it can be any value between 0 to around 90 degrees. 3-D nano object recognition has been performed through a conventional correlation technique that uses a matched filter. We will present some correlation results for the two input objects as shown in Fig. 23.3b and c to show that it can provide a practical capability for 3-D nano object recognition.

As the first step, we generate 0th order $\Psi$ and $\Delta$ spectrum for the rectangular shape by using RCWA. It corresponds to the reference data $u_i^R(\lambda)$ which is used for making the matched filter. Then, 0th order $\Psi$ and $\Delta$ spectrum for the eight different trapezoidal input objects are generated. The eight

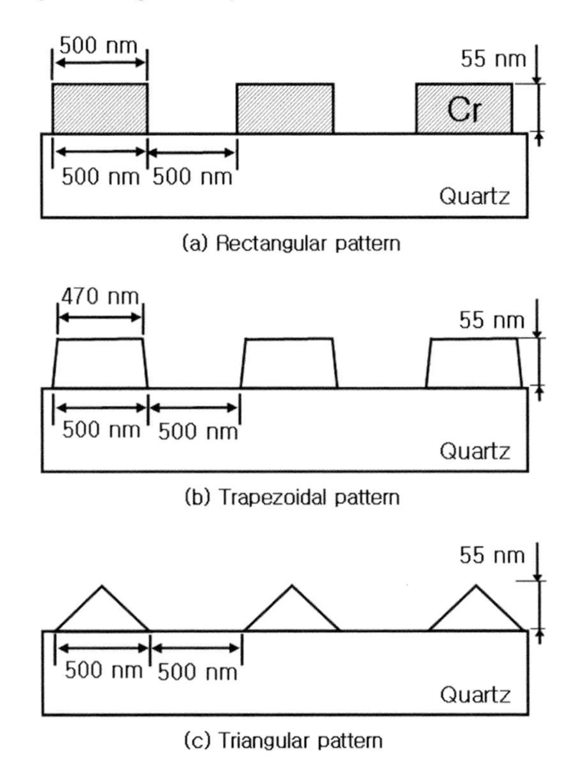


**Fig. 23.3.** Patterns used to validate scatterometry based recognition capability (**a**) rectangular pattern (reference pattern), (**b**) trapezoidal pattern (the third input target object) and (**c**) triangular pattern (the 8th input target object)

corresponding generated input target object spectrum data are denoted as ${u_i}^T(\lambda)$.

Figure 23.4a–b represent the 0th order $\Psi$ and $\Delta$ spectrum. Here, the solid line corresponds to the reference spectrum of the rectangular shape described in Fig. 23.3a, and the dotted line represents that of the trapezoidal input object in Fig. 23.3b. The correlation function $C(\lambda)$ between the reference spectrum and the input object spectrum can be represented as follows:

$$c(\lambda) = \left| \boldsymbol{F}^{-1}\{\boldsymbol{F}[u_i^T(\lambda)] \times \boldsymbol{F}^*[u_i^R(\lambda)]\} \right|^2 \tag{23.6}$$

Here, spectrum data can be either $\Psi$ or $\Delta$. Figure 23.4c represents the auto-correlation (solid line) and the cross correlation (dotted line) results when we used $\Psi$ spectrums as the reference and input target object. Similarly, Fig. 23.4d denotes the auto-correlation (solid line) and cross correlation (dotted line) results when we use $\Delta$ spectrums as the reference and input target object data. We can see that there is a strong correlation between the rectangular

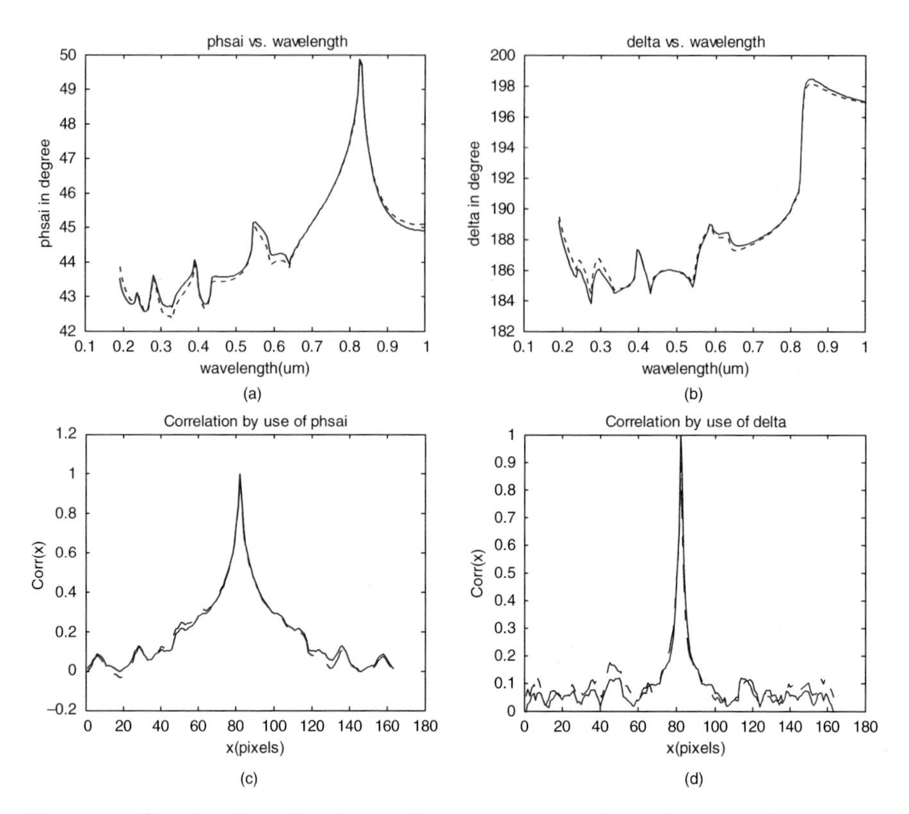


**Fig. 23.4.** Simulated results obtained by using rectangular and trapezoidal patterns (upper width: 470 nm) [(a)–(b) $\Psi(\lambda)$ and $\Delta(\lambda)$, respectively (*solid line*: rectangular pattern, *dotted line*: trapezoidal pattern), and (c)–(d) Amplitude correlation ($\lambda$) obtained by using the two $\Psi(\lambda)$ spectrums represented by *solid* and *dotted line* in Fig. 23.4a and phase correlation ($\lambda$) obtained by using the two $\Delta(\lambda)$ spectrums denoted by *solid* and *dotted* in Fig. 23.4b, respectively]

and trapezoidal pattern with an upper width of 470 nm, which is as expected for both the $\Psi$ and $\Delta$ cases.

Likewise, we use the same procedure with the triangular pattern shape that has an upper width of 0 nm as the second input target. As can be seen in Fig. 23.5c and d, there is still a strong correlation between the rectangular and triangular pattern when we use $\Psi$ spectrum, while $\Delta$ spectrum case does not. Figure 23.6b shows how the correlation peak values are varied as the side angle $\theta$ defined in Fig. 23.6a increases from 0 degrees to around 80 degrees for the $\Psi$ and $\Delta$ spectrum cases. This result clearly shows that the amplitude based scatterometry is much less likely to recognize input object 3-D shape correctly, while the phase sensitive approach can provide a moderate capability.

Only phase based $\Delta$ spectrum can provide a reasonable 3-D nano object recognition capability. The scatterometry based 3-D object recognition

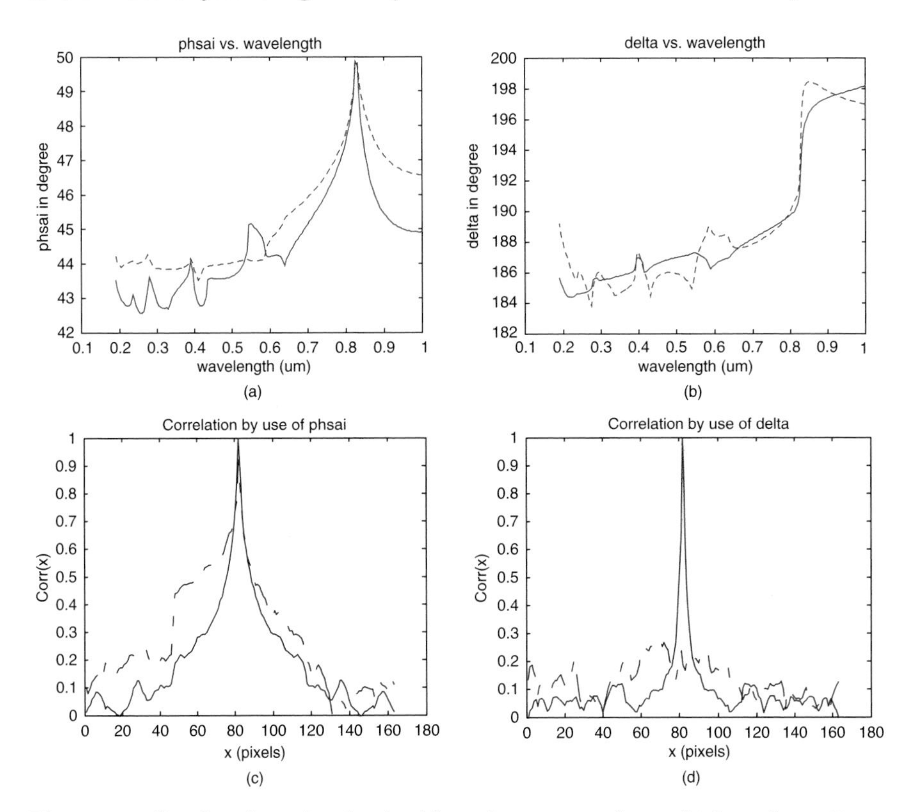


**Fig. 23.5.** Simulated results obtained by using rectangular and triangular patterns (upper width: 0 nm) [(a)–(b) $\Psi(\lambda)$ and $\Delta(\lambda)$, respectively (*solid line*: rectangular pattern, *dotted line*: trapezoidal pattern), and (c)–(d) Amplitude correlation ($\lambda$) obtained by using the two $\Psi(\lambda)$ spectrums represented by solid and dotted line in Fig. 23.5a and phase correlation ($\lambda$) obtained by using the two $\Delta(\lambda)$ spectrums denoted by solid and dotted in Fig. 23.5b, respectively]

theory seems to be very similar to that of digital holography based recognition in the sense that the phase information measurement capability of digital holographic technique gives much higher sensitive 3-D recognition capability than the amplitude based approach. The proposed scatterometry method enables us to recognize nano-patterns up to the size of around sub-100 nm with UV region scatterometry technology [10]. Also, we can say that it can provide a very fast recognition capability since such $\Delta$ spectrums can be measured almost in real-time with the current state-of-the-art SE technology. Although this kind of nano object recognition application is not yet that popular in the nano technology industry, we expect that it will play an important role in various nano technology fields such as photonic crystal, semiconductor and NEMS (Nano Electro-Mechanical System) applications in the near future.

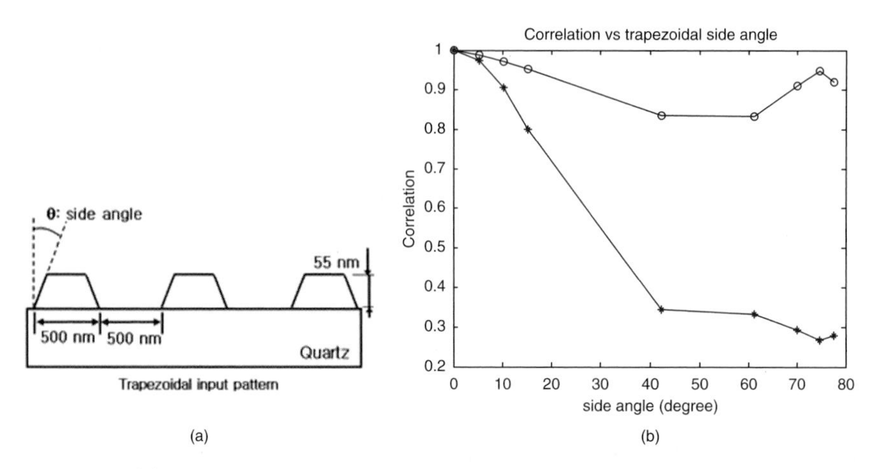


**Fig. 23.6.** (**a**) Definition of trapezoidal pattern side angle and (**b**) correlation peak value versus pattern side angle as defined in Fig. 23.6a [o: when we used Ψ spectrum, *: when we used Δ spectrum]

## 23.4 Conclusion

A novel method for nano size 3-D object recognition using phase sensitive scatterometry has been described. In this study, we have obtained a meaningful conclusion that the amplitude based scatterometry is much less likely to recognize a 3-D nano object correctly, while phase sensitive scatterometry can provide a reasonable moderate recognition capability. We expect that this kind of scatterometry based real-time nano object recognition technology will be more and more important as various nano technology fields gradually grow.

## References

[1] A. Mahalanobis, ."Correlation Pattern Recognition: An Optimum Approach," in *Image Recognition and Classification* (Marcel Dekker, New-York, 2002).
[2] J. Rosen, "Three dimensional electro-optical correlation," J. Opt. Soc. Am. A **15**, 430–436 (1998).
[3] J. W. Goodman, R. W. Lawrence, "Digital image formation from electronically detected holograms," Appl. Phy. Lett. **11**, 77–79 (1967).
[4] B. Javidi, E. Tajahuerce, "Three-dimensional object recognition by use of digital holography," Opt. Lett. **25**, 610–612 (2000).
[5] B. Javidi, D. Kim, "Three-dimensional-object recognition by use of single-exposure on-axis digital holography," Opt. Lett. **30**, 236–238 (2005).
[6] X. Niu, N. Jakatdar, J. Bao, C. Spanos, S. Yedur, Pro. SPIE **3677**, 159 (1999).
[7] H. Huang, F. Terry, "Normal incidence spectroscopic ellipsometry for critical dimension monitoring," App. Phy. Lett. **78**, 3983–3985 (2001).
[8] H. Gross et al., "Mathematical modeling of indirect measurements in scatterometry," Measurement, **39**, 782–794 (2006).

[9] M. G. Moharam, E. B. Grann, D. A. Pommet, T. K. Gaylord, "Formulation for stable and efficient implementation of the rigorous coupled-wave analysis of binary gratings," J. Opt. Soc. Am. A **12**, 1068–1076 (1995).
[10] X. Niu, N. Jakatdar, J. Bao, C. J. Spanos, "Specular Spectroscopic Scatterometry," IEEE Transactions on Semiconductor Manufacturing, **14**(2), 97–111 (2001)

# Index